Alcoholism

A Molecular Perspective

NATO ASI Series

Advanced Science Institutes Series

A series presenting the results of activities sponsored by the NATO Science Committee, which aims at the dissemination of advanced scientific and technological knowledge, with a view to strengthening links between scientific communities.

The series is published by an international board of publishers in conjunction with the NATO Scientific Affairs Division

A	**Life Sciences**	Plenum Publishing Corporation
B	**Physics**	New York and London
C	**Mathematical and Physical Sciences**	Kluwer Academic Publishers
D	**Behavioral and Social Sciences**	Dordrecht, Boston, and London
E	**Applied Sciences**	
F	**Computer and Systems Sciences**	Springer-Verlag
G	**Ecological Sciences**	Berlin, Heidelberg, New York, London,
H	**Cell Biology**	Paris, Tokyo, Hong Kong, and Barcelona
I	**Global Environmental Change**	

Recent Volumes in this Series

Volume 202—Molecular Aspects of Monooxygenases and Bioactivation of Toxic Compounds
edited by Emel Arinç, John B. Schenkman, and Ernest Hodgson

Volume 203—From Pigments to Perception: Advances in Understanding Visual Processes
edited by Arne Valberg and Barry B. Lee

Volume 204—Role of Melatonin and Pineal Peptides in Neuroimmunomodulation
edited by Franco Fraschini and Russel J. Reiter

Volume 205—Developmental Patterning of the Vertebrate Limb
edited by J. Richard Hinchliffe, Juan M. Hurle, and Dennis Summerbell

Volume 206—Alcoholism: A Molecular Perspective
edited by T. Norman Palmer

Volume 207—Bioorganic Chemistry in Healthcare and Technology
edited by U. K. Pandit and F. C. Alderweireldt

Volume 208—Vascular Endothelium: Physiological Basis of Clinical Problems
edited by John D. Catravas, Allan D. Callow, C. Norman Gillis, and Una S. Ryan

Series A: Life Sciences

Alcoholism
A Molecular Perspective

Edited by

T. Norman Palmer

The University of Western Australia
Nedlands, Perth, Australia

Plenum Press
New York and London
Published in cooperation with NATO Scientific Affairs Division

Proceedings of a NATO Advanced Study Institute on the
Molecular Pathology of Alcoholism,
held August 26–September 6, 1990,
in Il Ciocco, Italy

Library of Congress Cataloging-in-Publication Data

NATO Advanced Study Institute on the Molecular Pathology of Alcoholism (1990 : Il Ciocco, Italy)
Alcoholism : a molecular perspective / edited by T. Norman Palmer.
p. cm. -- (NATO ASI series. Series A, Life sciences ; v. 206)
"Proceedings of a NATO Advanced Study Institute on the Molecular Pathology of Alcoholism held August 26-September 6, 1990, in Il Ciocco, Italy"--T.p. verso.
"Published in cooperation with NATO Scientific Affairs Division."
Includes bibliographical references and index.
ISBN-13: 978-1-4684-5948-7 e-ISBN-13: 978-1-4684-5946-3
DOI: 10.1007/978-1-4684-5946-3
1. Alcoholism--Molecular aspects--Congresses. 2. Alcohol--Mechanism of action--Congresses. I. Palmer, T. Norman. II. North Atlantic Treaty Organization. Scientific Affairs Division. III. Title. IV. Series.
[DNLM: 1. Alcohol, Ethyl--metabolism--congresses. 2. Alcoholism--genetics--congresses. 3. Alcoholism--metabolism--congresses. WM 274 N281a 1990]
RC564.7.N38 1990
616.86'107--dc20
DNLM/DLC
for Library of Congress 91-21253
CIP

ISBN-13: 978-1-4684-5948-7

Softcover reprint of the hardcover 1st edition 1991
A Division of Plenum Publishing Corporation
233 Spring Street, New York, N.Y. 10013

PREFACE

This book contains selected proceedings from the NATO Advanced Study Institute (ASI) "The Molecular Pathology of Alcoholism" held at the Hotel Il Ciocco in Tuscany during 26th August - 6th September 1990. Alcoholism remains one of the most challenging problems in medical care, with far-reaching medical, social and economic consequences. For example in the U.S., estimates indicate that 18 million people have a serious drinking problem and that the total cost to the economy of alcohol abuse is $117 billion. Treatment of alcohol dependence and other alcohol-related disorders accounts for almost 15% of the total health bill of the United States.

Despite the scale of the medical problem, biomedical research on alcoholism remains something of a 'Cinderella science'. Research funding from government and other bodies is relatively poor and the number of medical scientists working in the field remains small. The Organizing Committee for this NATO ASI, comprising Charles Lieber (New York), Timothy Peters (London), Mario Dianzani (Torino), Emanuele Albano (Torino) and Norman Palmer (Perth, Director), were therefore particularly grateful to the NATO Scientific Affairs Division for their active support of this ASI, the first dealing with a topic related to alcohol abuse. We moreover hope that this support will continue. The theme of the ASI was an in depth discussion of the molecular events initiated by alcohol abuse that culminate in onset of alcohol-related disease. The last few years have seen major advances in our knowledge of the molecular pathology of alcoholism in a number of key areas. These include the molecular genetics and enzymology of the alcohol metabolising enzymes, the role of genetic factors in alcoholism, the interaction of alcohol with the central nervous system, and the role of acetaldehyde, acetaldehyde-protein adducts and free radicals in the pathogenesis of alcohol-related disease. The scale of these advances in knowledge may herald new advances in diagnosis and therapy.

I would like to express my appreciation to Charles Lieber, Timothy Peters, Mario Dianzani and Emanuele Albano for their invaluable help in the organisation of the ASI. My particular thanks go to Bobbie Ward and Bruno Giannasi for their enthusiastic assistance in the practical management of the Institute. Aside from the NATO Scientific Affairs Division, the ASI was generously supported in a variety of ways by The Greek Ministry of Industry, Energy and Technology, The Scientific and Technical Research Council of Turkey, The National Council for Scientific Investigation and Technology of Portugal, The Brewer's Society (London), SA Sanofi Labaz (Bruxelles), Du Pont (U.K.), the Ram Brewery (London), Scottia Pharmaceuticals and the Dow Chemical Company (Midland, Michigan). Finally I would like to thank Jenny Gillet and especially Raelene Bacon for their enormous assistance in the preparation of this book.

March 1991 Norman Palmer

CONTENTS

PATHWAYS OF ETHANOL METABOLISM AND RELATED PATHOLOGY

Charles S. Lieber

Bronx VA Medical Center and Mount Sinai School of Medicine
130 West Kingsbridge Road
Bronx, New York 10468 USA

METABOLIC AND PATHOLOGIC EFFECTS ASSOCIATED WITH ALCOHOL DEHYDROGENASE-MEDIATED ETHANOL METABOLISM

In ADH-mediated oxidation of ethanol, hydrogen is transferred from the substrate to the cofactor nicotinamide adenine dinucleotide (NAD), converting it to the reduced form (NADH) (Fig. 1). As a net result, ethanol oxidation by ADH generates an excess of reducing equivalents as free NADH in hepatic cytosol, primarily because the metabolic systems involved in NADH removal are not able to fully offset the accumulation of NADH. The acetaldehyde produced in this reaction is converted to acetate by aldehyde dehydrogenase, which is also associated with reduction of NAD to NADH. The large amounts of reducing equivalents produced overwhelm the hepatocyte's ability to maintain redox homeostasis and a number of metabolic disorders ensue, including changes in protein, carbohydrate (hypoglycemia) and uric acid metabolism (Lieber, 1982). This redox mechanism also affects lipids: 1) ethanol oxidation provides reducing equivalents and two carbon units for lipid synthesis, 2) the more reduced redox state inhibits the oxidation of fatty acids and diverts them into esterification which is further enhanced by the increased concentration of sn-glycerol-3-phosphate, 3) ethanol affects the amount of fatty acids transported from the adipose tissue into the liver (Lieber et al., 1959; Lieber and Schmid, 1961; Lieber, 1982). A characteristic feature of liver injury in the alcoholic is the predominance of steatosis and other lesions in the perivenular (also called centrilobular) zone or zone 3 of the hepatic acinus. The mechanism for this zonal selectivity of the toxic effects involves several distinct and not mutually exclusive mechanisms: one hypothesis postulates that ethanol can produce hypoxic damage of perivenular hepatocytes whereas another postulates that relative hypoxic conditions normally prevailing in the perivenular zone enhance the metabolic toxicity of ethanol. Furthermore, the selective presence and induction of enzymes of alcohol metabolism in the perivenular zone is also contributory.

The hypoxia hypothesis originated from the observation that liver slices from rats fed alcohol chronically consume more oxygen than those of controls (Videla and Israel, 1970). It was then postulated that the increased consumption of oxygen would increase the gradient of oxygen tensions along the sinusoids to the extent of producing anoxic injury of perivenular hepatocytes (Israel et al., 1975). Indeed both in human alcoholics (Kessler et al., 1954) and in animals fed alcohol chronically (Sato et al., 1983; Jauhonen et al., 1982), decreases in either hepatic venous oxygen saturation (Kessler et al., 1954) or PO_2 (Jauhonen et al., 1982) and in tissue oxygen tensions (Sato et al., 1983) have been found during the

Alcoholism: A Molecular Perspective
Edited by T.N. Palmer, Plenum Press, New York, 1991

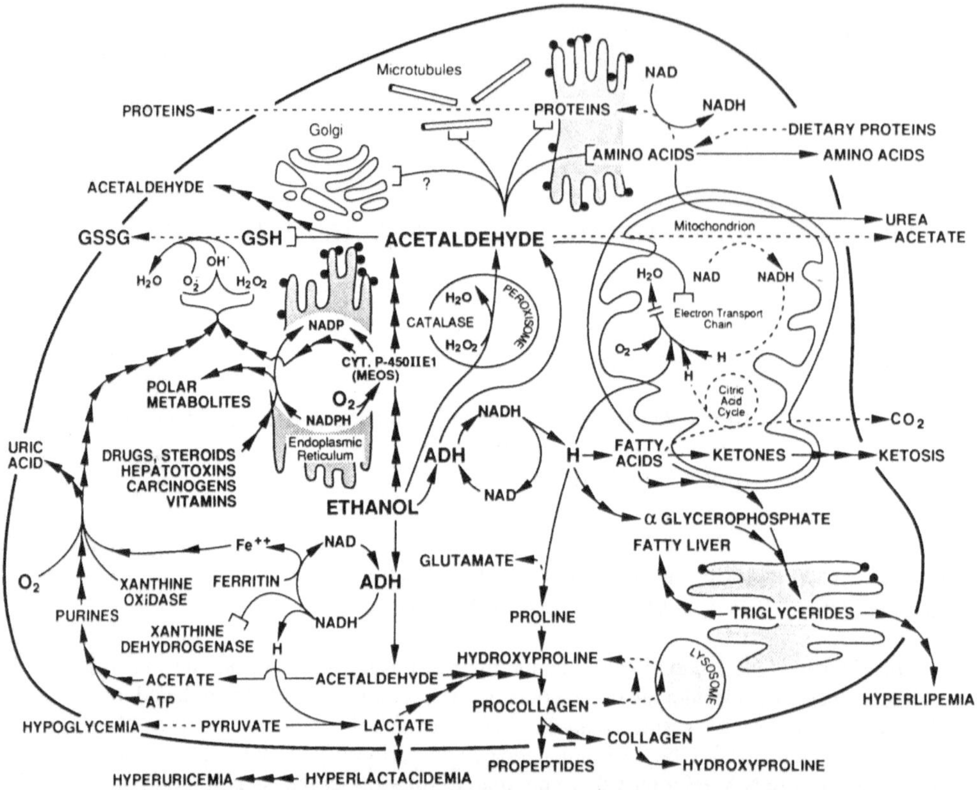

Fig. 1. Oxidation of ethanol in hepatocyte. Many disturbances in intermediary metabolism and toxic effects can be linked to 1). alcohol dehydrogenase (ADH) mediated generation of NADH 2). the induction of microsomal enzymes, especially P450IIE1 and 3). acetaldehyde, the product of ethanol oxidation. NAD, nicotinamide adenine dinucleotide; NADH, reduced NAD; GSH, reduced glutathione; GSSG, oxidized glutathione. The broken lines indicate pathways that are depressed by ethanol, whereas repeated arrows reflect stimulation or activation. The symbol -[denotes interference or binding. (From Lieber, 1990a).

withdrawal state. However, the changes in hepatic oxygenation found during the withdrawal state disappeared (Jauhonen et al., 1982; Shaw et al., 1977) or decreased (Sato et al., 1983) when alcohol was present in the blood. Acute ethanol administration increased splanchnic oxygen consumption in naive baboons, but the consequences of this effect on oxygenation in the perivenular zone were offset by increased blood flow resulting in unchanged hepatic venous oxygen tension (Jauhonen et al., 1982). Indeed, ethanol in fact induces an increase in portal hepatic blood flow (Stein et al., 1963; Shaw et al., 1977; Jauhonen et al., 1982; Carmichael et al., 1987). In baboons fed alcohol chronically, defective O_2 utilization rather than lack of O_2 blood supply characterized liver injury produced by high concentration of ethanol (Lieber et al., 1989). We postulated that the low oxygen tensions normally prevailing in perivenular zones could exaggerate the redox shift produced by ethanol (Jauhonen et al., 1982). To study the magnitude of such a shift in the baboon, the effects of ethanol on the lactate/pyruvate ratio in hepatic venous blood (an approximation of that in perivenular hepatocytes) were compared with the ratio in total liver. Ethanol increased the lactate/pyruvate ratio and decreased pyruvate more in hepatic venous blood than in total liver. In isolated rat hepatocytes, the ethanol-induced redox shift was markedly exaggerated by lowering the oxygen to a tension similar to those found in centrilobular zones. The process was also assessed in the isolated perfused rat liver, by varying the oxygen supply, to produce the oxygen tensions prevailing in vivo along the sinusoid (Jauhonen et al., 1985). It is noteworthy that hypoxia increases NADH which in turn inhibits the activity of NAD+-dependent xanthine dehydrogenase (XD), thereby favoring that of oxygen-dependent xanthine oxidase (XO) (Kato et al., 1990) (Fig. 1). It has been postulated that, due to the acetate derived from ethanol, purine metabolites accumulate and could be metabolized via XO. This process may lead to the production of oxygen radicals which may mediate toxic effects towards liver cells, including peroxidation. Physiological substrates for XO, hypoxanthine and xanthine, as well as AMP, significantly increased in the liver after ethanol, together with an enhanced urinary output of allantoin (a final product of xanthine metabolism). Allopurinol pretreatment resulted in 90% inhibition of XO activity, and also significantly decreased ethanol induced lipid peroxidation (Kato et al., 1990). Zonal distribution of some enzymes can influence the selective perivenular toxicity. There might be more ADH activity in the perivenular zone and, as discussed subsequently, proliferation of the smooth endoplasmic reticulum after chronic ethanol consumption is maximal in the perivenular zone, with associated enzyme induction and related effects. By means of immunohistochemical techniques, human ADH has now been demonstrated mainly in hepatocytes around the terminal hepatic venule (Buehler et al., 1982). Thus, a presumably higher level of ethanol metabolism in the perivenular zone, could contribute to the increased hepatotoxicity of ethanol, for instance by providing (together with the "induced" microsomal pathway: vide infra) an increased amount of the toxic metabolite acetaldehyde (Lieber, 1985). One must, however, also take into account that after chronic ethanol consumption, unlike the activity of MEOS, which is induced, that of ADH may not change or even may decrease (Lieber and DeCarli, 1970b; Brighenti and Pancaldi, 1970; Salaspuro et al., 1981). Alcoholics may display decreased hepatic ADH activity even in the absence of liver damage (Ugarte et al., 1967).

Although it is generally recognized that the liver is the main site of ethanol metabolism, extrahepatic metabolism occurs: ethanol oxidation in the digestive tract of the rat has been previously reported (Lamboeuf et al., 1981, 1983) and has been related to the ADH present in this tissue (Hempel and Pietruszko, 1979; Pestalozzi et al., 1983). However, the magnitude of gastrointestinal ethanol metabolism was assumed to be small (Lamboeuf et al., 1981, 1983). Some authors (Lin and Lester, 1980) could not demonstrate any significant gastrointestinal ethanol oxidation when they gave an acute high dose to rats; they concluded that this process was of negligible quantitative significance. The issue was reopened when it was shown that a significant fraction of alcohol ingested in doses in keeping with usual "social drinking" does not enter the systemic circulation in the rat and is oxidized mainly in the stomach (Julkunen et al., 1985a, b). This process was also shown to occur in man (Di Padova et al., 1987). Furthermore, Julkunen et al. (1985a) found an increase of acetate in portal blood after oral administration of ethanol to the rat, which clearly shows that ethanol must have been oxidized and therefore that some first pass metabolism had occurred locally. Moreover, gastrectomy was associated with an abolition of the first pass metabolism (Hernandez-Munoz et al., 1990). The magnitude of this process was assessed in the rat to amount to about 20% of the ethanol administered when given at a low dose (Caballeria et al., 1987). This process determines, in part, the bioavailability of alcohol and thus modulates its potential toxicity. This gastric barrier is low in females (Frezza et al., 1990), thus contributing to their increased susceptibility to ethanol. Commonly used H_2-antagonists, such as cimetidine, decrease the activity of gastric ADH (Caballeria et al., 1989), thereby enhancing peripheral blood levels of ethanol, a potentially hazardous complication if it occurs in an unsuspecting driver. Other equally potent H_2-antagonists, however, such as famotidine, do not have this undesirable side-effect

(Hernandez-Munoz et al., 1990). In alcoholics, first pass metabolism was much smaller and after fasting, the first pass metabolism had virtually disappeared in these alcoholics (Di Padova et al., 1987). The reduced first pass metabolism in alcoholics was found to be due, at least in part, to diminished gastric ADH activity.

METABOLISM OF ALCOHOL VIA THE MICROSOMAL ETHANOL OXIDIZING SYSTEM AND ITS INTERACTIONS WITH OTHER DRUGS AND HEPATOTOXIC AGENTS

Characterization of the Microsomal Ethanol Oxidizing System (MEOS) and its Role in Ethanol Metabolism

The observation in rats and man (Lane and Lieber, 1966) that chronic ethanol consumption is associated with proliferation of hepatic microsomal membranes prompted the suggestion that liver microsomes could be the site for a distinct and adaptive system of ethanol oxidation. Such a system has been demonstrated in liver microsomes in vitro (Lieber and DeCarli, 1968) and was named the microsomal ethanol oxidizing system (MEOS) (Lieber and DeCarli, 1968, 1970b). It was concluded that the MEOS was distinct from ADH and catalase and dependent on cytochrome P-450 because: a) isolation of a P-450-containing fraction from liver microsomes which, although devoid of any ADH or catalase activity, could still oxidize ethanol as well as higher aliphatic alcohols (e.g., butanol which is not a substrate for catalase) (Teschke et al., 1972, 1974; Mezey et al., 1973) and b) reconstitution of ethanol-oxidizing activity using NADPH-cytochrome P-450 reductase, phospholipid, and either partially-purified or highly-purified microsomal P-450 from untreated (Ohnishi and Lieber, 1977) or phenobarbital-treated (Miwa et al., 1978) rats. That chronic ethanol consumption results in the induction of a unique P-450 was shown by Ohnishi and Lieber (1977) using a liver microsomal P-450 fraction isolated from ethanol-treated rats. An ethanol-inducible form of P-450 (LM3a), purified from rabbit liver microsomes (Koop et al., 1982), catalyzed ethanol oxidation at rates much higher than other P-450 isozymes, and also had an enhanced capacity to oxidize 1-butanol, 1-pentanol, and aniline (Morgan et al., 1982), acetaminophen (Morgan et al., 1983), CCl_4 (Ingelman-Sundberg and Johansson, 1984), acetone (Koop and Cassaza, 1985), and N-nitrosodimethylamine (NDMA) (Yang et al., 1985). Similar results have been obtained with cytochrome P-450j, a major hepatic P-450 isozyme purified from ethanol- or isoniazid-treated rats (Ryan et al., 1985, 1986). Others have also provided evidence for the existence of a P-450j-like isozyme in humans (Wrighton et al., 1986; Song et al., 1986); however, its catalytic activity toward ethanol was not described. We have succeeded in obtaining the purified human protein (now called P450IIE1) in a catalytically active form, with a high turnover rate for ethanol and other specific substrates (Lasker et al., 1987). Compounds other than ethanol (e.g., acetone) can also serve as P450IIE1 inducers, but it has been shown that the activity of P450IIE1 can be induced after short-term and relatively light consumption of ethanol, in the absence of increased acetonemia or hepatic steatosis (Lieber et al., 1988).MEOS has a relatively high Km for ethanol (8-10 mM compared with 0.2-2 mM for ADH) and thus normally ADH accounts for the bulk of ethanol oxidation at low blood ethanol levels (Fig. 2A), but not necessarily at high ethanol levels (Fig. 2B) or during long-term use of alcohol (Fig. 2C), especially in view of the inducibility of the MEOS (Lieber and DeCarli, 1968, 1970b). Although data obtained with inhibitors are suggestive of the involvement (Lieber and DeCarli, 1970b, 1972; Teschke et al., 1976; Matsuzaki et al., 1981), they cannot be considered conclusive, since the inhibitors are not sufficiently specific. However, a mutant deermouse strain that lacks the hepatic low Km ADH (ADH^-) nevertheless actively oxidizes ethanol (Burnett and Felder, 1980; Shigeta et al., 1984; Kato et al., 1987, 1988), and studies with stable isotopes indicated that this effect

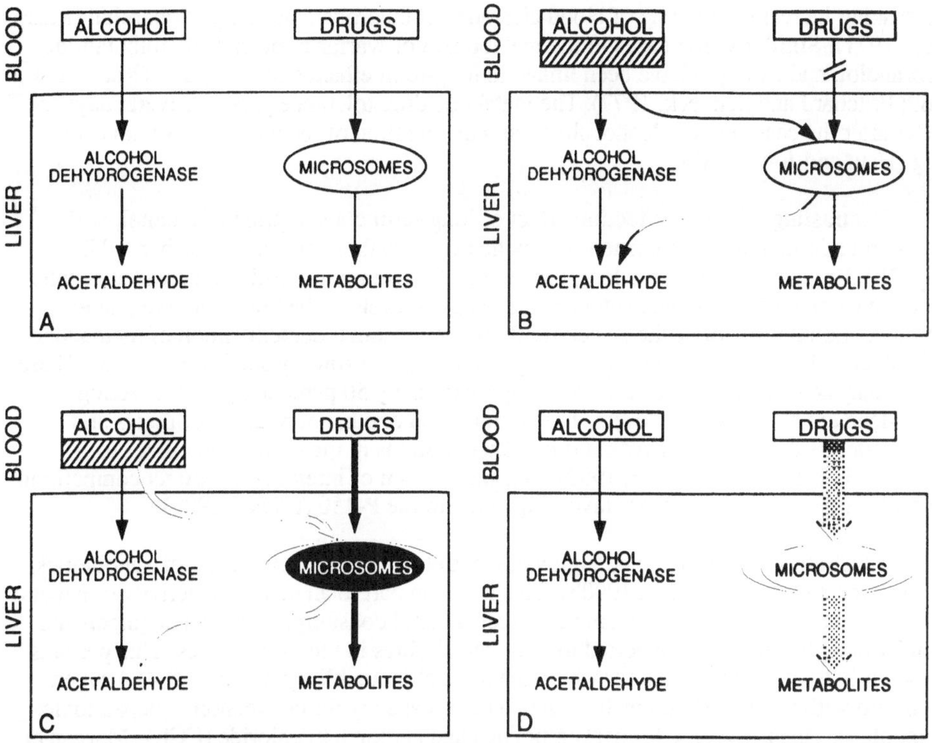

Fig. 2. Alcohol is metabolized by alcohol dehydrogenase, and drugs by microsomes (A). Microsomal drug metabolism is inhibited in the presence of high concentrations of ethanol, in part through competition for a common microsomal detoxification process (B). Microsomal induction after long-term alcohol consumption contributes to accelerated ethanol metabolism at high blood ethanol levels (C). Increased drug metabolism and activation of xenobiotics (due to microsomal induction) persist after cessation of long-term alcohol consumption (D). Hatching indicates high blood alcohol levels. (From Lieber, 1988).

is mediated principally by the MEOS (Alderman et al., 1987). Even when alcohol dehydrogenase is present, alternative pathways (mostly the MEOS) participate in ethanol metabolism at all concentrations tested in vivo and play a major part when blood ethanol levels are high (Alderman et al., 1987, 1989).

More recently it was suggested that 50% or more of ethanol elimination in ADH-deermice was caused by a mitochondrial dehydrogenase (Ingelman-Sundberg et al., 1988; Norsten et al. 1989). However, the activity was measured only at pH 10 and no indication was given of whether any such activity was observed at physiological pH, nor was this activity well characterized. Using the same strain of ADH- deermice, we could not find any activity at physiological pH and, in our hands, the oxidation found at pH 10 is minimal and does not appear to be due to an ADH activity (Inatomi et al., 1990).

Interactions with Drugs, Including Drug Tolerance and Enhanced Hepatotoxicity and Carcinogenicity

The administration of ethanol to volunteers under metabolic-ward conditions resulted in a striking increase in the rate of blood clearance of meprobamate and pentobarbital (Misra et al., 1971). Similarly, increases in the metabolism of warfarin, phenytoin, tolbutamide, propranolol, and rifampin have been linked to long-term ethanol consumption (Kater et al., 1969; Pritchard and Schneck, 1977). The metabolic drug tolerance persists several days to weeks after the cessation of alcohol abuse, and the duration of recovery varies with each drug (Hetu and Joly, 1985).

Contrasting with the induction effect of long-term consumption of ethanol, with short-term administration inhibition of hepatic drug metabolism is seen (Lieber, 1982) (Fig. 2B). For instance, whereas long-term ethanol consumption leads to increased hepatic microsomal metabolism of methadone and decreased levels in the brain and liver, short-term administration has the opposite effect: it inhibits microsomal demethylation of methadone and enhances brain and liver concentrations of the drug (Borowsky and Lieber, 1978). These effects may be of clinical relevance, since approximately 50 percent of patients taking methadone are also alcohol abusers. The combination of ethanol with tranquilizers and barbiturates also results in increased drug concentrations in the blood, sometimes to dangerously high levels (Lieber, 1982). One mechanism of interaction is direct competition for a common metabolic process involving cytochrome P-450 (Lieber, 1982).

Alcoholics are known to be highly susceptible to the adverse effects of xenobiotics, as reviewed elsewhere (Lieber, 1990b). This is due in part to an increased activity of hepatic microsomal enzymes resulting from prolonged alcohol consumption which augments the toxicity of agents that are converted to toxic metabolites in the microsomes. This pertains particularly to those substrates for which the alcohol inducible cytochrome P-450, when compared with other P-450s, displays an enhanced capacity for conversion to hepatotoxic metabolites. This is the case for acetaminophen and carbon tetrachloride (CCl_4) (Morgan et al., 1983; Johansson and Ingelman-Sundberg, 1985).

It is known that CCl_4 exerts its toxicity after conversion to an active compound in the microsomes, and we found that alcohol pretreatment remarkably stimulates the toxicity of CCl_4 (Hasumura et al., 1974). The conversion of $^{14}CCl_4$ to $^{14}CO_2$ and covalent binding of CCl_4 metabolites to protein were significantly accelerated in microsomes of ethanol-pretreated rats (Hasumura et al., 1974). Furthermore, pretreatment of rats with ethanol or rabbits with either imidazole or pyrazole, agents known to induce the ethanol-specific form of liver microsomal cytochrome P-450 (P450IIE1), caused 3 to 25-fold enhanced rates of CCl_4-dependent lipid peroxidation or chloroform production in isolated liver microsomes (Johansson and Ingelman-Sundberg, 1985).

Enhanced liver toxicity after chronic alcohol consumption also occurs with various other industrial solvents such as *bromobenzene*. Indeed, hepatotoxicity of bromobenzene was increased after chronic ethanol treatment (Hetu et al., 1983). The enhanced vulnerability of the alcoholic also pertains to *anaesthetic agents*. Hepatic microsomal enflurane defluorination increased 10-fold following chronic treatment (Tsutsumi et al., 1990) and this was inhibited with antibodies against P450IIE1. The same mechanism of hepatotoxicity also pertains to some "over the counter" medications. *Acetaminophen* (paracetamol, N-acetyl-p-aminophenol) intake in doses much smaller than those taken for suicidal attempts has been shown to produce hepatic injury (Prescott and Wright, 1973). Even smaller amounts well within the accepted tolerable range (2.5-6 g) have been incriminated as the cause of hepatic injury in alcoholic patients (Black, 1984; Seef et al., 1986). Experimentally, after chronic ethanol feeding, enhanced urinary excretion and covalent binding of reactive metabolite(s) of acetaminophen to microsomes from ethanol-fed rats was observed by Sato et al. (1981) and potentiation of acetaminophen hepatotoxicity occurred after ethanol withdrawal (Sato et al., 1981). In addition, depletion of reduced glutathione after both ethanol and acetaminophen may contribute to the toxicity of each compound, as discussed in greater detail subsequently. Unlike pretreatment with alcohol, which accentuates toxicity, the presence of ethanol in fact prevented the acute acetaminophen-induced hepatotoxicity, most likely because of the inhibition of the biotransformation of acetaminophen to reactive metabolites (C. Sato and Lieber, 1981; Altomare et al., 1984a,b) (Fig. 2B). The selective perivenular induction of P450IIE1, as visualized by the proliferation of the smooth endoplasmic reticulum (Iseri et al., 1966; Lane and Lieber, 1966) and the specific induction of P450IIE1 (Tsutsumi et al., 1989) in that zone after ethanol, is one of the mechanisms for the selective perivenular zonal toxicity of ethanol as well as that of various xenobiotic agents activated by the induced P450IIE1.

The increased activity of microsomal NADPH oxidase after ethanol consumption (Lieber and DeCarli, 1970c) may result in enhanced superoxide and hydrogen peroxide production, thereby theoretically favoring liver injury through lipid peroxidation (*vide infra*). Furthermore, it has been shown that long-term ethanol consumption is accompanied by increased formation of *hydroxyl radicals* (Krikun et al., 1984) and that P450IIE1 may be involved in this process in rabbits (Ingelman-Sundberg and Johansson, 1984). However, when the effect of long-term alcohol feeding on lipid peroxidation was studied in rat liver microsomes, no correlation with the generation of hydroxyl radicals was observed (Shaw et al., 1984a). Other radicals, such as the hydroxyethyl radical (Reinke et al., 1987), are also produced in liver microsomes, and some of these may initiate lipid peroxidation. Ethanol pretreatment clearly leads to enhancement of N-*nitrosodimethylamine* (NDMA) injury of the liver (Lorr et al., 1984), reflected by a 6-fold increase in ALT elevation (Maling et al., 1975). Ethanol has a unique effect on NDMA because P450IIE1 functions as a NDMA demethylase at low NDMA concentrations, both in rodents and in humans (Song et al., 1986). NDMA is not only a hepatotoxin, but it is also a carcinogen (*vide infra*).

There is an association between alcohol abuse and an increased incidence of upper alimentary and respiratory tract cancers (Lieber et al., 1986). Many factors have been incriminated, one of which is the effect of ethanol on enzyme systems involved in the cytochrome P-450-dependent activation of carcinogens. This effect has been demonstrated with the use of microsomes derived from a variety of tissues, including the liver (the principal site of xenobiotic metabolism) (Garro et al., 1981; Farinati et al., 1985), the lungs (Seitz et al., 1981; Farinati et al., 1985), and intestines (Seitz et al., 1978, 1981) (the chief portals of entry for tobacco smoke and dietary carcinogens, respectively), and the esophagus (Farinati et al., 1985) (where ethanol consumption is a major risk factor in the development of cancer). Alcoholics are commonly heavy smokers, and a synergistic effect of alcohol consumption and smoking on cancer development has been described (Lieber et al., 1986). Long-term ethanol consumption was found to enhance the mutagenicity of tobacco-derived products (Garro et al., 1981). Carcinogenesis may also be affected by altered vitamin A status. Even at the early fatty-liver stage, alcoholics commonly have a very low hepatic concentration of vitamin A, despite normal levels of circulating vitamin A and the absence of obvious dietary vitamin A deficiency (Leo and Lieber, 1982). Ethanol administration in animals has been shown to depress hepatic levels of vitamin A, even when administered with diets containing large amounts of vitamin A (M. Sato and Lieber, 1981), reflecting in part accelerated microsomal degradation of the vitamin. New hepatic pathways of microsomal retinol metabolism, inducible by either ethanol or drug administration, have been discovered (Leo and Lieber, 1985; Leo et al., 1987). Furthermore, reconstituted systems with purified forms of cytochrome P-450 have been used to show that retinoic acid (Leo et al., 1984) and retinol (Leo and Lieber, 1985; Leo et al., 1989) can serve as substrates for microsomal oxidation. The hepatic depletion of vitamin A was strikingly exacerbated when ethanol and other drugs were given together (Leo et al., 1987), mimicking a rather common clinical combination. Furthermore, ethanol contributes to hepatic retinoid depletion by enhanced mobilization of the vitamin (Leo et al., 1986a). Hepatic vitamin A depletion is associated with lysosomal lesions (Leo et al., 1983), decreased detoxification of NDMA (Leo et al., 1986b), and probably scores of other adverse effects, possibly including the formation of hyaline bodies (Mallory bodies) (French, 1981). Thus, vitamin A supplementation should be given to the alcoholic patient not only to correct problems of night blindness and sexual inadequacies, but also to alleviate potential liver dysfunction. Such therapy, however, is complicated by the fact that excessive amounts of vitamin A are hepatotoxic (Leo and Lieber, 1988). Long-term ethanol consumption enhances this effect, resulting in striking morphologic and functional alterations of the mitochondria (Leo et al., 1982), along with hepatic necrosis and fibrosis (Leo and Lieber, 1983). Hypervitaminosis A itself can induce fibrosis and even cirrhosis, as reviewed elsewhere (Leo and Lieber, 1988).

ADVERSE EFFECTS OF ACETALDEHYDE

Chronic ethanol consumption results in a significant reduction of the capacity of rat mitochondria to oxidize acetaldehyde (Hasumura et al., 1975). The decreased capacity of mitochondria of alcohol-fed subjects to oxidize acetaldehyde, associated with unaltered or even enhanced rates of ethanol oxidation (and therefore acetaldehyde generation) may result in an imbalance between production and disposition of acetaldehyde. Such a mechanism may result in the elevated acetaldehyde levels observed after chronic ethanol consumption in rats (Koivula and Lindros, 1975), in man (Korsten et al., 1975; Di Padova et al., 1987), and in baboons (Pikkarainen et al., 1981).

Acetaldehyde-Protein Adduct Formation, Immune Response and Microtubular Alterations

Nomura and Lieber (1981) have reported covalent binding of exogenously added acetaldehyde to the proteins of liver microsomes and an even greater effect when acetaldehyde was generated from ethanol in situ. This binding was significantly increased after long-term ethanol consumption and paralleled the induction of MEOS activity. We now found that this process occurs in vivo and that acetaldehyde selectively forms a stable adduct in rat liver microsomes with the ethanol-inducible P450IIE1 (Behrens et al., 1988). Using an animal model, Israel et al. (1986) demonstrated that acetaldehyde adducts may serve as neoantigens generating an immune response in mice. Our own studies (Hoerner et al., 1986) have shown that antibodies against acetaldehyde adducts (produced in vitro) are present in the serum of most alcoholics. Hepatic injury initially may promote the release of significant amounts of acetaldehyde-altered proteins and the severity of liver disease plays a role in the appearance of circulating antibodies against acetaldehyde adducts, not only in alcoholics, but also in some non-alcoholics with liver injury (Hoerner et al., 1988). In turn, complement-binding acetaldehyde-adduct-containing immune complexes may contribute to the perpetuation or exaggeration of liver disease. This could represent one of the immune mechanisms - cell mediated as well as humoral - which can play a role in the pathogenesis of alcoholic liver injury. The sulfhydryl groups of cysteine residues, which are present in tubulin, are involved in polymerization, resulting in microtubule formation. Acetaldehyde has a high affinity for sulfhydryl groups, and its binding to these, as well as to lysine, may alter the capacity of tubulin to polymerize. One of the key functions of microtubules is to promote the intracellular transport of proteins and their secretion. Long-term alcohol feeding seriously delayed the secretion of proteins into the plasma, with a corresponding retention in the liver (Baraona et al., 1977) thereby contributing to the hepatomegaly. In animals fed alcohol-containing diets, it was shown that lipids account for only half the increase in liver dry weight (Lieber et al., 1965), while the other half is almost totally accounted for by an increase in proteins (Baraona et al., 1975) possibly secondary to acetaldehyde induced impairment of microtubular mediated protein secretion. Indeed, hepatic microtubules are decreased in alcoholic liver disease (Matsuda et al., 1979, 1985). Experimental ethanol feeding results in enlargement of the hepatocytes in association with a decrease in the concentration of total tubulin (Baraona et al., 1975, 1977) and a decreased polymerization of tubulin to microtubules (Matsuda et al., 1979; Baraona et al., 1981) -- an effect that has been confirmed by morphometric studies (Matsuda et al., 1979). One suspects that ballooning and associated gross distortion of the volume of the hepatocytes may result in severe impairment of key cellular functions. In alcoholic liver disease, some cells not uncommonly have a diameter which is increased 2-3 times, and thereby volume is increased 10 fold or more. One may wonder to what extent this type of cellular disorganization, with protein retention and ballooning, may promote progression of the liver injury in the alcoholic.

Structural and Functional Alterations of the Mitochondria and Promotion of Lipid Peroxidation: Interaction with Cysteine, Glutathione and Vitamin E

Controlled studies in animals and man (Iseri et al., 1966; Lane and Lieber, 1966; Rubin and Lieber, 1967a,b; Lieber and Rubin, 1968) have revealed striking morphologic alterations, including swelling and abnormal cristae, in the liver mitochondria. Mitochondria of alcohol fed animals have a reduction in cytochrome a and b content (Rubin et al., 1970; Koch et al., 1977). The respiratory capacity of the mitochondria was observed to be depressed (Kiessling and Pilstrom, 1968; Rubin et al., 1972; Gordon, 1973; Hasumura et al., 1975). Oxidative phosphorylation was found to be altered (Cederbaum et al., 1974a). It is noteworthy that high concentrations of acetaldehyde mimic the defects produced by chronic ethanol consumption on oxidative phosphorylation (Cederbaum et al., 1974b). One may wonder to what extent chronic exposure to acetaldehyde is the cause for the defect observed after chronic ethanol consumption. Indeed, after chronic alcohol consumption, the liver mitochondria are unusually susceptible to the toxic effects of acetaldehyde, and a variety of important mitochondrial functions, such as fatty acid oxidation, are depressed, even in the presence of relatively low acetaldehyde concentrations (Matzusaki and Lieber, 1977). This mitochondrial injury is probably responsible for the impairment in oxygen utilization by the liver observed after chronic ethanol administration (Lieber et al., 1989). Binding of acetaldehyde with cysteine and/or glutathione (Fig. 1) may contribute to a depression of liver glutathione (Shaw et al., 1981). Acute ethanol administration inhibited GSH synthesis and produced an increased loss from the liver (Speisky et al., 1985). A severe reduction in glutathione favors peroxidation (Wendel et al., 1979), and the damage may possibly be compounded by the increased generation of active radicals by the "induced" microsomes following chronic ethanol consumption (*vide supra*). The glutathione depletion can be offset, in part, by the administration of S-adenosylmethionine, with an associated attenuation of liver lesions (including mitochondrial alterations) and enzyme leakage into the bloodstream (Lieber et al., 1990a). Ethanol induced lipid peroxidation was even more striking in the baboon (Shaw et al., 1981): administration of relatively small doses of ethanol (1-2g/kg) after 5-6 hours produced lipid peroxidation and GSH depletion. In the baboon chronically fed alcohol (50% of total calories for 1-4 years) alcoholic liver disease, including cirrhosis in some, developed and such animals exhibit evidence of enhanced hepatic lipid peroxidation and GSH depletion. These changes were observed following an overnight withdrawal from ethanol and were exacerbated by the readministration of ethanol. Evidence for GSH depletion and lipid peroxidation (enhanced diene conjugates) was found in liver biopsies of alcoholics who were withdrawn from alcohol (Shaw et al., 1983). α-Tocopherol, the major antioxidant in the membrane, is viewed as the "last line" of defense against membrane lipid peroxidation. Kawase et al. (1989) found that hepatic lipid peroxidation was significantly increased after chronic ethanol feeding in rats receiving a low vitamin E diet, indicating that dietary vitamin E is an important determinant of hepatic lipid peroxidation induced by chronic ethanol feeding. Both low dietary vitamin E and ethanol feeding significantly reduced hepatic α-tocopherol content. Together with acetaldehyde this may potentiate lipid peroxidation. Another major adverse effect of acetaldehyde pertains to fibrogenesis (*vide infra*).

Precursor Lesions and Cells Associated with Fibrogenesis

The presently prevailing view is that alcoholic cirrhosis develops in response to alcoholic hepatitis, characterized by ballooning of the hepatocytes, extensive necrosis and polymorphonuclear inflammation. It is understandable that necrosis and inflammation may trigger the scarring process of cirrhosis, but one must question whether this is the only mechanism involved. In some populations in the U.S. (Nakano et al., 1982; Worner and Lieber, 1985) as well as in Japan (Takada et al., 1982; Minato et al., 1983; Hasumura et al., 1985) cirrhosis commonly develops in alcoholics without an apparent intermediate stage of florid alcoholic hepatitis . This observation raised the question of whether alcohol can promote development of cirrhosis without a stage of alcoholic hepatitis. Indeed, in baboons fed alcohol, fatty liver developed, the hepatocytes increased in size, and there was some obvious ballooning. This ballooning was associated with some mononuclear inflammation but very few of the polymorphonuclear cells so characteristic of human alcoholic hepatitis (Popper and Lieber, 1980). Although some clumping was apparent in the cytoplasm, by electron microscopy there was no alcoholic hyaline. Thus there was no florid picture of alcoholic hepatitis; yet, in one-third of the animals, typical cirrhosis developed. Thus, it appears that full blown alcoholic hepatitis may not be a necessary intermediate step in the development of alcoholic cirrhosis. Sequential clinical observations support this contention (Worner and Lieber, 1985) (Fig. 3). This hypothesis in turn raises the question of which process, in the absence of alcoholic hepatitis, may initiate cirrhosis upon chronic consumption of ethanol. It is possible that cirrhosis develops in the presence of minimal inflammation and necrosis, which may suffice to trigger the fibrosis. On the other hand, it is also conceivable that, independent of the necrosis and inflammation, alcohol may have some more direct effects on the metabolism of collagen. In alcoholic liver injury there is a great variability in the magnitude of collagen disposition.

At the earlier stages, in the so-called simple or uncomplicated fatty liver, collagen is detectable by chemical means only (Feinman and Lieber, 1972). When collagen deposition is sufficient to become visible by light microscopy, usually it appears first around the central (also called terminal) venules, resulting in so-called "pericentral" or "perivenular" fibrosis or sclerosis, discussed before. Sequential biopsies in alcohol fed baboons revealed that in some animals, already at the early fatty liver stage, an increased number of mesenchymal cells, particularly myofibroblasts, appear in the perivenular areas (Nakano and Lieber, 1982). Myofibroblasts are present in normal liver and after alcohol feeding (Lieber et al., 1981). After ethanol, proliferation of these cells is eventually accompanied by deposition of abundant collagen bundles, first in the perivenular areas, leading to perivenular and pericellular fibrosis and ultimately to diffuse fibrosis and cirrhosis. Myofibroblasts have been isolated from baboon surgical liver biopsies by collagenase digestion and Percoll density gradient centrifugation (Savolainen et al., 1984a). By immunofluorescence , the cells synthesized collagen types I, III, and IV and laminin. Myofibroblasts processes also extended into the Disse space. There, however, the most common mesenchymal cell is the lipocyte (also called fat storing or Ito cell).

Lipocytes, myofibroblasts and fibroblasts may belong to the same cell family. Transformation of Ito cells into myofibroblasts has been suggested (Farrell et al., 1977). Lipocytes are characterized electron microscopically by abundant lipid droplets (Mak et al.,

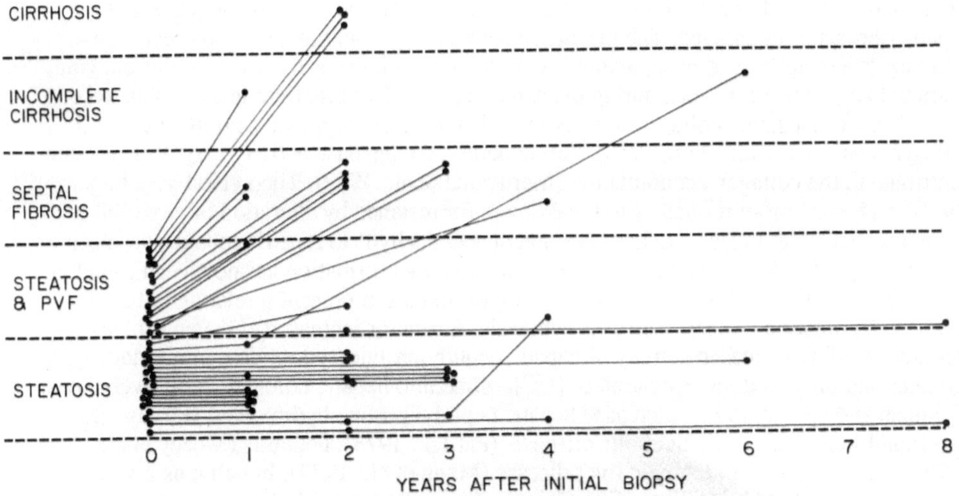

Fig. 3. Progression of fibrosis in alcoholics without hepatitis. Follow-up to 8 years after the initial biopsy. The presence of perivenular fibrosis (PVF) in the initial biopsy is a harbinger of rapid development of fibrosis to more severe stages, including cirrhosis (From Worner and Lieber, 1985).

1984; Mak and Lieber, 1988), microfilament bundles, dense bodies and pinocytic vesicles. After chronic alcohol consumption, about half the lipocytes are replaced by transitional cells between lipocytes and fibroblasts; these transitional cells were arbitrarily defined by a lipid droplet volume of less than 20%, measured by morphometry (Mak et al., 1984; Mak and Lieber, 1988). The area of the rough endoplasmic reticulum in transitional cells is greater than in lipocytes. Transitional cells have abundant microfilaments, dense bodies and pinocytic vesicles. They resemble myofibroblasts in the perivenular zones but they can be differentiated by the lack of surrounding basal laminae. Moreover, myofibroblasts typically have an indented nucleus. Transitional cells are surrounded by abundant collagen fibers and can be seen in the perisinusoidal spaces throughout the lobule associated with a net-like fibrosis, sometimes linking up with the perivenular lesion. There was good correlation between hepatic fibrosis, the percent of transitional cells and the area of their rough endoplasmic reticulum (Mak et al., 1984). Already at the fatty liver stage, these non parenchymal cells participate in the production of collagen but we do not know the respective role of parenchymal and non parenchymal cells in this process. In addition to the extensive evidence concerning the role of myofibroblasts and Ito cells (with transitional cells) in the production of collagen, several studies reported evidence that hepatocytes can also be contributory, as discussed in more detail elsewhere (Lieber, 1990a).

Mechanisms of Collagen Accumulation

Even in the presence of an adequate diet, alcohol can produce a fatty liver and associated hepatic fibrosis and cirrhosis in the baboon (Lieber and DeCarli, 1974), thereby clearly indicating the capacity of ethanol to produce fibrosis, even when associated with diets containing all nutrients presently known to be required. The accumulation of hepatic collagen during the development of cirrhosis could theoretically be accomplished by increased synthesis, decreased degradation, or both. The rate of hepatic fibrous tissue degradation has never been directly measured in human alcoholic liver disease or in any of the animal models of alcoholic liver disease. Therefore its role in the pathogenesis of hepatic fibrosis is unresolved. The mechanisms of collagen degradation in the liver are complex. There appears to be a paradoxical increase in collagenase activity of this enzyme in animals fed ethanol, at least during the early stage of alcoholic liver injury (Okazaki et al., 1977). At that stage, collagenolytic activity is increased, together with the increased collagen synthesis (Kato et al., 1985). Subsequently, collagenase activity may decrease and contribute to the collagen accumulation (Maruyama et al., 1982). Theoretically, collagen may also become more refractory to breakdown, for instance by alteration of cross linking due to acetaldehyde. Indeed, covalent binding of acetaldehyde to type III collagen has been described by Jukkola and Niemelä (1989). Similar adduct formation has now been found in vivo (Behrens et al., 1989). Thus, cirrhosis might, in part, represent a relative failure of collagen degradation to keep pace with synthesis. A role for increased collagen synthesis was suggested by increased activity of hepatic peptidylproline hydroxylase in rats and primates and increased incorporation of [^{14}C]proline into hepatic collagen in rat liver slices (Feinman and Lieber, 1972). Increased hepatic peptidyl proline hydroxylase activity was also found in patients with alcoholic cirrhosis (Patrick, 1973), hepatitis (Mezey et al., 1979) or in all stages of alcoholic liver disease (Mann et al., 1979). In baboons given ethanol that developed significant fibrosis, type I procollagen mRNA content was significantly higher (per liver RNA) as determined by hybridization analysis (Zern et al., 1985). Similarly in vitro, mRNA for type I collagen was increased by acetaldehyde in lipocytes in vitro (Casini et al., 1990) in association with increased collagen production (Moshage et al., 1990). Acetaldehyde results in increased collagen production in myofibroblasts (Savolainen et al., 1984a). Acetaldehyde also increases collagen synthesis in fibroblasts (Holt et al., 1984) with increased messenger RNA (Brenner and Chojkier, 1987). Another possible mechanism whereby alcohol consumption may be linked to collagen formation is the increase in tissue lactate secondary to alcohol metabolism. Elevated concentrations of lactate increase peptidylproline hydroxylase activity, and lactate may also play a role through inhibition of proline oxidase and expansion of the hepatic free proline pool size, which has been incriminated in the regulation of collagen synthesis. However, a high hepatic concentration of free proline did not induce collagen synthesis, at least in rat liver (Forsander et al., 1983). Lactate also stimulated collagen synthesis in myofibroblast cultures (Savolainen et al., 1984a). The mechanism whereby acetaldehyde and other putative mediators of the ethanol effects act on collagen production and increased mRNA for collagen have not been elucidated as yet. Protein synthesis is involved, whereas TGF was not incriminated (Casini et al., 1990). A number of other factors may modulate the effects of alcoholic fibrogenesis including nutritional factors, as discussed in detail elsewhere (Lieber and Leo, 1986).

Blood Tests for the Detection of Hepatic Fibrosis and Precursor Lesions of Cirrhosis

As mentioned before, the detection of perivenular fibrosis seems to indicate that a subject has already entered the fibrotic process and therefore is likely, upon continuation of drinking, to rapidly progress to more severe stages, including cirrhosis. At present, liver biopsy is required to obtain this information. Efforts are being conducted to develop easier and more widely applicable techniques to achieve detection of alcoholic liver injury at an early and still reversible stage.

Hyperprolinemia has been reported as a specific and significant feature of alcoholic liver fibrosis (Mata et al., 1975). However, in a more recent study (Shaw et al., 1984b), hyperprolinemia occurred in only a minority of patients with alcoholic liver disease, including cirrhosis, if they were withdrawn from alcohol. Furthermore, patients with severe liver disease unrelated to alcohol may have marked elevations in plasma proline. Other measurements used in the search for blood markers of liver fibrogenesis include serum procollagen peptides. Propeptides are cleaved (in the process of collagen formation) from the procollagen and are released into the blood stream. Serum propeptides of type III collagen measured by radioimmunoassay are elevated in alcoholic liver disease including steatosis, portal fibrosis, cirrhosis, and alcoholic hepatitis (Rhode et al., 1979). The serum concentration of this peptide correlated with the fibrotic activity in the liver in chronic liver diseases (Ackerman et al., 1981). In another study, both propeptides I and III were markedly elevated in alcoholic hepatitis and cirrhosis (Savolainen et al., 1984b). Moderately increased values were found less frequently in patients with fatty liver. These tests, however, did not differentiate patients with fatty liver from those with fatty liver and early fibrosis. The difference in serum procollagen type III propeptide between fatty liver and both alcoholic hepatitis and cirrhosis was significant. The propeptide concentrations seemed to relate to inflammation, and the highest values were found in a subgroup of patients with alcoholic hepatitis and numerous Mallory bodies. The results suggest that in alcoholic liver disease serum collagen propeptide concentrations give information on the activity of the disease process in the liver , but do not detect early stages of fibrosis. If the Fab fragments of the antibody against procollagen III peptides are used, however, elevated blood PIIIP levels within 1 week of alcohol withdrawal can be diagnostic of hepatic fibrosis: in more than half of the subjects, significant increases occurred at the perivenular fibrosis stage and thus could be used to detect a majority of those subjects prone to develop cirrhosis with continued drinking (Sato et al., 1985). If the test can be further perfected, Fab-PIIIP blood levels might become a realistic and more practical alternative to liver biopsy for the early recognition of precirrhotic stages.

Therapeutic Approaches

Colchicine, which inhibits collagen synthesis and procollagen secretion in embryonic tissue, may provide a new approach for the treatment of alcoholic liver injury (Kershenobich et al., 1988). Experimentally, various compounds, such as prostaglandin derivatives (Ruwart et al., 1988) and proline hydroxylase inhibitors (Fujiwara et al., 1988) have been shown to delay collagen formation. None of these have had clinical applications as yet. Polyunsaturated lecithin (PUL) has recently shown promising results in the prevention of alcohol induced fibrosis (Lieber et al., 1990b). This was demonstrated in a 10 year study conducted in baboons: 12 were fed dilinoleoylphosphatidylcholine-rich PUL with either ethanol (50% of total energy) or isocaloric carbohydrate, up to 8 years. They were compared to another 18 baboons fed an equivalent amount of the same diet (with or without ethanol) but devoid of PUL. In the two groups, the ethanol-fed animals developed comparable increases in lipids, but striking differences in the degree of fibrosis. Whereas 7/9 baboons fed the regular diet with ethanol developed septal fibrosis (with cirrhosis in 2), none of the animals fed PUL reached the stage of septal fibrosis ($p<.005$). The most severe lesions they developed were perivenular or interstitial fibrosis; they also had significantly less transformation of lipocytes to transitional cells (55 vs 81%; $p<.01$), as determined by electron microscopy. Furthermore, when 3 of these animals were taken off PUL, but continued on the same amount of the ethanol-containing diet, all progressed (within 18-21 months) to cirrhosis, with an increased transformation of lipocytes to transitional cells (from 55 to 82%). These results indicate that some component of PUL exerts a protective action against the fibrogenic effects of ethanol. Since we previously found that choline, in amounts present in PUL, has no comparable effect (Lieber et al., 1985), the polyunsaturated phospholipids themselves might be responsible, because of their high bioavailability and selective incorporation into liver membranes . Their effect on fibrogenesis was then verified in rat liver lipocytes and similar cells isolated from baboons (Lieber et al., 1990c). PUL prevented the acetaldehyde stimulated collagen production in vitro and it also stimulated collagenase activity 2-3 fold, thereby suggesting the following possible mechanism of protection: by activating collagenase, PUL may offset the stimulatory effect of acetaldehyde on fibrogenesis, whereas the latter may be unleashed upon PUL withdrawal.

The capacity of ethanol to produce cirrhosis in the baboon despite an adequate diet has been challenged by Ainley et al. (1988). The diet administered by Ainley et al.(1988), however, was not fully defined. Furthermore, there is some question concerning the amount of alcohol administered to Ainley's animals: surprisingly high and actually lethal amounts of alcohol (25 g/kg of body weight per day) were reportedly administered, but the animals survived. Their blood ethanol levels were not higher than those observed with four to five times less alcohol (namely 5-6 g/kg/day) (Lieber et al., 1975; Jauhonen et al., 1982). In fact, as acknowledged by Ainley et al. (1988), it is not known how much alcohol (or diet) was actually consumed by their animals. Moreover, as pointed out by French (1989), the diet was administered only to 2 control animals, 4 ethanol-fed baboons and 4 animals fed ethanol with a zinc supplement, without pair-feeding and with mixed gender. It is therefore difficult to make significant comparisons between the experimental groups. In addition, only two of these baboons were given ethanol with the regular alcohol containing diet for a period exceeding 18 months. Since the results of other studies (Lieber and DeCarli, 1974; Popper and Lieber, 1980; Lieber et al., 1990a) showed that a longer period of treatment is required to produce septal fibrosis or cirrhosis, it is obvious that the number of animals studied by Ainley et al. (1988) do not allow for definitive conclusions. By contrast, in the aggregate of our various studies (Lieber et al., 1975; Popper and Lieber, 1980; Lieber et al., 1985; Lieber et al., 1990b), production of cirrhosis was observed in 13 of a total of 63 baboons fed ethanol for three years or more, with septal fibrosis developing in an additional 13 animals. The capacity of ethanol to produced fibrosis in well fed baboons was confirmed by Porto et al. (1989a,b); these lesions were less severe than those produced by Lieber and DeCarli (1974) and Lieber et al. (1975) but the amount of alcohol administered was also smaller and the duration of the study was shorter.

NUTRITIONAL FACTORS

Until two decades ago, primary malnutrition (due to dietary deficiencies) was considered the main cause of liver disease in the alcoholic. As the overall nutrition of the population improved, more emphasis was placed on secondary malnutrition resulting from the maldigestion and malabsorption caused by long-term consumption of alcohol (Lieber, 1982). This includes malabsorption of thiamine and folate as well as maldigestion due to pancreatic insufficiency. At the same time, toxic effects of ethanol were determined that provided insight into processes whereby ethanol alters the degradation of key nutrients, such as pyridoxine, as discussed in detail elsewhere (Lieber, 1988), as well as vitamin A (*vide supra*). Such enhanced degradation results in increased vitamin requirements.

Similarly, protein requirements may be increased by alcohol (Bunout et al., 1987), but protein tolerance may be reduced in alcoholic liver disease (Lieber, 1982). Nevertheless, protein deficiency, when present, should be corrected, in view of the adverse effects on the metabolism of ethanol (Wilson et al., 1986) and the potentiating effect of protein deficiency on alcohol-induced liver injury (Lieber, 1982). There seems to be no comparable indication for the supplementation with lipotropic agents. Indeed, massive doses of choline not only failed to prevent the development of alcoholic cirrhosis but also exerted some toxic effects (Lieber et al., 1985). Thus, treatment with various nutritional factors must take into account a narrowed therapeutic window in alcoholic patients who, in spite of an increased need for some nutrients, nevertheless have an enhanced susceptibility to the adverse effects of nutrient excess.

CONCLUSIONS

Three decades of research in ethanol metabolism have established the hepatotoxicity of alcohol including the capacity to produce cirrhosis, even in the absence of dietary deficiencies. These studies have also defined alterations produced by the oxidation of ethanol by alcohol dehydrogenase and the associated oxidation-reduction change, including impairments in carbohydrate, lipid, and protein metabolism. Another key accomplishment was the ethanol-inducible microsomal cytochrome P-450 (P450IIE1) which was found to play a significant role, not only in terms of ethanol metabolism and ethanol tolerance, but also with regard to the activation of a number of xenobiotics to toxic metabolites. The unique potency of P450IIE1 to activate xenobiotic agents now provides a better understanding for the increased vulnerability of the heavy drinker to the hepatotoxicity of industrial solvents, commonly used drugs, over-the-counter medications, chemical carcinogens and even nutritional factors such as vitamin A. The induction of this new pathway may also result in energy wastage and leads to increased production of acetaldehyde, a very reactive compound. Acetaldehyde, in turn, causes injury through the formation of adducts with proteins, resulting in antibody formation, enzyme inactivation, decreased DNA repair, and alterations in microtubules, mitochondria, and plasma membranes. Acetaldehyde also promotes glutathione depletion, toxicity mediated by free radicals, lipid peroxidation, and hepatic collagen synthesis. The involvement of myofibroblasts and transitional cells derived from interstitial fat-storing cells was demonstrated. Precursor lesions have been recognized that might be helpful in the early detection of the process which ultimately leads to the development of cirrhosis. Experimentally several agents have shown promise in the prevention and treatment of fibrosis. A better understanding of the biochemical alterations produced by ethanol in the body, furthermore, provided insight into processes whereby ethanol alters both the activation and the degradation of key nutrients. Thus, the original dichotomy between nutritional and toxic effects of ethanol has now been bridged. Unraveling the molecular biology of ethanol metabolism may foster our ultimate goal: the detection, at an even earlier stage, of subjects who are biologically predisposed to the most severe complications of alcoholism.

ACKNOWLEDGEMENTS

Original studies reviewed were supported by the Department of Veterans Affairs and DHHS grants AA03508, AA05934, AA07802 and DK 32810, and the Alcoholic Beverage Medical and Kingsbridge Research Foundations, presented at the ISBRA Meeting, Toronto, Canada, June 1990, and published, in part, in the corresponding proceedings.

REFERENCES

Ackerman, W., Pott, G., Voss, B., Muller, K-M., and Gerlach, U. 1981. Serum concentrations of procollagen peptide in comparison with the serum activity of N-acetylglucosaminidase for diagnosis of the activity of liver fibrosis in patients with chronic active liver disease. **Clin. Chim. Acta**, 112, 365-369.

Ainley, C.C., Senapati, A., Brown, I.M.H., Slavin, B.M., Davies, D.R., Keeling, P.W.N., and Thompson, R.P.H. 1988. Is alcohol hepatotoxic in the baboon. **J. Hepatology**, 7, 8-92.

Alderman, J. Takagi, T., and Lieber, C.S. 1987. Ethanol metabolizing pathways in deermice:estimation of flux calculated from isotope effects. **J. Biol. Chem.**, 262, 7497-7503.

Alderman, J. Takagi, T., and Lieber, C.S. 1987. Ethanol metabolizing pathways in deermice:estimation of flux calculated from isotope effects. **J. Biol. Chem.,** 262, 7497-7503.
Alderman, J., Kato, S., and Lieber, C.S. 1989. The microsomal ethanol oxidizing system mediates metabolic tolerance to ethanol in deermice lacking alcohol dehydrogenase. **Arch. Biochem. Biophys.,** 271, 33-39.
Altomare, E., Leo, M.A., and Lieber, C.S. 1984a. Interaction of acute ethanol administration with acetaminophen metabolism and toxicity in rats fed alcohol chronically. **Alcoholism: Clin. Exp. Res.**, 8, 405-408.
Altomare, E., Leo, M.A., Sato, C., Vendemiale, G., and Lieber, C.S. 1984b. Interaction of ethanol with acetaminophen metabolism in the baboon. **Biochem. Pharmacol.**, 33, 2207-2212.
Baraona, E., Leo, M.A., Borowsky, S.A., and Lieber, C.S. 1975. Alcoholic hepatomegaly:Accumulation of protein in the liver. **Science**, 190, 794-795.
Baraona, E., Leo, M.A., Borowsky, S.A., and Lieber, C.S. 1977. Pathogenesis of alcohol-induced accumulation of protein in the liver . **J. Clin. Invest.**, 60, 546-554.
Baraona, E., Matsuda, Y., Pikkarainen, P., Finkelman, F., and Lieber, C.S. 1981. Effects of ethanol on hepatic protein secretion and microtubules. Possible mediation by acetaldehyde. In **Currents in Alcoholism**, (M. Galanter, Ed.) Vol. VIII, pp. 421-434, Grune and Stratton, New York.
Behrens, U.J., Hoerner, M., Lasker, J.M., and Lieber, C.S. 1988. Formation of acetaldehyde adducts with ethanol-inducible P450IIE1 in vivo. **Biochem. Biophys. Res. Commun.**, 154, 584-590.
Behrens, U.J., Ma, X-L., Baraona, E., and Lieber, C.S. 1989. Acetaldehyde-collagen adducts in CCl4-induced liver injury in rats. **Hepatology**, 10, 608.
Black, M., 1984. Acetaminophen hepatotoxicity. **Ann. Rev. Med.**, 35, 577-593.
Borowsky, S.A., and Lieber, C.S. 1978. Interaction of methadone and ethanol metabolism. **J. Pharmacol. Exp. Ther.**, 207, 123-129.
Brenner, D.A., and Chojkier, M. 1987. Acetaldehyde increases collagen gene transcription in cultured human fibroblasts. **J. Biol. Chem.**, 262, 17690-17695.
Brighenti, L., and Pancaldi, G. 1970. Effecto della simministrazione di alcohol etillico sualcune attivita enzimatiche del fegato di ratto. **Boll. Soc. Ital. Biol. Sper.**, 46, 1-5.
Buehler, R., Hess, M., and von Wartburg, J-P. 1982. Immunohistochemical localization of human liver alcohol dehydrogenase in liver tissue, cultured fibroblasts and hela cells. **Am. J. Pathol.**, 108, 89-99.
Bunout, D., Petermann, M., Ugarte, G., Barrera, G., and Iturriaga, H. 1987. Nitrogen economy in alcoholic patients without liver disease. **Metabolism**, 36, 651-653.
Burnett, E.G., and Felder, M.R. 1980. Ethanol metabolism in Peromyscus genetically deficient in alcohol dehydrogenase. **Biochem. Pharmacol.**, 29, 125-130.
Caballeria, J., Baraona, E., and Lieber, C.S. 1987. The contribution of the stomach to ethanol oxidation in the rat. **Life Sci.,** 41, 1021-1027.
Caballeria, J., Baraona, E., Rodamilans, M. and Lieber, C.S. 1989. Effects of cimetidine on gastric alcohol dehydrogenase activity and blood ethanol levels. **Gastroenterology,** 96, 388-392.
Carmichael, F.J., Saldivia, V., Israel, Y., McKaigney, J.P., and Orrego, H. 1987. Ethanol-induced increase in portal hepatic blood flow: interference by anesthetic agents. **Hepatology**, 7, 89-94.
Casini, A., Greenwel, P., Rojkind, M., and Lieber, C.S. 1990. Acetaldehyde increases procollagen type I and fibronectin mRNAs in cultured rat fat-storing cells through a protein synthesis-dependent mechanism. **Gastroenterology,** 98, 574.
Cederbaum, A.I., Lieber, C.S., and Rubin, E. 1974. Effects of chronic ethanol treatment on mitochondrial functions. **Arch. Biochem . Biophys.**, 165, 560-569.

Cederbaum, A.I., Lieber, C.S., and Rubin, E. 1974b. The effect of acetaldehyde on mitochondrial function. **Arch Biochem. Biophys.**, 161, 26-39.

Di Padova, C., Worner, T.M., Julkunen, R.J.K., and Lieber, C.S. 1987. Effects of fasting and chronic alcohol consumption on the first pass metabolism of ethanol. **Gastroenterology**, 92, 1169-1173.

Farinati, F., Zhou, Z., Bellah, J., Lieber, C.S., and Garro, A.J. 1985. Effect of chronic ethanol consumption on activation of nitrosopyrrolidine to a mutagen by rat upper alimentary tract, lung, and hepatic tissue. **Drug Metab. Dispos.**, 13, 210-214.

Farrell, G.C., Bathal, P.S., and Powell, L.W. 1977. Abnormal liver function in chronic hypervitaminosis A. **Dig. Dis. Sci**. 22, 724-7 28.

Feinman, L., and Lieber, C.S. 1972. Hepatic collagen metabolism: Effect of alcohol consumption in rats and baboons. **Science** 176, 79 5.

Forsander, O.A., Pikkarainen, J.A.J., Salaspuro, M. 1983. A high hepatic concentration of free proline does not induce collagen synthesis in rat liver. **Hepatogastroenterology**, 30, 6-8.

French, S.W. 1981. The Mallory body: structure, composition and pathogenesis. **Hepatology**,1, 76-83.

French, S.W. 1989. Alcoholic hepatotoxicity. **J. Hepatology**, 9, 134-135.

Frezza, M., Di Padova, C., Pozzato, G., Terpin, M., Baraona, E., and Lieber C.S. 1990. High blood alcohol levels in women: Role of decreased gastric alcohol dehydrogenase activity and first pass metabolism. **New Engl. J. Med.**, 322, 95-99.

Fujiwara, K., Ogata, I., Ohta, Y., Hayashi, S., Mishiro, S., Takatsyki, K., Yuzuru, S., Yamada, S., Hirata, K., Oka, H., Oda, T., Kawaji, H., Matsuda, S., Niiyama, Y., and Tsukuda, R. 1988. Decreased collagen accumulation by a prolylhydroxylase inhibitor in pig serum-induced fibrotic rat liver. **Hepatology,** 8, 804-807.

Garro, A.J., Seitz, H.K., and Lieber, C.S. 1981. Enhancement of dimethylnitrosamine metabolism and activation to a mutagen following chronic ethanol consumption. **Cancer Res.**, 41, 120-124.

Gordon, E.R. 1973. Mitochondrial functions in an ethanol-induced fatty liver. **J. Biol. Chem.**, 248, 8271-8280.

Hasumura, Y., Teschke, R., and Lieber, C.S. 1974. Increased carbon tetrachloride hepatotoxicity, and its mechanism, after chronic ethanol consumption. **Gastroenterology**, 66, 415-422.

Hasumura, Y., Teschke, R., and Lieber, C.S. 1975. Acetaldehyde oxidation by hepatic mitochondria: Its decrease after chronic ethanol consumption. **Science**, 189, 727-729.

Hasumura, Y., Minato, Y., Nishimura, M., Kaku, Y., Tozuka, S., Koyama, W., Tanaka, Y., and Takeuchi, J. 1985. Hepatic fibrosis in alcoholics: Morphologic characteristics, clinical diagnosis, and natural course. **Pathobiol. Hepatic Fibrosis,** 7, 13-24.

Hempel, J.D., and Pietruszko, R. 1979. Human stomach alcohol dehydrogenase: Isoenzyme composition and catalytic properties. **Alcoholism: Clin. Exp. Res.**, 3, 95-98.

Hernandez-Munoz, R., Caballeria, J., Baraona, E., Uppal, R., Greenstein, R., and Lieber, C.S. 1990. Human gastric alcohol dehydrogenase: its inhibition by H2-receptor antagonists, and its effect on the bioavailability of ethanol. **Alcoholism: Clin. Exp. Res.,** 14, 946-950.

Hetu, C., and Joly, J.G. 1985. Differences in the duration of the enhancement of liver mixed-function oxidase activities in ethanol-fed rats after withdrawal. **Biochem. Pharmacol**. 34, 1211-1216.

Hetu, C., Dumont, A., and Joly, J-G. 1983. Effect of chronic ethanol administration on bromobenzene liver toxicity in the rat. **Toxicol. Appl. Pharmacol.**, 67, 166-167.

Hoerner, M., Behrens, U.J., Worner, T., and Lieber, C.S. 1986. Humoral immune responses to acetaldehyde adducts in alcoholic patients. **Res. Commun. Chem. Pathol. Pharmacol.**, 54, 3-12.

Hoerner, M., Behrens, U.J., Worner, T., and Lieber, C.S. 1988. The role of alcoholism and liver disease in the appearance of serum antibodies against acetaldehyde adducts. **Hepatology**, 8, 569-574.

Holt, K., Bennett, M., and Chojkier, M. 1984. Acetaldehyde stimulates collagen and noncollagen protein production by human fibroblasts. **Hepatology** 4, 843-848.

Inatomi, N., Ito, D., and Lieber, C.S. 1990. Ethanol oxidation by deermice mitochondria under physiological conditions. **Alcoholism: Clin. Exp. Res.**, 14:130-133.

Ingelman-Sundberg, M., and Johansson, I. 1984. Mechanisms of hydroxyl radical formation and ethanol oxidation by ethanol-inducible and other forms of rabbit liver microsomal cytochromes P-450. **J. Biol. Chem.**, 259, 6447-6458.

Ingelman-Sundberg, M., Norsten C., Ekström G., Johansson, I., Eliasson, E., Handler, J.A.,Terelius, Y., Tindberg, N., Thurman, R.G., and Cronholm, T. 1988. Ethanol effects and metabolism. A study using stable isotopes. In **Biomedical and Social Aspects of Alcohol and Alcoholism.** (K. Kuriyama. A. Takada, H. Ishii, Ed.), pp. 119-122.

Iseri, O.A., Lieber, C.S., and Gottlieb, L.S. 1966. The ultrastructure of fatty liver induced by prolonged ethanol ingestion. **Am. J. Pathol.**, 48, 535-555.

Israel, Y., Kalant, H., Orrego, H., Khanna, J.M., Videla, I., and Phillips, J.M. 1975. Experimental alcohol-induced hepatic necrosis: Suppression by propylthiouracil. **Proc. Natl. Acad. Sci. USA**, 72, 1137-1141.

Israel, Y., Hurwitz, E., Niemela, O., and Arnon, R. 1986. Monoclonal and polyclonal antibodies against acetaldehyde-containing epitopes in acetaldehyde-protein adducts. **Proc. Natl. Acad. Sci. USA**, 83, 7923-7927.

Jauhonen, P., Baraona, E., Miyakawa, M., and Lieber, C.S. 1982. Mechanism for selective perivenular hepatotoxicity of ethanol. **Alcoholism: Clin. Exp. Res.**, 6, 350-357.

Jauhonen, P., Baraona, E., Lieber, C.S., and Hassinen, I.E. 1985. Dependence of ethanol-induced redox shift on hepatic oxygen tensions prevailing in vivo. **Alcohol**, 2, 163-167.

Johansson, I., and Ingelman-Sundberg, M. 1985. Carbon tetrachloride-induced lipid peroxidation dependent on an ethanol-inducible form of rabbit liver microsomal cytochrome P-450. **FEBS Lett.** 185, 265-269.

Jukkola, A., and Niemelä, O. 1989. Covalent binding of acetaldehyde to type III collagen. **Biochem. Biophys. Res. Commun.**, 159, 163-169.

Julkunen, R.J.K., Tannenbaum, L., Baraona, E., and Lieber, C.S. 1985a. First pass metabolism of ethanol: An important determinant of blood levels after alcohol consumption. **Alcohol**, 2, 437-441.

Julkunen, R.J.K., Di Padova, C., and Lieber, C.S. 1985b. First pass metabolism of ethanol - a gastrointestinal barrier against the systemic toxicity of ethanol. **Life Sci.**, 37, 567-573.

Kater, R.M., Tobon, J., and Iber, F.L. 1969. Increased rate of tolbutamide metabolism in alcoholic patients. **J. Am. Med. Assoc.**, 20 7, 363-365.

Kato, S., Murawaki, Y., and Hirayama, C. 1985. Effects of ethanol feeding on hepatic collagen synthesis and degradation in rats. **Res. Commun. Chem. Path. Pharmacol.**, 47, 163-180.

Kato, S., Alderman, J., and Lieber, C.S. 1987. Respective roles of the microsomal ethanol oxidizing system (MEOS) and catalase in ethanol metabolism by deermice lacking alcohol dehydrogenase. **Arch. Biochem. Biophys.**, 254, 586-591.

Kato, S., Alderman, J., and Lieber, C.S. 1988. In vivo role of the microsomal ethanol oxidizing system in ethanol metabolism by deermice lacking alcohol dehydrogenase. **Biochem. Pharmacol.**, 37, 2706-2708.

Kato, S., Alderman, J., Kawase, T., and Lieber, C.S. 1990. Role of xanthine oxidase in ethanol-induced lipid peroxidation in rats. **Gastroenterology**, 98:203-210.

Kawase, T., Kato, S., and Lieber, C.S. 1989. Lipid peroxidation and antioxidant defense systems in rat liver after chronic ethanol feeding. **Hepatology**, 10:815-821.

Kershenobich, D., Vargas, F., Garcia-Tsao, G., Tomayo, P.R., Gent, M., and Rojkind, M. 1988. Colchicine in the treatment of cirrhosis of the liver. **N. Engl. J. Med.**, 318, 1709-1713.

Kessler, B.J., Lieber, J.B., Bronfin, G.J., and Sass, M. 1954. The hepatic blood flow and splanchnic oxygen consumption in alcoholic fatty liver. **J. Clin. Invest.**, 33, 1338-1345.

Kiessling, K.H., and Pilstrom, L. 1968. Effect of ethanol on rat liver. V. Morphological and functional changes after prolonged consumption of various alcoholic beverages. **Quart. J. Stud. Alcohol,** 29, 819-827.

Koch, O.R., Boveris, A., Sirotzky De Favelukes, S., Schwarcz De Tarlovsky, and Stoppani,A.O.M. 1977. Biochemical lesions of liver mitochondria from rats after chronic alcohol consumption. **Exp. Mol. Pathol.**, 27, 213-220.

Koivula, T., and Lindros, K.O. 1975. Effects of long term ethanol treatment on aldehyde and alcohol dehydrogenase activities in rat liver. **Biochem. Pharmacol.**, 24, 1937-1942.

Koop, D.R., and Cassaza, J.P. 1985. Identification of ethanol-inducible P-450 isozyme 3a as the acetone and acetol monooxygenase of rabbit microsomes. **J. Biol. Chem.**, 260, 13607-13612.

Koop, D.R., Morgan, E.T., Tarr, G.E., and Coon, M.J. 1982. Purification and characterization of a unique isozyme of cytochrome P-450 from liver microsomes of ethanol-treated rabbits. **J. Biol. Chem.**, 257, 8472-8480.

Korsten, M.A., Matsuzaki, S., Feinman, L., and Lieber, C.S. 1975. High blood acetaldehyde levels after ethanol administration in alcoholics. Difference between alcoholic and non-alcoholic subjects. **N. Engl. J. Med.**, 292, 386-389.

Krikun, G., Lieber, C.S., and Cederbaum, A.I. 1984. Increased microsomal oxidation of ethanol by cytochrome P-450 and hydroxyl radical-dependent pathways after chronic ethanol consumption. **Biochem. Pharmacol.**, 33, 3306-3309.

Lamboeuf, Y., DeSaint, B.G., and Derache, R. 1981. Mucosal alcohol dehydrogenase - and aldehyde dehydrogenase-mediated ethanol oxidation in the digestive tract of the rat. **Biochem. Pharmacol.**, 30, 542-545.

Lamboeuf, Y., la Droitte, P., and DeSaint, D.B. 1983. The gastro-intestinal metabolism of ethanol in the rat. Effect of chronic alcoholic intoxication. **Arch. Int. Pharmacodyn. Ther.,** 261, 157-169.

Lane, B.P., and Lieber, C.S. 1966. Ultrastructural alterations in human hepatocytes following ingestion of ethanol with adequate diets. **Am. J. Pathol.**, 49, 593-603.

Lasker, J.M., Raucy, J., Kubota, S., Bloswick, B.P., Black, M., and Lieber, C.S. 1987a. Purification and characterization of human liver cytochrome P-450ALC. **Biochem. Biophys. Res. Commun.**, 148, 232-238.

Leo, M.A., and Lieber, C.S. 1982. Hepatic vitamin A depletion in alcoholic liver injury. **N. Engl. J. Med.**, 307, 597-601.

Leo, M.A., and Lieber, C.S. 1983. Hepatic fibrosis after long term administration of ethanol and moderate vitamin A supplementation in the rat. **Hepatology**, 3, 1-11.

Leo, M.A., and Lieber, C.S. 1985. New pathway of retinol metabolism in liver microsomes. **J. Biol. Chem.**, 260, 5228-5231.

Leo, M.A., and Lieber, C.S. 1988. Hypervitaminosis A: A liver lover's lament. **Hepatology**, 8, 412-417.

Leo, M.A., Arai, M., Sato, M., and Lieber, C.S. 1982. Hepatotoxicity of moderate vitamin A supplementation in the rat. **Gastroenterology**, 82, 194-205.

Leo, M.A., Sato, M., and Lieber, C.S. 1983. Effect of hepatic vitamin A depletion on the liver in men and rats. **Gastroenterology**, 84, 562-572.

Leo, M.A., Iida, S., and Lieber, C.S. 1984. Retinoic acid metabolism by a system reconstituted with cytochrome P-450. **Arch. Biochem. Biophys.**, 234, 305-312.

Leo, M.A., Kim, C., and Lieber, C.S. 1986a. Increased vitamin A in esophagus and other extrahepatic tissues after chronic ethanol consumption in the rat. **Alcoholism: Clin. Exp. Res.,** 10, 487-492.

Leo, M.A., Lowe, N., and Lieber, C.S. 1986b. Interaction of drugs and retinol. **Biochem. Pharmacol.**, 35, 3949-3953.

Leo, M.A., Lowe, N., and Lieber, C.S. 1987. Potentiation of ethanol-induced hepatic vitamin A depletion by phenobarbital and butylated hydroxytoluene. **J. Nutr.**, 117, 70-76.

Leo, M.A., Lasker, J.M., Raucy, J.L., Kim, C., Black, M., and Lieber, C.S. 1989. Metabolism of retinol and retinoic acid by human liver cytochrome P450IIC8. **Arch. Biochem. Biophys.**, 269, 305-312.

Lieber, C.S. 1982. **Medical disorders of alcoholism: pathogenesis and treatment**. pp. 589 WB Saunders Company, Philadelphia.

Lieber, C.S. 1985. Alcohol and the liver: Metabolism of ethanol, metabolic effects and pathogenesis of injury. **Acta Med. Scand.**, Suppl 703, 11-55.

Lieber, C.S. 1988. Biochemical and molecular basis for alcohol-induced injury to liver and other tissues. **N. Engl. J. Med.,** 319, 1639-1650.

Lieber, C.S. 1990a. **Medical and Nutritional Complications of Alcoholism: Mechanism and Management**. (CS Lieber, ed.), Plenum Press, New York, NY (in press).

Lieber, C.S. 1990b. Interaction of ethanol with drugs, hepatotoxic agents, carcinogens and vitamins. **Alcohol Alcoholism**, 25, 157-176.

Lieber, C.S., and DeCarli, L.M. 1968. Ethanol oxidation by hepatic microsomes: adaptive increase after ethanol feeding. **Science**, 162, 917-918.

Lieber, C,S., and DeCarli, L.M. 1970a. Quantitative relationship between the amount of dietary fat and the severity of the alcoholic fatty liver. **Am. J. Clin. Nutr.**, 23, 474-478.

Lieber, C.S., and DeCarli, L.M. 1970b. Hepatic microsomal ethanol-oxidizing system: In vitro characteristics and adaptive properties in vivo. **J. Biol. Chem.**, 245, 2505-2512.

Lieber, C.S., and DeCarli, L.M. 1970c. Reduced nicotinamide-adenine dinucleotide phosphate oxidase: Activity enhanced by ethanol consumption. **Science**, 170, 78-80.

Lieber, C.S., and DeCarli, L.M. 1972. The role of the hepatic microsomal ethanol oxidizing system (MEOS) for ethanol metabolism in vivo. **J. Pharmacol. Exp. Ther.,** 181, 279-287.

Lieber, C.S., and DeCarli, L.M. 1974. An experimental model of alcohol feeding and liver injury in the baboon. **J. Med. Primatol.**, 3, 153-163.

Lieber, C.S., and Leo, M.A. 1986. Role of vitamin A deficiency and excess in liver disease. In **Metabolism and Nutrition in Liver Disease** (H. Popper and F. Schaffner, eds.), Grune and Stratton, New York, Vol. III. Chap. 14, pp. 253-272.

Lieber, C.S., and Rubin, E. 1968. Alcoholic fatty liver in man on a high protein and low fat diet. **Am. J. Med.,** 44, 200-206.

Lieber, C.S., and Schmid, R. 1961. Stimulation of hepatic fatty acid synthesis by ethanol. **Am. J. Clin. Nutr.**, 9, 436-438.

Lieber, C.S., and Schmid, R. 1961. The effect of ethanol on fatty acid metabolism: stimulation of hepatic fatty acid synthesis in vitro. **J. Clin. Invest.,** 40, 395-399.

Lieber, C.S., DeCarli, L.M., and Schmid, R. 1959. Effects of ethanol on fatty acid metabolism in liver slices. **Biochem. Biophys. Res. Commun.,** 1, 302-306.

Lieber, C.S., Jones, D.P., and DeCarli, L.M. 1965. Effects of prolonged ethanol intake:Production of fatty liver despite adequate diets. **J. Clin. Invest.**, 44, 1009-1020.

Lieber, C.S., DeCarli, L.M., and Rubin, E. 1975. Sequential production of fatty liver, hepatitis and cirrhosis in sub-human primates fed ethanol with adequate diets. **Proc. Natl. Acad. Sci. USA,** 72, 437-441.

Lieber, C.S., Nakano, M., and Worner, T.M. 1981. Ultrastructure of the initial stages of hepatic perivenular fibrosis after alcohol. **Trans. Ass. Am. Physicians**, 94, 292-330.

Lieber, C.S., Leo, M.A., Mak, K.M., DeCarli, L.M., and Sato, S. 1985. Choline fails to prevent liver fibrosis in ethanol-fed baboons but causes toxicity. **Hepatology**, 5, 561-575.

Lieber, C.S., Garro, A., Leo, M.A., Mak, K.M., and Worner, T. 1986. Alcohol and cancer. **Hepatology**, 6, 1005-1019.

Lieber, C.S., Lasker, J.M., DeCarli, L.M., Saeli, J., and Wojtowicz, T. 1988. Role of acetone, dietary fat and total energy intake in the induction of the hepatic microsomal ethanol oxidizing system. **J. Pharmacol. Exp. Ther**., 247, 792-795.

Lieber, C.S., Baraona, E., Hernandez-Munoz, R., Kubota, S., Sato, N., Kawano, S., Matsumura,T., and Inatomi, N. 1989. Impaired oxygen utilization: A new mechanism for the hepatotoxicity of ethanol in sub-human primates. **J. Clin. Invest.**, 83, 1682-1690.

Lieber, C.S., Casini, A., DeCarli, L.M., Kim, C., Lowe, N., Sasaki, R., and Leo, M.A. 1990a. S-adenosyl-L-methionine attenuates alcohol-induced liver injury in the baboon. **Hepatology**, 11, 165-172.

Lieber, C.S., DeCarli, L.M., Mak, K.M., Kim, C-I., and Leo, M.A. 1990b Attenuation of alcohol-induced hepatic fibrosis by polyunsaturated lecithin. **Hepatology**, 12, 1390-1398.

Lieber, C.S., Li, J-J., DeCarli, L.M., Mak, K.M., Kim, C-I., and Leo, M.A. 1990c. Polyunsaturated lecithin (PUL) protects against alcoholic cirrhosis in the baboon. **Hepatology**, 12, 871.

Lin, G.W.J., and Lester, D. 1980. Significance of the gastrointestinal tract in the in vivo metabolism of ethanol in the rat. **Adv. Exp. Med. Biol.,** 131, 281-286.

Lorr, N.A., Miller, K.W., Chung, H.R., and Yang, C.S. 1984. Potentiation of the hepatotoxicity of N-nitrosodimethylamine by fasting, diabetes, acetone and isopropanol. **Toxicol. Appl. Pharmacol.**, 73, 423-431.

Mak, K.M., and Lieber, C.S. 1988. Lipocytes and transitional cells in alcoholic liver disease: A morphometric study. **Hepatology**, 8, 1027-1033.

Mak, K.M., Leo, M.A., and Lieber, C.S. 1984. Alcoholic liver injury in baboons:transformation of lipocytes to transitional cells. **Gastroenterology**, 87, 188-200.

Maling, H.M., Stripp, B., Sipes, I.G., Highman, B., Waul, W., and Williams, M.A. 1975. Enhanced hepatotoxicity of carbon tetrachloride, thioacetamide, and dimethylnitrosamine and some comparisons with potentiation by isopropanol. **Toxicol. Appl. Pharmacol.**, 33, 291-308.

Mann, S.W., Fuller, G.C., Rodil, J.V., and Fidins, E.I. 1979. Hepatic prolyl hydroxylase and collagen synthesis in patients with alcoholic liver disease. **Gut**, 20, 825-832.

Maruyama, K., Feinman, L., Fainsilber, Z., Nakano, M., Okazaki, I., and Lieber, C.S. 1982. Mammalian collagenase increases in early alcoholic liver disease and decreases with cirrhosis. **Life Sci.**, 30, 1379-1384.

Mata, J.M., Kershenobich, D., Villarreal, E., and Rojkind, M. 1975. Serum free proline and free hydroxyproline in patients with chronic liver disease. **Gastroenterology**, 68, 1265-1269.

Matsuda, Y., Baraona, E., Salaspuro, M., and Lieber, C.S. 1979. Effects of ethanol on

liver microtubules and Golgi apparatus. **Lab. Invest.**, 41, 455-463.
Matsuda, Y., Takada, A., Sato, H., Yasuhara, M., and Takase, S. 1985. Comparison between ballooned hepatocytes occurring in human alcoholic and nonalcoholic liver diseases. **Alcoholism: Clin. Exp. Res.**, 9, 366-370.
Matsuzaki, S., and Lieber, C.S. 1977. Increased susceptibility of hepatic mitochondria to the toxicity of acetaldehyde after chronic ethanol consumption. **Biochem. Biophys. Res. Commun.**, 75, 1059-1065.
Matsuzaki, S., Gordon, E., and Lieber, C.S. 1981. Increased alcohol dehydrogenase independent ethanol oxidation at high ethanol concentrations in isolated rat hepatocytes: the effect of chronic ethanol feeding. **J. Pharmacol. Exp. Ther.**, 217, 133-137.
Mezey, E., Potter, J.J., and Maddrey, W.C. 1973. Collagen turnover in alcoholic liver disease. **Gastroenterology**, 65, 560 (abstract).
Mezey, E., Potter, J.J., Iber, F.L., and Maddrey, W.C. 1979. Hepatic collagen proline hydroxylase activity in alcoholic hepatitis effect of d-penicillamine. **J. Lab. Clin. Med.**, 93, 92-100.
Minato Y., Hasumura, Y., and Takeuchi, J. 1983. The role of fat-storing cells in Disse space fibrogenesis in alcoholic liver disease. **Hepatology**, 3, 559-566.
Misra, P.S., Lefevre, A., Ishii, H., Rubin, E., and Lieber, C.S. 1971. Increase of ethanol, meprobamate and pentobarbital metabolism after chronic ethanol administration in man and in rats. **Am. J. Med.**, 51, 346-351.
Miwa, G.T., Levin, W., Thomas, P.E., and Lu, A.Y.H. 1978. The direct oxidation of ethanol by catalase- and alcohol dehydrogenase-free reconstituted system containing cytochrome P-450. **Arch. Biochem. Biophys.**, 187, 464-475.
Morgan, E.T., Koop, D.R., and Coon, M.J. 1982. Catalytic activity of cytochrome P-450 isozyme 3a isolated from liver microsomes of ethanol-treated rabbits. **J. Biol. Chem.**, 257, 13951-13957.
Morgan, E.T., Koop, D.R., and Coon, M.J. 1983. Comparison of six rabbit liver cytochrome P-450 isozymes in formation of a reactive metabolite of acetaminophen. **Biochem. Biophys. Res. Commun.**, 112, 8-13.
Moshage, H., Casini, A., and Lieber, C.S. 1990. Acetaldehyde selectively stimulates collagen production in cultured rat liver fat-storing cells but not in hepatocytes. **Hepatology**, 12, 511-518.
Nakano, M., and Lieber, C.S. 1982. Ultrastructure of initial stages of perivenular fibrosis in alcohol-fed baboons. **Am. J. Path.**, 106, 145-155.
Nakano, M., Worner, T., and Lieber, C.S. 1982. Perivenular fibrosis in alcoholic liver injury: ultrastructure of histologic progression. **Gastroenterology**, 83, 777-785.
Nomura, F., and Lieber, C.S. 1981. Binding of acetaldehyde to rat liver microsomes:Enhancement after chronic alcohol consumption. **Biochem. Biophys. Res. Commun.**, 100, 131-137.
Norsten, C., Cronholm, T., Ekstrom, G., Handler, J.A., Thurman, R.G., and Ingelman-Sundberg M. 1989. Dehydrogenase-dependent ethanol metabolism in deermice (Peromyscus maniculatus) lacking cytosolic alcohol dehydrogenase. **J. Biol. Chem.**, 264, 5593-5597.
Ohnishi, K., and Lieber, C.S. 1977. Reconstitution of the microsomal ethanol-oxidizing system: qualitative and quantitative changes of cytochrome P-450 after chronic ethanol consumption. **J. Biol. Chem.**, 252, 7124-7131.
Okazaki, I., Feinman, L., and Lieber, C.S. 1977. Hepatic mammalian collagenase: Development of an assay and demonstration of reversal activity after ethanol consumption. **Gastroenterology**, 73, 1236.
Patrick, R.S. 1973. Alcohol as a stimulus to hepatic fibrogenesis. **J. Alcoholism**, 8, 13-27.

Pestalozzi, D.M., Buhler, R., von Wartburg, J.P., and Hess, M. 1983. Immunohistochemical localization of alcohol dehydrogenase in the human gastrointestinal tract. **Gastroenterology**, 85, 1011-1016.
Pikkarainen, P.H., Gordon, E.R., Lebsack, M.E., and Lieber, C.S. 1981. Determinants of plasma free acetaldehyde level during the steady state oxidation of ethanol: Effects of chronic ethanol feeding. **Biochem. Pharmacol.**, 30, 799-802.
Popper, H., and Lieber, C.S. 1980. Histogenesis of alcoholic fibrosis and cirrhosis in the baboon. **Am. J. Pathol.,** 98, 695-716.
Porto, L.C., Chevallier, M., and Crimaud, J-A. 1989a. Morphometry of terminal hepatic veins. 1. Comparative study in man and baboon . **Virchows Archiv. A Pathol. Anat.**, 414, 129-134.
Porto, L.C., Chevallier M., and Crimaud, J-A. 1989b. Morphometry of terminal hepatic veins.2. Follow up in chronically alcohol-fed baboons. **Virchows Archiv. A Pathol. Anat.**, 414, 299-307.
Prescott, L.R., Wright, N. 1973. The effects of hepatic renal damage on paracetamol metabolism and excretion following overdosage. A pharmacokinetic study. **Br. J. Pharmacol.**, 49, 602-613.
Pritchard, J.F., and Schneck, D.W. 1977. Effects of ethanol and phenobarbital on the metabolism of propanolol by 9000g rat liver supernatant. **Biochem. Pharmacol.**, 26, 2453-2454.
Reinke, L.A., Lai, E.K., DuBose, C.M., and McCay, P.B. 1987. Reactive free radical generation in vivo in heart and liver of ethanol-fed rats: correlation with radical formation in vitro. **Proc. Natl. Acad. Sci. USA**, 84, 9223-9227.
Rhode, H., Vargas, L., Hahn, E., Kalbfeisch, H., Bruguera, M. and Timpl, R. 1979.Radioimmunoassay for type III procollagen peptide and its application to human liver disease. **Eur. J. Clin. Invest.**, 9, 451-459.
Rubin, E., and Lieber, C.S. 1967a. Early fine structural changes in human liver induced by alcohol. **Gastroenterology**, 52, 1-37.
Rubin, E., and Lieber, C.S. 1967b. Experimental alcoholic hepatic injury in man:Ultrastructural changes. **Fed. Proc.**, 26, 1458-1467 .
Rubin, E., Beattie, D.S., and Lieber, C.S. 1970. Effects of ethanol on the biogenesis of mitochondrial membranes and associated mitochondrial functions. **Lab. Invest.**, 23, 620-627.
Rubin, E., Beattie, D.S., Toth, A., and Lieber, C.S. 1972. Structural and functional effects of ethanol on hepatic mitochondria. **Fed. Proc.**, 31, 131-140.
Ruwart, M.J., Rush, B.D., Synder, K.F., Peters, K.M., Appelman, H.D., and Henley, K.S.1988. 16,16-Dimethyl prostaglandin E2 delays collagen formation in nutritional injury in rat liver. **Hepatology**, 8, 61-64.
Ryan, D.E., Ramanthan, L., Iida, S., Thomas, P.E., Haniu, M., Shively, J.E., and Lieber, C.S.1985. Characterization of a major form of rat hepatic microsomal cytochrome P-450 induced by isoniazid. **J. Biol. Chem.**, 260, 6385-6393.
Ryan, D.E., Koop, D.R., Thomas, P.E., Coon, M.J., and Levin, W. 1986. Evidence that isoniazid and ethanol induced the same microsomal cytochrome P-450 in rat liver, and isozyme homologous to rabbit liver cytochrome P-450 isozymes 3a. **Arch. Biochem. Biophys.**, 246 , 633-644.
Salaspuro, M.P., Shaw., S., Jayatilleke, E., Ross, W.A., and Lieber, C.S. 1981. Attenuation of the ethanol induced hepatic redox change after chronic alcohol consumption in baboons: metabolic consequences in vivo and in vitro. **Hepatology**, 1, 33-38.
Sato, C., and Lieber, C.S. 1981. Mechanism of preventive effect on acetaminophen-induced hepatotoxicity. **J. Pharmacol. Exp. Ther.**,218, 811-815.
Sato, M., and Lieber, C.S. 1981. Hepatic vitamin A depletion after chronic ethanol consumption in baboons and rats. **J. Nutr.,** 111 , 2015-2023.

Sato, C., Nakano, M., and Lieber, C.S. 1981. Increased hepatotoxicity of acetaminophen after chronic ethanol consumption in the rat . **Gastroenterology**, 80, 140-148.
Sato, N., Kamada, T., Kawano, S., Hayashi, N., Kishida, Y., Meren, H., Yoshihara, H., and Abe, H. 1983. Effect of acute and chronic ethanol consumption on hepatic tissue oxygen tension in rats. **Pharmacol. Biochem. Behav.**, 18, 443-447.
Sato, S., Nouchi, T., Worner, T., and Lieber, C.S. 1985. Fab radioimmunoassay of blood procollagen-III peptides detects precirrhotic lesions. **Gastroenterology**, 88, 1692.
Savolainen, E.R., Leo, M.A., Timple, R., and Lieber, C.S. 1984a. Acetaldehyde and lactate stimulate collagen synthesis of cultured baboon liver myofibroblasts. **Gastroenterology**, 87, 777-787.
Savolainen, E.R., Goldberg, B., Leo, M.A., Velez, M., and Lieber, C.S. 1984b. Diagnostic value of serum procollagen peptide measurements in alcoholic liver disease. **Alcoholism: Clin. Exp. Res.**, 8, 384-389.
Seef, L.B., Cuccherini, B.A., Zimmerman, H.J., Alder, E., and Benjamin, S.B. 1986.Acetaminophen hepatotoxicity in alcoholics (Clinical review). **Ann. Int. Med.**, 104, 399-404.
Seitz, H.K., Garro, A.J., and Lieber, C.S. 1978. Effect of chronic ethanol ingestion on intestinal metabolism and mutagenicity of benzo(a)pyrene. **Biochem. Biophys. Res. Commun.**, 85, 1061-1066.
Seitz, H.K., Garro, A.J., and Lieber, C.S. 1981. Enhanced pulmonary and intestinal activation of procarcinogens and mutagens after chronic ethanol consumption in the rat. **Eur. J. Clin. Invest.**, 11, 33-38.
Shaw, S., Heller, E., Friedman, H., Baraona, E., and Lieber, C.S. 1977. Increased hepatic oxygenation following ethanol administration in the baboon. **Proc. Soc. Exp. Biol. Med.**, 156, 509-513.
Shaw, S., Jayatilleke, E., and Lieber, C.S. 1981. Hepatic lipid peroxidation: Potentiation by chronic alcohol feeding and attenuation by methionine. **J. Lab. Clin. Med.**, 98, 417-425.
Shaw, S., Rubin, K.P., Lieber, C.S. 1983. Depressed hepatic glutathione and increased diene conjugates in alcoholic liver disease: evidence of lipid peroxidation. **Dig. Dis. Sci.**, 28, 585-589.
Shaw, S., Jayatilleke, E., and Lieber, C.S. 1984a. The effect of chronic alcohol feeding on lipid peroxidation in microsomes: lack of relationship to hydroxyl radical generation. **Biochem. Biophys. Res. Commun.**, 118, 233-238.
Shaw, S., Worner, T., and Lieber, C.S. 1984b. Frequency of hyperprolinemia in alcoholic liver cirrhosis: Relationship to blood lactate. **Hepatology**, 4, 295-299.
Shigeta, Y., Nomura, F., Iida, S., Leo, M.A., Felder, M.R., and Lieber, C.S. 1984. Ethanol metabolism in vivo by the microsomal ethanol oxidizing system in deermice lacking alcohol dehydrogenase (ADH). **Biochem. Pharmacol.**, 33, 807-814.
Song, B.J., Gelboin, H.V., Park, S.S., and Gonzales, F.J. 1986. Complementary DNA and protein sequences of ethanol-inducible rat and human cytochrome P-450s. **J. Biol. Chem.**, 261, 16689-16697.
Speisky, H., Macdonald, A., Giles, G., Orrego, H., and Israel, Y. 1985. Increased loss and decreased synthesis of hepatic glutathione after acute ethanol administration. **Biochem. J.**, 225, 565-572.
Stein, W.S., Lieber, C.S., Leevy, C.M., Cherrick, G.R., and Abelman, W.H. 1963. The effect of ethanol upon systemic and hepatic blood flow in man. **Am. J. Clin. Nutr.**, 13, 68-74.
Takada, A., Nei, J., Matsuda, Y., and Kanayama, R. 1982. Clinicopathological study of alcoholic fibrosis. **Am. J. Gastroenterology**, 77, 660-666.
Teschke, R., Hasumura, Y., Joly, J.G., Ishii, H., and Lieber, C.S. 1972. Microsomal ethanol-oxidizing system (MEOS).: Purification and properties of a rat liver system

free of catalase and alcohol dehydrogenase. **Biochem. Biophys. Res. Commun.**, 49, 1187-1193.

Teschke, R., Hasumura, Y., and Lieber, C.S. 1974. Hepatic microsomal alcohol oxidizing system. Solubilization, isolation and characterization. **Arch. Biochem. Biophys.**, 163, 404-415.

Teschke, R., Hasumura, Y., and Lieber, C.S. 1976. Hepatic ethanol metabolism: respective roles of alcohol dehydrogenase, the microsomal ethanol oxidizing system and catalase. **Arch. Biochem. Biophys.**, 175, 635-643.

Tsutsumi, M., Shimizu, M., Lasker, J.M., and Lieber, C.S. 1989. The intralobular distribution of ethanol-inducible P450IIE1 in rat and human liver. **Hepatology**, 10, 437-446.

Tsutsumi, R., Leo, M.A., Kim, C., Tsutsumi, M., Lasker, J., Lowe, N., and Lieber, C.S. 1990.Interaction of ethanol with enflurane metabolism and toxicity: Role of P450IIE1. **Alcoholism: Clin. Exp. Res.**, 14, 174-179.

Ugarte, G., Pinto, M.E., and Insunza, J. 1967. Hepatic alcohol dehydrogenase in alcoholic addicts with and without hepatic damage. **Am. J. Dig. Dis.**, 12, 589-599.

Videla, L., and Israel, Y., 1970. Factors that modify the metabolism of ethanol in rat liver and adaptive changes produced by its chronic administration. **Biochem. J.**, 118, 275-281.

Wendel, A., Fenerstein, S., and Konz, K.H. 1979. Acute paracetamol intoxication of starved mice leads to lipid peroxidation in vivo . **Biochem. Physiol.**, 28, 2051-2055.

Wilson, J.S., Korsten, M.A., and Lieber, C.S. 1986. The combined effects of protein deficiency and chronic ethanol administration on rat ethanol metabolism. **Hepatology**, 6, 823-829.

Worner, T.M., and Lieber, C.S. 1985. Perivenular fibrosis as precursor lesions of cirrhosis.**JAMA**, 254, 627-630.

Wrighton, S.A., Campanile, C., Thomas, P.E., Maines, S.L., Watkins, P.B., Parker, G., Mendez-Picon, G., Haniu, M., Shively, J.E., Levin, W., and Guzelian, P.S. 1986. Identification of a human liver cytochrome P-450 homologous to the major isosafrole-inducible cytochrome P-450 in the rat. **Mol. Pharmacol.**, 29, 405-410.

Yang, C.S., Tu, Y.Y., Koop, D.R., and Coon, M.J. 1985. Metabolism of nitrosamines by purified rabbit liver cytochrome P-450 isozyme s. **Cancer Res.**, 45, 1140-1145.

Zern, M.A., Leo, M.A., Giambrone, M.A., and Lieber, C.S. 1985. Increased type I procollagen mRNA levels and in vitro protein synthesis in the baboon model of chronic alcoholic liver disease. **Gastroenterology**, 89, 1123-1131.

REGULATION OF RATES OF ETHANOL METABOLISM AND LIVER $[NAD^+]/[NADH]$ RATIO

Michael J. Hardman, Rachel A. Page, Mark S. Wiseman and Kathryn E. Crow

Department of Chemistry and Biochemistry
Massey University
Palmerston North
New Zealand

INTRODUCTION

Factors that control the rate of alcohol metabolism in mammals have been the subject of debate for many years (for detailed reviews, see Crow, 1985; Crow and Hardman, 1989). There have been two main theories as to the major rate limitation on the ethanol metabolic pathway. The first theory was that the rate at which NADH (generated in the alcohol and aldehyde dehydrogenase reactions) could be reoxidised to NAD^+ was limiting (Hawkins and Kalant, 1972; Khanna and Israel, 1980; Thurman et al, 1989). This theory arose from the observation that the ratio of free $[NAD^+]/[NADH]$ in liver cytosol decreased during ethanol metabolism. It was assumed either that the liver ran out of NAD^+, because the rate of reoxidation of NADH was limiting, and this lack of NAD^+ then limited the rate of the alcohol dehydrogenase (ADH) reaction (Khanna and Israel, 1980) or that NADH accumulated and caused product inhibition of ADH (Thurman et al, 1989). The second theory was that the amount of ADH present in the liver was the main rate-determining factor for the pathway (Crow et al, 1977; Cornell et al, 1979; Braggins and Crow, 1981; Cornell, 1983; Bosron et al, 1983). This theory arose in part from observations that the activity of liver ADH measured *in vitro* was only slightly more than necessary to explain rates of ethanol metabolism observed *in vivo* (Crow et al, 1977; Cornell et al, 1979; Braggins and Crow, 1981). ADH was not present 'in excess' as had sometimes been claimed (Hawkins and Kalant, 1972; Kalant et al, 1975). It was also observed that variations in ADH activity induced by castration (Rachamin et al, 1980; Mezey et al, 1980; Cicero et al, 1980, 1982), starvation (Braggins and Crow, 1981; Lumeng et al, 1979, 1980) or stress (Mezey et al, 1979) were associated with corresponding changes in rates of ethanol metabolism. Although there was considerable evidence to support the theory that the activity of ADH was the major rate-determining factor for the pathway, this theory did not provide any explanation for the decrease in free $[NAD^+]/[NADH]$ ratio that occurs during ethanol metabolism. Indeed it appeared to be incompatible with this observation. The development of methods for the quantitative assessment of metabolic regulation (Kacser and Burns, 1973, 1979; Westerhoff et al, 1984) indicated that asking the question "What is **the** rate determining step for this pathway?" was no longer a tenable approach. Several steps can make significant contributions to the overall control of a pathway.

In this paper we describe research that has explained how a decrease in the ratio of free cytosolic $[NAD^+]/[NADH]$ can occur in the liver even if the activity of ADH plays a major rate-determining role in the pathway. We also discuss a method for quantitative assessment of metabolic regulation. Using this method we have determined the contribution of ADH to regulation of the ethanol metabolic pathway. The experimental data discussed in this paper were mainly obtained using rats. The regulation of the ethanol metabolising pathway in humans is more complex as there are a large number of ADH isozymes with varying kinetic parameters and different people have different combinations of isozymes (Burnell and Bosron, 1989). A full analysis of the regulation of the pathway in humans would be even more difficult than in experimental animals - indeed given the limits imposed by ethical experimentation it is probably impossible. However, the data from studies using rats can provide useful indications as to the likely regulatory factors in humans.

THE ROLE OF MALATE DEHYDROGENASE IN REGULATION OF NADH CONCENTRATIONS DURING ETHANOL METABOLISM

The decrease in free cytosolic $[NAD^+]/[NADH]$ that occurs during ethanol metabolism does not result from a decrease in the free NAD^+ concentration. The ratio of $[NAD^+]/[NADH]$ is about 1000 in the absence of ethanol and decreases to about 200 to 500 in the presence of ethanol. This change is caused by an increase in the free [NADH] from about 0.5 μM to 1-2 μM and is too small to have any significant effect on the free $[NAD^+]$ (Crow et al, 1983a). Therefore the rate of alcohol metabolism is not limited by the liver running out of NAD^+. The increase in [NADH] is not sufficient to cause more than about 10-20% inhibition of ADH (Cornell et al, 1979; Cornell, 1983; Bosron et al, 1983) suggesting that the reoxidation of NADH is not likely to be the major rate-determining factor for ethanol metabolism. Can we explain the increase in free cytosolic [NADH] if, as we have suggested, the activity of ADH is a major rate-determining factor for ethanol metabolism? Under physiological conditions cytosolic malate dehydrogenase (MDH), which is the first enzyme involved in NADH reoxidation, is not saturated with NADH (Crow et al, 1982). Therefore the rate of NADH reoxidation increases as the [NADH] in the cytosol increases. When ethanol is being metabolised, the free cytosolic [NADH] increases until the rate of removal of NADH by cytosolic MDH equals the rate of NADH formation by ADH. This means that increasing the activity of ADH increases both the rate of ethanol metabolism and the free cytosolic NADH concentration. The kinetic parameters of cytosolic MDH, in particular the K_m of the enzyme for NADH, are therefore important in determining the degree of change in the free cytosolic $[NAD^+]/[NADH]$ ratio during ethanol metabolism (Crow et al, 1982, 1983a). The parameters for the human cytosolic MDH are similar to those for rats (Crow et al, 1983b).

Our determination of kinetic parameters for cytosolic MDH under conditions similar to those *in vivo* allowed us to carry out simulations of the change in $[NAD^+]/[NADH]$ ratio that should occur during ethanol metabolism, using previously-published parameters for alcohol and aldehyde dehydrogenases. The changes in free cytosolic $[NAD^+]/[NADH]$ that we were able to generate using simulations agreed very closely with those observed experimentally in rat liver (Crow et al, 1982). In order to extend these studies we have now also determined the kinetic parameters for mitochondrial MDH under conditions similar to those *in vivo* . The results of this study (manuscript in preparation) show that for mitochondrial MDH to operate at the required rate the concentration of free mitochondrial oxaloacetate must be very low (<1 μM). The mitochondrial malate concentration, which has been shown to increase during ethanol metabolism (Bobyleva-Guarriero et al, 1986), is probably regulated by the activity of mitochondrial MDH in a manner similar to the regulation of cytosolic NADH concentration by cytosolic MDH. Mitochondrial MDH has a K_m for malate of 1.9 mM, and the mitochondrial malate concentration in the absence of ethanol is probably about 1-2 mM (Siess et al, 1982).

QUANTITATIVE ANALYSIS OF METABOLIC REGULATION

1. The Approach

Traditionally biochemists have assumed that a metabolic pathway should have a single rate determining step. This view has been challenged over the last few years by a number of theories which have allowed the quantitative assessment of the contribution of individual steps to the overall metabolic control of a pathway. The approach which has been most widely used experimentally, and which we have chosen to use for this study, is that of Kacser and Burns (1973, 1979). This approach has been discussed and expanded by a number of other workers since its original publication (Westerhoff et al, 1984; Derr, 1985, 1986; Kacser and Porteous, 1987).

Only a brief description of the parameters important to this approach to quantitative analysis will be given here. For more details readers are referred to the references given above. The theory proposes that a number of steps may share in the regulation of a metabolic pathway. A parameter known as the flux control coefficient gives a quantitative measure of the degree of control that each step exerts. The flux control coefficient for a given enzyme is determined by varying the activity of that enzyme, in a physiological setting, and monitoring the resultant variation in flux through the metabolic pathway of interest. The ratio of the fractional change in flux (J) to the fractional change in enzyme activity (E) gives the flux control coefficient, C:

$$C^{J}_{E_i} = \left(\frac{dJ/J}{dE_i/E_i}\right)$$

The required changes in enzyme activity are usually generated by the use of specific enzyme inhibitors, and a method called modulation is used. This involves making a series of very small changes in enzyme activity, using a range of inhibitor concentrations. The flux control coefficient is then given by the limiting value, when I = 0, of the following equation:

$$C^{J}_{E_i} = \frac{\left(\frac{\partial J/J}{\partial I/I}\right)}{\left(\frac{\partial v_i/v_i}{\partial I/I}\right)}$$

where v is the activity of the enzyme *in vitro* and I is the concentration of inhibitor. In the case of a non-competitive inhibitor this equation simplifies to:

$$C^{J}_{E_i} = -K_i\, \partial J/J\partial I$$

The summation theorem states that the sum of all the flux control coefficients for any pathway is 1. Thus if one enzyme in a pathway has a flux control coefficient close to 1, it is a major rate controlling factor and all the other enzymes will have small coefficients. Alternatively, a pathway may have a number of enzymes which all have small but significant flux control coefficients; therefore there is no single major rate-controlling step.

Flux control coefficients are not a property of an individual enzyme, but of the pathway as a whole, and they vary with changes in metabolic state. A given enzyme may have different flux control coefficients under different experimental conditions.

2. Results and Discussion

We have carried out experiments using isolated hepatocytes and specific inhibitors of ADH to obtain flux control coefficients for ADH in cells isolated from fed and starved rats. The rate of ethanol disappearance was determined over a range of inhibitor concentrations and the flux control coefficient was calculated using values of K_i determined *in vitro* (Chadha et al, 1983; Plapp et al, 1984).

In cells from starved rats, where acetaldehyde concentrations are low, flux control coefficients of 0.41 ± 0.03 (using tetramethylene sulfoxide as inhibitor) and 0.60 ± 0.03 (using isobutyramide as inhibitor) were obtained. The average value of about 0.5 shows that the activity of ADH is an important rate-controlling factor under these conditions. The other potential control steps should have flux control coefficients which total about 0.5 since the summation theorem states that the sum of all the coefficients should be 1.0. Therefore no other individual step is likely to have a flux control coefficient as high as that of ADH.

In cells from fed rats, acetaldehyde concentrations vary widely, from as low as 5 μM to as high as 200 μM (Crow et al, 1983c). This variation has also been observed in perfused rat liver (Braggins et al, 1980) and in rats *in vivo* (Braggins and Crow, 1981). In the presence of low acetaldehyde concentrations, the flux control coefficients for ADH were similar to those obtained for starved rats. With high acetaldehyde concentrations, however, the flux control coefficients for ADH were much lower (Table 1). These results indicate that where acetaldehyde concentrations are low, even in fed rats, ADH is a major rate-controlling factor for ethanol metabolism. When acetaldehyde concentrations are higher, however, ADH becomes less important in regulation and some other enzyme must play a more important role.

The acetaldehyde concentration is governed by a balance between the activities of alcohol and aldehyde dehydrogenases (Eriksson et al, 1975; Braggins et al, 1980; Crow et al, 1982;

Table 1. Flux Control Coefficient for ADH and Acetaldehyde Concentration

C^J_{ADH}	[Acetaldehyde] μM
0.14	143
0.02	138
0.07	81
0.25	42
0.21	32
0.41	16
0.43	11
0.54	6.0

Lindros, 1983; Crow and Hardman, 1989); steady state concentrations of acetaldehyde may vary widely as a result of small changes in the relative activities of these two enzymes. The relationship between flux control coefficient for ADH and acetaldehyde concentration strongly suggests that the activity of aldehyde dehydrogenase becomes an important rate-determining factor when [acetaldehyde] is high. Flux control coefficients for aldehyde dehydrogenase need to be determined to confirm this suggestion.

CONCLUSIONS

The activity of ADH is an important rate-controlling factor for ethanol metabolism in hepatocytes from starved rats. In hepatocytes from fed rats, the importance of ADH is often less; a lower importance of ADH is associated with a higher steady state [acetaldehyde]. Some other step, probably aldehyde dehydrogenase, becomes more important in these circumstances.

Therefore we conclude that alcohol dehydrogenase and aldehyde dehydrogenase share a major part of the regulation of the pathway of ethanol metabolism.

The degree of control exerted by individual reactions of the malate-aspartate shuttle and the electron transport chain remains to be determined. The relative contributions to control of the steps of the electron transport chain are likely to vary between different metabolic conditions, as has been shown by quantitative analysis of the regulation of this pathway (Duszynski et al, 1982; Tager et al, 1983). The cytosolic and mitochondrial malate dehydrogenases are unlikely to make an important contribution to regulation of flux, since their maximum activities are much greater than the flux through the pathway (Crow et al, 1982 (cytosolic); manuscript in preparation (mitochondrial)). Enzymes that make only a minor contribution to control of flux may, however, contribute significantly to regulation of metabolite concentrations, and the malate dehydrogenases have a role in regulating the concentrations of NADH (in the cytosol) and malate (in the mitochondria) during ethanol metabolism.

REFERENCES

Bobyleva-Guarriero, V., Wehbie, R.S. and Lardy, H.A., 1986, The role of malate in hormone-induced enhancement of mitochondrial respiration, **Arch. Biochem. Biophys.,** 245: 477.

Bosron, W.F., Crabb, D.W. and Li, T.-K., 1983, Relationship between kinetics of liver alcohol dehydrogenase and alcohol metabolism, **Pharmacol. Biochem. Behav.,** 18 (suppl. 1): 223.

Braggins, T.J. and Crow, K.E., 1981, Effects of high ethanol doses on rates of ethanol oxidation in rats, **Eur. J. Biochem.,** 119: 633.

Braggins, T.J., Crow, K.E. and Batt, R.D., 1980, Acetaldehyde and acetate production during ethanol metabolism in perfused rat liver, **in:** Alcohol and Aldehyde Metabolising Systems, Vol 4, R.G. Thurman, ed., pp.441-449, Plenum Press, New York.

Burnell, J.C. and Bosron, W.F., 1989, Genetic polymorphism of human liver alcohol dehydrogenase and kinetic properties of the isozymes, **in**: Human Metabolism of Alcohol, Vol. 2, K.E. Crow and R.D. Batt, eds., pp. 65-75, CRC Press Inc., Boca Raton, Florida.

Chadha, V.K., Leidal, K.G. and Plapp, B.V., 1983, Inhibition by carboxamides and

sulfoxides of liver alcohol dehydrogenase and ethanol metabolism, **J. Med. Chem.,** 26: 916.
Cicero, T.J., Bernard, J.D. and Newman, K., 1980, Effects of castration and chronic morphine administration on liver alcohol dehydrogenase and the metabolism of ethanol in the male Sprague-Dawley rat, **J. Pharmacol. Exp. Ther.,** 215: 317.
Cicero, T.J., Newman, K.S., Schmoeker, P.F. and Meyer, E.R., 1982, Role of testosterone in ethanol- and morphine-induced increases in the alcohol dehydrogenase dependent metabolism of ethanol in the male rat, **J. Pharmacol. Exp. Ther.,** 222: 20.
Cornell, N.W., 1983, Properties of alcohol dehydrogenase and ethanol oxidation *in vivo* and in hepatocytes, **Pharmacol. Biochem. Behav.,** 18 (suppl. 1): 215.
Cornell, N.W., Crow, K.E., Leadbetter, M.G. and Veech, R.L., 1979, Rate-determining factors for ethanol oxidation *in vivo* and in isolated hepatocytes, **in**: Alcohol and Nutrition, T.-K. Li., S. Schenker and L. Lumeng, eds., pp. 315-330, U.S. Govt. Printing Office, Washington D.C.
Crow, K.E., 1985, Ethanol metabolism by the liver, **Reviews on Drug Metabolism and Drug Interactions,** 5:113.
Crow, K.E., Braggins, T.J., Batt, R.D. and Hardman, M.J., 1982, Rat liver cytosolic malate dehydrogenase: purification, kinetic properties, role in control of free cytosolic NADH concentration, **J. Biol. Chem.,** 257: 14217.
Crow, K.E., Braggins, T.J., Batt, R.D. and Hardman, M.J., 1983a, Kinetics of malate dehydrogenase and control of rates of ethanol metabolism in rats, **Pharmacol. Biochem. Behav.,** 18 (suppl. 1): 233.
Crow, K.E., Braggins, T.J. and Hardman, M.J., 1983b, Human liver cytosolic malate dehydrogenase: Purification, kinetic properties and role in ethanol metabolism, **Arch. Biochem. Biophys.,** 225, 621.
Crow, K.E., Newland, K.M. and Batt, R.D., 1983c, Factors influencing rates of ethanol oxidation in isolated rat hepatocytes, **Pharmacol. Biochem. Behav.,** 18 (suppl. 1): 237.
Crow, K.E., Cornell, N.W. and Veech, R.L., 1977, The role of alcohol dehydrogenase in governing rates of ethanol metabolism, **in**: Alcohol and Aldehyde Metabolising Systems, Vol. 3, R.G. Thurman, J.R. Williamson, H.R. Drott and B. Chance, eds., pp. 335-342, Academic Press, New York.
Crow, K.E. and Hardman, M.J., 1989, Regulation of rates of ethanol metabolism, in: Human Metabolism of Alcohol, Vol. 2, K.E. Crow and R.D. Batt, eds., pp. 3-16, CRC Press Inc., Boca Raton, Florida.
Derr, R.F., 1985, Modern metabolic control theory. I. Fundamental theorems, **Biochem. Arch.,** 1: 239.
Derr, R.F., 1986, Modern metabolic control theory. II. Determination of flux control coefficients. **Biochem. Arch.,** 2: 31.
Duszynski, J., Groen, A.K., Wanders, R.J.A., Vervoorn, R.L. and Tager, J.M., 1982, Quantification of the role of the adenine nucleotide translocator in the control of mitochondrial respiration in isolated rat liver cells, **FEBS Letters,** 146: 262.
Eriksson, C.J.P., Marselos, M. and Koivula, T., 1975, The role of cytosolic rat liver aldehyde dehydrogenase in the oxidation of acetaldehyde during ethanol metabolism *in vivo*, **Biochem. J.,** 152: 709.
Hawkins, R.D. and Kalant, H., 1972, The metabolism of ethanol and its metabolic effects, **Pharmacol. Rev.,** 24: 67.
Kacser, H. and Burns, J.A., 1973, The control of flux, **Symp. Soc. Exp. Biol.,** 32: 65.
Kacser, H. and Burns, J.A., 1979, Molecular democracy: who shares the controls, **Biochem. Soc. Trans.,** 7: 1149.
Kacser, H. and Porteous, J.W., 1987, Control of metabolism: what do we have to measure? **Trends Biochem. Sci.,** 12: 5.

Kalant, H., Khanna, J.M. and Endrenyi, L., 1975, Effect of pyrazole on ethanol metabolism in ethanol-tolerant rats, **Can. J. Physiol. Pharmacol.,** 53: 416.

Khanna, J.M. and Israel, Y., 1980, Ethanol metabolism, **Int. Rev. Physiol.,** 21: 275.

Lindros, K.O., 1983, Human blood acetaldehyde levels: with improved methods a clearer picture emerges, **Alcoholism, Clin. Exp. Res.,** 6: 70.

Lumeng, L., Bosron, W.F. and Li, T.-K., 1979, Quantitative correlation of ethanol elimination rates *in vivo* with liver alcohol dehydrogenase activity in fed, fasted and food restricted rats, **Biochem Pharmacol.,** 28: 1547.

Lumeng, L., Bosron, W.F. and Li, T.-K., 1980, Rate-determining factors for ethanol metabolism *in vivo* during fasting, **in**: Alcohol and Aldehyde Metabolising Systems, Vol. 4, R.G. Thurman, ed., pp. 489-496, Plenum Press, New York.

Mezey, E., Potter, J.J., Harmon, S.M. and Tsitouras, P.D., 1980, Effects of castration and testosterone administration on rat liver alcohol dehydrogenase activity, **Biochem. Pharmacol.,** 29: 3175.

Mezey, E., Potter, J.J. and Kvetnansky, R., 1979, Effect of stress by repeated immobilisation on hepatic alcohol dehydrogenase activity and ethanol metabolism, **Biochem. Pharmacol.,** 28: 657.

Plapp, B.V., Leidal, K.G., Smith, R.K. and Murch, B.P., 1984, Kinetics of inhibition of ethanol metabolism in rats and the rate-limiting role of alcohol dehydrogenase, Arch. **Biochem. Biophys.,** 230: 30

Rachamin, G., Macdonald, J.A., Wahid, S., Clapp, J.J., Khanna, J.M. and Israel, Y., 1980, Modulation of alcohol dehydrogenase and ethanol metabolism by sex hormones in the spontaneously hypertensive rat, **Biochem. J.,** 186: 483.

Siess, E.A., Brocks, D.G. and Wieland, O.H., 1982, Subcellular distribution of adenine nucleotides and of metabolites of the tricarboxylate cycle and gluconeogenesis in hepatocytes, **in** Metabolic Compartmentation, H. Sies, ed., pp. 235-257, Academic Press, New York.

Tager, J.M., Groen, A.K., Wanders, R.J.A., Duszynski, J., Westerhoff, H.V. and Vervoon, R.C., 1983, Control of mitochondrial respiration, **Biochem. Soc. Trans.,** 11: 40.

Thurman, R.G., Glassman, E.B., Handler, J.A. and Forman, D.T., 1989, The swift increase in alcohol metabolism (SIAM): A commentary on the regulation of alcohol metabolism in mammals, **in**: Human Metabolism of Alcohol, Vol. 2, K.E. Crow and R.D. Batt, eds., pp. 17-30, CRC Press Inc., Boca Raton, Florida.

Westerhoff, H.V., Groen, A.K. and Wanders, R.J.A., 1984, Modern theories of metabolic control and their applications, **Biosci. Rep.,** 4: 1.

FREE RADICAL PATHOLOGY IN ALCOHOL-INDUCED LIVER INJURY

Mario Umberto Dianzani

Department of Experimental Medicine and Oncology
Section of General Pathology
University of Turin
10125 Turin, Italy

INTRODUCTION

Increasing importance is attributed to the damage provoked by free radicals during both acute and chronic alcohol-induced liver injury. The first insight into the problem came from the experiments by Di Luzio and Costales (1964), showing a protective action of antioxidants on fatty liver following acute ethanol treatment in the rat. As lipid peroxidation had been shown to increase in CCl_4-induced liver damage (Comporti et al., 1965; Recknagel and Ghoshal, 1966), Comporti et al. (1967) studied the behaviour of this parameter in ethanol-treated rats and found an increase in the production of malonyldialdehyde. Hashimoto et al.(1968), however, found no appearance of the diene conjugation band in microsomal lipids of treated rats. This started a long-standing controversy, that may be considered today as concluded by several demonstrations of increased lipid peroxidation after acute, and especially after chronic, treatment of rats (Diazani, 1985; Yamada et al., 1988). Clinical demonstration of increased lipid peroxidation in chronic alcoholics has been reported (Moscarella et al., 1984).

It is clear now that several types of free radicals are produced during ethanol poisoning. Among the oxygen species, anion superoxide O_2^- has been shown to be produced by cytochrome P-450 and this formation is greatly increased following chronic alcohol feeding of rats (Ekstrom et al., 1986). Hydrogen peroxide (H_2O_2), originating from the dismutation of superoxide anion, in the presence of either reduced or oxidized iron is degraded through the Fenton and the Haber-Weiss reactions giving rise to hydroxyl radicals which can participate in the oxidation of ethanol (Klein et al., 1983). Moreover, it has been shown that thiyl free radicals are produced in the presence of anion superoxide from substances containing free SH-groups (Cederbaum and Dicker, 1983). Other free radicals arise from lipid peroxidation (lipodienyl free radicals and lipoperoxy-free radicals). Moreover, it has been recently shown that the ethanol free radical is produced by microsomes in the presence of iron (Ross et al., 1984; Lai and Piette, 1977; Finkelstein et al., 1980). Other free radicals involved are those arising from the interaction of lipid free radicals with proteins or lipids (Albano et al., 1988; Roubal and Tappel, 1966a). One can also consider as related to the free radical pathology the damage provoked by products of lipid peroxidation, as aldehydes, ketones and possibly other substances.

Alcoholism: A Molecular Perspective
Edited by T.N. Palmer, Plenum Press, New York, 1991

The present paper presents a review of the present knowledge in this field, as well as a critical consideration of the possible types of damage due to free radicals in vivo.

OXYGEN SPECIES

The problem of the formation of free radical oxygen species was first examined by Cederbaum and collaborators from the Lieber group (Roubal and Tappel, 1966b; Cederbaum et al., 1979; Klein et al., 1983; Ohnishi and Lieber, 1978). These authors have shown that, in the presence of microsomes and iron, ethanol becomes oxidized to acetaldehyde in a nonenzymic way. Moreover, the iron chelating agent desferrioxamine is able to inhibit ethanol oxidation by liver microsomes (Klein et al., 1983). The authors therefore postulated the formation of the $OH^{\cdot}$ free radical, that would be the real promoter of ethanol oxidation to acetadehyde, possibly with the intermediate formation of the ethanol free radical. As the life span of this free radical (never measured as yet) is presumably longer than the highly reactive $OH^{\cdot}$, ethanol would act in this way as a scavenger of this dangerous free radical.

$OH^{\cdot}$ would be formed through the Fenton reaction involving hydrogen peroxide in the presence of Fe^{2+}

$$Fe^{2+} + H_2O_2 \text{-------}> Fe^{3+} + OH^{\cdot} + OH^{-}$$

Hydrogen peroxide needed for this reaction may arise from the reduction of the anion superoxide $O_2^{\cdot}$. This would be formed in the smooth endoplasmic reticulum chain, possibly at the level of FAD, with the intermediate formation of a semiquinone free radical. This would transform O_2 present in the medium into the superoxide by electron transfer. The origin of O_2^{-} would be therefore related to the oxidation of ethanol in the drug metabolizing system, i.e. by an enzymic reaction. The reasons for the discharge of the electron from the semiquinone to O_2 at this level, instead of progressing through the usual cytochrome P450 pathway, are still unclear. In principle, this might occur when the rate and amount of FAD reduction during the first steps of the chain exceed the ability of cytochrome P450 to be reduced by oxygen. So, an obstacle in the second part of the chain might cause a bypass of the electron directly on oxygen at the level of the FAD semiquinone. Of course, this might depend on either excess ethanol entering the chain or decreased ability of cytochrome P450 to release the ethanol-derived electrons. It is noteworthy that oxidation of ethanol in the cytosol by alcohol dehydrogenase, that represents by far the major oxidative pathway for ethanol, seems to be unable to produce free radicals in the same manner.

It is well known that anion superoxide is rather long-living in water phase. So it could diffuse from the production sites. At the same time, it might act as an hydrogen acceptor, so forming hydrogen peroxide. This might contribute in a nonenzymic way to ethanol oxidation to acetaldehyde. The reaction is facilitated in the presence of catalase, mainly present intracellularly inside peroxysomes.

It is still discussed whether $OH^{\cdot}$ or/and anion superoxide may be produced at the level of cytochrome P450 or of other heme-proteins. According to several authors (Ohnishi and Lieber, 1978; Bando and Aki, 1990; Monteiro and Winterbourn, 1989; Aruoma and Halliwell, 1987; Puppo and Halliwell, 1988), this would happen under certain conditions, releasing iron from heme. $OH^{\cdot}$ would be produced also in this way, but it is still impossible to measure the amounts of $OH^{\cdot}$ formed at the different sites in the chain either in normal or pathological conditions.

Hydroxyl radicals might be formed even by the Haber-Weiss reaction:

$$O_2^- + H_2O_2 + Fe^{3+} \text{ -------> } O_2 + OH^{\cdot} + OH^-$$

The reaction requires the presence of iron in the oxidized form, but it is known that the transformation of Fe^{2+} into Fe^{3+} and vice versa is quite easy in the intracellular medium. The local concentration of O_2^- is decreased by the activity of superoxide dismutase. It is known, however, that this enzyme is mostly present in the cytosol. So, it is questionable whether it plays a role in the inactivation of superoxide formed in the endoplasmic reticulum membranes.

THIYL FREE RADICAL

The formation of this radical was firstly demonstrated by Ross et al. (1984) during the peroxidase-catalyzed metabolism of acetaminophen in the presence of thiols. Thiols were able to reduce the acetaminophen-generated radical so producing thiyl radicals that were trapped by the spin trap 5,5-dimethyl-1,pyrrolidine-N-oxide (DMPO).

In the system used, the production of thiyl free radicals, especially of glutathione, $GS^{\cdot}$, may account for a protective action of thiols against the possibly pure toxic acetaminophen free radicals. As far as ethanol is concerned, the formation of thiyl free radicals is probably through no interaction with $OH^{\cdot}$. So, even in this case the formation of thiyl free radicals may be as a protective reaction destroying the more dangerous $OH^{\cdot}$. $GS^{\cdot}$ may then be transformed into G-S-S-G by a reaction involving the condensation of two $GS^{\cdot}$. Another possibility is, however, the formation of other types of adducts involving only one molecule of $GS^{\cdot}$, and displaying the characteristics of a more complex free radical. It is impossible for the moment to establish if these reactions may provoke any type of damage in the cell. Of course, the involvement of GSH in these reactions produces a decrease in the local concentration of this antioxidant, but usually the cell replaces the lost amount very quickly by direct synthesis of new GSH molecules or by reduction of G-S-S-G.

LIPODIENYL- AND LIPOPEROXY-FREE RADICALS

The problem concerning the presence of increased lipid peroxidation after ethanol has been debated for a long time, and is now resolved in the sense that it really exists, especially in chronically treated animals. Several reviews on this problems have appeared (Torrielli et al., 1978; Dianzani, 1985). I will deal here only with the mechanisms producing the onset of lipid peroxidation after ethanol. A possible hypothesis is that this radical reaction is initiated by reaction of polyunsaturated fatty acids with some oxygen free radical species. The extremely short life span of $OH^{\cdot}$ might be an argument against its direct involvement since it is expected to react immediately after formation. $OH^{\cdot}$ is thought to be formed after ethanol administration at the level of the smooth endoplasmic reticulum. This hypothesis that $OH^{\cdot}$ is directly responsible presumes that it may be formed inside the microsomal phospholipids. As $OH^{\cdot}$ may be formed by the microsomal chain by either the Fenton or the Haber-Weiss reactions, this condition would be apparently fulfilled if iron, either in the reduced or oxidized form, were to be present at sufficient concentration.

Iron, however, may be set free from heme compounds, such as cytochrome P450 in certain conditions. This might explain production of $OH^{\cdot}$ at the level of this cytochrome, but not at that of the flavoprotein, where most of O_2^- is thought to be formed. The alternative hypothesis is that the free radical involved in the onset of lipid peroxidation is anion superoxide (O_2^-). This is also formed inside the membranes and is known to be much more reactive in lipid than in aqueous phase. A point in favour of the intervention of oxygen free radical species in the onset of lipid peroxidation is the fact the malonyldialdehyde production by liver microsomes is definitely decreased in the presence of added catalase (Dianzani, 1985). It is also well known that chelation of iron by EDTA or by desferroxiamine results in a prevention of lipid peroxidation.

Another possibility is that lipid peroxidation is started by free radical species other than oxygen species. The possibility that thiyl free radicals are involved has been put forward by Schoneich et al. (1989). These authors have demonstrated this possibility in vitro and proposed therefore the reaction:

$$RS^{\cdot} + PUFA \text{ ------> } PUFA^{\cdot} + RSH$$

The rate constant for this process is thought to be rather high.

According to Munday (1989), thiyl free radicals would be involved in hemolysis caused by several types of poisons. Hemolysis would be the consequence of the thiyl attack of membrane alpha-beta unsaturated acids.

Another sulphur-containing free radical involved in lipid peroxidation might be, however, sulfite radical anion (Erben-Russ et al., 1987). This is usually formed from one sulfite anion by interaction with $OH^{\cdot}$ but it is possible that even other oxygen species

$$SO_3^{2} + OH^{\cdot} \text{ ------> } SO_3^{-}$$

may produce the same result. Lizada and Yang (1981) have shown that sulfite, through the intermediate formation of free radicals in the presence of oxygen, can induce peroxidation of both linoleic and linolenic acids. There is no evidence, however, that either thiyl or sulfite anion free radicals are really produced after ethanol-administration.

The intracellular level of GSH decreases during lipid peroxidation by microsomes or homogenates (Dianzani, 1985), most of pre-existing G-SH being transformed into G-S-S-G. The decline of this water-soluble antioxidant has been considered among the causes facilitating lipid peroxidation in the presence of ethanol, but this may be due to the fact that G-SH is used for regeneration of vitamin E from its free radical, so recycling this lipid soluble antioxidant that is considered as the most powerful chain-breaker naturally occurring inside membranes.

The types of damage provoked by lipid peroxidation in membranes have been extensively discussed. It is only necessary here to say that the known mechanisms of its damaging effects are: i) anatomical and physiological damage to the membranes; ii) attack by lipodienyl- or lipoperoxy-free radicals on other types of structures; iii) distant damage provoked by products of lipid peroxidation, and especially by aldehydes (Dianzani, 1982). Among these aldehydes, those belonging to the 4-hydroxy-2,3-trans unsaturated series, and especially 4-hydroxynonenal, are the most toxic. Such substances are produced in rather high amounts during ethanol-induced lipid peroxidation. The mechanism of aldehyde-induced damage is related to the high reactivity of such substances with either SH or $-NH_2$ groups. In consequence, the aldehydes inhibit, when present at sufficient concentration, the activities of -SH-enzymes. Moreover, the aldehydes produce impairment of microtubular function (Dianzani, 1982; Gabriel et al., 1985) and can inhibit lipoprotein secretion when added at relatively high concentration (10^{-4}-10^{-5}M) to the system (Dianzani, 1982). Such concentrations may be reached in vivo in conditions of stimulated lipid peroxidation, for instance in the late phases of CCl_4-poisoning.

There are, however, some activities of the aldehydes that are displayed at very low concentration (less than 10^{-6}M). These are a stimulation of chemotaxis in rat polymorphonuclear leukocytes (Curzio, 1988; Curzio et al., 1987a,b,c), strong activation of plasma membrane adenylate cyclase of both hepatocytes (Dianzani et al., 1990) and granulocytes, stimulation of hepatocyte plasma membrane guanylate cyclase (Dianzani et al., 1990), and strong stimulation of hepatocyte membrane phospholipase C (Dianzani et al., 1990; Rossi et al., 1988,1989). These three enzymes contain G-proteins in their complex systems, in the adenylate cyclase system there being at least two G-proteins. Experiments done in the presence or absence of known stimulators of either Gi (inhibitory) or Gs (stimulatory) show that the aldehydes act by inhibiting the Gi protein. In fact, 4-hydroxynonenal displays an additive effect in the presence of the pertussis toxin that is thought to be mediated by inhibition of the G inhibitory protein in the adenylate cyclase system. Through these effects very low amounts of aldehydes are able to act as signal transmitters, as well as to affect the intracellular distribution of Ca^{2+}. Whether these facts have some relevance not only in the pathogenesis of cell damage, but even in cell physiology, is still open to debate. In fact, there are several arguments to support the view that lipid peroxidation may work at a very limited rate in normal cells, at least in those whose organelles (mitochondria, smooth endoplamic reticulum etc.) have a life-span shorter than that of the whole cell. No clearcut evidence exists, however, for the formation of such amounts of 4-hydroxynonenal in normal cells. The amounts of this aldehyde usually found in freshly prepared liver homogenates are in the range of the active concentrations, but one cannot exclude that these have been formed during the homogenization.

The aldehydes are able to give chromogenic adducts with proteins, so to account for the chemoluminescent structures (lipofuscin and others) often found in tissues having undergone in vivo lipid peroxidation.

It is noteworthy that, as far as it regards CCl_4-poisoning, it is now possible to distinguish between the biochemical lesions provoked by lipid peroxidation (possibly through the formation of the aldehydes) and those which are related to the covalent binding of CCl_4-derived free radicals to cellular proteins, lipids or nucleotides. In fact, pretreatment with the antioxidant promethazine, or preloading of the rat with vitamin E, results in the prevention of lipid peroxidation and at the same time block of the inhibition of microsomal glucose-6-phosphatase, of the Golgi apparatus galactosyl-transferase and of the release from cells into the surrounding medium of lactate dehydrogenase. These pretreatments do not inhibit covalent binding, and cannot protect against loss in protein synthesis, accumulation of fat inside the cells, the block in lipoprotein secretion, inhibition of ornithine decarboxilase, and the block in the discharge of Ca^{2+} from mitochondria and from microsomal vesicles into the cytosol. These results show that it is possible to separate the causes of fat infiltration in CCl_4-treated cells from those provoking acute cell death. In fact, fat accumulation seems to be related to covalent binding, or at least to be lipid peroxidation-independent, whereas acute cell death following CCl_4 administration is related to lipid peroxidation (Dianzani and Poli, 1986). This type of separation has not been done as yet in the case of ethanol poisoning, both in vivo and in vitro. It is important to remember, however, that acute ethanol treatment produces fat accumulation, but no cell death, whereas the opposite is found during chronic treatment. This represents an obvious experimental complication in this type of experiment.

It is noteworthy that lipid peroxidation in vitro is more stimulated in the presence of acetaldehyde than in that of ethanol itself (Stege, 1981; Muller and Sies, 1982; Dianzani,

1985). It is possible that this is due to the fact that acetaldehyde is removed by aldehyde oxidase, a flavoenzyme that forms anion superoxide. This mechanism might work even in vivo, but clear evidence is still lacking.

It has been also shown that anion superoxide is produced during the xanthine oxidase reaction (Bolann and Ulvik, 1987; Kato et al, 1990). Xanthine oxidase is a flavoenzyme, and is able to stimulate lipid peroxidation when added together with its substrate to microsomes. By contrast, uric acid, the product of the xanthine oxidase reaction, has some antioxidant properties as does allopurinol, the well-known inhibitor of xanthine oxidase. It is well known that uric acid production is increased after ethanol. So, the xanthine oxidase-uric acid system might play a role in the peroxidative balance during ethanol poisoning. It is noteworthy that xanthine oxidase can use acetaldehyde as a substrate.

ETHANOL FREE RADICAL

The formation of an ethanol free radical during ethanol intoxication was first hypothesized by Slater (1972). Cederbaum et al. (Roubal and Tappel, 1966b; Cederbaum et al., 1979; Klein et al., 1983; Ohnishi and Lieber, 1978) have shown the oxidation of ethanol by liver microsomes depends on the presence of reactive oxygen species, as stated

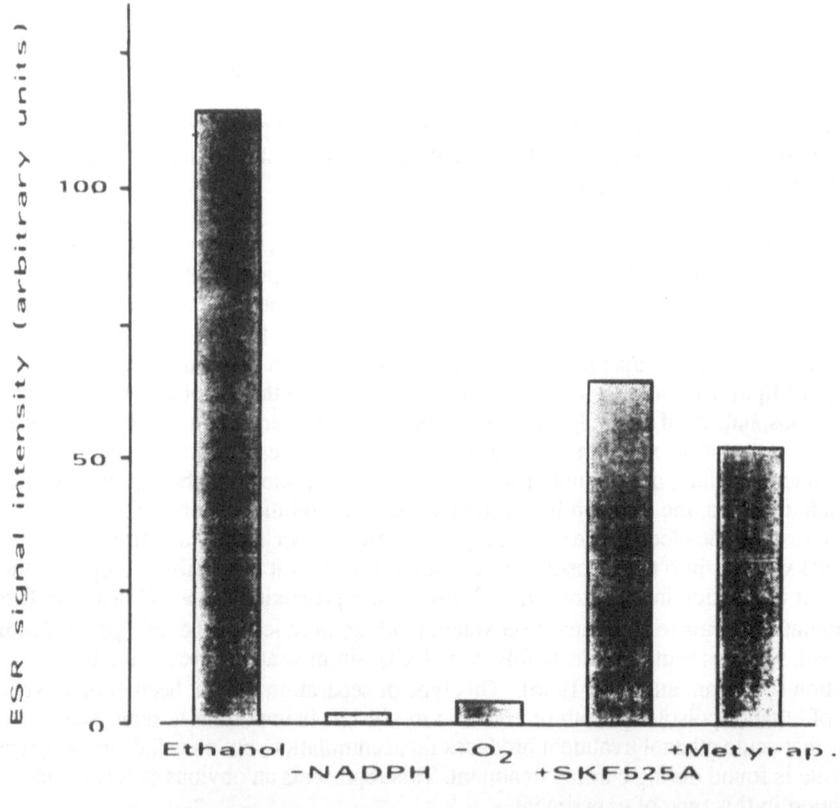

Figure 1. Intensity of the ESR signals produced in rat liver microsomes incubated with 20 mM ethanol plus or minus oxygen or NADPH and in the presence of cytochrome P-450 inhibitors SKF 525A (0.5 mM) and metyrapone (0.5 mM).

above, and have hypothesized that an ethanol free radical might be formed from ethanol and OH·. Clearcut evidence for the formation of hydroxyethyl free radical has been produced, however, by Albano et al. (1988), who used 4-pyridyl-1-oxide-t-butylnitrone (4-POBN) as spin trap (Fig 1). The formation of this free radical depends upon the presence of oxygen and NADPH and was blocked by heat inactivation of microsomes indicating that the microsomal monooxygenase system was involved. When a mixture of ^{12}C- and of $^{13}C^{1}$-ethanol was used, the authors observed a splitting of the signal seen with ^{12}C-ethanol alone, lying exactly on the same position of the pattern.

Free radicals are produced even when alcohols higher than ethanol are used. The height of the signal increases from ethanol to propanols, butanols and pentanols. It is noteworthy that in ethanol-induced animals, the height of the signals obtained with these alcohols is strongly increased. This means that the drug metabolizing system is presumably the same for all the used alcohols. Moreover, this experiment may have clinical implications: the expected damage might be higher when a drinker uses distillates that contain certain amounts of these alcohols than when he drinks wine.

The site for the formation of the ethanol free radical is now sufficiently clear (Albano et al., 1988). The addition of catalase, destroying H_2O_2 and preventing in this way its use in the Fenton's or Haber-Weiss' reactions, decreases alcohol radical formation, whereas the addition of azide, inhibiting catalase, increases the formation of the signal. Moreover, the use in vitro of OH· scavengers, such as mannitol or benzoate, has a protective action upon the formation of the signal. All these results indicate the importance of the non-enzymic fórmation of the ethanol free radical through the interaction of ethanol with OH·.

The OH· scavengers, however, never produced complete prevention of the formation of the signal. This led Albano et al (1988) to study further the possibility of ethanol free radical formation in an enzymic way at the level of cytochrome P450. The block of the endoplasmic reticulum oxidative systems by inhibitors of cytochrome P450 such as metyrapone and SKF525A provokes a strong inhibition of the signal formation (Fig. 1).

Recent experiments using monoclonal antibodies against the cytochrome P450 isoform that is known to metabolize specifically ethanol, usually referred to as cytochrome P450j, or P450IIE1, have shown that these antibodies were able not only to decrease the oxidation of ethanol but also inhibit ESR signal formation by more than 50%. These experiments suggest that ethanol free radical can be produced even at the level of cytochrome P450 and that these pathways account in vitro for about 50% of total hydroxyethyl free radical production.

The problem of the formation of the ethanol free radical in vivo is completely open. In previous studies, in fact, this species has never been found in the liver of the whole animal (Reinke et al., 1987) or in isolated hepatocytes in vitro. Only very recently, Knecht et al. (1990) have succeeded in detecting the adduct of hydroxyethyl radical with the spin trap 4-POBN in the gall bladder of deermice previously given ^{13}C-ethanol. In the case of ethanol, it is very difficult to study covalent binding in comparison to lipid peroxidation-provoked damage, as covalent binding occurs even in the case of acetaldehyde, and it is practically impossible to distinguish between a free radical-derived and an acetadehyde-derived adduct by this method. Moreover, acetaldehyde is by itself a toxic substance able to block -SH-groups, even if this happens with minor effect and at higher concentration than in the case of the lipid peroxidation-derived aldehydes.

So, for the moment we do not have any idea about the real toxicological importance of the ethanol free radical. It has been stated above that in appropriate concentration ethanol may act as a scavenger of the more dangerous OH· free radical, but it is not clear at all if this protective effect becomes a damaging effect when the ethanol concentration exceeds a certain level. What is certain, is that in vitro the hydroxyethyl free radical is produced only during ethanol metabolism in the smooth endoplasmic reticulum, and that its formation is much higher in ethanol-pretreated animals. This may indicate participation of the free radical reaction in the chronic alcoholic, without any indication about the species of free radical mainly involved in cell damage. Attempts to protect alcoholics by long-term vitamin E treatment gave some results (Kawase et al., 1989), showing the participation of lipid peroxidation to the production of the chronic damage, but the fact that protection is never complete seems to indicate the existence of mechanisms of damage other than lipid peroxidation. How many of these mechanisms are related to the major metabolite of ethanol, i.e. acetaldehyde, or to free radicals other than those arising during lipid peroxidation is also completely unknown. One has to wait for new technical approaches to the problem, therefore, before the last word can be said on this problem.

REFERENCES

Albano, E., Tomasi, A., Goria-Gatti, L. and Dianzani, M.U., 1988, Spin trapping of free radical species produced during the microsomal metabolism of ethanol. **Chem-Biol. Interacts.**, 65:223.

Aruoma, O.I. and Halliwell, B., 1987, Superoxide-dependent and ascorbate-dependent formation of hydroxyl radical from hydrogen peroxide in the presence of iron. Are lactoferrin and transferrin promoters of hydroxyl-radical generation? **Biochem. J.**, 241:273.

Bando, Y. and Aki, K., 1990, Superoxide-mediated release of iron from ferritin by some flavoproteins. **Biochem. Biophys. Res. Commun.**, 168:389.

Bolann, B.J. and Ulvik,R.J., 1987, Release of iron from ferritin by xanthine oxidase. Role of the superoxide radical. **Biochem. J.**, 243:55.

Cederbaum, A.I. and Dicker, E., 1983, Inhibition of microsomal oxidation of alcohols and hydroxyradical-scavenging agents by the iron-chelating agent desferrioxamine. **Biochem. J.,** 210:107.

Cederbaum, A.I., Miwa, G., Cohen, G. and Lu, A.Y.H., 1979, Production of hydroxyl radicals and their role in the oxidation of ethanol by a reconstituted microsomal systems containing cytochrome P_{450} purified from phenobarbital-treated rats. **Biochem. Biophys. Res. Commun.**, 91:747.

Comporti, M., Hartman, A. and Di Luzio, N.R., 1967, Effect on "in vivo" and "in vitro" ethanol administration on liver lipid peroxidation. **Lab. Invest.**, 16:616.

Comporti, M., Saccocci, C. and Dianzani, M.U., 1965, Effect of CCl_4 in vitro and in vivo on lipid peroxidation on rat liver homogenates and subcellular fractions. **Enzimologia**, 29:185.

Curzio, M., 1988, Interaction between neutrophils and 4-hydroxy-alkenals and consequences in neutrophils motility. **Free Rad. Res. Commun.**, 5:55.

Curzio, M., Di Mauro, C., Esterbauer, H. and Dianzani, M.U., 1987a, Chemotactic activity of aldehydes. Structural requirements. Role in the inflammatory process. **Biomed. Pharmacol.**, 41:304.

Curzio, M., Esterbauer, H., Di Mauro, C. and Dianzani, M.U., 1987b, Chemotactic activity of hydroxyalkenals in rat neutrophils. **Int. J. Tissue Reacts.**, 9:295.

Curzio, M., Esterbauer, H., Poli, G., Biasi, F., Cecchini, G., Di Mauro, C., Cappello, N., and Dianzani, M.U., 1987c, Possible role of aldehydic lipid peroxidation products as chemo-attractants. **Int. J. Tissue Reacts.**, 9:295.

Di Luzio, N.R. and Costales, F., 1964, Inhibition of the ethanol and carbon tetrachloride-induced fatty liver by antioxidants. **Exp. Mol. Pathol.**, 4:141.

Dianzani , M.U. and Poli, G., 1986, Lipid peroxidation and haloalkylation in CCl_4-induced liver injury. **In**: Free Radicals in Liver Injury, Poli, G., Cheeseman, K.H., Dianzani, M.U. and Slater, T.F., eds. pp. 149-158, IRL Press, Oxford.

Dianzani, M.U., 1982, Biochemical effects of saturated and unsaturated aldehydes. **In**: Free Radicals, Lipid Peroxidation and Cancer, ed. by D.C.H. McBrien, and T.F. Slater, pp. 129-158, Academic Press, London.

Dianzani, M.U., 1985, Lipid peroxidation in ethanol poisoning. A critical reconsideration. **Alcohol, Alcoholism**, 20:161.

Dianzani, M.U., Paradisi, L., Barrera, G., Rossi, M.A. and Parola, M., 1990, The action of 4-hydroxynonenal on the plasma membrane enzyme from rat hepatocytes. **In**: Free Radicals, Metal Ions and Biopolymers, P.C. Beaumont, D.C. Parsons, and C. Rice-Evans, eds., pp. 329-346, Richelieu Press, London.

Ekstrom, G. and Ingelman-Sundberg, M., 1989, Rat liver NADPH-supported oxidase activity and lipid peroxidation dependent on ethanol-inducible cytochrome P_{450} (P_{450}IIE1). **Biochem. Pharmacol.** 38:1313.

Ekstrom, G., Chroholm, T. and Ingelman-Sundberg, M., 1986, Hydroxyl-radical production and ethanol oxidation by liver microsomes isolated from ethanol-treated rats. **Biochem. J.** 233:755.

Erben-Russ, M., Michel, C., Bors, W. and Saran, M. 1987 The reaction of sulfite radical anion with nucleic acid components. **Free Rad. Res. Commun**. 2:285.

Finkelstein, E., Rosen, G.M. and Rauckman, E.J., 1980, Spin trapping of superoxide and hydroxyl radical: practical aspects. **Arch. Biochem. Biophys.** 200:1.

Gabriel, L., Miglietta, A. and Dianzani, M.U., 1985, 4-hydroxy-alkenals interaction with purified microtubular proteins. **Chem-Biol. Interact.** 56:201.

Hashimoto, S. and Recknagel, R.O., 1968, No chemical evidence for hepatic lipid peroxidation in acute ethanol toxicity. **Exp. Mol. Pathol.** 8:224.

Kato, S., Kawase, T., Alderman, J., Triatomi, N. and Lieber, C.S., 1990, Role of xanthine oxidase in ethanol-induced lipid peroxidation in rats. **Gastronterol**. 98: 203.

Kawase, T., Kato, S. and Lieber, C.S., 1989, Lipid peroxidation and antioxidant defence systems in rat liver after chronic ethanol feeding. **Hepatol**. 10:815.

Klein, S.M., Cohen, G., Lieber, C.S. and Cederbaum, A.I., 1983, Increased microsomal oxidation of hydroxyl radical scavenging agents and ethanol after chronic consumption of ethanol. **Arch. Biochem. Biophys.** 223:425.

Knecht, K.T., Bradford, B.U., Mason, R.P. and Thurman, G.R., 1990, Detection of the alpha-hydroxyethyl free radical from ethanol *in vivo*. 5th Congress of the International Society for Biomedical Research on Alcoholism, Toronto, Abs. 158.

Lai, C.S. and Piette, L.H., 1977, Hydroxyl radical production involved lipid peroxidation of rat liver microsomes. **Biochem. Biophys Res. Commun.**, 78:51.

Lizada, M.C.C. and Yang, S.F., 1981, Sulfite-induced lipid peroxidation. **Lipids**, 16:189.

Monteiro, H.P. and Winterbourn, C.C., 1989, 6-hydroxydopamine releases iron from ferritin and promotes ferritin-dependent lipid peroxidation. **Biochem. Pharmacol.**,38:4177.

Moscarella, S., Laffi, G., Coletta, D., Arena, U., Cappellini, A.P. and Gentilini, P., 1984, Volatile hydrocarbons in the breath of patients with chronic alcoholic disease: a possible marker of ethanol-induced lipid peroxidation. **In**: Frontiers of Gastrointestinal Research. Liver Cirrhosis., P. Gentilini and M.U. Dianzani eds., pp. 208-216, Karger, Basel.

Muller, A. and Sies, H., 1982, Role of alcohol dehydrogenase activity and of acetadehyde in ethanol-induced ethane and pentane production by isolated perfused rat liver. **Biochem. J.**, 206:153.

Munday, P.R., 1989, Toxicity of thiols and disulphides: involvement of free-radical species. **Free Rad. Biol. Med.**, 7:659.

Ohnishi, K. and Lieber, C.S., 1978, Respective role of superoxide and hydroxyl-radical in the activity of the reconstituted microsomal ethanol-oxidizing system. **Arch. Biochem. Biophys.**, 191:798.

Puppo, A. and Halliwell, B., 1988, Formation of hydroxyl radicals in biological systems. Does myoglobin stimulate hydroxyl radical formation from hydrogen peroxide? **Free Rad. Res. Comms.**, 4:415.

Recknagel, R.O. and Ghoshal, A.K., 1966, Lipoperoxidation as a vector in carbon tetrachloride hepatotoxicity. **Lab. Invest.**, 15:132.

Reinke, L.A., Dubose, C.M. , McCay, P.B., 1987, Reactive free radical generation in vivo in heart and liver of ethanol-fed rats: correlation with radical formation *in vitro*. **Proc. Natl. Acad. Sci. USA**, 84:9223.

Ross, D., Albano, E., Nilsson, U. and Moldèus, P., 1984, Thyil radicals formation during peroxidase-catalysed metabolism of acetaminophen in the presence of thiols. **Biochem. Biophys. Res. Commun.**, 125:109.

Rossi, M.A., Garramone, A. and Dianzani, M.U., 1988, Stimulation of phospholipase C activity by 4-hydroxy-nonenal: influence of GTP and calcium concentration. **Int. Tissue React.**, 5:321.

Rossi, M.A., Garramone, A. and Dianzani, M.U. 1989 Effect of 4-hydroxy-2,3-trans-nonenal, a lipid peroxidation product, on hepatic phospholipase C. **Med. Sci. Res.**, 17: 257.

Roubal, W.T. and Tappel, A.L., 1966a, Damage to proteins, enzymes and aminoacids by peroxidizing lipids. **Arch. Biochem. Biophys.**, 113:5.

Roubal, W.T. and Tappel, A.I., 1966b, Polymerization of protein produced by free-radical lipid peroxidation. **Arch. Biochem. Biophys.**, 113:150.

Schoneich, C., Asmus, K.D., Dillinger, U. and Buchhausen, F., 1989, Thyil radical attack on polyunsaturated fatty acids: a possible route to lipid peroxidation, **Biochem. Biophys. Res. Commun.**, 161:113.

Stege, T.E., 1981, Acetadehyde-induced lipid peroxidation in isolated hepatocytes. **Res. Comm. Chem. Pathol.Pharmacol.**, 36:287.

Torrielli, M.V., Gabriel, L. and Dianzani, M.U., 1978, Ethanol-induced hepatotoxicity; experimental observations on the role of lipid peroxidation. **J. Pathol.**, 126:11 .

Yamada, S., Fujiwara, K., Masaki, N., Olita, Y., Sato, Y. and Oka, H., 1988, Evidence for potentiation of lipid peroxidation in the rat liver after chronic ethanol feeding. **Scand. J. Clin. Lab. Invest.**, 48:627.

FREE RADICAL MEDIATED REACTIONS AND ETHANOL TOXICITY: SOME CONSIDERATIONS ON THE METHODOLOGICAL APPROACHES

Emanuelle Albano, Magnus Ingelman-Sundberg, A. Tomasi and G. Poli

Dept. of Experimental Medicine and Oncology
University of Turin
Turin, Italy
Dept. of Physiological Chemistry
Karolinska Institutet
Stockholm, Sweden
Institute of General Pathology
University of Modena
Modena, Italy

In the last few years the occurrence of free radical-mediated reactions, including lipid peroxidation, as a result of ethanol intoxication have received new attention, and it is now generally accepted that oxidative damages might play a role in the pathogenesis of tissue damage due to alcohol abuse (Lieber, 1988). Nonetheless, it should be realized that since the early work of Di Luzio, who first suggested an involvement of lipid peroxidation in causing liver lesions following ethanol poisoning (Di Luzio and Hartman, 1967; Comporti et al., 1967), this topic has been a matter of a long standing debate. The main reason of such a controversy is the fact that the results obtained in different laboratories have often been contradictory because of methodological differences (Videla and Valenzuela, 1982; Dianzani, 1985), and also because of the use of analytical techniques not always suitable for the detection of low levels of radical-associated tissue damage.

The aim of this paper is to give a rapid overview of the methods so far employed in studying the role of free radicals in ethanol toxicity discussing the capabilities as well as the limits of the different assays.

DETECTION OF ETHANOL FREE RADICALS BY ESR SPECTROSCOPY

Electron spin resonance (ESR) spectroscopy is the most specific technique to study free radicals since the unpaired electrons characteristic of free radical molecules give them paramagnetic properties which can be exploited by this spectroscopic technique (Janzen, 1980). Keller and co-workers (1971) were the first to apply ESR spectroscopy to investigate free radical formation in the liver of ethanol-treated rats. Their lack of success can now be explained because the radical species they were looking for have rates of reaction so high that their actual steady-state concentrations in the liver are well below the detection range of

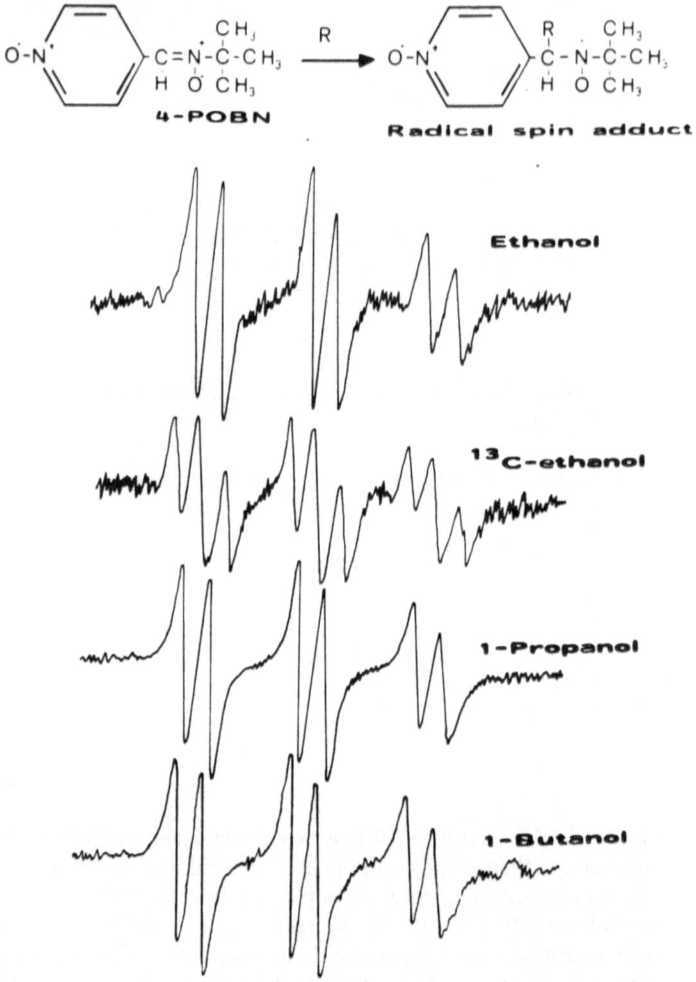

Figure 1. Reaction of the spin trapping agent 4-pyridyl-1-oxide-t-butyl nitrone (4-POBN) with free radicals and ESR spectra of the spin adducts between this spin trap and alcohol-derived free radical species produced during the metabolism of ethanol, ^{13}C-ethanol, 1-propanol and 1-butanol, respectively, by rat liver microsomes.

the ESR spectrometer, of about 10^{-6}M. This limitation can be now partially overcome by complexing highly reactive free radicals with nitrone or nitroso compounds (Fig. 1), named spin traps, to form relatively stable adducts, which can be then analyzed by ESR spectroscopy (Janzen, 1980).

Using the spin trapping agent 4-pyridyl-1-oxide-t-butyl nitrone (4-POBN) we succeeded in detecting the formation of a carbon centred free radical in rat liver microsomes incubated in the presence of ethanol and a NADPH-regenerating system (Albano et al., 1986). This species was positively identified as 1-hydroxyethyl radical on the basis of the similarities of the ESR spectra with those of a chemically prepared 4-POBN-hydroxyethyl spin adduct as well as by means of ^{13}C isotope substitution in the ethanol molecule (Albano et al., 1988a), as shown in Figure 1. We also found that alcohol free radicals were similarly produced by liver microsomes during the metabolism of various aliphatic alcohols (Fig. 1), indicating the existence of a common activation pathway for these compounds (Albano et al., 1987; Albano et al., 1991).

Although spin trapping is mainly a qualitative technique it can be used with some precautions for investigating the site and mechanisms of free radical formation. Previous studies by the group of Cederbaum have demonstrated that ethanol can be oxidized to acetaldehyde by a non-enzymatic mechanism involving the presence of hydroxyl ($OH^{\cdot}$) radicals, originating from iron-catalyzed degradation of hydrogen peroxide (Cederbaum et al., 1978; Cederbaum and Dikers, 1983). Spin trapping experiments have confirmed that such non-enzymatic pathway can be partially responsible for the formation of hydroxyethyl radicals, since the addition of desferrioxamine, catalase and $OH^{\cdot}$ scavengers decreased the yield of the radical trapped. However, when iron contaminating the incubation system was carefully removed it was possible to demonstrate about 50% of the free radical production was still taking place and was unaffected by the presence of H_2O_2, desferrioxamine and $OH^{\cdot}$ scavengers (Albano et al., 1988a). Moreover, in these conditions blockers of cytochrome P-450 were able to inhibit the free radical activation of ethanol, suggesting the involvement of a metabolic process independent of the presence of H_2O_2 and iron and probably involving cytochrome P-450 (Albano et al., 1988a). The role of cytochrome P-450 in catalyzing the free radical activation of alcohol has been recently confirmed by demonstrating that reconstituted membrane vesicles containing cytochrome P-450 reductase do not form alcohol-derived free radicals when incubated with NADPH, 4-POBN and ethanol or 1-butanol, unless cytochrome P-450 was incorporated in the membranes (Albano et al., 1991). Moreover, the ethanol-inducible cytochrome P-450IIE1 was twice as active as the phenobarbital-inducible isoenzyme P-450IIB1.

The ability of cytochrome P-450IIE1 to catalyze the radical activation of alcohols were further confirmed using antibodies raised against cytochrome P-450IIE1 which inhibited by more than 50% the spin trapping of ethanol and 1-butanol derived radicals in microsomes from ethanol fed rats, but had minimal effects in those from untreated controls (Albano et al., 1991).

Another problem investigated using the spin trapping technique has been the mechanism responsible for the free radical formation by cytochrome P-450IIE1. We have observed that both in microsomal fractions and in reconstituted membrane systems the production of 1-butanol radicals correlated with the capacity of cytochrome P-450IIE1 to generate superoxide anion and that inhibitors of the NADPH oxidase activity of cytochrome P-450IIE1 affected alcohol radical production (Albano et al., 1991). This led to the hypothesis that ferric cytochrome P-450-superoxide complex might be the reactive form of the enzyme catalyzing the free radical conversion (Albano et al., 1991).

In conclusion the results above show that spin trapping can be applied successfully not only to demonstrate the presence of alcohol free radical, but can be also used to elucidate the biochemical mechanisms responsible for the formation of such reactive species. Unfortunately, previous attempts to spin trap hydroxyethyl free radicals in the liver of ethanol-treated rats have failed (Reinke et al., 1987) and only very recently a report showing the detection of this radical in the bile of ADH-negative deermice has appeared (Knecht et al., 1990). This finding is of great importance since it indicates that hydroxyethyl radicals can be formed under the conditions present *in vivo* and supports their possible involvement in causing ethanol toxicity.

DETECTION OF ETHANOL-INDUCED LIPID PEROXIDATION

The demonstration of lipid peroxidation is the most widely used index for assessing the presence of free radical-mediated reactions either *in vitro* or *in vivo*, largely because several assays of lipid peroxidation are easy to perform. It should also be considered that peroxidative reactions might contribute to the spreading of a free radical-induced damage within the tissue. Therefore the detection of lipid peroxidation following a toxic injury, as in the case ethanol intoxication, is not only an indication for the presence of free radical reactions, but helps in elucidating the pathogenetic mechanisms of such injury.

Lipid peroxidation is initiated by the attack of any free radical species at the methylene carbon of an unsaturated fatty acid followed by the abstraction of an hydrogen atom which leaves behind a radical on the lipid molecule (Fig. 2). This lipid radical undergoes a molecular rearrangement to the conjugated diene configuration, followed by the reaction with oxygen to give a peroxyl radical. Peroxyl radicals are highly reactive species which can abstract an hydrogen from an adjacent fatty acid molecule giving an hydroperoxide and, at the same time propagating the chain reaction of lipid peroxidation (Fig. 2). The length of such self sustaining chain reaction will depend upon many factors and particularly upon the presence of the so-called chain-breaking antioxidants, as for instance vitamin E (α-tocopherol). Vitamin E is able to donate an hydrogen atom to the peroxyl radical forming an α-tocopheryl radical, which is probably then reduced back to tocopherol by the action of ascorbic acid (Niki, 1987). Although lipid peroxides are relatively stable, in the presence of traces of transition metals (iron, copper), they can be converted to alkoxyl radicals which then undergo a complex series of decomposition reactions to produce a variety of carbonyl compounds (alkanals, alkenals, ketones) and hydrocarbons (ethane and pentane).

As shown by Figure 2, lipid peroxidation can be monitored at different stages by measuring the loss of unsaturated fatty acids, or by detecting the formation of peroxidation products generated during either the propagation phase (lipid radicals, conjugated dienes, lipid hydroperoxides) or the degradation process (malonildialdehyde and other carbonyls, hydrocarbon gases). Nonetheless, it should be considered that peroxidative reactions occurring in biological membranes are extremely complicated processes, since they involve a mixture of different lipids in matrices also containing proteins, carbohydrates, catalytically-active enzymes and traces of metals and antioxidants. Thus, no single assay method can fully describe what is happening in a system undergoing lipid peroxidation.

Several of the assays shown in Figure 2 have been used in detecting lipid peroxidation induced by either acute or chronic alcohol administration. In the next few pages these tests will be critically discussed in relation to their applications in studies concerning ethanol.

A decrease in the content of unsaturated fatty acids and particularly of arachidonic acid has been observed in the livers of ethanol-fed rats as compared to pair fed controls (Reitz, 1975; Peters et al., 1986) and in hepatic biopsies from patients with alcoholic cirrhosis (Peters et al., 1986). These findings, although suggestive of the presence of lipid peroxidation, should be considered very carefully since ethanol is known to affect fatty acid desaturase activities in the rodent liver, therefore interfering with the endogenous synthesis of unsaturated fatty acids (Umeki et al., 1984). Moreover, dietary imbalances and alterations in enteral lipid absorption are possible causes for the lowering of arachidonic acid content in cirrhotic livers.

Reinke and coworkers (1987) have reported that lipid free radicals can be detected by ESR spectroscopy in the liver and heart of rats chronically fed with alcohol after the *in vivo* administration of the spin trap 2,4,6-trimethoxyphenyl-t-butyl nitrone. This is an unequivocal indication that peroxidative reactions are occurring during chronic ethanol intoxication not only in the liver but also in the heart, an organ which is also a target of oxidative damage by alcohol (Nordmann et al., 1990). However, the spin trapping technique can not be used as a general purpose technique for investigating lipid peroxidation, since it requires expensive instrumentation and some experience in the interpretation of results. Moreover, it can not be applied to humans because of the toxicity of the spin trapping agents (Albano et al., 1986).

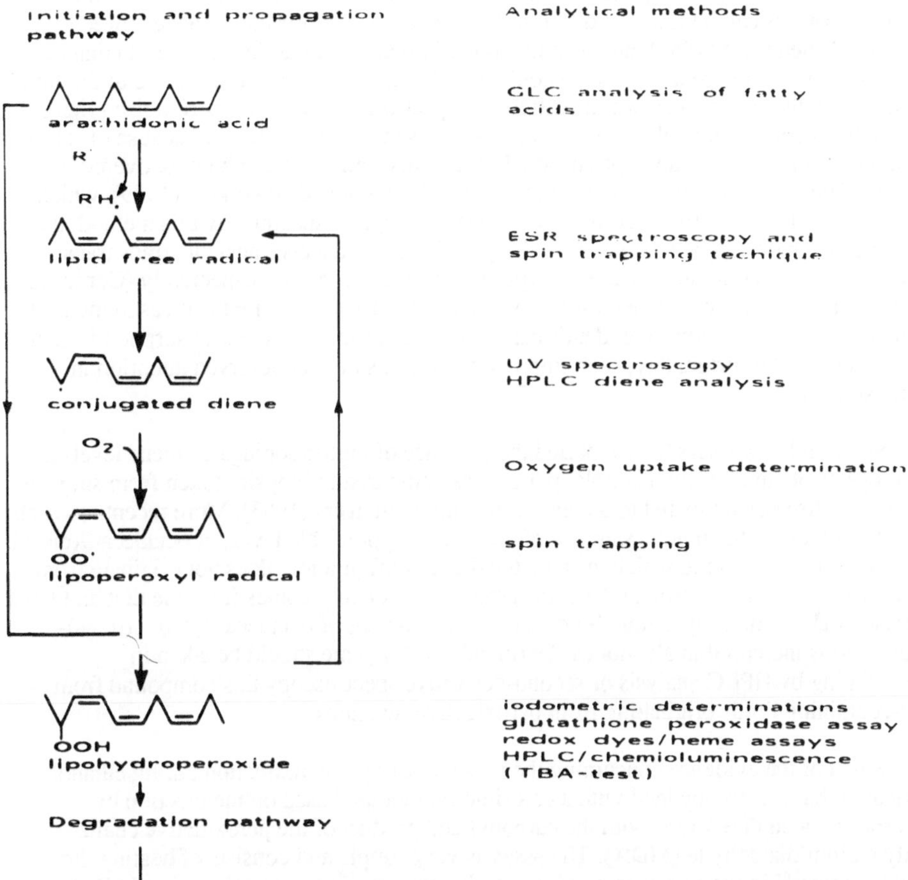

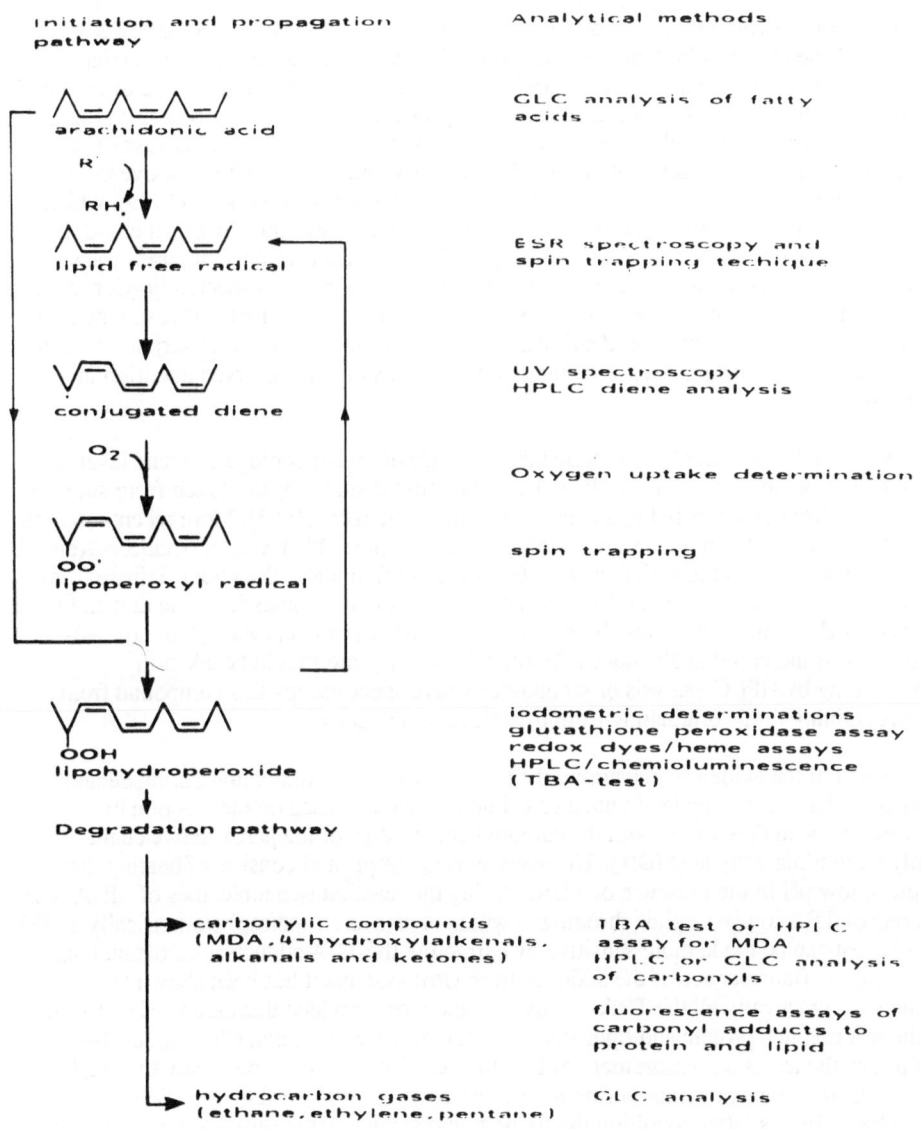

Figure 2. Reactions involved in the initiation and propagation of lipid peroxidation and methods currently used for the measurement of lipid peroxidation.

The most widely used test for measuring the early phases of lipid peroxidation is the assay of conjugated dienes. As outlined in Figure 2, the abstraction of an hydrogen atom from an unsaturated fatty acid molecule is immediately followed by the rearrangement of the double bonds to a diene configuration which has a strong absorption around 233 nm. The determination of conjugated dienes is simple, requiring only the extraction of the lipids in a chloroform-methanol mixture followed by re-dissolution of the chloroform extract in cyclohexane for the spetroscopic analysis. It is also a rather sensitive method for detecting lipid peroxidation and can be applied to *in vivo* experiments. An increase in the conjugated diene absorbances has been observed following acute (MacDonald, 1973; Valenzuela et al., 1980; Kera et al., 1985) and chronic (Peters et al., 1986) alcohol intoxication of rats. However, negative results were also obtained by several groups (Hashimoto and Recknagel., 1968; Torrielli et al., 1978; Speisky et al., 1985; Inomata et al., 1987). The reasons of these discrepancies can be found in differences in experimental procedures as reviewed by Videla and Velenzuela (1982), and also by taking into account that in the case of ethanol intoxication the diene "peak" is just a shoulder on the large absorption peak due to the lipids themselves. In these cases performing difference spectroscopy with non-peroxidized lipids obtained from control animals does not increase the sensitivity of the test because of intra-individual variation in the absorptions at 233 nm. This limitation can be now overcome by a recent refinement of diene analysis, proposed by Corongiu and coworkers (1989), which consists of second derivative recording of the UV-absorption spectra. By this method, the shoulder absorption can be resolved into sharp peaks and often cis,trans- and trans,trans-isomers can be differentiated by the absorption at 233 and 245 nm, respectively (Corongiu et al., 1989). In addition, second-derivative spectroscopy does not require the measurement of difference spectra and allows the identification of small changes in diene absorption in each single specimen. Unfortunately, such improved assay has not yet deserved attention in ethanol studies.

Shaw and coworkers have reported the presence of higher conjugated diene levels in liver biopsies obtained from alcoholic patients as compared to biopsies taken from subjects having liver diseases unrelated to alcohol or healthy volunteers (1983). More recent research indicates, however, that human samples often contain appreciable levels of octadeca-9(cis)11 (trans)-dienoic acid, a lipid which, despite the diene configuration, does not originate from lipid peroxidation (Thompson and Smith, 1985), but probably comes from the diet and from bacterial metabolism in the colon. Since the plasma content of octadeca-9(cis)11(trans)-dienoic acid is increased in alcoholics (Dormandy, 1988), care should be taken in differentiating by HPLC analysis or second-derivative spectroscopy this compound from diene-containing lipids originating from peroxidative reactions.

Much of the evidence concerning the presence of lipid peroxidation during ethanol intoxication have been obtained with a colorimetric method based on the reaction of thiobarbituric acid (TBA test) with the carbonyl end product of the peroxidative chain, mainly malonildialdehyde (MDA). The assay is very simple and consists of heating the sample at low pH in the presence of TBA. During the reaction two molecules of MDA react with one of TBA to give a pink chromogen which is measured spetrophotometrically at 533 nm. The test can be made more sensitive by extracting the TBA adducts with butanol and measuring the fluorescence of the adducts. In *in vitro* systems it has been shown that MDA accounts for practically all the TBA-reactive compounds, provided that care is taken to carry out the reaction in a protein and lipid-free supernatant (Esterbauer and Cheeseman, 1990). In such a case the test is a measurement of free MDA. The same does not apply to samples obtained from organs or serum during *in vivo* experiments where MDA is rapidly metabolized. In this latter condition the tissue homogenates or the serum are usually boiled for long time (30-45 min) in the presence of TBA and the substances forming the chromogen are generated by the decomposition of lipid hydroperoxides catalyzed by traces of transition metals (Gutteridge and Halliwell, 1990). This method is often improperly referred as estimation of the lipoperoxide content of tissues. However, since also sugars and amino acids react with TBA forming a chromogen identical to that generated by MDA (Gutteridge and Halliwell, 1990), the term TBA reactive substances (TBARS) is more appropriate to describe the compounds that are effectively measured. A further problem in the application of the TBA test as an index of lipid peroxidation *in vivo* concerns the heterogeneous composition of the samples where variation in the concentration of antioxidants and transition metals might have a large influence on chromogen formation. These points should be taken into account when considering the many reports showing that either acute (Videla and Valenzuela, 1982; Kera et al., 1985; Tanaka et al., 1970; Videla et al., 1980; Sippel, 1983; Kato et al., 1990) or chronic ethanol treatment of rats (Shaw et al., 1981; Kavase et al., 1989) increase TBARS in liver as well as in cerebellum and heart homogenates (Nordmann et al., 1990). The same applies to the observation that alcoholic patients have higher TBARS levels in the serum and in hepatic biopsies as compared to healthy subjects (Suematzu et al., 1981; Tanner et al., 1986). It has been shown, in fact, that acute ethanol intoxication decreases the hepatic glutathione content in experimental animals (Videla and Valenzuela, 1982; Kera et al., 1985; Videla et al., 1980), while lower levels of vitamin E are present both in the liver of chronically ethanol-fed rats and in the plasma of alcoholics (Kavase et al., 1989; Tanner et al., 1986). Furthermore, there is evidence that ethanol administration increases the amount of tissue free iron, which can substantially accelerate the formation of TBARS during the assay (Rouch et al., 1990; Shaw et al., 1989). On the other hand, *in vitro* experiments have shown the production of free MDA by isolated hepatocytes incubated with ethanol (Albano et al., 1988b). Moreover, in liver cells prepared from ethanol-fed rats ethanol-induced lipid peroxidation is greatly increased (Albano et al., 1988b), probably as a result of the combined stimulation in the formation of both hydroxyethyl free radical (Albano et al., 1988a) and of reactive oxygen species (Ekström and Ingelman-Sundberg, 1989) resulting from the induction of cytochrome P-450IIE1. Therefore, the results so far obtained *in vivo* by the TBA test require validation by more sophisticated and selective analytical techniques. This is particularly true in the case of human studies where the use of TBA test has been shown to overestimate by about seventy fold the plasma peroxide content when compared to direct measurements lipohydroperoxides by HPLC (Gutteridge and Halliwell, 1990).

In looking for a simple assay suitable for the detection of lipid peroxidation products in human serum we have studied in alcoholic patients the fluorescence between proteins and both malonildialdehyde (MDA) or 4-hydroxynonenal, an highly reactive aldehyde originating during lipid peroxidation (Esterbauer and Cheeseman, 1990), which can be detected in plasma by fluorescence analysis at 460 nm emission wave-length using the excitation wave-lengths at 390 and 356 nm, respectively (Fig. 3). Moreover, we have found that alcoholics with normal liver function show an increase over controls in levels of fluorescent protein adducts with the two aldehydes (Fig. 4). This assay has the advantage of being very simple and sufficiently sensitive and could be used as preliminary screening in research on oxidative injury in human diseases.

Another group of final products of lipid peroxidation which has been used for demonstrating the presence of ethanol-induced oxidative damage is that comprising the hydrocarbons ethane and pentane. These gases originate from the oxidative degradation of omega-3 and omega-6 polyunsaturated fatty acids, respectively (Burk and Ludden, 1989). Although they represent only a small proportion of the end products of lipid peroxidation, they have been detected by gas chromatography both *in vitro* in the perfusate of isolated

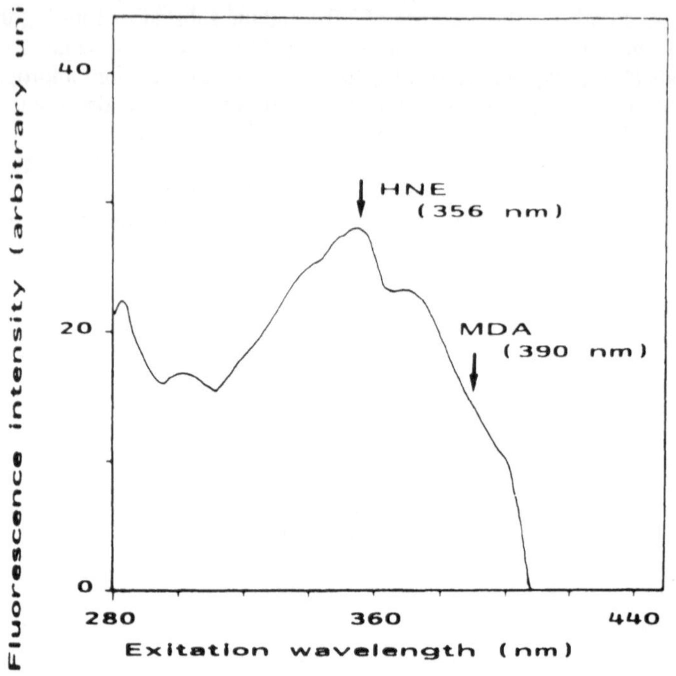

Figure 3. Fluorescence spectra at 460 nm of malonildialdehyde (MDA) and 4-hydroxynonenal (HNE) adducts with plasma proteins.

livers perfused with a solution containing ethanol (Litov et al., 1978) and *in vivo* in the expired air of rats acutely treated with ethanol (Muller and Sies, 1982).

The determination of ethane and pentane represents so far the only genuinely non-invasive method for determining lipid peroxidation *in vivo* and enables the continuous monitoring of animals and, even humans, for long time periods. Concerning this latter aspect, patients with chronic alcoholic liver diseases have been shown to exhale more pentane than normal subjects (Peters et al., 1986; Moscarella et al., 1984). Despite the sensitivity and the advantages of this technique it is now open to criticism on several grounds. Firstly when applied to the whole animal or human it does not give information about the organ where the two gases originate.

The second problem arises because, even avoiding contamination from fecal gases, it is difficult to determine how much of the ethane and pentane exhaled come from the absorption of hydrocarbons originating by bacterial fermentation in the gut. (Burk and Ludden, 1989). Finally, the stoichiometry with respect to peroxidized lipid is not fixed, depending upon the oxygen concentration, and pentane undergoes considerable metabolism by the liver microsomal monoxygenase system (Burk and Ludden, 1989). This latter aspect is particularly important for the measurement of ethane and pentane during alcohol intoxication because the ethanol-inducible cytochrome P-450IIE1 is particularly active in metabolizing pentane. Therefore, in alcoholic patients the amounts of these gases in the breath may vary considerably depending on the activity of the liver enzymes as well as on the degree of hepatic damage.

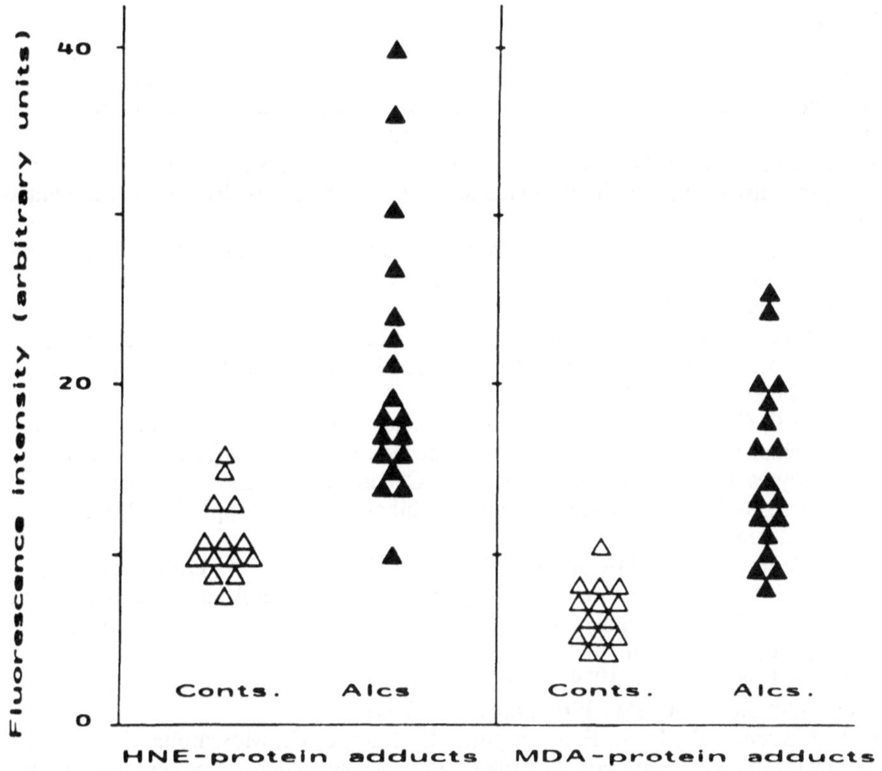

Figure 4. Fluorescence intensities of HNE-protein and MDA-protein adducts in plasma from control subjects and healthy alcoholics (Results from Carini et al., 1988).

CONCLUSIONS

Despite the many lines of evidence indicating that free radical-mediated oxidative damage occurs as a result of both acute and chronic alcohol intoxication, the methodologies employed for these assays have not always been properly evaluated. In many cases, in fact, the complexity of the interaction between ethanol and living tissues requires that the tests be carried out in *in vivo* experimental models or in human samples in which we are still lacking suitable analytical methods for demonstrating free radical damage (Gutteridge and Halliwell, 1990). Nonetheless, a number of new and very sensitive assays, such as the measurement of lipid hydroperoxides by HPLC coupled with photochemical detection (Miyazawa, 1989) and the analysis of carbonyl products of lipid peroxidation by HPLC or by combined gas-chromatography and mass spectrometry (Esterbauer and Cheeseman, 1990) have been developed in the last few years and now wait to be applied in the detection of alcohol-mediated peroxidative injury. The use of these techniques as well as of antibodies able to recognise aldehydes bound to proteins (Mowri et al., 1988) will be the methods of choice for investigating the role of lipid peroxidation in causing tissue damage in humans and to assess the advantages of antioxidant therapies.

REFERENCES

Albano, E. , Tomasi, A. , Goria-Gatti, L. and Dianzani, M.U., 1986, Free radical metabolism of ethanol, **in**: "Free Radicals Cell Damage and Disease," C. Rice-Evans, ed., pp.117-126, Richelieu Press, London.

Albano, E., Cheeseman, K.H., Carini, R. , Dianzani, M.U. and Slater, T.F., 1986, Effects of spin traps in isolated hepatocytes and liver microsomes, **Biochem. Pharmacol.** 35:257.

Albano, E., Tomasi, A., Goria-Gatti, L., Poli, G., Vannini, V. and Dianzani, M.U., 1987, Free radical metabolism of alcohols by rat liver microsomes, **Free Rad.Res. Communs.** 3:234.

Albano, E. , Tomasi, A. , Goria-Gatti, L. and Dianzani, M.U., 1988a, Spin trapping of free radical species produced during the microsomal metabolism of ethanol, **Chem. Biol. Interact.** 65:223.

Albano, E., Poli, G., Tomasi, A., Goria-Gatti, L. and Dianzani, M.U., 1988b, Study on the mechanisms responsible for ethanol-induced oxidative damage using isolated hepatocytes, **in**: Medical, Biochemical and Chemical Aspects of Free Radicals, Hayaishi, O., Niki, E., Kondo, M. and Yoshikawa, T. eds., pp. 1389-1392, Elsevier, Amsterdam.

Albano, E., Tomasi, A., Goria-Gatti, L., Persson, J.O., Terelius, Y., Goria-Gatti, L., Dianzani, M.U. and Ingelman-Sundberg, M., Role of ethanol-inducible cytochrome P-450 (P-450IIE1) in catalyzing the free radical activation of aliphatic alcohols, **Biochem.Pharmacol.**, in press.

Burk, R.F. and Ludden, T.M., 1989, Exhalated alkanes as indices of *in vivo* lipid peroxidation, **Biochem. Pharmacol.** 38:1029.

Carini, R., Mazzanti, R., Biasi, F., Chiarpotto, E., Marmo, G., Moscarella, S., Gentilini, P., Dianzani, M.U. and Poli, G., 1988, Fluorescence aldehyde protein adducts in the blood serum of healthy alcoholics, **in**: Alcohol Toxicity and Free Radical Mechanisms, Nordmann, R., Ribière, C. and Rouach, H. eds., pp. 61-64, Pergamon Press, Oxford.

Cederbaum, A.I. and Dikers, E., 1983, Inhibition of microsomal oxidation of alcohol and hydroxyl radical scavengers by the iron chelating agent desferrioxamine, **Biochem J.** 210:107.

Cederbaum, A.I., Dikers, E. and Cohen, G., 1978, Effect of hydroxyl radical scavengers on microsomal oxidation of alcohols and associated microsomal reactions, **Biochem.** 17:3558.

Comporti, M., Hartman, A.D. and Di Luzio, N.R., 1967, Effect of lipid peroxidation, **Lab.Invest.** 16:616.

Corongiu, F.P., Banni, S. and Dessì, M.A., 1989, Conjugated dienes detected in tissue lipid extracts by second derivative spectrophotometry, **Free Rad.Biol.Med.** 7:183.

Di Luzio, N.R. and Hartman, A.D., 1967, Role of lipid peroxidation in the pathogenesis of ethanol induced fatty liver, **Fed.Proc.** 26:1463.

Dianzani, M.U., 1985, Lipid peroxidation in ethanol poisoning: a critical reconsideration, **Alcohol Alcoholism** 20:161.

Dormandy, T.L., 1988, Diene conjugation in chronic alcoholics, **in**: Alcohol Toxicity and Free Radical Mechanisms, Nordmann, R., Ribière, C. and Rouach, H., eds., pp. 55-59, Pergamon Press, Oxford.

Ekström, G. and Ingelman-Sundberg, M., 1989, Rat liver microsomal NADPH-supported oxidase activity and lipid peroxidation dependent on ethanol-inducible cytochrome P-450 (P-450IIE1), **Biochem.Pharmacol.** 38:1313.

Esterbauer, H. and Cheeseman, K.H., 1990, Determination of aldehydic lipid peroxidation products with special attention to malonaldehyde and 4-hydroxynonenal, **Methods Enzymol.**, 186:407.

Gutteridge, J.M.C. and Halliwell, B., 1990, The measurement and the mechanism of lipid peroxidation in biological systems, **Trends Biochem. Sci.**, 15:129.

Hashimoto, S. and Recknagel, R.O., 1968, No chemical evidence of lipid peroxidation in acute ethanol toxicity, **Exp. Mol. Pathol,** 8:225 .

Inomata, T., Ananda Rao, G. and Tsukamoto, H., 1987, Lack of evidence for increased lipid peroxidation in ethanol-induced centrilobular necrosis of rat liver, **Liver**, 7:233.

Janzen, E., 1980, A critical review of spin trapping in biological systems, **in**: "Free Radicals in Biology", vol. 4 pp.116-150, W.A. Pryor, ed., Academic Press, New York.

Kato, S., Kavase, T., Alderman, J., Inatomi, N. and Lieber, C.S., 1990, Role of xanthine oxidase in ethanol-induced lipid peroxidation in rats, **Gastroenterol**. 98:203.

Kavase, T., Kato, S. and Lieber, C.S., 1989, Lipid peroxidation and antioxidant defense systems in rat liver after chronic ethanol feeding, **Hepatol**. 10:815.

Keller, F., Snider, A.B., Petracek, T.J. and Sancier, K.M., 1971, Hepatic free radical levels in ethanol-treated and carbon tetrachloride-treated rats, **Biochem. Pharmacol.** 20:2507 .

Kera, Y., Komura, S., Ohbora, Y., Kiriyama, T. and Inoue, K., 1985, Ethanol induced changes in lipid peroxidation and non protein sulphydryl content: different effects in rat liver and kidney, **Res.Comm.Chem.Pathol.Pharmacol.** 47:203 .

Knecht, K.T., Bradfort, B:U:, Mason, R.P. and Thurman, G.R., 1990, Detection of a-hydroxyethyl free radical from ethanol *in vivo*, **Mol.Pharmacol.**, 38:26.

Lieber, C.S., 1988, Biochemical and molecular basis of alcohol-induced injury to liver and other tissues, **New Engl.J. Med.**, 319:463.

Litov, R.E., Inving, O.H., Downey, J.F. and Tappel, A.L., 1978, Lipid peroxidation: a mechanism involved in acute ethanol toxicity as demonstrated by *in vivo* pentane production in the rat, **Lipids**, 13:305.

MacDonald, C.M., 1973, The effect of ethanol on hepatic lipid peroxidation and on the activity of glutathione reductase and peroxidase, **FEBS Lett**, 35:227 .

Miyazawa, T., 1989, Determination of phospholipid hydro-peroxides in human blood by chemioluminescence-HPLC assay, **Free Rad.Biol.Med.**, 7:209.

Moscarella, S., Laffi, G., Coletta, D., Arena, U., Cappellini, A.P. and Gentilini, P., 1984, Volatile hydrocarbons in the breath of patients with chronic alcoholic liver disease: a possible marker of ethanol induced lipo-peroxidation. **in**: Frontiers of Gastrointestinal Research, vol. 8, P. Gentilini and M.U. Dianzani, eds., pp.208-216, Krager, Basel.

Mowri, H., Ohkuma, S. and Takano, T., 1988, Monoclonal DLRIa/ 104G antibodies recognized peroxidized lipoproteins in atherosclerotic lesions, **Biochim.Biophys.Acta**, 963:208 .

Muller, A. and Sies, H., 1982, Role of ethanol dehydrogenase activity and of acetaldehyde in ethanol-induced ethane and pentane production by isolated perfused rat liver, **Biochem.J.**, 206:153.

Niki, E., 1987, Antioxidants in relation to lipid peroxidation, **Chem.Phys Lipids**, 44:227.

Nordmann, R., Ribière, C. and Rouach, H., 1990, Ethanol-induced lipid peroxidation and oxidative stress in extrahepatic tissues, **Alcohol Alcoholism**, 23:213 .

Peters, T.J., O'Connell, M.J., Venkatesan, S. and Ward, R.J., 1986 Evidence for free radical-mediated damage in experimental and human alcoholic liver disease, **in**: "Free Radicals Cell Damage and Disease," pp. 99-110, Rice-Evans, C. ed., Richelieu Press, London .

Reinke, L.A., Lai, E.K., Du Bose, C.M. and McCay, P.B., 1987, Reactive free radical generation *in vivo* in heart and liver of ethanol-fed rats: correlation with radical formation *in vitro*, **Proc.Natl.Acad.Sci.** 84:8223.

Reitz, R.C., 1975, A possible mechanism for the peroxidation of lipids due to chronic ethanol ingestion, **Biochim. Biophys. Acta**, 380:145.

Rouach, H., Houzè, P., Orfanelli, M.T., Gentil, M., Bourdon, R. and Nordmann, R., 1990, Effect of acute ethanol administration on the subcellular distribution of iron in rat liver and cerebellum, **Biochem. Pharmacol.** 39:1095.

Shaw, S., Jayatilleke, E. and Lieber, C.S., 1989, Lipid peroxidation as a mechanism of alcoholic liver injury: Role of iron mobilization and microsomal induction. **Alcohol**, 5:135 .

Shaw, S., Jayatilleke, E., Ross, W.A., Gordon, E.R. and Lieber, C.S., 1981, Ethanol induced lipid peroxidation: potentiation by long term alcohol feeding and attenuation by methionine, **J.Lab.Clin.Med**. 98:417.

Shaw, S., Rubin, K.P., Lieber, C.S., 1983, Depressed hepatic glutathione and increased diene conjugates in alcoholic liver disease. Evidence of lipid peroxidation, **Dig.Dis. Sci.** 28:585.

Sippel, H.W., 1983, Effect of an acute dose of ethanol on lipid peroxidation and on the activity of glutathione S-transferase in rat liver, **Acta Pharmacol.Toxicol.**, 53:135.

Speisky, H., Bunout, D., Orrego, H., Giles, H.G., Gunasecara, A. and Israel, Y., 1985, Lack of changes in diene conjugate levels following ethanol induced glutathione depletion or hepatic necrosis, **Res.Commun.Chem.Pathol**. 48:77.

Suematzu, T., Matsumura, T., Sato, N., Miyamoto, T., Ooka, T., Kamada, T. and Abe, H., 1981, Lipid peroxidation in alcoholic liver disease in humans, **Alcohol.Clin.Exp. Res.** 5:427 .

Tanaka, A., Ikegami, F., Okumura, Y., Hasumura, Y., Kanayama, R. and Takeuki, J., 1970, Effect of alcohol in liver of rats-III. The role of lipid peroxidation and sulphydryl compounds in ethanol-induced liver injury, **Lab.Invest**. 23:421.

Tanner, A.R., Bantock, I., Hinks, L., Lloyd, B., Turner, N.R. and Wright, R., 1986, Depressed selenium and vitamin E levels in alcoholic population. Possible relationship to hepatic injury through increased lipid peroxidation, **Dig.Dis.Sci.** 31:1307.

Thompson, S. and Smith, M.T., 1985, Measurement of diene conjugated from linoleic acid in plasma by high performance liquid chromatography: a questionable non-invasive assay for free radical activity?, **Chem.-Biol. Interact**. 55:357 .

Torrielli, V., Gabriel, L. and Dianzani, M.U., 1978, Ethanol-induced hepatotoxicity: experimental observations on the role of lipid peroxidation, **J.Pathol.** 126:11 .

Umeki, S., Shiojiri, H. and Nozawa, Y., 1984, Chronic ethanol administration decreases fatty acyl-CoA desaturase activities in rat liver microsomes, **FEBS Lett**. 169: 274.

Valenzuela, A., Fernandez, N., Ugarte, G. and Videla, L.A., 1980, Effect of acute ethanol ingestion on lipid peroxidation and on the activity of the enzymes related to peroxide metabolism in the rat, **FEBS Lett.** 111:11.

Videla, L.A. and Valenzuela, A., 1982, Alcohol ingestion, liver glutathione and lipid peroxidation: metabolic inter-relations and pathological implications, **Life Sci.** 31: 2395.

Videla, L.A. Fernandez, V., Ugarte, G., Valenzuela, A. and Villanueva, A., 1980, Effect of acute ethanol intoxication on the content of reduced glutathione of the liver in relation to its lipoperoxidative capacity in the rats, **FEBS Lett**. 111:6.

THE MOLECULAR PATHOLOGY OF ALCOHOLIC LIVER DISEASE: AN OVERVIEW

S.W. French

Department of Pathology
Faculty of Medicine
University of Ottawa
Ottawa, Ontario, Canada

INTRODUCTION

The primary event which characterizes the beginning of ethanol alteration of hepatocytes is the effect ethanol has on the signal transduction systems. Through interaction of the receptor G protein and catalytic enzymes located in the plasma membrane ethanol alters the homeostasis of the cell. The signal transduction is either enhanced or inhibited by the action of alcohol, usually through the action of ethanol on the alpha subunit of the G protein. To understand these relationships it is important to have a basic concept of signal transduction (Fig. 1) (Gilman, 1989): agonist receptor interaction leads to replacement of GDP by GTP in the alpha sub segment of the G protein which then activates catalytic proteins such as adenylyl cyclase or phospholipase C or ion channels.

The catalytic units provide second messengers such as cyclic AMP, cyclic GMP and inositol trisphosphate or diacylglycerol (DAG). Consequently DAG activates protein kinase C (Fig 2) (Shenolikar, 1988). Increased cytosolic calcium activates protein kinase which is calcium dependent. Cyclic AMP activates protein kinase A and cyclic GMP activates protein kinase G.

A feedback mechanism through phosphorylation of receptor G proteins or catalytic units such as phosphodiesterase alters the signal transduction response. In this complex system it is possible for crosstalking between different signal transduction systems so that for instance phorbol ester (TPA), which activates protein kinase C, may lead to protein kinase C phosphorylation of the GS alpha subunit to enhance the beta adrenergic response initiated by isoproterenol (Aasheim et al., 1989) (Fig. 3). This may explain how ethanol augments the response of various tissues to ß agonists (Tabakoff et al., 1988) (Fig. 3). Another example is the effect of ethanol on the vasopressin response of the V1 receptor. Ethanol or phorbol ester triggers the activation of protein kinase C which inhibits the G protein linked to phospholipase C activation by the V1 vasopressin receptor in a way that higher doses of vasopressin are required to overcome this feedback inhibition mechanism (Hoek et al., 1988).

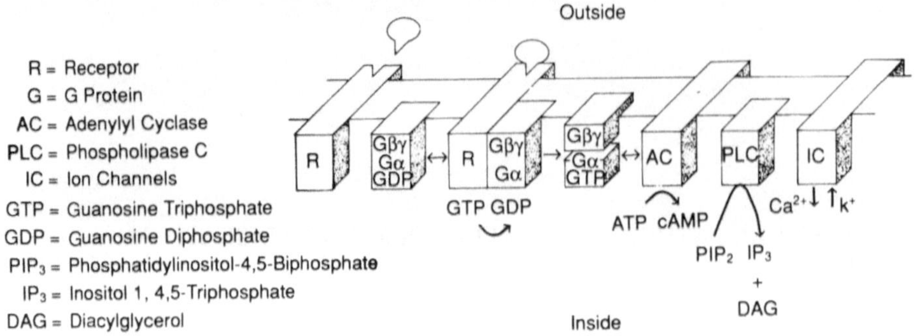

Fig 1. Signal Transduction, Modified from Gilman, 1989.

EFFECT OF ETHANOL ON LIVER FUNCTION

The response of the bile secretory apparatus to vasopressin V1 may be down regulated by the increased PKC inhibition of PLC stimulated by ethanol since the vasopressin V1 receptor stimulates bile flow through PLC activation which increases cytosolic Ca^{++} via IP_3 (Nathanson et al., 1990). This could account for the decrease in bile flow seen after ethanol infusion (Okanoue et al., 1984). Similarly, the activation of protein kinase C phosphorylation of enzymes such as phosphorylase a may account for the mobilization of glucose from glycogen stored in the liver in the acute response to ethanol (Hoek, Taraschi, 1988). Similarly, proteins such as intermediate filaments that constitute the cytoskeleton are phosphorylated by the ethanol-stimulated protein kinase C response similar to EGF (Baribault et al., 1989). Intermediate filaments when phosphorylated may disassemble and the liberated monomers characteristically are quickly degraded by calcium-dependent proteases in the cytosol. The peptide fragment products of this hydrolysis may regulate gene gating as postulated by Traub et al., (1987). It is possible that gene regulation as influenced

Hormone Receptor	Catalyst	Membrane Regulator Protein	Second Messenger	Phosporylation of receptors or G proteins or ion channels
β_2 Adrenergic Dopamine V_2 Vasopressin	Adenylyl cyclase	Gs	cAMP	PKA
α_2 Adrenergic Muscarinic / Cholinergic	Adenylyl cyclase	Gi	cAMP	PKA
ANF	Guanylate cyclase		cGMP	PKG
α_1 Adrenergic Muscarinic/ Cholinergic V_1 Vasopressin	Phospholipase C	G	IP_3 DAG	PKC

Fig 2. Signal Transduction: Routes and Cross Talk, Modified from Shenolikar, 1988.

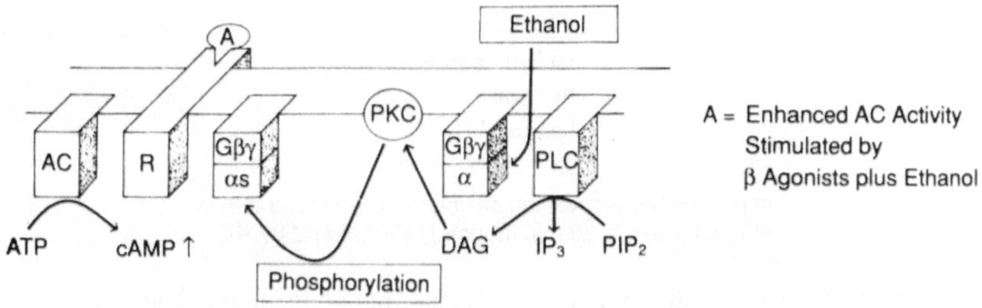

Fig 3. Proposed Mechanism for Ethanol-enhanced Receptor-activated Adenylyl Cyclase Activity in Liver Cells

by ethanol is a result of phosphorylation of the intermediate filaments triggered by acute ethanol exposure.

Alternatively ethanol stimulated increase in PKC activity could have long term effects by regulating oncogene expression (Alcon and Nelson, 1990) or differentiation (Ginsberg and Kimmel, 1989). Ethanol induces the gene expression of numerous proteins which are important in the defense of the hepatocyte to injury or aid in the elimination of ethanol by oxidation. In the former case, heat shock proteins are induced by the exposure of hepatocytes to ethanol (Zatoukal et al., 1988) (See Fig. 4).

Mallory Body Formation

One of the possible consequences of heat shock protein induction by alcohol exposure in the liver is the Mallory body, which is composed of an aggregate of intermediate filaments of the cytokeratin type. This aggregate is characterized by a marked increase in binding by ubiquitin (Ohta et al., 1988). Heat shock protein 70 Kd is present in abundance in liver cells in alcoholic hepatitis with Mallory bodies (Omar et al., 1990). Also involved in the process may be an enzyme transglutaminase which crosslinks intermediate filaments by forming glutamyl-lysine bridges. Zatoukal et al (1989) and Denk et al (1984) have shown that Mallory body formation is associated with increased transglutaminase in liver cells induced by griseofulvin feeding of mice. Catzavelos and colleagues have shown that Mallory body formation induced by griseofulvin feeding in mice is associated with a marked increase in messenger RNA for transglutaminase. Associated with this is the conversion of the normal lobular zonal distribution of transglutaminase message concentrated in the central lobular zone to a heterogenous cell population of liver cells throughout the lobule, some expressing a large amount of message and others expressing very little. Should this pattern be confirmed in alcoholic liver disease, such as alcoholic hepatitis, a new basic molecular explanation for Mallory body formation in alcoholic hepatitis would be evident.

Hypoxia Induced by Ethanol

The effect of ethanol on liver blood flow and oxygen extraction by liver cells has led to the concept that central lobular necrosis and fibrosis is a result of hypoxia caused by ethanol metabolism. The centrilobular hypoxia may be a consequent of a regional decrease

HSP induced by: ethanol
sodium arsenite
hypoxia
heat (41° C)

HSP = ubiquitin, 68 – 70 KD, 87 KD, 97 KD

In mouse hepatocytes in cell suspension incubated in ethanol 4 to 8% 30 min induced HSP 68 and 87 KD

Fig 4. Alcohol Triggers the Synthesis of "Heat Shock" Proteins (Thermal Tolerance), Zatoukal et al., 1988.

in blood flow as indicated by the study of Fukui et al (1990) using infrared spectroscopy by measuring oxygen saturation of hemoglobin in the sinusoids of the liver and also measuring blood flow by timing the movement of RBCs within the sinusoid by video recording. These investigators showed that ethanol in large amounts (4 g per kilogram) caused an increase in blood flow in some sinusoids and decreased blood flow in other sinusoids and the level of the oxygen tension in low flow areas and the central vein was decreased. Thus, the acute effect of ethanol was to produce focal hypoxia (See Fig. 5).

Israel et al (1973) has shown that alcohol stimulates 0_2 consumption by the liver by increasing the metabolic rate of the liver. They showed that this made the liver more vulnerable to further hypoxia. Necrosis in the centrilobular region resulted and this was prevented by propyl-thiouracil (PTU) by inhibiting thyroid function (Israel et al., 1975). Yuki and Thurman (1980) on the other hand, showed that the increase in oxygen consumption was mediated through catecholamine release caused by high blood alcohol levels. French et al (1983) showed that in vivo exposure to low oxygen tension for 5 hours in rats fed ethanol chronically caused central necrosis whereas controls did not develop liver damage. Miyamoto et al (1988) using the same chronic alcohol feeding intragastric tube feeding model showed that there was central necrosis leading to central fibrosis associated with chronic alcohol feeding and this was associated with a low liver cell ATP content. In this system the ATP levels dropped faster in alcohol fed animals than controls in the response to a three minute hypoxia episode and this was reversed to normal levels of ATP

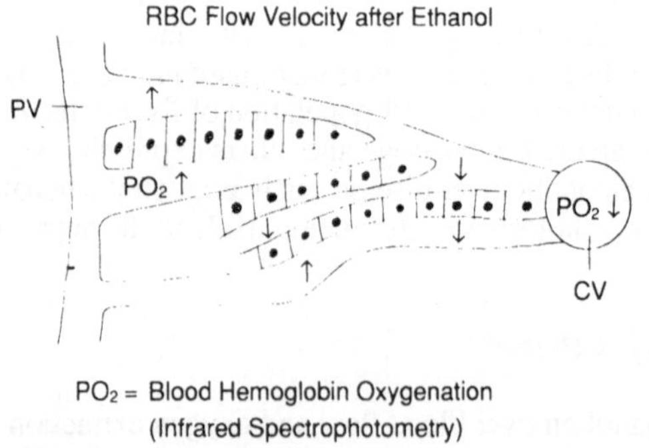

Fig 5. Effect of Ethanol on Liver Microcirculation in the Rat (Acute 4g/kg), Modified from Fukui et al, 1990.

by 100% oxygen in these animals. Takahashi et al (1990) confirmed the low energy state of the chronic ethanol-fed rat model using NMR spectroscopy in vivo noninvasively. In this system the low energy state was documented by calculating the ratio of Pi to beta-ATP. The vulnerability to hypoxia has recently been confirmed by Spach and Cunningham (1990). Tsukamoto and Xi (1989), using the chronic intragastric tube feeding rat model for alcohol liver disease, showed that there was an increase in oxygen extraction by the liver and this led to significantly reduced PO_2 levels in the central vein. Thus the increased blood flow did not compensate for the increased oxygen utilization by alcohol. In the baboon ALD model, there is reduced utilization of O_2 due to loss of mitochondrial function (Fig. 6).

One of the possible mechanisms for central necrosis caused by alcohol may be that hypoxia activates xanthine oxidase converting it from xanthine dehydrogenase to increase the superoxide produced from the oxidation of acetaldehyde (See Fig. 7) (Lewis and Paton, 1982).

The increased oxygen consumption, seen in this chronic alcohol-fed rat model, where high blood alcohol levels are maintained, may be due to the induction of the microsomal ethanol oxidizing system and the oxidation of ethanol by P450IIE1 (Ingelman-Sundberg et al., 1990). This enzyme is induced at least 10-fold by chronic ethanol feeding. The generation of superoxide may cause oxidative stress when ethanol is oxidized by P450IIE1 (Cederbaum, 1989b, Ingelman-Sundberg et al., 1990) (Fig. 8).

Oxidative Stress

Nordmann et al (1988) and Cederbaum (1989b) have shown that the superoxide generated by the P450 system maybe utilized through the Fenton/Haber-Weiss, iron catalyzed system to form the hydroxyethyl free-radical form of ethanol (Fig. 9). In this way, ethanol is a free-radical scavenger and is oxidized by the hydroxyl-radical generated by the P450 system. This has been confirmed using ESR spectroscopy using a spin trap both in vitro in microsomes, metabolizing ethanol (Albano et al., 1988) and in vivo by Krecht et al (1990). However the free-radicals formed by ethanol peroxidation may be quite varied and some of them are unstable and highly reactive and can produce oxidative damage to proteins and DNA. Thus it is very important to consider in alcohol liver disease the free-radical scavengers which protect the liver from oxidative stress (See Fig. 10). Free-radical scavengers such as vitamin E and glutathione (GSH) or enzymes which are selenium or

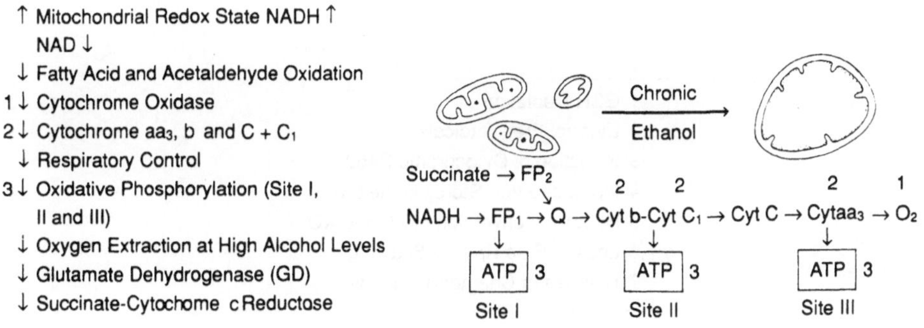

Fig 6. Impaired Oxygen Utilization by Liver Mitochondria (Chronic Ethanol-fed Baboons)

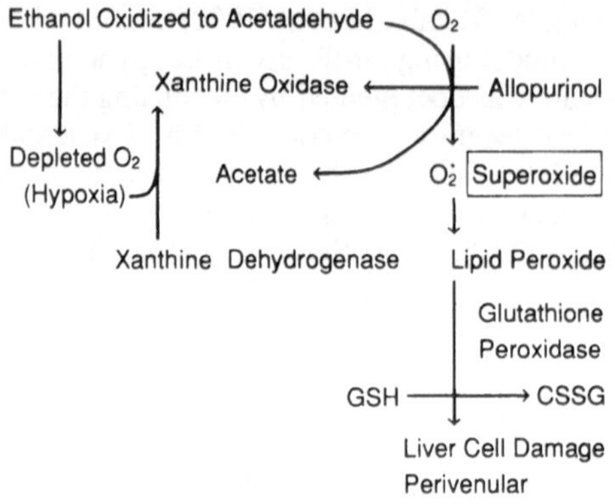

Fig 7. How Ethanol Oxidation Leads to Lipid Peroxidation in Perivenular Hepatocytes, Modified from Lewis and Paton, 1982.

copper-zinc based or enzyme inhibitors such as allopurinol which inhibits xanthine oxidase may be important in considering protective mechanisms for preventing the progression of alcoholic liver disease.

Israel et al (1990) has shown that the induction of gamma glutamyl transpeptidase (GGT) by ethanol may function to supply the centrilobular region of the liver with substrate for GSH synthesis and thus protect these cells from centrilobular hypoxia and oxidase stress induced by chronic ethanol metabolism. They have shown that GSH is broken down into glutamate and cysteinyl-glycine in the periportal zone and this supplies substrate for the synthesis of GSH in the centrilobular hepatocytes down stream (See Fig. 11).

Barak et al (1985) and Lieber et al (1990) have utilized the fact that glutathione is protective of free-radical damage in alcohol liver disease by augmenting glutathione synthesis in the chronic-alcohol fed baboon model. To do this, they have fed s-adenosyl methionine which is in short supply due to lack of energy in the form of ATP and the inhibition of methyl transferase by ethanol (See Fig. 12). This therapeutic approach has had an ameliorating effect on the progression of alcohol liver disease in baboons although it does not alter the basic histopathology.

1 GSH Depletion
2 Ethanol Free Radicals
3 Induction of Cytochrome P450
4 Increased Iron Stores in the Liver
5 Hypoxia Conversion of XDH to XO
6 Loss of Free Radical Scavengers
7 Release of Chemoattractants

Fig 8. Possible Mechanisms of Oxidative Stress by Ethanol, Modified from Cederbaum, 1989b.

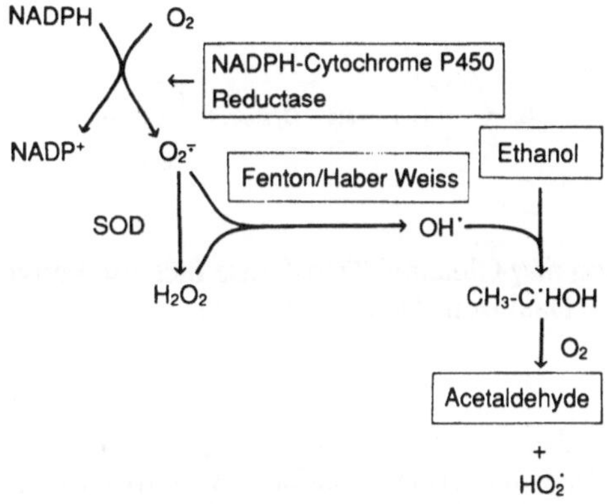

Fig 9. Ethanol Derived Hydroxyethyl Free Radicals in Microsomes Detected by ESR Spectroscopy Using a Spin Trap, Modified from Albano et al, 1988.

LIVER FIBROSIS: ROLE OF THE Ito CELL

The mechanism for fibrosis leading to cirrhosis in alcohol liver disease involves activation of Ito cells, the cell of origin of collagen located normally in the liver lobule in the perisinusoidal space. Hepatocytes, Ito cells and endothelial cells synthesize collagen type I in vivo and in vitro(Biagini and Ballardini,1989). Collagen type III is synthesized by hepatocytes, Ito cells and endothelial cells. Type IV collagen, located in the basement membrane and in the non-filamentous extracellular matrix, is synthesized by hepatocytes, ducts, Ito cells and endothelial cells. Fibronectin is synthesized by hepatocytes, Ito cells, endothelial cells and Kupffer cells. Laminin is synthesized by hepatocytes, bile duct cells, Ito cells, endothelial cells and Kupffer cells. Proteoglycans are synthesized by hepatocytes, Ito cells and Kupffer cells. Ito cells are stimulated to proliferate by growth factors including EGF, TGF alpha, fibroblast growth factor, platelet derived growth factor and TNF-alpha (Gressner and Bachem, 1990).

Glutathione (2GSH + $H_2O_2 \rightarrow$ GSSG + $2H_2O$

Ethanol

Vit E (Tocopherol-OH + $-C-O_2 \rightarrow -C-O_2H$ + Tocopherol-O)

Vit A

Vit C

Uric Acid

Enzymes

Glutathione Peroxidase (Selenium)

Superoxide Dismutase (Copper-Zinc)

Catalase

Inhibitors

Allopurinol (x anthine Oxidase)

Fig 10. Free Radical Scavengers Protecting the Liver From Oxidative Stress, Modified from Cederbaum, 1989b.

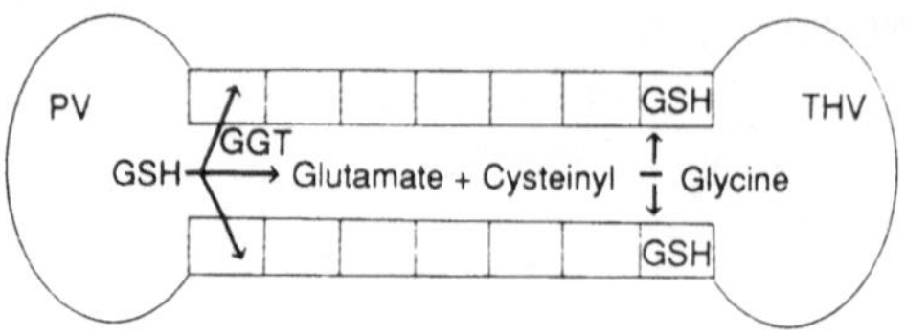

Fig 11. Ethanol-induction of γ-Glutamyl Transferase Increase Perivenular Hepatocyte Supply of GSH, Data from Israel et al, 1990.

TGF beta stimulates the synthesis of the collagens and proteoglycans such as heparin sulphate, chondroitin 6-sulfate and dermatan sulfate. As Tsukamoto et al (1990) and others have shown, cytokines are involved in the regulation of the synthesis of collagen and collagenase in alcohol liver disease. Maher (1990) has illustrated the sequence of events involved in fibrosis of the liver in alcohol liver disease (See Fig. 13). Following liver cell necrosis and activation of Kupffer cells which initiate the repair process, cytokines and the extracellular matrix both stimulate Ito cell activation to synthesize collagen.

This process can take place in central necrosis or in the perisinusoidal region. Studying alcohol liver disease in man and in the intragastric tube feeding rat model of alcohol liver disease Ito cell activation is found primarily in scars where necrosis has occurred. The perisinusoidal Ito cells appear to be inactive. They do not show the electron microscopy features of activation. These features include enlargement of the space within the Ito cell occupied by rough endoplasmic reticulum and filling of the endopasmic reticulum cisternae with floculent material which is distined for secretion to form the

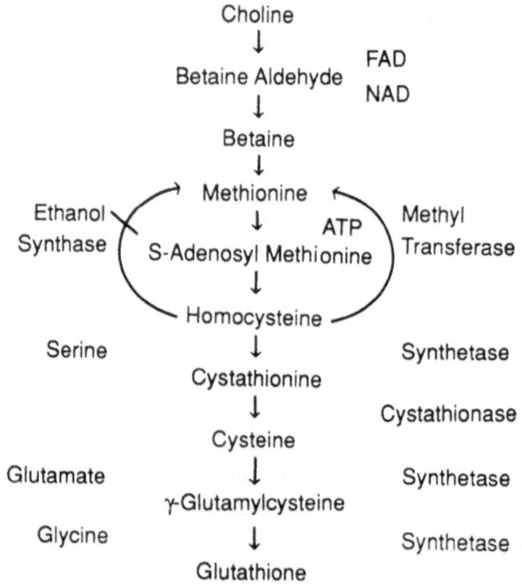

Fig 12. Gutathionine Deficiency Induced by Alcohol is Partially Reversed by S-Adenosyl Methionine (Lieber).

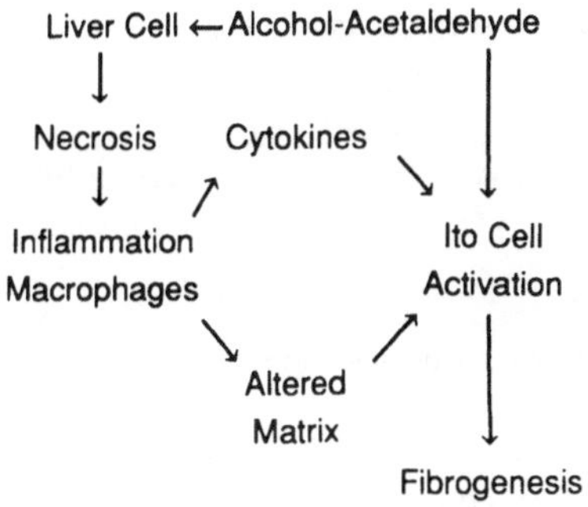

Fig 13. Schematic: Alcohol-induced Liver Fibrosis, Modified from Maher, 1990.

extracellular matrix and collagen. At the same time that the endoplasmic reticulum proliferates in the activated Ito cell the storage of lipid including retinoids diminishes.

In both alcoholic hepatitis in man and in the chronic ethanol-fed rat this activation of Ito cells is observed only in the scar except when ethanol is maintained at high alcohol levels in the rat for more than 4 months (Takahashi et al., 1990). When Ito cells and macrophages are isolated from the livers of the chronic ethanol-fed rats however, they show a functional change including secretion by the Kupffer cells of TGF-beta (Tsukamoto et al., 1990). The Ito cell response to stimulation by TGF β from macrophages was increased by ethanol and the high fat diet. Thus long before scar formation appears in the rat model Ito cells are activated when assessed by in vitro assays.

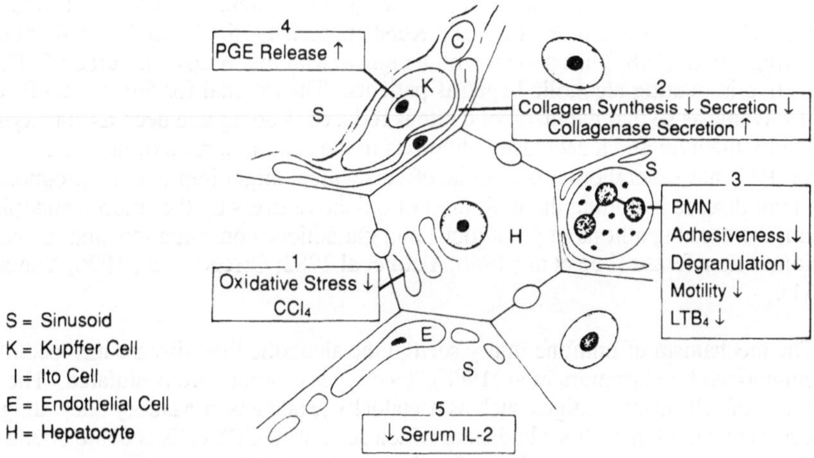

Fig 14. Possible Rationale for Colchicine Therapy for Alcoholic Liver Fibrosis.

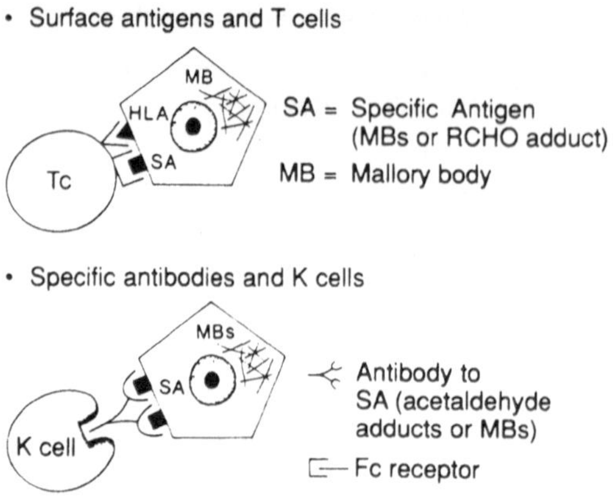

Fig 15. Immune Mechanisms of Hepatocellular Damage in Alcoholic Hepatitis by Lymphocytes, Modified from Hasumura et al., 1988.

THERAPEUTIC INTERVENTION

The therapeutic implications of the activated Ito cell production of fibrosis in alcohol liver disease is that any drug which would inhibit collagen synthesis or increase collagenase secretion or reduce the inflammatory role of granulocytes in alcohol liver disease or would inhibit oxidative stress derived from drug metabolism or would reduce growth factors which cause Ito cell proliferation should retard progression or decrease the fibrosis even after cirrhosis has developed. One drug which has all these actions is colchicine. It has been given a clinical trial in cirrhotic patients including alcohol liver disease with success (Kershenobich et al., 1988). The mechanisms of hypothetical action of colchicine are indicated in Fig. 14 (Ehrenfeldet al., 1980; Ehrlich and Bornstein 1972; Mourelle and Meza 1989; Mourelle et al 1987a, b; Rojkind and Kershenobich 1975; Taylor 1865; Wahl et al., 1978). Orrego et al (1987) has shown with a clinical trial that propyl-thiouracil (PTU) is also effective in treating alcoholic hepatitis patients. The rational for this, is that PTU blocks T4 synthesis so that the metabolic rate is reduced, leading to a decrease in oxygen extraction by the liver and a decrease in hypoxia in the central portion of the lobule. However, PTU has several other beneficial effects which might improve the prognosis of alcohol liver disease including the inhibition of oxidative stress by the smooth endoplasmic reticulum, increased synthesis of glutathione and glutathione concentration, and increase in glutathione s transferase (Lee et al., 1986; Yuki et al 1982; Orrego et al., 1976; Yamada et al., 1981).

The mechanism of immune injury seen in the alcoholic liver disease has recently been summarized by Hasumura et al (1988). Two types of injury are postulated. The first is the presence of cell surface antigen such as acetaldehyde adducts or Mallory body antigens of hepatocytes contained in Mallory bodies to which sensitized CD8 cells become bound and associated with HLA antigen (see Fig. 15). Presumably the T cell that is adherent to the hepatocyte secretes a membrane lytic component like perforin which leads to lethal damage to the involved hepatocyte. The second mechanism involves K cells which bind to the FC fragment of an antibody which is bound to the surface antigens.

REFERENCES

Aasheim, L.H., Kleine, L.P. and Franks, D.J., 1989, Activation of protein kinase C sensitizes the cyclic AMP signalling system of T51B rat liver cells. **Cellular Signal** 6: 617-625.

Alkon, D.L. and Nelson, T.J.,1990, Specificity of molecular changes in neurons involved in memory storage. **FASEB J** 4: 1567-1576.

Albano, E., Tomasi, A., Goria-Gatti, L. and Dianzani, M.U., 1988, Sign trapping of free radical species produced during the microsomal metabolism of ethanol. **Chem Biol Interactions** 65: 223-234.

Barak, A.J., Beckenhawa, H.C. and Tuma, D.J., 1985, Ethanol feeding inhibits the activity of hepatic NS-methyltetrahydrofolate homocysteine methyl transferase in the rat. **IRCS Med** 13: 760-761.

Baribault, H., Blouin, R., Bourgon, L. and Marceau, N., 1989, Epidermal growth factor-induced selective phosphorylation of cultured rat hepatocyte 55 kD cytokeratin before filament reorganization and DNA synthesis. **J Cell Biol** 109: 1665-1676.

Biagini, G. and Ballardini, G., 1989, Liver fibrosis and extracellular matrix. **J Hepatol** 8: 115-124.

Catzavelos, C., Leger, J., Tenniswood, M. and French, S.W., unpublished results.

Cederbaum, A., 1989a, Introduction: Role of lipid peroxidase and oxidative stress in alcohol toxicity. **Free Radical Biol Med** 7: 537-539.

Cederbaum, A., 1989b, Oxygen radical generation by microsomes: Role of iron and implications for metabolism and toxicity. **Free Radical Biol Med** 7: 559-567.

Denk, H., Bernklau, G. and Krepler, R., 1984, Effect of griseofulvin treatment and neoplastic transformation on transglutaminase activity in mouse liver. **Liver** 4: 208.

French, S.W., 1983, Present understanding of the development of the Mallory body. **Arch Pathol Lab Med** 107: 445-450.

French, S.W., Benson, N.C. and Sun, P.S., 1984, Centrilobular liver necrosis induced by hypoxia in chronic ethanol-fed rats. **Hepatology** 4: 912-917.

Fukui, H., Sato, N., Yoshihara, N., Kashio, S. and Hijioka, T., 1990, High dose of ethanol causes hepatic hypoxia. **Alcoholism** 14: 290.

Ginsburg, G. and Kimmel, A.R., 1989, Inositol trisphosphate and diacylglycerol can differentially modulate gene expression in Dictyostelium. **Proc Natl Acad Sci USA** 86: 9332-9336.

Gilman, A.G., 1989, G proteins and regulation of adenylyl cyclase. **JAMA** 262: 1819-1825.

Gressner, A.M. and Bachem, M.G., 1990, Cellular sources of non-collagenous matrix proteins: Role of fat-storing cells in fibrogenesis. **Semin Liver Dis** 10: 30-46.

Hasumura, Y., Izumi, N., Sakai, Y. and Takeuchi, J., 1988, Lymphocytic infiltration in the liver in patients with alcoholic hepatitis. **In**: Biomedical and Social Aspects of Alcohol and Alcoholism. Kuriyama, K., Takada, A., Ishii, H. (eds) Elsevier Science Publ, pp 779-782.

Hoek, J.B., Armstrong, J.J. and Crovello, M.T., 1988, Feedback control of ethanol-induced phospholipase C activation in rat hepatocytes. **In**: Biomedical and Social Aspects of Alcohol and Alcoholism, Kuriyama, K., Takada, A., Ishii, H. (eds) Elsevier Sci Publishers, pp 651-655.

Hoek, J.B. and Taraschi, T.F., 1988, Cellular adaptation to ethanol. **Trends Biochem Sci** 13: 269-274.

Hoek, J.B., Taraschi, T.F. and Rubin, E., 1988, Functional implications of the interaction of ethanol with biologic membranes: Actions of ethanol on hormonal signal transduction systems. **Semin Liver Dis** 8: 36-46.

Ingelman-Sundberg, M., Johansson, I., Eliasson, E. and Tindberg, N., 1990, Regulation and toxicology of ethanol-inducible cytochrome P450E1. **Alcoholism Clin Exp Res** 114: 299.

Israel, Y., Videla, Y., MacDonald, A. and Bernstein, J., 1973, Metabolic alterations produced in the liver by chronic ethanol administration. Comparison between the effects produced by éthanol and by thyroid hormones. **Biochem J** 134: 523-529.

Kershenobich, D., Vargas, F., Garcia-Tsao, G., Tamayo, R.P., Gent, M. and Rojkind, M., 1988, Colchicine in the treatment of cirrhosis of the liver. **N Engl J Med** 318: 1709-1713.

Knecht, K.T., Bradford, B.U., Mason, R.P. and Thurman, R.G, 1990, Detection of the hydroxyethelyl free radical from ethanol in vivo. **Alcoholism Clin Exp Res** 14: 304.

Lewis, K.O. and Paton, A. 1982. Could superoxide cause cirrhosis? **Lancet**, July 24, p. 188-189.

Lieber, C.S., Casini, L.M., De Carli, L. M., Kin, C.I., Lowe, N., Sasaki, R. and Leo, M.A., 1990, S-adenosyl-L-methionine attenuates alcohol-induced liver injury in the baboon. **Hepatology** ll: 165-172.

Maher, J.J., 1990, Hepatic fibrosis caused by alcohol. **Sem Liver Dis.** 10: 66-74.

Miyamoto, K. and French, S.W., 1988, Hepatic adenine nucleotide metabolism measured in vivo in rats fed ethanol and a high fat-low protein diet: Relation to pathogenesis of alcohol-induced liver injury in the rat. **Hepatology** 8: 53-60.

Nathanson, M.H., Gautam, A. and Boyer, J.L., 1990, Effects of vasopressin (VP) and phorbol dibutyrate (PDB) on bile secretion in isolated rat hepatocyte couplets (IRHC). **Gastroenterology** 98: A615.

Nordmann, R., Ribiere, C. and Rouach, H., 1988, Alcool et radicaux libres: donnees actvelles. **Med Sci** 6: 336-345.

Ohta, M., Marceau, N., Perry, G., Maneto, V., Gambetti, P., Metuzals, J., Cadrin, M., Kawahara, H. and French, S.W., 1988, Ubiquitin is present on the cytokeratin intermediate filaments and Mallory bodies of hepatocytes. **Lab Invest** 58: 848-856.

Okanoue, T., Kondo, I., Ihrig, T.J. and French, S.W., 1984, Effect of ethanol and chlorpromazine on transhepatic transport and biliary secretion of horseradish peroxidase. **Hepatology** 4: 253-260.

Omar, R., Papolla, M. and Bommankanti, S., 1990, Immunocytochemical detection of the 70 Kd heat shock protein in alcoholic liver disease. **Arch Pathol Lab Med** 114: 589-592.

Orrego, H., Blake, J.E., Blendis, L.M., Compton, K.V. and Israel, Y. 1987, Long-term treatment of alcoholic liver disease with propylthiouracil. **N Engl J Med** 317: 1421-1427.

Shenolikar, S., 1988, Protein phosphorylation: hormones, drugs, and bioregulation. **FASEB J** 2: 2753-2764.

Spach, P.I., Herbert, J.S. and Cunningham, C.C., 1990, Effect of oxygen tension on ethanol-induced alterations in hepatic energy state. **Alcoholism Clin Exp Res** 14: 341.

Speisky, H., Shackel, N., Varghese, G., Wade, D. and Israel, Y., 1990, Role of hepatic gamma-glutamyl transferase in the degradation of circulating glutathione: Studies in the intact guinea pig perfused liver. **Hepatology** 11: 843-849.

Tabakoff, B., Hoffman, P.L. and McLaughlin, A., 1988, Is ethanol a discriminating substance? **Semin Liver Dis** 8: 26-35.

Takahashi, H., Geoffrion, Y, Butler, K.W. and French, S.W., 1990, In vivo hepatic energy metabolism during the progression of alcoholic liver disease: A non invasive 31P nuclear magnetic resonance study in rats. **Hepatology** 11: 65-73.

Takahashi, H., Wong, K., Jui, L. and French, S.W., 1990, Effect of dietary fat on Ito cell activation by chronic ethanol intake. **Alcoholism: Clin Exp Res** 14: 344.

Traub, P., Pagens, U., Kuhn, S. and Perides, G., 1987, Function of intermediate filaments. A novel hypothesis. **In**: Zortschrite der Zoologie Band 34. Wohfarth-Bottermann (Hrsg): Nature and Function of Cytoskeletal Proteins in Motility and Transport. Gautav-Fischer Verlag Stuttgart, pp 275-287.

Tsukamoto, H., Matsuoka, M. and French, S.W., 1990, Experimental models of hepatic fibrosis: A review. **Semin Liver Dis** 10: 56-65.

Tsukamoto, H. and Xi, X.P., 1989, Incomplete compensation of enhanced hepatic oxygen consumption in rats with alcoholic centrilobular liver necrosis. **Hepatology** 9: 302-306.

Yuki, T and Thurman, R.G., 1980, The swift increase in alcohol metabolism. Time course for the increase in hepatic oxygen uptake and the involvement of glycolysis. **Biochem J** 186: 119-126.

Yuki, T., Israel, Y. and Thurman, R.G., 1982. The swift increase in alcohol metabolism. Inhibition by propylthiouracil. **Biochem Pharmacol** 31: 2403-2407.

Zatoukal, K., Denk, H, Lackinger, E. and Rainer, I., 1989, Hepatocellular cytokeratins as substrates of transglutaminase. **Lab Invest** 61: 603-608.

Zatoukal, K., Sohar, R., Lackinger, E. and Denk, H., 1988, Induction of heat-shock proteins in short-term cultured hepatocytes derived from normal and chronically griseofulvin-treated mice. **Hepatology** 8: 607-612.

ACETALDEHYDE ADDUCTS AND EXCESSIVE ALCOHOL CONSUMPTION

Onni Niemelä

Department of Clinical Chemistry, University of Oulu
SF-90220 Oulu, Finland

INTRODUCTION

Acetaldehyde, a product of ethanol metabolism, has been implicated in a number of adverse effects of ethanol (Lieber, 1988). A major reason for acetaldehyde toxicity may be its ability to form covalent adducts with various proteins and cell constituents. The primary site of acetaldehyde adduct formation should be the liver, where it is believed to inhibit microtubule assembly, decrease enzyme activities and increase protein catabolism (Sorrell and Tuma, 1987; Lieber, 1988). Recent findings indicating that acetaldehyde binding with proteins can also trigger immune responses have provided the basis for antibody-based detection of acetaldehyde condensates as biological markers of excessive alcohol consumption (Israel et al., 1986, 1988; Lin et al., 1988). The present contribution summarizes recent studies on the formation of acetaldehyde-protein adducts as a result of alcohol consumption and suggests a possible association of adduct formation with hepatic fibrogenesis. For more detailed accounts and references on the immunological and functional aspects of acetaldehyde-protein binding the reader is referred to Israel et al. (1988) and Sorrell and Tuma (1987), respectively. Acetaldehyde-promoted fibrogenesis has been discussed in greater depth by Lieber (1988).

ACETALDEHYDE ADDUCTS AS MARKERS OF ALCOHOL ABUSE

Detection of acetaldehyde condensates from red cells has been postulated as an approach to measure ethanol consumption analogous to the measurement of glucosylated hemoglobin to estimate glucose control in diabetes (Stevens et al., 1981; Israel et al., 1988), although to date, measurements of such condensates have been hampered by lack of sensitivity and specificity of the methods of analysis. Recently, new chromatographic methods with improved sensitivities have, however, been introduced (Peterson and Polizzi, 1987; Sillanaukee and Koivula, 1990). Peterson and coworkers (1987, 1989) using a fluorigenic high performance liquid chromatography reported increased amounts of hemoglobin- and plasma protein-associated acetaldehyde after ethanol ingestion. Studies by immunochemical methods have also indicated that increased concentrations of hemoglobin-acetaldehyde adducts occur in chronic alcoholics (Israel et al., 1988; Lin et al., 1990). Recently, immunoreactive adducts were also detected from the blood of volunteers following acute ethanol ingestion (Niemelä et al., 1990a). In these experiments, the adduct

Alcoholism: A Molecular Perspective
Edited by T.N. Palmer, Plenum Press, New York, 1991

values were found to increase rather slowly, the highest values being obtained from the samples at time points where blood ethanol had already reached zero. Subsequently, the immunoreactive adducts appear to mostly return to original levels within 1-3 weeks of follow-up following either acute ethanol dose or chronic alcohol consumption (Niemelä and Israel, unpublished observations). Accordingly, the adducts measured by fluorigenic high performance chromatography from mice blood after ethanol ingestion have been shown to peak at day 2 and decline by day 9 upon ethanol discontinuation (Peterson and Scott, 1989). Although not surprising, this would indicate a somewhat faster rate of decay than expected considering the life span of red cells. While the adduct formation in Caucasian subjects requires rather high doses of ethanol , a marked increase after a small ethanol dose has been observed in the case of an Oriental flusher (Niemelä et al., 1990a). Addition of acetaldehyde into the blood *in vitro* without a reducing agent does not, however, result in a significant increase in the immunoreactive material in short term incubations suggesting the necessity of adduct stabilization for the immunological recognition.

Chronic ethanol ingestion has been shown to lead to the appearance of circulating acetaldehyde adduct antibodies (Israel et al., 1986; Niemelä et al., 1987; Hoerner et al., 1988). The diagnostic value of antibody detection to mark alcohol consumption or liver disease of alcoholic origin appears, however, limited since rather low values often occur in patients with advanced alcoholic cirrhosis as well as in alcoholics with early stages of liver injury (Niemelä et al., 1987). Furthermore, elevated antibody titers have been observed in non-alcoholic liver diseases possibly due to excessive formation of endogenous acetaldehyde (Hoerner et al., 1988; Ma et al., 1989). Nevertheless, such antibodies may play a role in the aggravation of alcoholic liver disease in the presence of antigenic stimulation, i.e. acetaldehyde adducts as created by prolonged ethanol ingestion. (Israel et al., 1988). The extent of antigen formation in tissues has, however, remained unclear. *In vitro* and metabolic studies have described acetaldehyde adducts with various hepatic proteins and cell constituents (Sorrell and Tuma, 1987; Lieber. 1988). Using antibody-based detection from rat liver following ethanol consumption Lin and coworkers (1988) reported the existence of one, yet unidentified, cytosolic 37 kD acetaldehyde-modified polypeptide, while other studies have found that cytochrome P450IIE1 is a selective target of ethylation (Lieber, 1988). Recently, several intracellular acetaldehyde adducts have been observed by the immunochemical methods from the liver after ethanol consumption (Yokoyama et al., 1990, Niemelä et al., 1991). These findings suggest that an immunologic cascade against the acetaldehyde adducts can perhaps be initiated upon hepatocyte damage and intracellular antigen release.

Specificity of acetaldehyde adducts to mark ethanol consumption is expected to be high since they should represent true metabolites of ethanol (Israel et al., 1988). While comparisons of the cation exchange chromatographic and immunological methods for the adduct detection have revealed significant correlations between them (Sillanaukee et al., unpublished observation), the exact chemical nature of the adduct involved in these detection systems has not, however, been clearly defined. Apparently, at least some of the immunoreactive adducts are different (Lin and Lumeng, 1989). Immunogenic adducts have been shown to result from stabilization of acetaldehyde-modified lysines with appropriate reducing agents (Israel et al., 1986). Reaction of acetaldehyde with proteins may, however, also involve other amino acid residues including cysteine, valine and tyrosine (Stevens et al., 1981; San George and Hoberman, 1986; Sorrell and Tuma, 1987) and other types of adducts of which at least the acetaldehyde-tyrosine linkage may have immunoreactivity (Lin and Lumeng, 1989). This linkage may be similar to the cyclic imidazolidinone derivative from valine (San George and Hoberman, 1986). It is also possible that antibodies would recognize other types of arrangements from reactions of aldehyde groups with proteins, such as cross-linking. Other possible modifications of proteins include the glucosylation of diabetes and the carbamylation of uremia, which appear, however, not to cross react significantly with the antibodies raised against acetaldehyde-modified proteins (Niemelä and Israel, unpublished observation).

ACETALDEHYDE ADDUCTS AND FIBROGENESIS

Accumulation of collagens in the liver upon chronic alcohol consumption causes a major medical problem in many Western countries (Lieber, 1988). Induction of hepatic fibrogenesis in the presence of hepatocellular inflammation may be brought about by numerous mediators, including transforming growth factor-beta, fibrogenic factor or other regulatory proteins (Bissell et al., 1990). In alcoholic liver disease hepatitis appears, however, not to be a necessary precursor of cirrhosis (Lieber, 1988). Recent studies have indicated that acetaldehyde can induce collagen synthesis and collagen gene transcription in cell cultures (Brenner and Chojkier, 1987; Lieber, 1988). This effect is blocked by a scavenger of reducing equivalents (Brenner and Chojkier, 1987) and stimulated by ascorbic acid (Chojkier et al., 1989), which is also known as an enhancer of acetaldehyde adduct formation (Sorrell and Tuma, 1987).

The rate and specificity of gene transcription is regulated via specific transacting elements which bind to cisacting elements of the genes. The knowledge on the nature of such elements is rapidly growing and recent studies, for exampl, have indicated that histone H1 in addition to serving as a general repressor for large chromatin fragments may also be involved in the control of individual gene transcriptional activity (Wolffe, 1989). Histone H1 purified by a DNA recognition site affinity chromatography apparently binds to a CTF/NF-1 binding site at about -300 with respect to the start of the transcription in a mouse type I collagen promoter (Ristiniemi and Oikarinen, 1989). Recently, incubation of histone H1 with acetaldehyde *in vitro* has been shown to result in the formation of spontaneously stable adducts with the reactive lysine residues in its carboxyterminal tail (Niemelä et al., 1990b), which may be essential for its repressor function. The adduct formation with histone H1 and physiologically relevant concentrations of acetaldehyde markedly decreases histone H1 DNA binding, while equivalent concentrations of ethanol or acetaldehyde *per se* do not produce any effect.

The above findings have suggested that acetaldehyde adduct formation with DNA-binding proteins could interfere with hepatocyte gene expression. Since histone H1 appears to play a role in regulating collagen gene expression, functional impairment could result in the accumulation of collagen. This would occur upon excessive alcohol consumption when high concentrations of acetaldehyde may be continuously available for adduct formation especially in zone 3 of the hepatic acinus. This region is the predominant location of both ethanol-inducible cytochrome P-450IIE1 and alcohol dehydrogenase, the primary enzyme of alcohol metabolism (Lieber 1988). It is also the predominant site of precursor lesions in alcoholic liver disease. The acetaldehyde-mediated mechanism of excessive accumulation of collagen would primarily be dependent on the degree and duration of alcohol abuse and not the presence of hepatocellular inflammation. Consistent with this idea, the data by Lelbach (1974) has shown that every alcoholic will develop cirrhosis if they continue drinking long enough. Upon modification of the transacting proteins by acetaldehyde, impairment of the transcriptional regulation of not only collagen but a variety of other proteins is expected to occur possibly depending, however, on the type of protein and cell involved. It is tempting to speculate that acetaldehyde adduct formation would also interfere with other regulatory interactions between DNA and proteins, including transforming growth factor-beta, protein kinase C or steroid hormone receptors and thus account for various ethanol-induced changes in cellular activities, such as gene expression or signal transduction. Other aldehyde groups, such as malondialdehyde, a product of lipid peroxidation should also have similar reactivity (Chojkier et al., 1989).

CONCLUSIONS

Recent work has provided further evidence on the existence of acetaldehyde-modified epitopes in blood and tissues following ethanol consumption. While acetaldehyde adducts may have a pathophysiological role in creating the adverse effects of ethanol in the liver, such as fibrogenesis, there may also be potential diagnostic implications. Following acute ethanol ingestion, an increase in acetaldehyde conjugation to hemoglobin can be detected using sensitive immunological methods. Acetaldehyde-hemoglobin condensates remain elevated longer than blood ethanol is measurable suggesting the usefulness of the adduct measurements to detect heavy drinking in clinical settings. In addition, immunocytochemical detection of acetaldehyde-modified epitopes from the liver may prove to be of value in the assessment of ethanol-induced tissue damage.

ACKNOWLEDGEMENTS

The original studies in the author's laboratory were supported by a grant from the Finnish Foundation for Alcohol Studies.

REFERENCES

Bissell, D.M., Friedman, S.L., Maher, J.J., and Roll, F.J., 1990, Connective tissue biology and hepatic fibrosis: report of conference, **Hepatology**, 11:488.

Brenner, D.A., and Chojkier, M., 1987, Acetaldehyde increases collagen gene transcription in cultured human fibroblasts, **J. Biol. Chem.**, 262: 17690.

Chojkier, M., Houghlum, K., Solis-Herruzo, J., and Brenner, D.A., 1989, Stimulation of collagen gene expression by ascorbic acid in cultured human fibroblasts. A role for lipid peroxidation, **J. Biol. Chem.**, 264:16957.

Hoerner, M., Behrens, U.J., Worner, T.M., Blacksberg, I., Braley, R., Schaffner, F., and Lieber, C.S., 1988, The role of alcoholism and liver disease in the appearance of serum antibodies against acetaldehyde adducts, **Hepatology**, 8:569.

Israel, Y., Hurwitz, E., Niemelä, O., and Arnon, R., 1986, Monoclonal and polyclonal antibodies against acetaldehyde-containing epitopes in acetaldehyde-protein adducts, **Proc. Natl. Acad. Sci. USA**, 83:7923.

Israel, Y., Orrego, H., and Niemelä, O., 1988, Immune responses to alcohol metabolites: pathogenic and diagnostic implications. **Semin. Liver Dis.**, 8:81.

Lelbach, W.K., 1974, Organic pathology related to volume and pattern of alcohol use, **in:** "Research advances in alcohol and drug problems," R.J. Gibbins, Y. Israel, H. Kalant, eds., John Wiley and sons, New York.

Lieber, C.S., 1988, Biochemical and molecular basis of alcohol-induced injury to liver and other tissues, **N. Engl. J. Med.**, 319:1639.

Lin, R. C., Smith, S.R., and Lumeng, L., 1988, Detection of a protein-acetaldehyde adduct in the liver of rats fed alcohol chronically, **J. Clin. Invest.**, 81:615.

Lin, R.C., and Lumeng, L., 1989, Further studies on the 37 kD liver protein-acetaldehyde adduct that forms in vivo during chronic alcohol ingestion, **Hepatology**, 10:807.

Lin, R.C., Lumeng, L., Shahidi, S., Kelly, T., and Pound, D.C., 1990, Protein-acetaldehyde adducts in serum of alcoholic patients, **Alcoholism: Clin. Exp. Res.**, 14:438.

Ma, X. -L., Baraona, E., Hernandez-Munoz, R., and Lieber, C.S., 1989, High levels of acetaldehyde in non-alcoholic liver injury after threonine or ethanol administration, **Hepatology**, 10:933.

Niemelä, O., Klajner, F., Orrego, H., Vidins, E., Blendis, L., and Israel, Y., 1987, Antibodies against acetaldehyde-modified protein epitopes in human alcoholics, **Hepatology**, 7:1210.

Niemelä, O., Israel, Y., Mizoi, Y., Fukunaga, T., and Eriksson, C.J.P., 1990a, Hemoglobin-acetaldehyde adducts in human volunteers following acute ethanol ingestion. **Alcoholism: Clin. Exp. Res.**, 14:838.

Niemelä, O., Mannermaa, R.-M., and Oikarinen, J., 1990b, Impairment of histone H1 function by adduct formation with acetaldehyde, **Life Sci.**, 47:2241.

Niemelä, O., Juvonen, T. and Parkkila, S., 1991, Immunohistochemical detection of acetaldehyde - modified epitopes in human liver following alcohol consumption, **J. Clin. Inves.**, in press.

Peterson, C.M., and Polizzi, C.M., 1987, Improved method for acetaldehyde in plasma and hemoglobin-associated acetaldehyde: Results in teetotalers and alcoholics reporting for treatment, **Alcohol**, 4:477.

Peterson, C.M., and Scott, B.K., 1989, Studies of whole blood associated acetaldehyde as a marker of alcohol intake in mice, **Alcoholism: Clin. Exp. Res.**, 13:845.

Ristiniemi, J., and Oikarinen, J., 1989, Histone H1 binds to the putative nuclear factor I recognition sequence in the mouse $\alpha2(I)$ collagen promoter, **J. Biol. Chem.**, 264:2164.

San George, R.C., and Hoberman, H.D., 1986, Reaction of acetaldehyde with hemoglobin, **J. Biol. Chem.**, 261:6811.

Sillanaukee, P., and Koivula, T., 1990, Detection of a new acetaldehyde-induced hemoglobin fraction HbA1ach by cation exchange chromatography, **Alcoholism: Clin. Exp. Res.**, in press.

Sorrell, M.F., and Tuma, D.J., 1987, The functional implications of acetaldehyde Stevens, V.J., Fantl, W.F., Newman, C.B., Sims, R.V., Cerami, A., and Peterson, C.M., 1981, Acetaldehyde adducts with hemoglobin, **J. Clin. Invest.**, 67:364.

Wolffe, A.P., 1989, Dominant and specific repression of Xenopus oocyte 5S RNA genes and satellite I DNA by histone H1, **EMBO J.**, 8:527.

Yokoyama, H., Ishii, H., Nagata, S., Kato, S., and Tsuchiya, M., 1990, Detection of acetaldehyde adducts in hepatic microsomes and cytosol after ethanol consumption, **Alcoholism: Clin. Exp. Res.**, 14: 355a (Abstr.).

STUDIES OF A CHEMICAL MEASURE OF ACETALDEHYDE ADDUCT FORMATION

Charles M. Peterson

Sansum Medical Research Foundation
2219 Bath Street
Santa Barbara, CA 93105
USA

INTRODUCTION

Two of the major problems in alcohol research and treatment have been: 1) a lack of objective criteria with which to characterize the condition; and 2) the lack of a biochemical basis by which the numerous secondary sequelae of the disease can be explained. Recent studies make it feasible to address these problems using a biochemical approach since it appears that hemoglobin and other proteins combine with acetaldehyde to form a number of stable protein-acetaldehyde adducts.

The fact that acetaldehyde forms stable adducts with proteins suggests: 1) certain sequelae of alcoholism may be the result of post translational modification of particular proteins at specific sites of reaction and 2) the adducts might serve as a marker(s) for chronic alcohol consumption and/or alcoholism.

A biochemical marker would establish a rationale for evaluating social dysfunction in terms of a biochemical parameter. Such an assay would allow evaluation of groups of patients to determine whether alcohol consumption per se correlates with the level of "alcoholism" in social terms. It might be, for example, that such an assay would help dichotomize the population into those who are more vulnerable to alcoholism and those who are less so, following a specific alcohol challenge. Such an assay would also be invaluable in following individual patients. A reliable assay for alcohol consumption and abuse would enable the definition of the individual at risk and the prevalence of excessive alcohol intake in a way that has heretofore been impossible. Good prevalence data is mandatory before comprehensive intervention programs can be initiated.

In 1974, it was estimated that there were 9-10 million problem drinkers in the U.S., more than twice earlier estimates (Keller, 1974; Pernanen, 1974). Though federally-funded programs have attempted to reach problem drinkers in the U.S. and though a 70% success rate was established for those who did remain in the programs, 24% never really entered treatment, and an additional 45% dropped out before 30 days (Ruggels et al., 1975). A means whereby alcohol intake could be quantified is essential before the population at risk can be identified with confidence and effective treatment follow up programs can be devised and assessed (Holden, 1987).

Korsten and colleagues were the first to find significantly higher blood acetaldehyde levels in chronic alcoholics compared to nonalcoholic controls (Korsten et al., 1975). These findings have been confirmed by others utilizing improved methods of blood acetaldehyde measurement (Lindros et al., 1980; Nuutinen et al.,1980). It has been hypothesized for over a decade that some individuals have a high genetic-metabolic predisposition toward the development of alcoholism as a result of abnormal biochemical pathways for metabolism of acetaldehyde (Gallant, 1982; Schuckit and Vidamantas, 1979). Utilizing liver biopsies from normal controls and alcoholic patients who were biopsied acutely and after liver function tests had returned to normal, it was demonstrated that the activity of cytosolic acetaldehyde dehydrogenase in 15 alcoholic specimens was significantly lower than that in 6 controls; the activity of the mitochondrial acetaldehyde dehydrogenase was also slightly lower in the alcoholic specimens. Follow up biopsies showed that the cytosol acetaldehyde dehydrogenase remained low (Thomas et al., 1982). The authors concluded that the genetic propensity to alcoholism is associated with a relative reduction in cytosolic acetaldehyde dehydrogenase activity.

Reported blood acetaldehyde levels during ethanol oxidation have ranged from practically zero to 250 μM (Korsten et al., 1975; Stowell et al., 1972). By improved methods it has now been established that in nonalcoholic controls blood acetaldehyde levels during ethanol oxidation are very low (Lindros et al., 1980; Pikkarainen et al., 1979; Eriksson, 1983; our own data below).

Recent experiments indicate that acetaldehyde can form stable adducts with hemoglobin (Hoberman and Chiodo, 1982; Stevens et al., 1981; Hoberman, 1980; Tsuboi et al., 1981; Fantl et al., 1982; Nguyen and Peterson, 1984a,b; Peterson et al., 1985; Nguyen and Peterson, 1985; Nguyen and Peterson, 1986a,b; Peterson et al., 1986; Peterson and Polizzi, 1987; Peterson et al., 1988a,b; Israel et al., 1986), plasma proteins and albumin (Donahue et al., 1983), P450 (Behrens et al., 1988), nucleosides (Fraenkel-Conrat and Singer, 1988), plasma membrane proteins (Barry et al., 1987), microtubular proteins and tubulin (McKinnon et al., 1987; Jennett et al., 1987), low density lipoproteins (Savolainen et al., 1987), and ribonuclease (Mauch et al., 1987); and that quantification of such adducts might serve as a useful marker for alcohol consumption or the alcoholic state. The formation of these adducts provides an hypothesis whereby a number of the secondary sequelae of alcoholism might be explained including liver injury (Cederbaum et al., 1974; Matsuzake and Lieber, 1977; Cederbaum et al., 1975; Nomura and Lieber, 1981), cardiovascular pathology (Schreiber et al., 1974; Gaillis, 1975), neuropathology (Walsh, 1971), erythrocyte membrane dysfunction (Gaines et al., 1977), vascular smooth muscle dysfunction (Altura et al., 1978), and an increased risk of cancer through acetaldehyde-induced cross links in DNA with resultant sister chromatid exchanges in human cells (Ristowe and Obe, 1978).

INITIAL STUDIES WITH HEMOGLOBIN

In the course of clinical studies on minor hemoglobins, we found that certain patients who had a history of alcohol abuse appeared to have higher levels of minor hemoglobins than controls. We hypothesized that acetaldehyde might modify adult hemoglobin or hemoglobin A. A series of experiments were initiated to test this hypothesis (Stevens et al., 1981; Fantl et al., 1982; Nguyen and Peterson, 1984a,b; Peterson et al., 1985; Nguyen and Peterson, 1985; Nguyen and Peterson, 1986a,b; Peterson et al, 1986; Peterson and

Polizzi, 1987; Peterson et al., 1988a,b). Acetaldehyde adduct formation with hemoglobin was tested in red cells, hemolysate, and with HbA using labelled 1,2-^{14}C-acetaldehyde.

The in vitro "on-rate" of acetaldehyde adduct with hemoglobin was determined in the following manner. Destromatized hemolysate (2 mM hemoglobin) was reacted with 0.3 mM ^{14}C-acetaldehyde in a vaccine stoppered reaction vial at 22°C. Aliquots were withdrawn with a syringe at various times and reduced with 400 fold excess of $NaBH_4$. Globin was prepared (Clegg et al., 1966) from the reduced aliquots and the radioactivity associated with 1 mg protein determined. About 20% of the ^{14}C-acetaldehyde had reacted with hemoglobin after 60 minutes.

The "off rate" of acetaldehyde reacted with hemoglobin was determined by taking a portion of the reaction mixture described above at 60 minutes and dialyzing against 4 liters of sodium phosphate buffer, pH 7.0. Aliquots were withdrawn at time intervals and reduced with sodium borohydride. Globin was prepared from the aliquots and the radioactivity in 1 mg was measured. Under these conditions, the reaction was virtually complete within ten minutes at 22°C. The "off rate" experiment showed rapid loss of about 75% of the reducible adducts after 30 minutes of dialysis followed by a more gradual loss of counts to a relatively stable level of 15-20% of the total adducts formed.

The amount of adducts stable to dialysis was proportional to the concentration of acetaldehyde in the reaction mixture over 4 log units (0.003 to 3 mM). Incubation of ^{14}C-acetaldehyde with washed red cells also resulted in the incorporation of radioactivity into hemoglobin in amounts comparable to that of hemolysate. Exposure of hemoglobin to 15 µM pulses of acetaldehyde, each of which was separated by dialysis, led to a dose related increase in the amount of stable adducts formed with hemoglobin. This experiment would be analogous to serial alcohol consumption by an individual.

Amino acid residues of hemoglobin susceptible to modification by ^{14}C-acetaldehyde were investigated by amino acid analysis. Globin was prepared from hemoglobin which had been reacted with ^{14}C-acetaldehyde and reduced with cyanoborohydride. After amino acid analysis, radioactivity from the labeled adducts was found distributed in peaks which corresponded to derivatives of valine, lysine and tyrosine. The first peak which appeared at 18-20 min, was found to coelute with the synthesized, cyanoborohydride reduced adduct of acetaldehyde plus N- (1-deoxy-1-glucitolyl) valine. The second peak (30 min) coeluted with the synthesized, cyanoborohydride reduced adduct of acetaldehyde plus valine. The third peak coeluted with the cyanoborohydride-reduced adduct of N6- (1-deoxy-1-glucitolyl) lysine and acetaldehyde. Peak #4 coeluted with a cyanoborohydride-reduced alpha-t-Boc tyrosine adduct and peak #5 coeluted with an acetaldehyde adduct formed with the C-amino group of lysine. Acetaldehyde adduct formation with tyrosine residues may involve either a nucleophilic attack by the third or fifth carbon of the phenolic ring, by analogy to formaldehyde modification of proteins (Blass et al., 1966), or alternatively by reaction with the hydroxyl group of tyrosine.

It was necessary to reduce the modified globin protein with sodium borohydride or cyanoborohydride in order to recover any of the amino acid adducts with acetaldehyde after acid hydrolysis. However, even with reduction, the recovery of the acetaldehyde adducts after amino acid analysis, was only 5% of those which were stable to dialysis.

To assess whether modified hemoglobins might occur in patients chronically consuming ethanol, blood was drawn from patients admitted for alcohol detoxification. The total fast hemoglobin component determined by high performance chromatography was elevated when compared to controls ($p<0.001$). This suggested that the fast hemoglobin components might be enriched by non-glycosyl adducts. Subsequent experiments performed by Gordis and Hershkopf (1986) and by our group (Peterson et al., 1986) documented that the measured changes in charged hemoglobins seen in the blood of alcoholic individuals resulted from inhibition of glycolytic pathways and hence an artifactual increase in fructose 1,6-diphosphate adducts unrelated to acetaldehyde levels. Acetaldehyde itself appeared to function as an inhibitor of glycolysis at or below the aldolase step.

Therefore it became mandatory to devise a different approach for measurement of acetaldehyde adducts through either an immunological or chemical approach. We chose to concentrate on the latter because of the problems in multiple epitopes, the necessity for reduction of sample, and the difficulty in raising antibodies against a relatively well-conserved protein.

STUDIES WITH FLUORIGENIC HPLC

We developed a fluorigenic high performance liquid chromatographic (HPLC) assay which has now been utilized for the determination of acetaldehyde associated with whole blood, hemoglobin, plasma, and platelets. The assay is based on the reaction of acetaldehyde with two molecules of 1,3-cyclohexanedione in the presence of ammonium ion to form a fluorescent species followed by reverse phase HPLC separation of specific aldehyde derived compounds (Peterson and Polizzi, 1987). The assay is specific, has sensitivity in the picomole range, and intraassay precision of less than 3.5%. The total run time is less than 6 minutes on HPLC. Percent recovery of HBAA is 95% and PAA is 86%.

Figure 1 shows the plasma and HBAA results following the pulse of 0.3 gm/kg ethanol to volunteer subjects compared to results found in teetotallers and alcoholic individuals reporting for treatment. Plasma values peaked at the 30 minute time point with a value of 31 $\pm$ 16 nmol/g protein. HBAA levels peaked at the 30 minute time point with a level of 159 $\pm$ 48 nmol/g hemoglobin. Thereafter PAA and HBAA returned to levels not statistically different from the initial levels by 3.5 hours. PAA continued to decrease after discharge from the outpatient unit and reached levels found in teetotallers by day five. HBAA remained elevated above levels found in teetotallers for the duration of the experiment (Peterson et al., 1988a,b).

Acetaldehyde partitions preferentially to hemoglobin in vivo as evidenced by the five fold greater molar association per gram protein vs. plasma. In each volunteer subject, serum ethanol levels were at or below the limits of detection of the assay at the beginning of the experiment. Ethanol levels reached a peak at 30 minutes with a mean level of 0.042 $\pm$ 0.007 % and were below the level of detection of the assay by 24 hours. The mean HBAA level of the volunteers at entry into the study before the dose of ethanol (115 $\pm$ 32 S.D. nmol/g) was significantly ($p < 0.001$) greater than the mean level found in teetotaller subjects (74 $\pm$ 16 S.D. nmol/g) and less than the mean level found in alcoholic subjects (155 $\pm$ 109 nmol/g, $p = 0.12$). The mean PAA level of the volunteers at entry into the study (13 $\pm$ 9 S.D. nmol/g) was significantly ($p = 0.012$) greater than that found in teetotaller subjects (4.9 $\pm$ 0.6 S.D. nmol/g) and also greater than the mean level found in alcoholic subjects (9.9 $\pm$ 4.8 S.D. nmol/g, $p = 0.4$).

On entry into the study, the volunteers' levels of PAA were actually elevated above the levels seen in the alcoholic subjects. This finding is consistent with having a "last celebration" before the study, confirmed by informal questioning. The fact that the

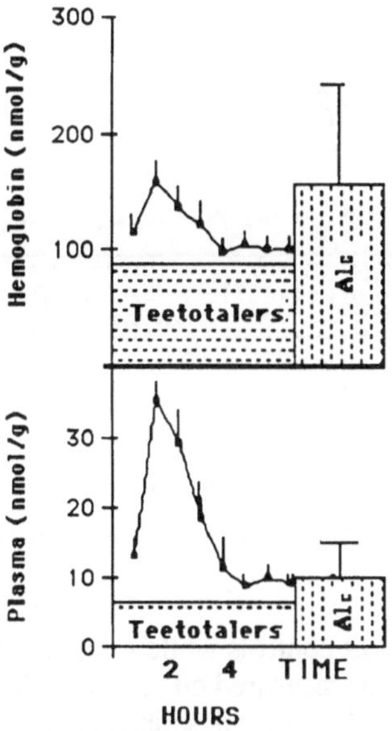

Fig. 1. Plasma and HBAA following the pulse of 0.3 gm/kg ethanol in teetotallers and alcoholic individuals. Mean levels are indicated with the bar representing the standard error of the mean. The lower and right stippled areas show the mean + 1 standard deviation values for teetotallers and alcoholic subjects respectively.

alcoholic subjects were voluntarily enrolling in a treatment program and attempting sobriety might also account in part for their lower PAA levels. Nevertheless, HBAA levels were less in the volunteers than in the alcoholic subjects, consistent with different kinetic responses of the two measurements and the ten fold greater lifespan of hemoglobin over plasma proteins.

For most subjects, HBAA levels did not decrease to the levels found in teetotaller subjects during the 28 day course of the study. PAA levels also rose intermittently during the study in many subjects. Some subjects who were well monitored showed a return of HBAA levels to those of teetotallers within 21 days as shown in figure 2 below.

From these experiments, it was apparent that the assay could be used to differentiate drinking from nondrinking humans well after the last observed ethanol intake (Peterson et al., 1988a,b). Nevertheless, ethanol consumption experiments in humans proved difficult to monitor in the outpatient setting and animal models were sought to further quantify the potential utility of the assay systems. Furthermore, the observation that 80-90% of acetaldehyde in blood was in the erythrocyte and bound to hemoglobin made it probable that measurement by our method of WBAA might provide a simple micromeasure of ethanol consumption. This hypothesis was tested in a rodent system.

The C57 black mouse has been documented to be a useful model of alcohol intake (Dole et al., 1988). Thirty C57Bl mice were randomized into two groups (Peterson and

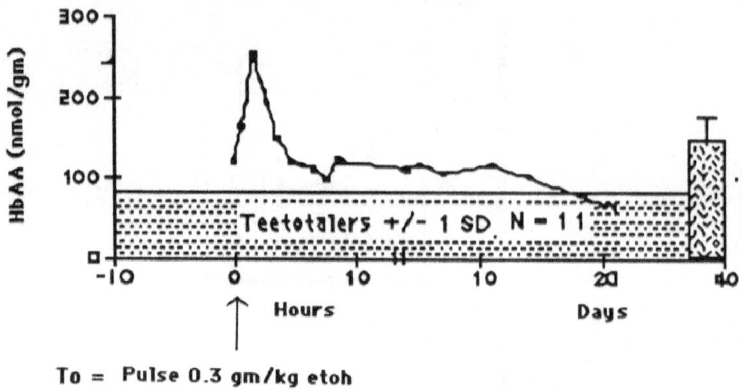

Fig. 2. Return of HBAA to levels found in teetotallers in a single subject.

Scott, 1989). Group 1 served as controls while group 2 was given 10% V/V ethanol with the drinking water. WBAA was measured on capillary blood samples using the fluorigenic high performance chromatographic assay described below. WBAA peaked at day 2. A stable mean plateau of 263 ± 71 S.D. with a range of 160-400 nmoles/gram hemoglobin WBAA was found in the group consuming ethanol compared with 122 ± 17 S.D. and a range of 88 - 150 nmoles/gram hemoglobin for controls ($p < 0.001$).

When ethanol was discontinued, levels of WBAA declined and became similar to those of controls by 9 days following cessation of ethanol. The quantitative difference between ethanol consuming and control animals in addition to the rapid rise of WBAA and the relatively slow decline following cessation of ethanol intake indicate that such a test might be a useful monitor of drinking behavior (Fig 3).

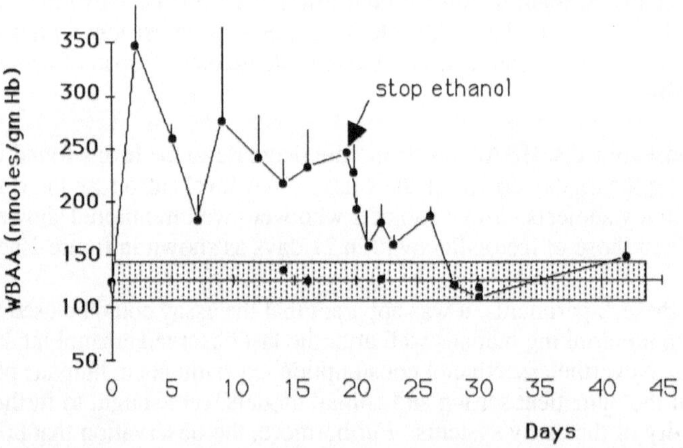

Fig. 3. The rapid rise, stable plateau and slow decline of whole blood associated acetaldehyde in C57 black mice given 10% ethanol which was discontinued at the time shown by the arrow. The hatched area represents the mean +/- 1 SD levels found in controls.

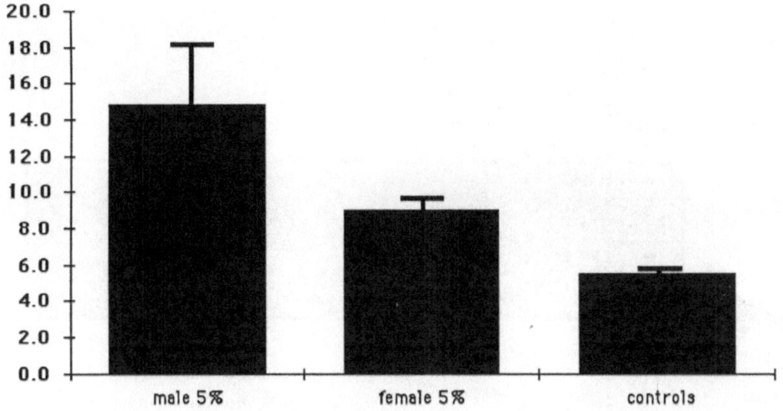

Figure 4 reveals a significant difference in WBAA levels between male and female mice maintained on 5% ethanol for 2 weeks. These data emphasize that norms for assays of ethanol consumption based on acetaldehyde adduct formation with proteins must consider gender(Peterson et al., 1990a).

Experiments in miniature swine have also proven to be useful (Peterson et al., 1990b). Blood samples were obtained from miniature swine maintained on 0, 2, or 6

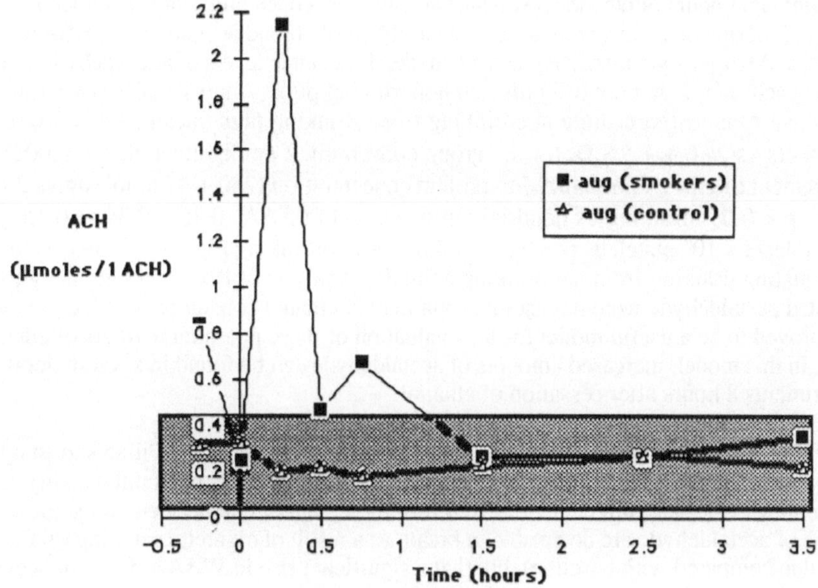

Fig. 5. Breath analysis of acetaldehyde as a function of time in 4 smokers.

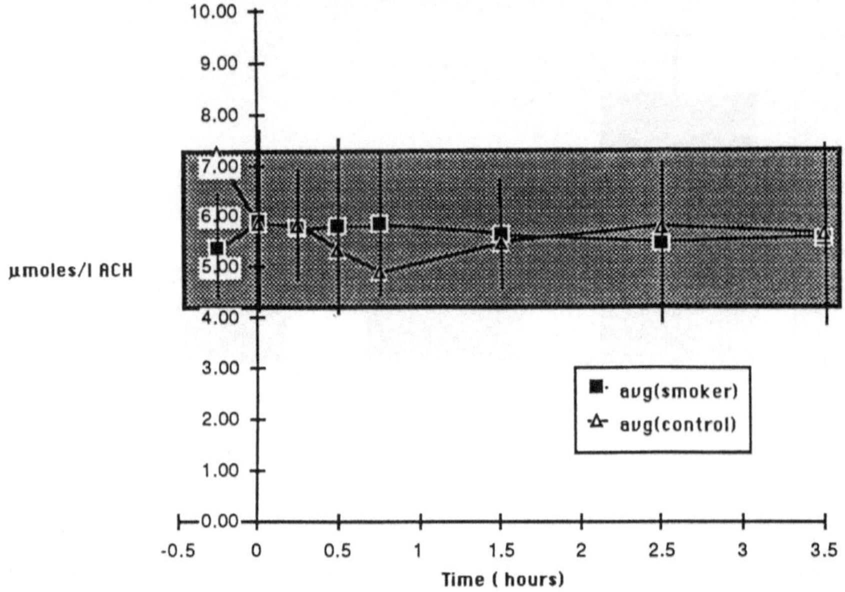

Fig. 6. Whole blood acetaldehyde values in 4 smokers following smoking 1-3 cigarettes over the 15 minutes after time 0. The shaded area represents the mean + 1 S.D. value for nonsmokers. The error bars represent the S.D.

gm/kg/24 hr ethanol for 8 months (N=6 in each group). Samples from drinking pigs were taken after eight hours of ethanol abstinence and all were coded and sent for "blinded" analysis. Fluorigenic assays were used to quantify whole blood acetaldehyde, HBAA, PAA, PLTAA, and lymphocyte associated acetaldehyde. Detectable levels of acetaldehyde were found in each sample in both drinking and nondrinking pigs. Analysis of WBAA was most discriminatory in distinguishing nondrinking from drinking pigs (mean 21.4 $\pm$ 1.0 µM for nondrinkers vs 24.6 $\pm$ 1.5 S.D. for the group consuming 2 gm/kg ethanol, $p = 0.001$). Measurement of HBAA normalized to protein concentration (250 $\pm$ 47 nmoles/g vs 203 $\pm$ 33 S.D., $p < 0.05$ drinking vs nondrinking pigs), and PLTAA (0.46 $\pm$ 0.34 vs 0.15 $\pm$ 0.16 nmoles/3 x 10^8 platelets, $p = 0.05$ drinking vs nondrinking pigs) were also useful in discriminating drinking from nondrinking animals. Analysis of PAA and lymphocyte associated acetaldehyde were not useful as markers of ethanol consumption. The miniature swine proved to be a useful model for the evaluation of these potential markers of ethanol intake. In this model, increased amounts of acetaldehyde can be found in several blood compartments 8 hours after cessation of ethanol.

Figures 5 and 6 show preliminary data obtained in 4 pairs of individuals to study the contribution of smoking to the acetaldehyde adducts under study as potential markers of ethanol intake (McLaughlin et al., 1990). From these data, it can be seen that significant amounts of acetaldehyde are detectable in breath as a result of cigarette smoking ($p<0.05$ peak value compared with baseline), but that a significant rise in WBAA does not occur. These findings can be compared with the results of ethanol consumption in Figures 1-3. Hence, these preliminary data lend credence to the hypothesis that acetaldehyde adducts with blood proteins will serve as useful and specific markers for ethanol intake in humans.

Studies documenting a correlation of HbAA with questionnaires which quantify alcoholism such as the SAAST (Peterson et al., 1990c), an instrument validated in two cultures (Davis et al., 1989) lend further weight to this hypothesis.

CONCLUSION

The assay system developed for the measurement of acetaldehyde adducts has been shown to be useful in discriminating ethanol consuming from nondrinking pigs, mice, and humans. The biological "on rate" for acetaldehyde in blood of mouse and man appears very rapid (minutes to hours) and the "off rate" appears to be sufficiently long (weeks) to allow for the clinical utility of these measurements as a monitor of ethanol consumption long after the last drink. Storage studies have now been completed up to 21 months with acceptable precision for PAA and HBAA. Further work is warranted to define the relative kinetics of all the measurements discussed above (WBAA, HbAA, PltAA, and PAA) and the potential utility of each as a marker for ethanol consumption and alcoholism in humans.

REFERENCES

Altura, B.M., Carella, A. and Altura, B.T., 1978, Acetaldehyde on vascular smooth muscle, possible role in vasodilator action of ethanol, **Eur. J. Pharmacol.** 5:73-83.

Barry, R.E., Williams, A.J. and McGivan, J.D., 1987, The detection of acetaldehyde/liver plasma membrane protein adduct formed in vivo by alcohol feeding, **Liver** 7:364-8.

Behrens,U.J., Hoerner M., Lasker, J.M. and Lieber C.S., 1988, Formation of acetaldehyde adducts with ethanol-inducible P450IIE1 in vivo. **Biochemical and Biophysical Research Communications**, 154:584-90.

Blass, J., Bizzini, B. and Raynaud, M., 1965, Mecanisme de la detoxification par le formol. **C.R. Acad. Sci. Paris.** 261:1448-1449.

Cederbaum, A.I., Lieber, C.S. and Rubin, E., 1974, The effect of acetaldehyde on mitochondrial function. **Arch. Biochem. Biophys.** 161:26-39.

Cederbaum, A.I., Lieber, C.S. and Rubin, E., 1975, Effect of acetaldehyde on fatty acid oxidation and ketogenesis by hepatic mitochondria. **Arch. Biochem. Biophys.** 169:29-41.

Clegg, J.B., Naughton, M.A, and Weatherall, D.J., 1966, Abnormal human hemoglobins. **J. Mol. Biol.** 19:91.

Davis, L.J., de la Fuente, J.R., Morse, R.M., Landa, E. and O'Brien, P.C., 1989, Self-administered alcoholism screening test (SAAST): Comparison of classificatory accuracy in two cultures. **Alcoholism: Clin Exp Res.** 13;224-8.

Dole, V.P., Ho, A., Gentry, R.T. and Chin, A., 1988, Toward an analogue of alcoholism in mice: Analysis of nongenetic variance in consumption of alcohol. **Proc Nat Acad Sci USA** 85:827-30.

Donohue, T.M. Jr, Tuma, D.J. and Sorrell, M.F. 1983, Acetaldehyde adducts with proteins: binding of ^{14}C-acetaldehyde to serum albumin. **Arch. Biochem. Biophys.** 220:239-246.

Eriksson, C.J., 1983, Human blood acetaldehyde concentration during ethanol oxidation (update 1982). **Pharm. Biochem. Behavior** 18 (Suppl. 1):141-150.

Fantl, W.J., Stevens, V.J. and Peterson, C.M., 1982, Reactions of biologic aldehydes with proteins. **Diabetes** 31 (Suppl.3):15-21.

Fraenkel-Conrat, H. and Singer, B., 1988, Nucleoside adducts are formed by cooperative reaction of acetaldehyde and alcohols: possible mechanism for the role of ethanol in carcinogenesis. **Proc Nat Acad Sci USA** 85:3759-61.

Gaillis, L., 1975, Cardiovascular effects of acetaldehyde: evidence for the involvement of tissue SH groups, in Lindros KO, Eriksson CJP (eds): **In:** The role of acetaldehyde in the actions of ethanol: Satellite symposim 6th Int. Congr. Pharm. Helsinki, pp. 135-147.

Gaines, K.C., Salhany, J.M., Tuma, D.J. and Sorrell, M.F., 1977, Reaction of acetaldehyde with human erythrocyte membrane proteins. **FEBS Let.** 75:115-119.

Gallant, D.M., 1982, New information on genetics of alcoholism. **Alcoholism: Clin. Exp. Res.** 6:130-131.

Gordis, E. and Herschkopf, S., 1986, Application of isoelectric focussing in immobilized pH gradients to the study of acetaldehyde-modified hemoglobin. **Alcoholism: Clin Exp Res** 10:311-19.

Hoberman, H.D., 1980, A possible hemoglobin marker of alcoholism, **in:** Thurman, R. (ed) Alcohol and Aldehyde Metabolizing Systems, vol. 3. New York, Plenum, p. 265.

Hoberman, H.D. and Chiodo, S.M., 1982, Elevation of the hemoglobin Al fraction in alcoholism. **Alcoholism: Clin. Exp. Res.**:260-266.

Holden, C., 1987, Is alcoholism treatment effective? **Science** 236:20-22.

Israel, Y., Hurwitz, E., Niemela, O. and Arnon, R., 1986, Monoclonal and polyclonal antibodies against acetaldehyde containing epitopes in acetaldehyde protein adducts. **Proc Nat Acad Sci USA** 83: 7923-7.

Jennett, R.B., Sorrell, M.F., Johnson, E.L. and Tuma, D.J:., 1987, Covalent binding of acetaldehyde to tubulin: evidence for preferential binding to the alpha-chain. **Archives of Biochem and Biophys.**256:10-18.

Keller, M., 1974, **ed:** Alcohol and Health: New Knowledge. Second special report to the U.S. Congress. DHEW Publication No. (ADM) 75-212. Washington, D.C.: U.S. Government Printing Office.

Korsten, M.A., Matsuzaki, S., Feinman, L. and Lieber, C.S., 1975, High blood acetaldehyde levels after ethanol administration. **N. Engl. J. Med.** 292:386-389.

Lindros, K.O., Stowell, A., Pikkarainen, P.H. and Salapuro, M., 1980, Elevated blood acetaldehyde in alcoholics with accelerated ethanol elimination. **Pharm. Biochem. Behavior** 13:119-124.

Matsuzake, S. and Lieber, C.S., 1977, Increased susceptibility of hepatic mitochondria to the toxicity of acetaldehyde after chronic ethanol consumption. **Biochem. Biophys. Res. Commun.** 75:1059-1065.

Mauch, T.J., Tuma, D.J. and Sorrell, M.F., 1987, The binding of acetaldehyde to the active site of ribonuclease: alterations in catalytic activity and effects of phosphate. **Alcohol and Alcoholism** 2:103-12.

McKinnon, G., de Jersey, J., Shanley, B. and Ward, L., 1987, The reaction of acetaldehyde with brain microtubular proteins: formation of stable adducts and inhibition of polymerization. **Neuroscience Letters** 79:163-8.

McLaughlin,S.D., Scott ,B.K. and Peterson, C.M., 1990, The effect of cigarette smoking on breath and whole blood associated acetaldehyde. **Alcohol** 7:285-287.

Nguyen, L.B. and Peterson, C.M., 1984a, The effect of acetaldehyde concentration on the relative rates of formation of acetaldehyde-modified hemoglobins. **Proc Soc Exp Biol and Med** 177:226-233.

Nguyen, L.B. and Peterson, C.M., 1984b, Modification of hemoglobin by acetaldehyde: A time course study by high pressure liquid chromatography. **Alcoholism: Clin Exp Res** 8:516-21.

Nguyen, L.B. and Peterson, C.M., 1985, Clinical implications of acetaldehyde adducts with hemoglobin. **In** Collins MA (ed). Aldehyde Adducts in Alcoholism. Alan R. Liss, New York, pp 19-30.

Nguyen, L.B, and Peterson, C.M., 1986a, Differential modification of hemoglobin chains by acetaldehyde. **Proc Soc Exp Biol and Med** 181:151-6.

Nguyen, L.B. and Peterson, C.M., 1986b, Stability of acetaldehyde fractions with various hemoglobin fractions. **Diabetes Research** 3:249-53.

Nomura,F. and Lieber, C.S., 1981, Binding of acetaldehyde to rat liver microsomes: enhancement after chronic alcohol consumption. **Biochem. Biophys. Res. Comm.** 100:131-133.

Nuutinen, H., Lindros, K.O. and Salaspuro, M., Determinants of blood acetaldehyde level during ethanol oxidation in chronic alcoholics. **Alcoholism: Clin. Exp. Res.** 7:163-168.

Pernanen, K., 1975, Validity of survey data on alcohol use. **Alcohol and drug problems,** Ruggels WL et al: A follow-up study of clients at selected alcoholism treatment centers funded by NIAAA: Final Report. SRI Project URU-3124. Menlo Park, Calif.: Stanford Research Institute.

Peterson., C.M., Nguyen, L., Fant, W., Stevens, B., Hawthorne, G. and Blackburn, P., 1985, Acetaldehyde Adducts with Hemoglobin. **In**: Chang NC, Chao HM (eds). Research Monograph NO. 17 Early Identification of Alcohol Abuse. U.S. Department of Health and Human Services Publication N. ADM, pp 85-1258, Washington D.C.

Peterson, C.M,, Polizzi, C.M. and Frawley, P.J., 1986, Artefactual increase in hemoglobins A1a-b in blood from alcoholic subjects. **Alcoholism: Clin. and Exp. Res.** 10:219-20.

Peterson, C.M. and Polizzi, C.M., 1987, Improved method for acetaldehyde in plasma and hemoglobin-associated acetaldehyde: Results in Teetotallers and Alcoholics Reporting for Treatment. **Alcohol** 4: 477-480.

Peterson, C.M. Jovanovic-Peterson,L. and Schmid-Formby, F., 1988, Rapid association of acetaldehyde with hemoglobin in human volunteers after low dose ethanol. **Alcohol** 5:371-4.

Peterson, C.M., Jovanovic-Peterson, L., Schmid-Formby, F. and Polizzi, C.M., 1988, Association of acetaldehyde with hemoglobin and plasma after ethanol: Potential markers of ethanol intake with different kinetics. **In**: Kuriyama K, Takada A, and Ishii H (eds) Biomedical and Social Aspects of Alcohol and Alcoholism. Elsevier Science Publishers. New York, 1988 pp321-324.

Peterson, C.M. and Scott, B.K., 1989, Studies of Whole Blood Associated Acetaldehyde as a Marker for Alcohol Intake in Mice. **Alcoholism : Clin. and Exp. Res.** 13: 845-848.

Peterson, C.M., Scott, B.K. and McLaughlin, S.D., 1990a, Studies of whole blood associated acetaldehyde as a marker for alcohol intake: Effect of gender in mice. **Alcohol**, in press.

Peterson, C.M., Scott, B.K., Sun, G.Y. and Sun, A.Y., 1990b, A comparative blinded study in miniature swine of whole blood, hemoglobin, platelet, plasma, and lymphocyte associated acetaldehyde as markers for ethanol intake, **Alcoholism: Clin. Exp. Research,** 14: 717-720.

Peterson, C.M., Ross, S.L. and Scott, B.K., 1990c, Correlation of self-administered alcoholism screening test with hemoglobin-associated acetaldehyde. **Alcohol** 7:289-294.

Pikkarainen,P.H., Salaspuro, M.P. and Lieber, C.S., 1979, A method for the determination of "free" acetaldehyde in plasma. **Alcoholism: Clin. Exp. Res.** 3:259-261.

Ristowe, H. and Obe, G., 1978, Acetaldehyde induces cross-links in DNA and causes sister chromatid exchanges in human cells. **Mut. Res.** 58:115-119.

Savolainen, M.J., Baraona, E. and Lieber, C.S., 1987, Acetaldehyde binding increases the catabolism of rat serum low-density lipoproteins. **Life Sciences,** 40:841-6.

Schreiber, S.S., Orak, M., Rothschild, M.A., Reff, F. and Evans, C., 1974, Alcoholic cardiomyopathy. II. The inhibition of cardiac microsomal protein synthesis by acetaldehyde. **J. Mol. Cell. Cardiol.** 6:207-213.

Schuckit, M.A. and Vidamantas, R., 1979, Ethanol ingestion: Differences in blood acetaldehyde concentrations in relatives of alcoholics and controls. **Science** 203:54-55.

Stevens, V.J., Fantl, W.J., Newman, C.B., Sims, R.V., Cerami, A. and Peterson C.M., 1981, Acetaldehyde adducts with hemoglobin. **J. Clin. Invest.** 67:362-369.

Stowell, A.R., Greenway, R.M. and Bat, R.D., 1972, Acetaldehyde formation during deproteinization of human blood samples containing ethanol. **Biochem. Med.** 8:392-401.

Thomas, M., Halsall, S and Peters, T.J., 1982, Role of hepatic acetaldehyde dehydrogenase in alcoholism: Demonstration of persistent reduction of cytosolic activity in abstaining patients. **Lancet** 2:1057-1059.

Tsuboi, K.K., Thompson, D.J., Rush, E.M. and Schwartz, H.C., 1981, Acetaldehyde-dependent changes in hemoglobin and oxygen affinity in human erythrocytes. **Hemoglobin** 5:241-250.

Walsh, M.J.,1971, Role of acetaldehyde in the interactions of ethanol with neuroamines, **in:** Roach MK, McIsaac WM, Creaven PJ (eds): Biological Aspects of Alcohol, Austin University of Texas Press, pp. 233-249.

ETHANOL-INDUCED PHOSPHORYLATION OF CYTOKERATINS IN PRIMARY CULTURED HEPATOCYTES

Samuel W. French, H. Kawahara and M. Cadrin

Department of Pathology
Faculty of Medicine
University of Ottawa
Ottawa, Ontario
Canada K1H 8M5

INTRODUCTION

Phosphorylation of specific proteins by protein kinase A (PKA) or protein kinase C (PKC) may play a role in the regulation of assembly and disassembly and stability of intermediate filaments including intermediate filaments composed of cytokeratins (CKs). Initiation of the phosphorylation is through signal transduction by agonist receptor G protein activation of a catalyst such as cyclic AMP or phospholipase C which activate PKA or PKC, respectively.

Data from other laboratories has demonstrated that phosphorylation of CKs occurs at the carboxyl and the amino terminal (Steinert, 1988). These endings of the CK molecule may be involved in the formation of intracellular attachments of the filaments (Alberts and Fuchs, 1989). Intermediate filament phosphorylation has also been associated with a reorganization of the intermediate filaments during mitoses (Celis et al., 1983) or in response to the stimulation by epidermal growth factor (Baribault et al., 1989). Intermediate filaments provide the substrate for PKA or PKC (Gard and Lazarides, 1982; Geisler et al., 1989). Geisler et al. (1989) showed that phorbol ester caused aggregation and clumping of intermediate filaments through the known associated PKC pathway. Ethanol is known to activate PKC phosphorylation through two different pathways. First is via the action of calcium-dependent protein kinase which results from IP3 generation from phospholipase C activation (Fig. 1) (Hock et al., 1988; Hock et al., 1987; Hock et al., 1989). Ethanol activates PLC through its action on the Gα subunit in the plasma membrane (Fig. 1). Protein kinase C is also activated by ethanol through diacylglycerol (DAG) generated from phospholipase C hydrolysis of PIP2. Thus, ethanol utilizes the same pathway as phorbol ester in activating protein kinase C phosphorylation.

Phosphorylation of intermediate filaments such as cytokeratin may lead to disassembly of the filament to liberate monomers which are degraded by calcium-dependent cytosolic proteases to generate polypeptide fragments which are postulated to enter the nuclear pores and regulate transcription through gene gating (Traub et al., 1987). Thus,

Alcoholism: A Molecular Perspective
Edited by T.N. Palmer, Plenum Press, New York, 1991

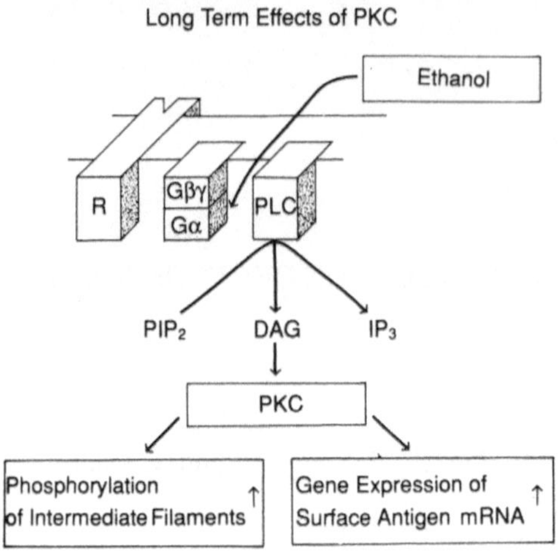

Fig. 1. Cellular adaptation to alcohol, modified from Hoek et al., 1989.

long-term cellular adaptative changes may occur in hepatocytes in response to the exposure of the cells to alcohol through the PKC phosphorylation pathway (Fig. 2). It is also possible that phosphorylation of microtubules may alter membrane traffic and activation of actin contraction of the bile canaliculus which may influence bile flow and secretion. It is already known that actin contraction influences the rate of bile flow and that ethanol in vivo reduces the rate of bile flow (Okanoue et al., 1984). Alcohol may interfere with the bile secretory function of the canaliculus through its known feedback mechanism where activated PKC inhibits phospholipase C-independent signal transduction to hormones such as vasopressin 1 which causes bile canalicular contraction by increasing cytosolic calcium (Nathanson et al., 1990; Hoek, 1988).

METHODS

Liver cell primary cultures were obtained from 14 day old male suckling rats of the Wistar strain using a two-step perfusion method (Seglen, 1976) with 0.05% collagenase via the portal vein. Cell viability was over 90% when tested with the trypan blue exclusion test. Hepatocytes were cultured in serum free William's E medium (1-5 X 10^5 cells/ml). The medium contained insulin (100 ng/ml) and dexamethasone (10 μM) to induce differentiation. Isolated hepatocytes were seeded on fibronectin coated dishes and 3 hrs after seeding the culture medium was exchanged to remove dead cells and non-parenchymal cells. The primary culture then was maintained for 48 hrs in the same medium. Labeling was accomplished with ^{32}P by incubating the hepatocytes for 4 hrs in a phosphorous-free α-minimal essential medium supplemented with insulin and dexamethasone containing ^{32}P-orthophosphate (50 μCi/ml) to label ATP pools with $^{32}P0_4$. Agonists were added as follows 30 mins before the end of the 4 hr incubation with $^{32}P0_4$: 1. Ethanol 300 mmol, 2. 10 nM vasopressin (VP), 150 nM 12-0-tetradecanoyl phorbol-13-acetate (TPA) or dibutyryl cyclic adenosine monophosphate (db-cAMP). The rationale for this was that

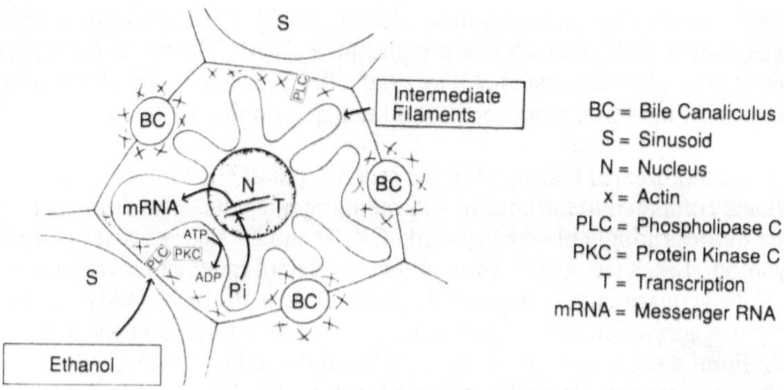

Fig. 2. Cellular adaptation to alcohol (liver cells).

phorbol ester and ethanol act through PKC to phosphorylate proteins as does vasopressin via a VP1 receptor, whereas db-cyclic AMP induces phosphorylation via PKA. Thus, it was anticipated that these agonists would phosphorylate isolated cytoskeletal filaments by activation of PKC or PKA. After 4 hours of incubation, the cytoskeletal preparation of CK filaments was prepared by extracting soluble proteins with a solution which contained EGTA, sodium pyrophosphate, sodium fluoride and sodium vanadate as well as the protease inhibitor, phenylmethylsulphonyl chloride and dithiothreitol. The detergent used to extract soluble proteins was 1% Triton-X100. The extraction was accomplished in 10 mins followed by a solution of 0.5% of Triton-X100 with 1.5 M KCl for 30 mins. Extraction was done at 4°C. Once the soluble proteins were removed and only the intermediate filament skeleton remained, the insoluble proteins were solubilized with 0.5% sodium dodecyl sulphate (SDS) then precipitated in ethanol at -20°C overnight and centrifuged and dried.

Two dimensional gel electrophoresis (2D) was performed as described by O'Farrell et al. (1977). The proteins were solubilized in 9.5 M urea in sample buffer. They were sonicated and the proteins were quantitated by Coomassie blue assay for samples in urea solubilized buffer (Ramagli and Rodriguez, 1985). For the first dimension, isoelectric focussing was used by the combination of 5-7 and 3-10 ampholites at a ratio of 4:1. For the second dimension, samples were run in 8% acrylamide gels. The gels were stained with Coomassie blue, photographed, dried and autoradiographed with Kodak XAR-5 film by a two hour exposure. All of the agonists were run together so as to allow comparison between the different treatments.

RESULTS

The cytoskeletal preparations in the control cultured hepatocytes showed major cytokeratin proteins, i.e. 1) 55 kDa (CK 8), 2) 52 kDa (possibly a major degradation product of CK 55) (Schuller and Franke, 1983) and 3) 49 kDa (CK 18) (Kawahara, Cadrin and French, 1990). Non-phosphorylated proteins were primarily actin which was present in considerable amount. The CKs 1, 2 and 3 had several isoelectric variants due to binding to different levels of phosphorylation (Sun and Green, 1978). The most basic variant of the CKs was a dominant protein which was not phosphorylated. The second and third more acidic variants were phosphorylated. Although the third variant of CK 55 was not detectable on the Coomassie blue stain, it appeared as a faint spot on the radiograph.

In the ethanol-treated hepatocytes, the second spot of CK 55 was markedly phosphorylated compared to the control autoradiograph results. The third spot of CK 55 also became more strikingly phosphorylated. CK 52 and 49 also showed enhanced phosphorylation. Likewise, results comparable to ethanol treatment were found with treatment of the primary culture hepatocytes with TPA or VP. Db-cAMP did not cause an increase in phosphorylation of the cytokeratin proteins and thus they appeared the same as the control. From these results, it can be concluded that ethanol stimulates phosphorylation of intermediate filament cytokeratins in cultured hepatocytes in a similar manner to TPA and vasopressin and, therefore, it can be surmised that ethanol induces phosphorylation through an activation of PKC by way of the G protein phospholipase C complex generation of DAG as reported by Hoek et al. (1989).

REFERENCES

Albers, K., and Fuchs, E., 1989, Expression of mutant keratin cDNAs in epithelial cells reveals possible mechanisms for initiation and assembly of intermediate filaments, **J. Cell. Biol.** 108: 1477-1493.

Baribault, H., Blouin, R., Bourgon, L., and Marceau, N., 1989, Epidermal growth factor-induced selective phosphorylation of cultured rat hepatocyte 55-KD cytokeratin before filament reorganization and DNA synthesis, **J. Cell. Biol.** 109: 1665-1676.

Celis, J.E., Larsen, S.J., Fey, S.J., and Celis, A., 1983, Phosphorylation of keratin and vimentin polypeptides in normal and transformed mitotic human epithelial amnion cells: Behaviour of keratin and vimentin filaments during mitosis, **J. Cell. Biol.** 97: 1429-1434.

Gard, D.L., and Lazarides, E., 1982, Cyclic AMP-modulated phosphorylation of intermediate filament proteins in cultured avian myogenic cells, **Mol. Cell. Biol.** 2: 1104-1114.

Geisler, N., Hatzfield, M., and Weber, K., 1989, Phosphorylation in vitro of vimentin by protein kinases A and C is restricted to the head domain. Identification of the phosphoserine sites and their influence on filament formation, **Europ. J. Biochem.** 123: 441-447.

Hoek, J.B., Thomas, A.P., Rubin, R., and Rubin, E., 1987, Ethanol-induced mobilization of calcium by activation of phosphoinositide-specific phospholipase C in intact hepatocytes, **J. Biol. Chem.** 262: 682-691.

Hoek, J.B., Rubin, R., and Thomas, A.P., 1988, Ethanol-induced phospholipase C activation is inhibited by phorbol esters in isolated hepatocytes, **Biochem. J.** 251: 865-871.

Hoek, J.B., Armstrong, J.J., and Crovello, M.T., 1988, Feedback control of ethanol-induced phospholipase C activation in rat hepatocytes, **in**: Biomedical and Social Aspects of Alcohol and Alcoholism, K. Kuriyama, A. Takada, and H. Ishii, eds., Elsevier Sci. Publ., pp. 651-655.

Hoek, J.B., Teraschi, T.F., Rubin, E., 1989, Functional implications of the interaction of ethanol with biologic membranes: Actions of ethanol on hormonal signal transduction systems, **Semin. Liver Dis.** 8: 36-46.

Kawahara, H., Cadrin, M., and French, S.W., 1990. Ethanol-induced phosphorylation of cytokeratin in cultured hepatocytes. **Life Sciences** 47: 859-863.

Nathanson, M.H., Gautam, A., and Boyer, J.L., 1990, Effects of vasopressin (VP) and phorbol dibutyrate (PDB) on bile secretion in isolated rat hepatocyte couplets (IRHC), **Gastroenterology** 98: A615.

O'Farrell, P.Z., Goodman, H.M., and O'Farrell, P.H., 1977, High-resolution two-dimensional electrophoresis of basic as well as acidic proteins, **Cell** 12: 1133-1142.

Okanoue, T., Kondo, I., Ihrig, T.J., and French, S.W., 1984, Effect of ethanol and chlorpromazine on transhepatic transport and biliary secretion of horseradish peroxidase, **Hepatology** 4: 253-260.

Ramagli, L.S., Capetillo, S., Becker, F.F., and Rodriguez, L.V., 1985, Quantitation of microgram amounts of protein in two-dimensional polyacrylamide gel electrophoresis sample buffer. **Electrophoresis** 6: 559-563.

Schiller, D.L., and Franke, W.W., 1983, Limited proteolysis of cytokeratin a by an endogenous protease: removal of positively charged terminal sequences, **Cell Biol. Int. Rep.** 7: 3.

Seglen, P.O., 1976, Preparation of isolated rat liver cells, **Methods Cell. Biol**. 13: 29-83.

Steinert, P.M., 1988, The dynamic phosphorylation of the human intermediate filament keratin 1 chain, **J. Biol. Chem.** 263: 13333-13339.

Sun, T.-T., and Green, H., 1978, Keratin filaments of cultured human epidermal cells: Formation of intermolecular disulfide bonds during terminal differentiation, **J. Biol. Chem.** 253: 2053-2060.

Traub, P., Pagens, V., Kuhn, S., and Perides, G., 1987, Function of intermediate filaments. A novel hypothesis, **in**: Zortschrite der Zoologie Brand 34. Wohfarth-Bottermann (Hrsg.): Nature and Function of Cytoskeletal Proteins in Motility and Transport, Gautav-Fisher Verlag, Stuttgart, pp. 275-287.

CLONING AND CHARACTERIZATION OF NEW ALCOHOL DEHYDROGENASE AND ALDEHYDE DEHYDROGENASE ISOZYMES

A. Yoshida, L. Hsu and M. Yasunami

Department of Biochemical Genetics
Beckman Research Institute of the City of Hope
Duarte, California, U.S.A.

INTRODUCTION

Large numbers of alcohol dehydrogenase (ADH) and aldehyde dehydrogenase (ALDH) isozymes have been found in humans and other mammals. These enzymes oxidize varieties of alcohols and aldehydes, and play a role, not only in alcohol detoxification, but also in the metabolism of drugs, neurotransmitters and other endogenous biologically active alcohols and aldehydes.

At present, five types of ADH subunits (α, ß, γ, π and χ), controlled by five non-allelic genes, are identified in humans. Primary amino acid sequences of these subunits were determined, and cDNAs for these subunits have been cloned and characterized. However, the existence of other ADH isozyme(s) has not been ruled out. For example, the testis-specific ADH isozyme found in rodents (Keung, 1988) has not yet been found in humans. Recently it was reported that a significant portion of orally administered ethanol was metabolized in the stomach of males, but not in females, and that ADH activity of stomach mucous membrane is substantially higher in males than in females (Frezza et al., 1990). It is not clear whether this sex-dependent stomach ADH isozyme is identical to the ADH_4 (ππ) isozyme or it is a hitherto unknown ADH isozyme.

At least nine human ALDH isozymes, which are distinguishable by tissue and subcellular distributions, kinetic properties and physico-chemical properties, have been described (see Yoshida et al., 1990). Among them, only a major liver cytosolic $ALDH_1$ and a major liver mitochondrial $ALDH_2$ were extensively studied, i.e. their primary amino acid sequences were determined and their genes and cDNAs were cloned and characterized. Our knowledge of the rest of the human ALDH isozymes is still rather limited. The brain and testis probably have other ALDH isozymes with unique physiological roles.

These ADH and ALDH isozymes are considered to have been diverged from their ancestral proteins by gene multiplication and evolution. These isozymes have a certain degree of similarity at the protein level and the coding nucleotide sequences. The ADH_1 (α subunit), ADH_2 (ß subunit) and ADH_3 (γ subunit) are highly similar to each other (>95%

Alcoholism: A Molecular Perspective
Edited by T.N. Palmer, Plenum Press, New York, 1991

positional identity) (Ikuta et al., 1986). These three isozymes and other isozymes, ADH_4 (π subunit) and ADH_5 (χ subunit), are also similar to each other (60~70% positional identity) (Jörnvall et al., 1987). The degree of resemblance between the cytosolic $ALDH_1$ and the mitochondrial $ALDH_2$ is 68% in their coding nucleotide sequences (Hsu et al., 1985). It is quite conceivable that such a similarity also exists between these known isozymes and hitherto unknown isozymes.

Thus far, ADH and ALDH isozymes have been identified through examination of proteins and enzymes. Recent development of molecular genetics allows us to identify and characterize hitherto-unknown isozymes through direct study of DNA.

METHODS FOR CLONING OF GENES FOR NEW ENZYMES

The related genes can be cloned by the following two approaches. The simple and straightforward method includes: a) screening a genomic DNA library (or cDNA library) with nucleotide probes which correspond to conserved coding sequences of known related isozymes under a low-stringent hybridization condition; and b) characterization of weakly positive clones by restriction mapping and partial nucleotide sequencing for identification. The obvious advantage of the use of genomic libraries over cDNA libraries is the equal (single copy) existence of all genes in the genomic library regardless of degree of expressions of individual genes in various tissues. Once a potentially new gene(s) is obtained by the above procedures, a corresponding cDNA may be cloned and characterized without difficulty.

The alternative method includes: a) amplification of polyA RNA by polymerase chain reaction (PCR) (Saiki et al., 1988) using a pair of sense and anti-sense oligonucleotide primers which are compatible to highly conserved coding nucleotide regions of known related isozymes; and b) subcloning of the PCR products and further characterization of candidate cDNA clones. Clone(s) with unique nucleotide sequence(s), which differs from that of known isozymes, may be obtained by the procedures.

Within the related isozymes from the same and different mammalian species, certain molecular regions, which are implicated in structural and/or functional importance, are particularly well conserved. In the case of ADH isozymes, exon 6 is highly conserved in all five human ADH subunits, and also in corresponding ADH isozymes of other mammals. In the case of ALDH isozymes, exon 7 is highly homologous, i.e. ~90% positional identity in this region versus 68% identity in their total coding sequences (Hsu et al., 1988; Hsu et al., 1989a). The conserved exon 7 sequence is also observed in the corresponding bovine and rat ALDH (Farrés et al., 1989). Complete or near complete identity exists within short (~20 bp) nucleotide stretches of a group of isozymes. Therefore, it is not difficult to select adequate cDNA probes and adequate primers based on the available protein and nucleotide sequence data of known isozymes of human and other mammals.

NEW ADH GENE

A cosmid genomic library (~40 Kb inserts) was screened with a full length ßADH cDNA probe, which includes the highly conserved exon 6 sequence. Several clones which exhibited relatively weak hybridization signals were picked up. One such clone (CosADH-96, insert size 38 Kb) contained 5.8 Kb BamHI-BamHI fragment which was hybridization positive. Partial nucleotide sequencing revealed that the 0.8 Kb XbaI fragment of the clone was similar, but not identical, to the exon 6 of the known ADH isozymes.

By screening another genomic DNA library (λDash vector, insert size ~15 Kb) using the XbaI fragment as a probe, several positive clones were obtained. Among them, λADH-61 and λADH-961 were overlapped at their 3'- and 5'-regions and contained the regions homologous to exon 6 and exon 5 of the known ADH isozymes.

Based on the sequence of these two clones, two 5'-side primers (20 mer and 21 mer) and a 3'-side primer (20 mer) with unique sequences were prepared, and human liver polyA$^+$ RNA was amplified by PCR method. The amplified 5' product and 3' product were sub-cloned in Bluescript KS vector and SK vector, respectively. A 5'-clone, 450 bp, was found to have an initiation codon and an upstream nonsense codon, and a 3'-clone, 950 bp, had a termination signal followed by a polyA tail. The two clones overlap at KpnI site as originally designed, and the two together can encode 368 amino acid residues. Using these two cDNA clones as probes, the genomic structure of the new ADH gene (designated as $\underline{ADH_6}$) was elucidated. The $\underline{ADH_6}$ gene consists of 8 coding exons interrupted by 7 introns, and spans about 15 Kb. By contrast to the three class I ADH genes ($\underline{ADH_1}$, $\underline{ADH_2}$, and $\underline{ADH_3}$), which have 9 exons and encode 375 amino acid residues, the $\underline{ADH_6}$ gene lacks exon 9 and encodes 368 amino acid residues. The exon 8 of $\underline{ADH_6}$ gene contains a polyadenylation signal, and the cDNA cloned a polyA tail. Therefore, the cDNA and the gene cloned are "full" not partial. However, an existence of an additional exon, and mRNA with extended coding sequence, which could be produced by an alternative splicing of primary transcripts, is not excluded.

Compared to the known five ADH subunits, the subunit encoded by the $\underline{ADH_6}$ gene is moderately similar (60~70% positional identity) to all of them, but not highly homologous to any one of these known ADHs (Yoshida et al., 1990). Thus, the ADH_6 does not belong to the class I, class II or class III ADH isozymes, i.e. the $\underline{ADH_6}$ gene is for an unique new class of ADH isozyme.

Northern blot hybridization, using a specific oligonucleotide probe, indicated that the $\underline{ADH_6}$ is far more strongly expressed in the stomach than in the liver. Whether or not the ADH_6 corresponds to the sex-dependent stomach ADH isozyme remains to be elucidated. Enzymatic properties of ADH_6 can be readily examined by expressing ADH_6 mRNA in the E.coli expression system. The physiological role of this isozyme may be discussed thereafter.

NEW ALDH GENE

A cosmid human genomic library was screened using a 29-mer oligonucleotide which partially matched a conserved region of the $\underline{ALDH_1}$ and $\underline{ALDH_2}$ under a low stringent hybridization condition. Positive clones thus isolated were hybridized with the $ALDH_1$ cDNA and $ALDH_2$ cDNA probes under a highly stringent hybridization condition, and strongly positive clones were rejected. The remaining seven clones were subjected to restriction mapping analysis. One of them (clone ALDH-98) containing an insert of 33.1 Kb was analyzed in detail. ALDH-98 contained a short, 9 bp, untranslated 5' sequence, a chain initiation codon followed by an uninterrupted open reading frame for 517 amino acid residues, a chain termination signal and two polyadenylation signals. The degree of resemblance between the deduced amino acid sequence of ALDH-98 clone and $ALDH_2$ is 72.5% (alignment of 517 residues), and that between the clone and $ALDH_1$ is 67.8% (alignment of 500 residues). Thus the gene cloned is designated as $\underline{ALDH_x}$. The $\underline{ALDH_x}$ gene encodes 517 amino acid residues, and the first 17 residues are compatible to the consensus signal sequence of other mitochondrial enzymes, i.e. the $\underline{ALDH_x}$ is for a new mitochondrial ALDH isozyme (Hsu et al., 1989b; Yoshida et al., 1990).

We examined expression of the $\underline{ALDH_x}$ gene in various human tissues by Northern blot hybridization of polyA$^+$ RNA prepared from the liver, brain and testis using a part of coding region (1.1 Kb SmaI fragment) of $\underline{ALDH_x}$ as a probe. A major 3.0 Kb and a minor 2.1 Kb bands were observed in the liver and more strongly in the testis. The major band did not hybridize with $ALDH_1$ or $ALDH_2$ probes, i.e. it is mRNA for the $\underline{ALDH_x}$. The minor 2.1 Kb band could either be superimposed with, or originate from cross-hybridization to $ALDH_2$. The existence of corresponding mRNA in the tissues indicates that the $\underline{ALDH_x}$ gene without an intron in its coding region is functional. The gene for human testis-specific phosphoglycerate kinase was the first intronless functional gene found in eukaryotes (McCarrey and Thomas, 1987). Several functional mammalian genes (various receptor genes, interferon genes, protamine genes, etc.) also have no intron in their coding regions.

A human testis cDNA library constructed in λgt11 was screened with a 1.1 Kb SmaI fragment of $\underline{ALDH_x}$ gene. Among eleven randomly selected independent clones, clone 69 had the longest insert of 2.6 Kb, and contained the longest extended 5' untranslated sequences. Clone 69 contained a chain initiation codon, entire coding sequence, a chain termination signal and a polyadenylation signal. Some of the other clones had longer 3' untranslated sequences containing two polyadenylation signals located at 130 nt and 1329 nt downstream from the chain termination codon.

The cDNA and genomic clone-98 were aligned to locate the exon-intron junctions. The 3'-untranslated region does not contain introns. However, the 5'-untranslated region of $\underline{ALDH_x}$ is interrupted by an intron of approximately 2.6 Kb. The sequences of intron-exon junction of $\underline{ALDH_x}$ is compatible to consensus splicing signals.

When the complete nucleotide sequences of the cDNA clone λ-69 and that of the coding exon of genomic clone-98 were compared, discrepancies were found in three nucleotide positions, i.e. C↔T at 183, C↔T at 257 and T↔G at 320, counting from the chain initiation adenine. The nucleotide changes create amino acid changes Ala↔Val at position 86 and Leu↔Arg at position 107 from the NH_2-terminal. The results indicate the existence of genetic polymorphism at the $\underline{ALDH_x}$ locus. In order to further confirm the polymorphism, genomic DNA samples prepared from several unrelated individuals were analyzed. The targeted region, which includes the three substitution sites, was amplified by PCR method using a pair of primers, and the PCR products were subcloned and sequenced. Seven samples were found to be -C-C-T- type and one was -T-T-G- type. Since the base changes result in producing two amino acid substitutions, the two types of allele synthesize two types of enzyme which may have different kinetic properties. The $\underline{ALDH_x}$ polymorphism may have some biological significance particularly in the testis. The possibility can be tested by examining the properties of the two types of enzymes produced in an adequate expression system by the two types of $ALDH_x$ mRNA.

The $\underline{ALDH_x}$ gene was assigned to chromosome 9 by Southern blot hybridization of DNA samples prepared from a panel of rodent-human hybrid cell lines.

More recently, we obtained other ALDH clones by the alternative cloning method, i.e. amplification of polyA$^+$ RNA by PCR. One of them had nucleotide sequences which are highly homologous to the mouse stomach ALDH, and thus it may be for the human stomach $ALDH_{3b}$ isozyme. Genes and cDNAs for new ADH and ALDH isozymes may be cloned and characterized by similar approaches. Exploration of isozymes expressed in brain tissues may have a particular importance in the study of alcoholism.

ACKNOWLEDGEMENTS

This work was supported by the U.S. Public Health Service Grant AA05763.

REFERENCES

Farrés, J., Guan, K., and Weiner, H., 1989, Primary structures of rat and bovine liver mitochondrial aldehyde dehydrogenase deduced from cDNA sequences, **Eur.J.Bioch.,** 180:67.

Frezza, M., di Padova, C., Possato, G., Terpin, M., Baraona, E., and Lieber, C. S., 1990, High blood alcohol levels in women: The role of decreased gastric alcohol dehydrogenase activity and first-pass metabolism, **New Eng.J.Med.,** 322:95.

Hsu, L. C., Tani, K., Fujiyoshi, T., Kurachi, K., and Yoshida, A., 1985, Cloning of cDNAs for human aldehyde dehydrogenase 1 and 2, **Proc. Natl.Acad.Sci.U.S.A.,** 82:3771.

Hsu, L. C., Bendel, R. E., and Yoshida, A., 1988, Genomic structure of the human mitochondrial aldehyde dehydrogenase gene, **Genomics**, 2:57.

Hsu, L. C., Chang, W-C., and Yoshida, A., 1989a, Genomic structure of the human cytosolic aldehyde dehydrogenase gene, **Genomics**, 5:857.

Hsu, L. C., Chang, W-C., and Yoshida, A., 1989b, Cloning of a new human aldehyde dehydrogenase gene and comparison with liver cytosolic $ALDH_1$ and mitochondrial $ALDH_2$ genes, **Am.J.Hum.Genet.,** 45:A196.

Ikuta, T., Szeto, S., and Yoshida, A., 1986, Three human alcohol dehydrogenase subunits: cDNA structure and molecular and evolutionary divergence, **Proc.Natl.Acad.Sci.U.S.A.,** 83:634.

Jörnvall, H., Höög, J-O., von Bahr-Lindström, H., Johansson, J., Kaizer, R., and Persson, B., 1987, Alcohol dehydrogenases and aldehyde dehydrogenases, **Bioch.Soc.Trans.,** 16:223.

Keung, W-M., 1988, A genuine organ specific alcohol dehydrogenase from hamster testes: Isolation, characterization and developmental changes, **Bioch.Biophys.Res.Comm.,** 156:38.

McCarrey, J. R., and Thomas, K., 1987, Human testis-specific PGK gene lacks introns and possesses characteristics of a processed gene, **Nature**, 326:501.

Saiki, R. K., Gelfand, D. H., Stoffel, S., Scharf, S. J., Higuchi, R., Horn, G. T., Mullis, K. B., and Erlich, H. A., 1988, Primer-directed enzymatic amplification of DNA with a thermostable DNA polymerase, **Science**, 239:487.

Yoshida, A., Hsu, L. C., and Yasunami, M., 1990, Genetics of human alcohol-metabolizing enzymes, **in**: "Progress in Nucleic Acid Research and Molecular Biology," W. Cohn and K. Moldave, eds., Academic Press, Orlando, in press.

PHYSIOLOGICAL ROLE OF ALDEHYDE DEHYDROGENASE (EC 1.2.1.3)

Regina Pietruszko, Gloria Kurys and Wojciech Ambroziak

Center of Alcohol Studies
Rutgers University
Piscataway, NJ 08855-0969 USA

INTRODUCTION

In living organisms, aldehydes occur as common metabolites; some are metabolized by substrate-specific enzymes, others by enzymes with broader substrate specificity. Both specific and non-specific enzymes are active with acetaldehyde as substrate (Pietruszko, 1989). Aldehyde dehydrogenase (EC 1.2.1.3) from human liver and brain as well as some specific enzymes including glutamic-γ-semialdehyde dehydrogenase (EC 1.5.1.12) from human liver (Forte-McRobbie and Pietruszko, 1986) and glyceraldehyde-3-phosphate dehydrogenase (Ryzlak and Pietruszko, 1988a) (EC 1.2.1.12) and succinic semialdehyde dehydrogenase (Ryzlak and Pietruszko, 1988b) (EC 1.2.1.24) from human brain have been recently purified to homogeneity in our laboratory. Aldehyde dehydrogenase (EC 1.2.1.3) which metabolizes ethanol-derived acetaldehyde has a broad substrate specificity which includes a variety of aldehyde structures (Pietruszko, 1989) Among physiologically occurring aldehydes, biogenic aldehydes derived from dopamine, serotonin and norepinephrine (MacKerrell et al., 1986) as well as aldehyde metabolites of corticosteroids (Martin and Monder, 1978; Monder et al., 1982) are substrates for this enzyme. The Michaelis constants are low micromolar values, suggesting that the enzyme may be important in biogenic amine and corticosteroid metabolism. More recently it has been established that γ-aminobutyraldehyde, is also a substrate (Ambroziak and Pietruszko, 1987; Kurys et al., 1989). All the above compounds are important physiological intermediates: biogenic aldehydes arise from biogenic amines which are hormones and/or neurotransmitters and aldehydes themselves are thought to have some hormonal activity (Tipton et al., 1977). γ-Aminobutyraldehyde is an oxidative metabolite of an amine, putrescine, which via aldehyde dehydrogenase, is converted to γ-aminobutyric acid, a known inhibitory neurotransmitter. Putrescine itself is an important intermediate in the synthesis of the polyamines spermidine and spermine which are essential for development and differentiation (Tabor and Tabor, 1984).

METABOLISM OF γ-AMINOBUTYRALDEHYDE

It has been known for some time that putrescine can be readily metabolized to γ-aminobutyric acid outside the central nervous system, suggesting that peripheral γ-

Edited by T.N. Palmer, Plenum Press, New York, 1991

aminobutyric acid may be involved in neuromodulation of local hormones and neurotransmitters. It has been also known that aldehyde dehydrogenase functioned in this metabolic pathway but the identity of the enzyme was unknown. Only recently was the enzyme purified to homogeneity from human liver and characterized (Kurys et al., 1989). This was quickly followed in another laboratory by the purification and characterization of a similar enzyme from rat brain (Abe et al., 1990). What was extremely interesting was that the enzyme metabolizing γ-aminobutyraldehyde that we purified did not appear to be a separate enzyme but an isozyme of aldehyde dehydrogenase (EC 1.2.1.3).

The presence of this isozyme was never previously observed because it's major component superimposed with the E1 isozyme on isoelectric focusing on agarose gels routinely employed for aldehyde dehydrogenase isozyme identification. In both the pI value and its Km value with acetaldehyde (see Table 1 and Figure 1), the enzyme resembled the cytoplasmic E1 isozyme. Its low Km value with γ-aminobutyraldehyde relative to that of E1 and E2 isozymes, which was recognized before purification, permitted purification which could not be otherwise accomplished. Thus, γ-aminobutyraldehyde at a low concentration could be routinely used for enzyme detection in crude liver homogenates and during various purification steps. Up to the chromatography on 5'-AMP Sepharose 4B the enzyme was found to copurify with the E1 and E2 isozymes; its separation occurred on the 5'-AMP Sepharose affinity column which absorbed two other isozymes and allowed E3 isozyme to pass through. Bound coenzyme (see Table 1) appears to be the reason why this isozyme does not bind to the 5'-AMP Sepharose affinity column. Three additional columns were necessary to purify E3 isozyme to homogeneity.

ALDEHYDE DEHYDROGENASE (EC 1.2.1.3): COMPARISON OF ISOZYMES

Similarities between E1, E2 and E3 isozymes are listed in part 1 of Table 1. Their molecular properties appear to be identical. Their amino acid composition is similar; also E3 interacts with antibodies developed in the rabbit to E1 and E2 isozymes. All three isozymes have low (μM) Km values with acetaldehyde and thus have a potential to participate in metabolism of ethanol-derived acetaldehyde. All three isozymes have very similar Km values with imidazoleacetaldehyde and 3,4-dihydroxyphenylacetaldehyde and are active with γ-aminobutyraldehyde and aldehyde metabolites of the polyamines spermidine and spermine.

Properties that, at the present stage of investigation, distinguish the isozymes are listed in part 2 of Table 1. The bound coenzyme is the major distinguishing characteristic in which the E3 isozyme differs markedly from both E1 and E2 isozymes. Although all three isozymes are active with γ-aminobutyraldehyde, the Km value of the E3 isozyme is considerably smaller than that of E1 or E2 isozymes. Magnesium has no effect on catalytic activity of this isozyme: it inhibits the activity of the cytoplasmic E1 isozyme and activates the mitochondrial E2 isozyme. In the isoelectric point of its major component and Km for acetaldehyde, the E3 isozyme resembles the cytoplasmic E1 isozyme; in its insensitivity to disulfiram it resembles the mitochondrial E2 isozyme.

BIOGENIC ALDEHYDES ARE NOT SUBSTRATES FOR OTHER ALDEHYDE DEHYDROGENASES

Other aldehyde dehydrogenases isolated from human tissues in our laboratory were found to be inactive with 3,4 dihydroxyphenylacetaldehyde as substrate. Glutamic-γ-semialdehyde dehydrogenase from human liver (Forte-McRobbie and Pietruszko, 1986) was also tested for activity with γ-aminobutyraldehyde (Ambroziak and Pietruszko, unpublished results) and found to be inactive with this substrate as well. Thus, these experiments suggested that activity with biogenic aldehydes and neurotransmitters may be the property which is exclusive to aldehyde dehydrogenase (EC 1.2.1.3).

Table 1. Molecular and Catalytic Properties of Human Aldehyde Dehydrogenase Isozymes E1, E2 and E3

	Isozymes		
Property	E1	E2	E3
SIMILARITIES BETWEEN ISOZYMES			
Native MW (daltons)	240,000	220,000	250,000
Subunit MW (daltons)	54,000	54,000	54,000
Amino Acid Composition	Similar	Similar	Similar
Rabbit Anti-E1 Interaction	+	+	+
Rabbit Anti-E2 Interaction	+	+	+
Km 3,4-dihydroxyphenylacetaldehyde (μM)	0.4	1	1
Km imidazoleacetaldehyde (μM)	40	35	60
Activity with γ-aminobutyraldehyde	+	+	+
μM Km with acetaldehyde	+	+	+
Activity with aldehyde metabolites of spermidine and spermine	+	+	+
DIFFERENCES BETWEEN ISOZYMES			
Bound coenzyme	-	-	+
Effect of Mg^{++}	Inhibition	Activation	No Effect
Disulfiram	Sensitive	Partly Sensitive	Partly Sensitive
Km γ-aminobutyraldehyde (μM)	760	512	14
Km acetaldehyde (μM)	50	1	60
Km NAD (μM)	10	70	15
Isoelectric Point (pI)	5.3	5.0	5.3, 5.45*

*E3 isozyme has been isolated as two components: pI 5.3 for the major component and pI 5.45 for the minor component (Kurys et al., 1989). Data taken from MacKerell et al., 1986, Ambroziak and Pietruszko, 1987, Kurys et al., 1989, Greenfield and Pietruszko, 1977, and Vallari and Pietruszko, 1981.

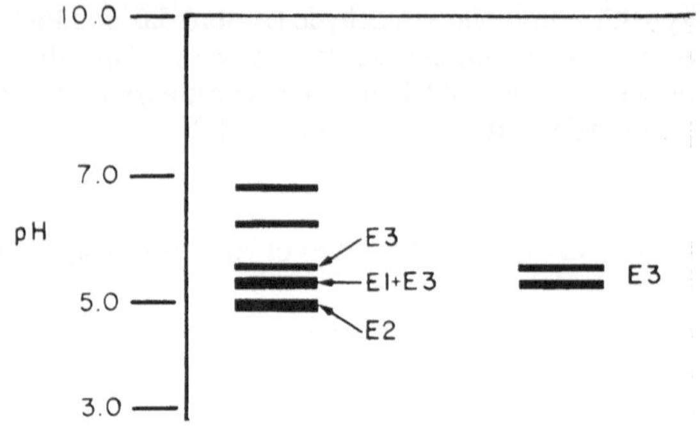

Fig.1. Isoelectric focusing pattern of aldehyde dehydrogenase in human liver homogenate and of purified E3 isozyme. Left hand side pattern - aldehyde dehydrogenase bands visualized in human liver homogenate with 20 mM acetaldehyde. The band with pI of approximately 6.8 represents glutamic-γ-semialdehyde dehydrogenase; the pI 6.4 band has not yet been fully characterized. Only pI 5.0, 5.3 and 5.45 bands are visualized at low concentrations (50 - 100 mM) acetaldehyde. Right hand side pattern - homogeneous E3 isozyme.

Another way of testing this possibility was via gel staining. The isoelectric focusing pattern of aldehyde dehydrogenases visualized on gels of human liver homogenates developed with 20 mM acetaldehyde is shown in Figure 1. When low concentrations of acetaldehyde (50-100 μM) were employed, only three bands were visualized which were localized in the pI 5.0 - 5.5 area. In order to find out if any of the gel areas other than those containing the E1 and E2 isozymes interacted with E1 and E2 antibodies, the isoelectric focusing gels were overlaid with anti-E1 and anti-E2 antibodies and examined for cross-reactivity. The cross-reactivity with both antibodies was seen only in the pI 5.0 - 5.5 gel area. In addition, the gels were developed with several substrates. These substrates included 50 μM propionaldehyde, 50 μM 3,4-dihydroxyphenylacetaldehyde, 50 μM 5-hydroxyindoleacetaldehyde and 100 μM γ-aminobutyraldehyde. In all cases, the same bands as those seen with 50 - 100 mM acetaldehyde were visualized. No other bands were detected in any other region of the gel, suggesting that aldehyde dehydrogenase (EC 1.2.1.3) was the only enzyme reacting with these substrates.

DISCUSSION

Aldehyde dehydrogenase (EC 1.2.1.3) is universally distributed in mammalian organisms. The presence of cytoplasmic and mitochondrial isozymes has been reported in a variety of species (Pietruszko, 1989). In man, absence of the mitochondrial isozyme associated with the aversive flush reaction to alcohol has been reported in Oriental individuals (Agarwal et al., 1981). However, although absence of the mitochondrial isozyme in Orientals, and absence of the cytoplasmic isozyme in one individual has been reported, there has not been a single report of the absence of both isozymes, suggesting that the enzyme may be essential for survival.

Because of its wide substrate specificity, aldehyde dehydrogenase (EC 1.2.1.3) is often regarded as a general, non-specific enzyme which functions in removal of various aldehydes ingested in foodstuffs. Our recent purification and characterization of the third isozyme (Kurys et al., 1989) and its identification as γ-aminobutyraldehyde dehydrogenase suggest that this view may be incorrect. Recent studies in insects (Prestwich, 1987), where aldehyde dehydrogenase functions in pheromone inactivation, essential in sex recognition and survival also argue against such a role. The likelihood of involvement of aldehyde dehydrogenase in the metabolism of putrescine has been an unexpected and exciting finding. Unlike other reactions catalyzed by aldehyde dehydrogenase, where chemically reactive or hormonally active aldehyde is converted to relatively unreactive or inactive acid, γ-aminobutyraldehyde (a metabolite of putrescine) is converted to γ-aminobutyric acid which is a neurotransmitter. Since all three isozymes dehydrogenate biogenic aldehydes and there appear to be no other enzymes in the human liver that perform this function, it is proposed that aldehyde dehydrogenase (EC 1.2.1.3) is a biogenic aldehyde dehydrogenase whose physiological role is in metabolism of biogenic aldehydes and neurotransmitters.

Since aldehyde dehydrogenase (EC 1.2.1.3) is also involved in the metabolism of ethanol-derived acetaldehyde, the possibility of interference of ethanol metabolism with the metabolism of hormones has to be considered. In the presence of acetaldehyde, levels of natural metabolites could considerably increase. Increased levels of neurotransmitters or hormones could produce a variety of undesirable messages to the organism which could result in a large variety of complications affecting not only liver but also other organs. It is well established that alcohol can cause fetal abnormalities documented in man and animals. Thus, interaction of ethanol metabolism with that of putrescine and polyamines could be of particular importance; the putrescine cycle is especially active in developing organisms.

REFERENCES

Abe, T., Takada, K., Ohkawa, K., and Matsuda, M., 1990, Purification and characterization of a rat brain aldehyde dehydrogenase able to metabolize γ-aminobutyraldehyde to γ-aminobutyric acid, **Biochem. J.** 269:25.

Agarwal, D.P., Harada, S. and Goedde, H.W., 1981, Racial differences in biological sensitivity to ethanol: The role of alcohol dehydrogenase and aldehyde dehydrogenase isozymes, **Alcohol. Clin. Exp. Res.** 5:12.

Ambroziak, W. and Pietruszko, R., unpublished results.

Ambroziak, W. and Pietruszko, R., 1987, Human aldehyde dehydrogenase: Metabolism of putrescine and histamine, **Alcohol. Clin. Exp. Res.** 11:528.

Forte-McRobbie, C.M. and Pietruszko, R., 1986, Purification and characterization of human liver "high Km" aldehyde dehydrogenase and its identification as glutamic-γ-semialdehyde dehydrogenase, **J. Biol. Chem.** 261:2154.

Greenfield, N.J. and Pietruszko, R., 1977, Two aldehyde dehydrogenases from human liver. Isolation via affinity chromatography and characterization of the isozymes, **Biochem. Biophys. Acta** 483:35.

Kurys, G. Ambroziak, W. and Pietruszko, R., 1989, Human aldehyde dehydro-genase. Purification and characterization of a third isozyme with low Km for γ-aminobutyraldehyde, **J. Biol. Chem.** 264:4715.

MacKerell, A.D. Jr., Blatter, E.E. and Pietruszko, R., 1986, Human aldehyde dehydrogenase: Kinetic identification of the isozyme for which biogenic aldehydes and acetaldehyde compete, **Alcohol. Clin. Exp. Res.** 10:266.

Martin, K.O. and Monder, C., 1978, Oxidation of steroids with the 20β-hydroxy-21-oxo side chain to 20β-hydroxy-21-oic acids by horse liver aldehyde dehydrogenases, **J. Steroid Biochem.** 9:1233.

Monder, C., Purkaystha, R. and Pietruszko, R., 1982, Oxidation of 17-aldo (20β-hydroxy-21-aldehyde) intermediates of corticosteroid metabolism to hydroxy acids by

homogeneous human liver aldehyde dehydrogenase, **J. Steroid Biochem.** 17:41.
Pietruszko R., 1989, Aldehyde dehydrogenase (EC 1.2.1.3), **in**: "Biochemistry and Physiology of Substance Abuse, Vol. I," CRC Press, Boca Raton, Florida, pp 89-127.
Prestwich, G.D., 1987, Chemistry of pheromone and hormone metabolism in insects, **Science** 237:999.
Ryzlak, M.T. and Pietruszko, R., 1987, Purification and characterization of aldehyde dehydrogenase from human brain, **Arch. Biochem. Biophys.** 255:409.
Ryzlak, M.T. and Pietruszko, R., 1988a, Heterogeneity of glyceraldehyde-3-phosphate dehydrogenase from human brain, **Biochim. et Biophys. Acta** 954:309.
Ryzlak, M.T. and Pietruszko, R., 1988b, Human brain high Km aldehyde dehydrogenase: Purification and identification as NAD+dependent succinic semialdehyde dehydrogenase, **Arch. Biochem. Biophys.** 266:386.
Ryzlak, M.T. and Pietruszko, R., 1989, Human brain glyceraldehyde-3-phosphate dehydrogenase, succinic semialdehyde dehydrogenase and aldehyde dehydrogenase isozymes: substrate specificity and sensitivity to disulfiram, **Alcohol. Clin. Exp. Res.** 13:755.
Tabor, W.C. and Tabor, H., 1984, Polyamines, **Ann. Rev. Biochem.** 53:749.
Tipton, K.F., Housley, M.D., and Turner, A.J., 1977, Metabolism of aldehydes in brain, **Essays Neurochem. Neuropharmacol.** 1:103.
Vallari, R.C. and Pietruszko, R., 1981, Kinetic mechanism of human cytoplasmic aldehyde dehydrogenase E1, **Arch. Biochem. Biophys.** 212:9.

THE ALCOHOL DEHYDROGENASE SYSTEM IN THE RAT: COMPARISON WITH THE HUMAN ENZYME

M. Dolors Boleda, Pere Julià, Alberto Moreno,
Narcís Saubi and Xavier Parés

Departament de Bioquímica i Biologia Molecular
Universitat Autònoma de Barcelona
08193, Bellaterra, Spain

INTRODUCTION

The rat is the most used animal model in studies on the metabolism and pharmacology of ethanol. The research on rat alcohol dehydrogenase (ADH; EC 1.1.1.1) started early (Markovic et al., 1971; Arslanian et al., 1971) and continued more recently (Crabb et al., 1983; Lad and Leffert, 1983, Mezey and Potter, 1983). These reports focused on the liver form that is most active with ethanol. During the last decade a great amount of information has been available on the human enzyme, which is now recognised to be a complex system of isoenzymes. The human ADH isoenzymes were grouped in three classes (I, II, III) according to their kinetic characteristics (Vallee and Bazzone, 1983). The rat, with only one ADH isoenzyme clearly described, appeared to be a model too simple for ethanol metabolism studies. We reinvestigated the rat enzyme in different organs and reported the existence of three distinct isoenzymes: ADH-3 is the liver enzyme with properties similar to human class I; ADH-2, present in all the organs examined, appeared homologous to human class III; ADH-1, isolated from stomach, exhibited characteristics of class II (Julià et al., 1987). Structural studies of classes I and III supported this classification (Julià et al., 1988). In the present report we summarize our work on the rat ADH isoenzymes and compare the rat and the human enzymes, taking into account the new human ADH that we have recently described in human stomach. We use the nomenclature of classes for the rat isoenzymes to allow a clear comparison with the human ADH.

STARCH GEL ELECTROPHORESIS OF RAT ADH ISOENZYMES

Sprague-Dawley rats have been generally used. No differences in ADH activity and isoenzyme pattern were found in the experiments (distribution in organs) performed with Wistar rats.

Starch gel electrophoresis of rat liver homogenate reveals class I ADH as a group of 3-4 bands in the cathodic part of the gel when it is stained with ethanol or 2-buten-1-ol as substrate (Fig. 1). The different forms of class I have very similar kinetic characteristics

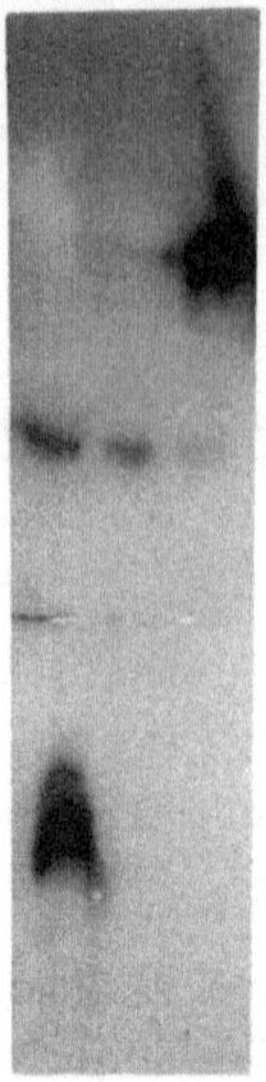

Fig. 1. Starch gel electrophoresis of rat tissue homogenates. The gel was stained for activity using 2-buten-1-ol as substrate. Lane 1, liver. Lane 2, brain. Lane 3, stomach.

(Mezey and Potter, 1983) and have been suggested to originated by changes in the oxidation of sulfhydryl groups (Jörnvall, 1973).

Class III ADH is detected in the anodic part of the gel when it is stained with 2-buten-1-ol or with a long chain alcohol as pentanol. Class II ADH is not present in the liver but can be observed as an anodic band in the electrophoresis of stomach homogenate when the gel is stained with 2-buten-1-ol, pentanol or with a high ethanol concentration (250 mM).

PURIFICATION

Classes I and III can be isolated from the same preparation of rat liver homogenate (Julià et al., 1987; Parés and Julià, 1990). The first step consists of chromatography on a DEAE-Sepharose column, equilibrated in 10 mM Tris-HCl, 0.5 mM dithiothreitol, pH 8.3. Class I ADH elutes in the buffer wash, while class III ADH appears in a 0-150 mM NaCl gradient. Class I can be further purified by chromatography on a CapGapp-Sepharose column equilibrated with 50 mM sodium phosphate, 0.5 mM dithiothreitol, 2 mM NAD, pH 7.5. The purified enzyme is eluted with 0.5 M ethanol, and the ethanol is finally removed by ammonium sulfate precipitation and gel filtration of the resuspended material on a Sephadex G-25 column.

The class III enzyme eluted from the DEAE-Sepharose column can be purified to homogeneity by application on an AMP-Sepharose column equilibrated with 10 mM sodium phosphate, 0.5 mM dithiothreitol, pH 7.4. The enzyme elutes in a 0-66 μM NADH gradient.

Homogeneous class II ADH can be obtained from rat stomach homogenate. The first step consists of chromatography on DEAE-Sepharose equilibrated with 10 mM Tris-HCl, pH 7.9. The enzyme is eluted in a 0-150 mM NaCl gradient and applied to an AMP-Sepharose column, equilibrated with 10 mM Tris-HCl, pH 7.9. The enzyme appears in the buffer wash and is loaded again on the same column and eluted under the same conditions. Final purification is obtained by passage through a Sephacryl S-300 column.

MOLECULAR PROPERTIES

SDS-polyacrylamide gel electrophoresis indicates slight but significant differences in the subunit Mr of rat ADH isoenzymes. The estimated Mr values are: 40,000 for class I, 43,000 for class II and 39,000 for class III. Gel filtration experiments show an apparent Mr of 80,000 for the three molecules. Moreover the three isoenzymes contain 4 mol Zn/mol protein. These are common properties of all mammalian ADHs known. The isoelectric point differentiates the rat isoenzymes; isoelectric focusing experiments yield a pI of 8.25-8.4 for class I, 5.1 for class II and 5.95-6.3 for class III.

KINETIC PROPERTIES

The three isoenzymes show a strong preference for NAD(H) over NADP(H). Moreover, pH-activity profiles of the three classes indicate a much more efficient alcohol oxidation above pH 10.0 than at neutral pH. It is because of this property, common to many ADHs, that many kinetic studies have been performed at pH 10.0 or 10.5. Nevertheless, we indicate the kinetic constants for several alcohols and aldehydes at pH 7.5 (Table I), which has more physiological interest. The three isoenzymes exhibit better kinetic constants for long chain alcohols and aldehydes than those for ethanol and acetaldehyde. Retinol, however, is a good substrate for class I, is less efficiently oxidized by class II, and is not a substrate for class III. The kinetic constants also show that only class I has a role in ethanol metabolism at the experimental blood concentrations of this alcohol. The rat ADH classes show much better kinetic constants for aldehyde reduction than those for the oxidation of the corresponding alcohols.

The rat isoenzymes exhibit very different sensitivities towards pyrazole, a classic inhibitor of ADH. Thus, the inhibition constants (Ki) at pH 10.0 are 0.4 μM for class I, 0.56 mM for class II and 79 mM for class III, using ethanol as substrate (Julià et al., 1987). At pH 7.5 the Ki values increase by more than 10 fold (Boleda et al., 1989).

The kinetic properties with ethanol and inhibition constants with pyrazole are the main criteria used for the human ADH isoenzyme classification (Vallee and Bazzone, 1983). On the basis of the same criteria, each rat ADH isoenzyme can be assigned to an ADH class, in analogy with the human enzyme. Thus, the low Km for ethanol and very low Ki for pyrazole clearly distinguish class I. Lack of saturation with ethanol and insensitivity to pyrazole are characteristic of class III. Class II shows intermediate properties. The three isoenzyme classes are also present in baboon, monkey, horse, mouse etc., suggesting a general occurrence in mammals (see Julià et al., 1987 for references).

DISTRIBUTION IN THE RAT ORGANS AND CONTRIBUTION TO ETHANOL METABOLISM

The ADH isoenzyme pattern of the rat organs were determined by starch gel electrophoresis of the tissue homogenates, followed by activity staining of the gel with 2-

buten-1-ol and densitometry analysis (Boleda et al., 1989). Class I is the typical liver enzyme, also present in intestine, lung, kidneys, testes and uterus. Class II is characteristic of the external epithelia and organs in contact with the external environment: skin and cornea; nose, ear and mouth mucosas; digestive, respiratory and sexual tracts. Class II is typically located in the most external mucosas, while class I appears in more internal organs. Thus, in the digestive system, class II shows high activity in mouth, esophagus, stomach and rectum, while class I is present in small intestine, colon and cecum. The respiratory and reproductive tissues of both sexes are also good examples of the selective distribution of ADH isoenzymes in correspondence with the proximity to external environment. Finally, class III is detected in all organs examined. Organs such as brain (Fig. 1), pancreas, spleen, heart and skeletal muscle show only class III.

Alcohol dehydrogenase activity with 33 mM ethanol and pH 7.5, is concentrated in liver, with 3500±640 milliunits per male liver, which represents about 90% of the total activity in the rat. The second organ in importance is the skin, with 88±15 milliunits. Intestine, kidneys, testes, epididymis and stomach exhibit 20 to 60 milliunits per organ, while other organs present smaller activities. We estimate that class I contributes about 96% of the total activity (90% hepatic and 6% extrahepatic), while the contribution of class II (extrahepatic) is only 4%.

Table 1. Kinetic constants of rat ADH isoenzymes at pH 7.5[a]

Substrate	Class I			Class II			Class III		
	Km (mM)	kcat (min^{-1})	kcat/Km ($mM^{-1}min^{-1}$)	Km (mM)	kcat (min^{-1})	kcat/Km ($mM^{-1}min^{-1}$)	Km (mM)	kcat (min^{-1})	kcat/Km ($mM^{-1}min^{-1}$)
Ethanol	1.4	39	28	5000	1000	0.2	n.s.[c]	–	–
1-Octanol	0.06	13	217	0.5	300	600	1.6	4	2.5
12-HDA[b]	0.06	23	380	0.42	240	570	0.04	12	300
Retinol	0.03	12	400	0.02	0.2	10	n.a.[d]	–	–
NAD	0.17			0.2			0.04		
Acetal-dehyde	0.17	170	1000	n.s.[c]	–	–	n.s.[c]	–	–
Octanal	0.05	440	8800	0.3	23400	78000	3.5	250	70
NADH	0.01			0.05			0.0017		

[a]Activities were determined with 0.1 M sodium phosphate, pH 7.5 (Julià et al., 1987).
[b]12-HDA: 12-hydroxydodecanoic acid.
[c]n.s.: no saturation was detected.
[d]n.a.: no activity

Class II exhibits a very large Km towards ethanol but also a very high kcat (Table 1). Therefore, class II activity can be very high at large ethanol concentrations, increasing practically in linear proportionality with this concentration. The class II enzyme of the digestive tract can be in contact with large ethanol concentrations. Ethanol feeding experiments frequently involve liquid diets with 5-10% (0.8-1.7 M) ethanol. At 1 M ethanol, pH 7.5, activity of stomach homogenate is 216±32 milliunits while homogenate from small intestine exhibits 113±37 milliunits. The whole gastrointestinal activity corresponds to about 10% of the total hepatic activity with 33 mM ethanol. In conclusion, when ethanol is orally administered the contribution of extrahepatic ADH is probably much higher than in the case of parenteral injection because of the class II (stomach and intestine) and class I (intestine) activities of the digestive tract.

PHYSIOLOGICAL ROLE

The physiological role of the rat ADH isoenzymes can be estimated from the kinetic properties and their specific tissue distribution. Classes I and II, located in organs with some relation with the external environment, may represent metabolic barriers to external alcohols or aldehydes. Class II of cornea, skin and the epithelia lining the respiratory, digestive and sexual tracts, would represent the most external barrier. In terms of kcat/Km, class II is the most efficient isoenzyme towards many alcohols and aldehydes of medium and long hydrophobic chain, and would be, therefore, effective in the metabolic transformation of such toxic compounds. Class I would represent a second barrier, very effective because of the high enzyme content in liver and its wide substrate specificity. The distribution of class III in all rat organs suggests a different role for this isoenzyme. It has been recently demonstrated that class III is highly specific for the oxidation of formaldehyde bound to glutathione (Koivusalo et al., 1989). In fact, it is now believed that the main physiological function of class III is the detoxification of formaldehyde.

The activity of ADH with alcohols and aldehydes of physiological interest (Table 1; Pietruszko, 1979; Julià et al., 1987) and the wide distribution of ADH isoenzymes in the organism in significant amounts, strongly suggest that the enzyme is involved in the transformation of endogenous compounds. Thus, the good kinetic constants with the ω-hydroxy fatty acids support a role for ADH in the ω-oxidation of fatty acids. Interestingly, class III exhibits a significant activity with these compounds (Table 1; Giri et al., 1989) which suggests that besides its function as a glutathione-dependent formaldehyde dehydrogenase, class III is also an ω-hydroxy fatty acid dehydrogenase.

It is well accepted that liver ADH (class I) is involved in retinol metabolism (Mezey and Holt, 1971; Frolik, 1984). Although class II shows lower activity with retinol than class I (Table 1), its distribution in epithelia also suggests a role in the transformation of retinoids. Studies *in vivo* support the involvement of class II ADH in the oxidation of retinol in mouse skin (Connor and Smit, 1987).

COMPARISON WITH THE HUMAN ENZYME

The isoenzyme pattern of the human liver ADH, obtained by starch gel electrophoresis, is indicated in Fig. 2 (lanes 3 and 4). The kinetic properties of the isolated isoenzymes allowed the definition and identification of three isoenzymatic classes: χ-ADH corresponds to class III, π-ADH is a class II isoenzyme, and the group of most cathodic forms constitutes class I (Vallee and Bazzone, 1983). A remarkable difference between the human and the rat ADH is the complex isoenzymatic pattern of human class I as compared with the single class I isoenzyme in the rat. However, the human class I ADHs are very similar molecules, with sequence identity greater than 90% (Jörnvall et al., 1987) and similar general kinetic features.

The kinetic properties of rat class I (Table 1) indicate a clear homology with human class I isoenzymes. This is confirmed by the structural studies that show a 82% identity with the human enzyme ($\beta_1\beta_1$ isoenzyme) (Julià et al., 1988), and the conservation in the rat enzyme of all residues typical of class I.

There is no doubt of the homology between human χ-ADH and rat class III. The substrate specificity and kinetic constants are practically the same for both enzymes (Table 1; Parés and Vallee, 1981). Moreover, the primary structure of rat class III shows a 94% identity with the χ-ADH amino acid sequence (Julià et al., 1988). Finally, both enzymes are glutathione-dependent formaldehyde dehydrogenases (Koivusalo et al., 1989).

As we have previously indicated (Julià et al., 1987) the only ambiguity in the correspondence between the human and the rat ADHs is in the class II isoenzyme. Certainly π-ADH and rat class II isoenzymes share general class II properties, such as high Km for ethanol oxidation and a medium sensitivity to pyrazole inhibition. The actual kinetic constant values are, however, quite distinct and, importantly, the tissue distribution is completely different. Thus, π-ADH has been described only in liver but rat class II is present in many organs (see above) but not in liver. We have recently reported the existence of a new ADH in human stomach mucosa (Moreno and Parés, 1990). This isoenzyme, named σ-ADH, has a slow cathodic mobility in starch gel electrophoresis which is clearly different from that of the liver isoenzymes (Fig. 2, lanes 1 and 2). The kinetic properties of σ-ADH are typical of a class II isoenzyme. Thus, it exhibits high Km (41 mM) and kcat (280 min^{-1}) values for ethanol at pH 7.5. The inhibition constant for pyrazole (20 μM) is higher than for the class I isoenzymes. At present and in the absence of primary structure studies, σ-ADH appears as a class II ADH. We have to conclude, therefore, that there are, probably, two class II forms in human ADH, the liver type and the stomach type. The liver type corresponds to π-ADH, and does not have an homologous isoenzyme in the rat.

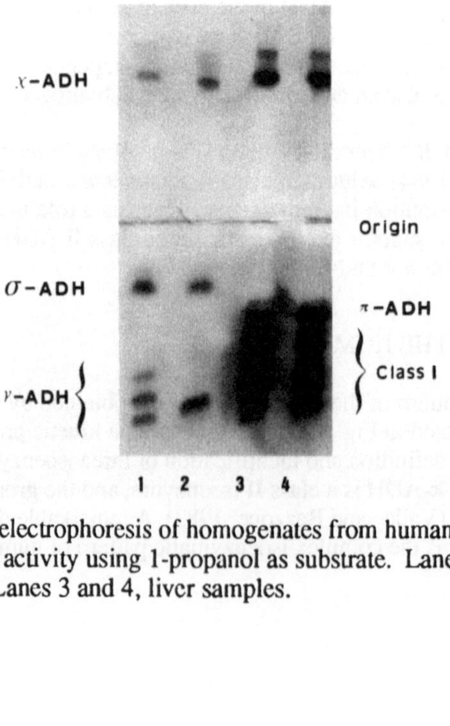

Fig 2. Starch gel electrophoresis of homogenates from human tissues. The gel was stained for activity using 1-propanol as substrate. Lanes 1 and 2, stomach samples. Lanes 3 and 4, liver samples.

The stomach type corresponds to σ-ADH which would be the homologous isoenzyme of rat class II. An additional property that supports this homology is the tissue distribution. Thus, it has been recently reported that σ-ADH is also present in human cornea (Holmes, 1989), a tissue that is also rich in class II ADH in the rat species.

In conclusion, the rat ADH system is more similar to the human enzyme than previously thought. In both species alcohol dehydrogenase exhibits three isoenzyme classes, with many similarities in tissue distribution (although more studies are required with the human tissues). However, two important differences have to be considered when comparing the enzymatic systems of alcohol oxidation of both species: the isoenzyme complexity and polymorphism of the human class I, and the absence of the liver class II type in the rat.

NOTE ADDED IN PROOF

The recent report on the partial amino acid sequence of the rat stomach ADH, here referred as class II, demonstrates an important variation (32-40% residue differences) from the structures of the three classes previously characterized (Parés et al., 1990). Consequently, the rat stomach ADH constitutes a new alcohol dehydrogenase class (class IV).

ACKNOWLEDGEMENTS

Supported by grant PB86-156 of Dirección General de Investigación Científica y Técnica (Spain).

REFERENCES

Arsalanian, M.J., Pascoe, E. and Reinhold, J.G., 1971, Rat liver alcohol dehydrogenase. Purification and properties, **Biochem. J.,** 125:1039.

Boleda, M.D., Julià, P., Moreno, A., and Parés, X.,1989, Role of extrahepatic alcohol dehydrogenase in rat ethanol metabolism, **Arch. Biochem. Biophys.** 274:74.

Connor, M.J. and Smit, M.H., 1987, Terminal-group oxidation by mouse epidermis. Inhibition *in vitro* and *in vivo*, **Biochem. J.,** 244:489.

Crabb, W.D., Bosron, W.F. and Li, T.-K., 1983, Steady state kinetic properties of purified rat liver alcohol dehydrogenase: Application to predicting alcohol elimination rates *in vivo*. **Arch. Biochem. Biophys.** 224: 299.

Frolik, C.A., 1984, Metabolism of retinoids, **in**: "The retinoids," Vol.2, M.B. Sporn, A.B. Roberts, and Goodman, D.S., eds. Academic Press, Orlando.

Giri, P.R., Linnoila, M., O'Neill, J.B. and Goldman, D., 1989, Distribution and possible metabolic role of class III alcohol dehydrogenase in the human brain. **Brain Res.,** 481:131.

Holmes, R.S., 1988, Alcohol dehydrogenases and aldehyde dehydrogenases of anterior eye tissues from humans and other mammals, **in**: "Biomedical and social aspects of alcohol and alcoholism,"K. Kuriyama, A. Takada and H. Ishii, eds., Elsevier, Amsterdam.

Jörnvall, H., 1973, Differences in thiol groups and multiple forms of rat liver alcohol dehydrogenase, **Biochem. Biophys. Res. Commun.** 53:1096.

Jörnvall, H., Hempel, J., and Vallee, B.L., 1987, Structures of human alcohol and aldehyde dehydrogenases **Enzyme**, 37:5.

Julià, P., Farrés, J., and Parés, X., 1987, Characterization of three isoenzymes of rat alcohol dehydrogenase. Tissue distribution and physical and enzymatic properties., **Eur. J. Biochem.,**162:179.

Julià, P., Parés, X. and Jörnvall, H., 1988, Rat liver alcohol dehydrogenase of class III. Primary structure, functional consequences and relationships to other dehydrogenases., **Eur. J. Biochem.**, 172:73.

Koivusalo, M., Baumann, M., and Uotila, L., 1989, Evidence for the identity of glutathione-dependent formaldehyde dehydrogenase and class III alcohol dehydrogenase, **FEBS Let.**, 257:105.

Lad, J.P., Leffert, H.L., 1983, Rat liver alcohol dehydrogenase. I. Purification and characterization, **Anal. Biochem.** 133:350

Markovic, O. Theorell, H., and Rao, S., 1971, Rat liver alcohol dehydrogenase. Purification and properties, **Acta Chem. Scand.** 25:195.

Mezey, E. and Holt, P.R., 1971, The inhibitory effect of ethanol on retinol oxidation by human liver and cattle retina, **Exp. Mol. Pathol.**, 15:148.

Mezey, E., Potter, J.J., 1983, Separation and partial characterization of multiple forms of rat liver alcohol dehydrogenase, **Arch. Biochem. Biophys.**, 225:787.

Moreno, A. and Parés, X., 1990, Purification and characterization of a new alcohol dehydrogenase from human stomach, **J. Biol. Chem.**, 266:1128.

Parés, X. and Vallee, B.L., 1981, New human liver alcohol dehydrogenase forms with unique kinetic characteristics, **Biochem. Biophys. Res. Commun.**, 119:1047.

Parés, X. and Julià, P., 1990, Isoenzymes of alcohol dehydrogenase in retinoid metabolism, **Methods Enzymol.**, 189:436.

Parés, X., Moreno, A., Cederlund, E., Höög, J-O. and Jörnvall, A., 1990, Class IV Mammalian alcohol dehydrogenase. Structural data of the rat stomach enzyme reveal a new class well separated from those already characterized, **FEBS Let.**, 277:115.

Pietruszko, R., 1979, Nonethanol substrates of alcohol dehydrogenase, **in**: "Biochemistry and pharmacology of ethanol," Vol. 1, E. Majchrowicz and E.P. Noble, eds., Plenum Press, New York.

Vallee, BL. and Bazzone, T.J., 1983, Isoenzymes of human liver alcohol dehydrogenase, **Isozymes, Curr. Top. Biol. Med. Res.**, 8:219.

CHANGES IN THE INDUCIBILITY OF A HEPATIC ALDEHYDE DEHYDROGENASE

P. Pappas, V. Vasiliou, M. Karageorgou and M. Marselos

Department of Pharmacology
Medical School, University of Ioannina
GR 451 10 Ioannina, Greece

INTRODUCTION

Rat liver contains at least two cytosolic aldehyde dehydrogenases (ALDH 1.2.1.3) which are inducible by inducers of drug metabolism (Deitrich, 1971; Marselos and Hänninen, 1974; Deitrich et al., 1977).Phenobarbital (PB) type inducers induce the φ-isozyme in certain rat strains with a genetic predisposition (RR) (Deitrich, 1971; Deitrich et al., 1972; Marselos, 1976; Nakanishi et al., 1978). Polycyclic aromatic hydrocarbons or other type II inducers of drug metabolism can induce another isozyme (τ-ALDH), an isozyme which has been detected in all rat strains tested to date (Deitrich et al., 1978; Marselos et al., 1979; Vasiliou et al., 1988). An increase in ALDH activity was also found in primary cultures of human or rat hepatocytes, after *in vitro* exposure to PB or methylcholanthrene (Marselos and Michalopoulos, 1986; Marselos et al., 1987).

A great deal of interest has focussed on τ-ALDH, because of its apparent similarity to a tumor-associated ALDH detected in many chemically induced hepatocellular neoplasms (Feinstein and Cameron, 1972; Lindahl and Feinstein, 1976; Ritter and Eriksson 1985). The physiological function of this isozyme and its induction is not fully understood but seems to be influenced by several factors (Marselos et al., 1987; Vasiliou, 1988; Vasiliou and Marselos 1989).

In the present study, we examined the possible role of glutathione depletion on the induction of ALDH by benzo(a)pyrene (BaP) and on the ALDH itself. For depleting glutathione we used phorone (PH), which does not affect the hepatic microsomal enzymes (Younes et al., 1986).

MATERIALS AND METHODS

Chemical All chemicals used were reagent grade. Pyrazole was obtained from Fluka AG (W. Germany) and propionaldehyde from Ferak (W. Germany). The other chemicals were obtained from Sigma Chem. Co, St. Louis, Mo (USA).

Animals Male albino rats (weighing 200-250 g) of the Wistar/Af/Han/Kuo/Io strain

Alcoholism: A Molecular Perspective
Edited by T.N. Palmer, Plenum Press, New York, 1991

were used. These animals have been inbred as two separate substrains (RR and rr), according to their response to PB, as far as the induction of ALDH is concerned (Marselos 1976). In the present experiments we used only rr-animals, which have been shown to lack the φ-isozyme (Marselos et al., 1979).

The rats were kept in plastic cages (Makrolon) with bedding of wood (Populus sp.) and had free access to tap water and pellet chow (Elviz, Greece).

Experiment A. Treatment with phorone Phorone was dissolved in olive oil and was administered to the animals intraperitoneally in a daily dose of 250 mg/Kg for two days. Control rats received olive oil (10 ml/Kg, i.p.); The animals were killed 18 h after the last PH injection.

Experiment B. Cotreatment with PH and BaP Twenty animals were divided into groups of five. The control group received olive oil (10 ml/Kg, i.p.). The first experimental group (PH) was treated with PH for 4 days (50 mg/Kg/day, i.p., in olive oil). The second group (PH-BaP) was treated with PH as above and with a single dose of BaP (20 mg/Kg, i.p., in olive oil) on the first day of the experiment. The third group was given a single dose of BaP (20 mg/Kg, i.p., in olive oil).

Experiment C. Post-treatment with PH Twelve animals were injected for 7 days with BaP (25 mg/Kg/day, i.p., in olive oil). After that, they were divided into two groups. The first group (BaP) was not given any further treatment. The second group (PH) was treated with PH for 3 days (50 mg/Kg/day). All the animals were killed 24 h after the last injection.

Tissue preparation The excised livers were weighed and placed immediately in an ice-cold solution used in the homogenization. Any extra non-hepatic tissue was cut off, and the livers were chopped with scissors into pieces (in solution). The livers were homogenized in a Potter Elvehjem homogenizer with a teflon pestle glass homogenizer in a 3 vol. (v/w) of ice-cold 0.25 M sucrose solution. The homogenate was then centrifuged at 10000 x g for 30 min. The supernatant was centrifuged at 10000 x g for 30 min. The new supernatant was carefully removed for the preparation of the cytosolic fraction, as described below. The remaining pellets (mitochondria) was resuspended in 3 ml of the sucrose medium and centrifuged as above. The final mitochondrial pellets were resuspended in 2 ml of 0.25 M sucrose solution, containing 2% sodium deoxycholate. An equal volume of 0.024 M $CaCl_2$ in 0.25 M sucrose solution was added to the supernatant of the 10000 x g centrifugation (Kamath et al. 1971; Kamath and Narayan 1972). The diluted supernatant was stirred and left to stand in ice for 10 min. The microsomal fraction was sedimented by centrifugation at 10000 x g for 30 min. The final supernatant was considered to correspond to the cytosolic fraction.

Aldehyde dehydrogenase was measured at 37° C, by following the reduction of NAD at 340 nm in a Hitachi 100-80 A spectrophotometer. The reaction cuvette contained sodium pyrophosphate buffer (0.075 M, pH 8.0), 1 mM pyrazole (to inhibit any possible activity of alcohol dehydrogenase), 1 mM NAD and 5 mM propionaldehyde (P/NAD activity). For measuring the B/NADP activity, the substrate was benzaldehyde (5 mM) and the coenzyme was NADP (2.5 mM). In either enzyme assay, the reaction was started by adding the substrate after a 5-min pre-incubation. A blank was run without the substrate (Marselos and Michalopoulos, 1986). The high- and the low-Km ALDH were measured according according to the methods described by Vasiliou et al., 1986.

Protein determination was carried out with the biuret method (Gornall et al. 1949), using bovine serum albumin as the standard.

Statistical analysis of the results was performed by Student's t-test.

RESULTS

Experiment A. Treatment with PH

Acute treatment of Wistar rats with phorone (PH) (250 mg/Kg, i.p., daily, x2) caused a statistically significant inhibition (30%, $p < 0.001$) of the hepatic mitochondrial low-Km P/NAD ALDH activity. The high-Km P/NAD ALDH activity was found to be inhibited by 25% ($p < 0.001$) in the mitochondrial,cytosolic fractions and also by 15% ($p < 0.01$) in the microsomal fraction (Table 1).

Experiment B. Cotreatment with PH and BaP

When phorone was administered in a daily dose of 5Q mg/Kg, i.p. x4, it did not affect cytosolic ALDH activities (Table 2). On the contrary, BaP (20 mg/Kg, i.p. once) increased the ALDH activity measured with benzaldehyde as the substrate and NADP as the coenzyme (B/NADP) by 25-fold. Co-treatment with PH and BaP showed a great inhibition (75%) of the B/NADP ALDH activity compared to the respective controls (Table 2).

Table 1. Effect of Phorone (250 mg/Kg, x2, i.p.) on the hepatic ALDH activities.

Enzyme	Control	Experimental
mitochondria		
Low-Km ALDH	5.5 ± 0.3	3.3 ± 0.4 *
High-Km ALDH	15.8 ± 1.1	12.5 ± 0.9 *
cytosol		
P/NAD	10.7 ± 1.1	8.0 ± 0.7 *
B/NADP	1.4 ± 0.7	2.1 ± 0.6 *
microsomes		
P/NAD	28.0 ± 3.8	23.3 ± 1.2 **
B/NADP	3.0 ± 0.3	3.2 ± 0.7

* Statistically different from control group at $p < 0.001$
** Statistically different from control group at $p < 0.025$

Table 2. Effects of cotreatment with phorone and BaP on the cytosolic ALDH activities.

	ALDH activity	
Groups	**P/NAD**	**B/NADP**
Control	10.2 ± 0.8	2.1 ± 1.0
Phorone	10.1 ± 1.1	1.5 ± 0.6
BaP	12.5 ± 1.2	46.3 ± 6.1 *
Phorone-BaP	11.5 ± 0.8	11.8 ± 1.9 *, †

* Statistically different from control group at $p < 0.001$
† Statistically different from BaP group at $p < 0.001$

Experiment C. *In vivo* effect of PH on the induced ALDH activity

The results were somewhat different when PH was administered to animals already treated with BaP (25 mg/Kg, i.p., daily x8). In this case PH did not affect the already induced ALDH activity

DISCUSSION

Mammalian cells possess protective mechanisms to minimize events that result from toxic chemicals and normal oxidative products of cellular metabolism. A major endogenous

Table 3. Effect of post-treatment with phorone on the induction of ALDH by benzo(a)pyrene.

Groups	**P/NAD**	**B/NADP**
Control	10.1 ± 0.8	3.8 ± 0.7
BaP	168.9 ± 23.9	1117.9 ± 169.3 *
BaP-Phorone	148.7 ± 32.3	1293.6 ± 367.8 *

* Statistically different from control group at $p < 0.001$

protective system is the glutathione redox cycle (Reed, 1986). Glutathione is present in high concentrations as GSH in most mammalian cells (generally in the millimolar range), with minor fractions being GSSG, mixed disulfides of GSH and other cellular thiols, and minor amounts of thioethers (Kosower and Kosower, 1978). GSH acts both as a nucleophilic "scavenger" of numerous compounds and their metabolites, via enzymatic and chemical mechanisms, converting electrophilic centers to thioether bonds, and as a substrate in the GSH peroxidase-mediated destruction of hydroperoxides. GSH depletion to about 20-30% of total glutathione levels can impair the cell's defense against the toxic actions of several compounds and may lead to cell injury and death (Reed and Fariss, 1984; Moldeus and Quanguan, 1987).

The mechanism of τ-ALDH induction by PAHs is not fully understood. Even with large doses of MC, given to the rat *in vivo* (Törrönen et al., 1981), or applied to hepatocyte cultures *in vitro* (Marselos et al., 1987), a lag time of at least 3 days is necessary before the induction of B/NADP activity becomes maximal. It has been previously reported by us that the induction of τ-ALDH is enhanced in rats and also in cultures of HepG2 cells already pretreated with phenobarbital. These observations suggest that the eventual effect of MC on the ALDH activity is mediated through an active metabolite (Vasiliou and Marselos 1989; Marselos et al., 1987). In order to study further this hypothesis, we examined the induction of τ-ALDH in glutathione depleted animals. It is possible that pretreatment with PH decreases the expected high activity of the τ-isozyme, not because of a direct action on the enzyme itself, but rather by affecting the process of induction. Although PH is not an inhibitor of the hepatic microsomal drug metabolizing enzymes (Younes et al. 1986), it may still influence the metabolism of PAH's, either by a non-specific competition for the same enzymes or by leading to deterioration in overall hepatocyte function. The latter alternative seems more probable, because in the absence of glutathione the accumulation of many toxic compounds will disturb the homeostasis of the hepatocytes.

ACKNOWLEDGEMENTS

This work was supported by a grant from the Greek Ministry of Health.

REFERENCES

Deitrich, R.A., 1971, Genetic aspects of increase in rat liver aldehyde dehydrogenase induced by phenobarbital. **Science** 173:334.

Deitrich, R.A., Collins, A.C.and Erwin, V.G., 1972, Genetic influence upon phenobarbital-induced increase in rat liver supernatant aldehyde dehydrogenase activity. **J. Biol. Chem.,** 247:7232-7236.

Deitrich, R.A., Bludeau, P., Stock, T. and Roper, M., 1977, Induction of different rat liver supernatant aldehyde dehydrogenases by phenobarbital and tetrachlorodibenzo-p-dioxin. **J. Biol. Chem.,** 252:6169.

Deitrich, R.A., Bludeau, P., Roper, M. and Schmuck, J., 1978, Induction of aldehyde dehydrogenase. **Biochem. Pharmacol.,** 27:2343.

Feinstein, R.N. and Cameron, E.C., 1972, Aldehyde dehydrogenase activity in rat hepatomas. **Biochem. Biophys. Res. Commun.,** 48:1140.

Gornall, A.G., Bardawill, C.J. and David, M.M., 1949, Determinations of serum proteins by means of the biuret reaction. **J Biol Chem.,** 177:751.

Kamath, S.A., Kummerow, F.A. and Narayan, K.N., 1971, A simple procedure for the isolation of rat liver microsomes. **FEBS Lett.,** 17:90-92.

Kamath, S.A.and Narayan, K.N., 1972, Interaction of Ca with endoplasmic reticulum of rat liver: a standardized procedure for the isolation of rat liver microsomes. **Anal. Biochem.,** 48:59-61.

Kosower, N.S. and Kosower, E.M., 1978, The glutathione status of the cell. **Int.Rev. Cytol.,** 54:109-60.

Lindahl, R.and Feinstein, E.C., 1976, Purification and immunochemical characterization of aldehyde dehydrogenase from 2-acetylaminofluorene-induced rat hepatomas. **Biochem. Biophys. Acta,** 452:345-355.

Marselos, M.and Hänninen, 0., 1974, Enhancement of D-glucuronolactone and aldehyde dehydrogenase activity in the rat liver by inducers of drug metabolism. **Biochem. Pharmacol.,** 24:1457.

Marselos, M., 1976, Genetic variation of drug metabolising enzymes in the Wistar rat. **Acta Pharmacol. et Toxicol.,** 39:186.

Marselos, M., Törrönen, R., Koivula, T. and Koivusalo, M., 1979, Comparison of phenobarbital- and carcinogen- induced aldehyde dehydrogenase in the rat. **Biochem. Biophys. Acta,** 583:110.

Marselos, M. and Michalopoulos, G., 1986, Phenobarbital enhances the aldehyde dehydrogenase activity of rat hepatocyte *in vitro* and *in vivo*. **Acta Pharmacol. et Toxicol.,** 59:405.

Marselos, M., Strom, S. and Michalopoulos, G., 1987, Effect of phenobarbital and 3-methylcholanthrene on aldehyde dehydrogenase activity in cultures of HepG2 and normal human hepatocytes. **Chem. Biol. Interactions,** 62:75.

Moldeus, P. and Quanguan, J., 1987, Importance of the glutathione cycle in drug metabolism, **Pharmacol. Ther.,** 33:37-40.

Nakanichi, S., Shiobara, E. and Tsukada, M., 1978, Rat liver aldehyde dehydrogenases: strain differences in response of the enzymes to phenobarbital treatment. **Jpn. J. Pharmacol.,** 28:653-659.

Reed, D.J. and Fariss, M.W., 1984, Glutathione depletion and susceptibility. **Pharmacol. Rev.,** 36:25S-33S.

Reed, D.J., 1986, Regulation of reductive processes by glutathione. **Biochem. Pharmacol.,** 35:7-13.

Ritter, E. and Eriksson, L.C., 1985, Aldehyde dehydrogenase activities in hepatocyte nodules and hepatocellular carcinomas from Wistar rats. **Carcinogenesis** 6:1683-1687.

Törrönen, R., Nousiainen, U. and Hänninen, 0., 1981, Induction of aldehyde dehydrogenase by polycyclic aromatic hydrocarbons in rats. **Chem. Biol.Interactions,** 36:33.

Younes, M., Sharma, S.C. and Siegers, C.-P., 1986, Glutathione depletion by phorone: Organ specificity and effect on hepatic microsomal mixed-function oxidase system. **Drug Chem. Toxicol.,** 9:67-73.

Vasiliou, V., Malamas, M. and Marselos, M., 1986, The mechanism of alcohol intolerance produced by various therapeutic agents. **Acta Pharmacol Toxicol.,** 58:305-310.

Vasiliou, V., 1988, The effect of xenobiotics on the hepatic aldehyde dehydrogenase activity. **Academic dissertation**, University of Ioannina, Greece.

Vasiliou, V., Athanasiou, K. and Marselos, M., 1988, The use of ALDH induction as a carcinogenic risk marker in comparison with a typical *in vitro* system. **In**: Travis C.C. (ed): "Biologically Based Methods for Cancer Risk Assessment" New York: Plenum Press, 159:231-240.

Vasiliou, V. and Marselos, M., 1989, Tissue distribution of inducible aldehyde dehydrogenase activity in the rat after treatment with phenobarbital or 3-methylcholanthrene. **Pharmacol. Toxicol.,** 64:39-42.

CLASS III ALCOHOL DEHYDROGENASES: EVIDENCE FOR THEIR IDENTITY WITH THE GLUTATHIONE-DEPENDENT FORMALDEHYDE DEHYDROGENASES

Martti Koivusalo and Lasse Uotila

Department of Medical Chemistry
University of Helsinki
SF-00170 Helsinki, Finland

INTRODUCTION

Alcohol dehydrogenases (EC 1.1.1.1) from human (Vallee and Bazzone, 1983), rat (Julià et al., 1987b), mouse (Algar et al., 1983) and other mammalian species (Julià et al., 1987a) are divided into three classes, I, II and III, which are characterized by their specific, differing electrophoretic mobilities, reactivities with different alcohols and sensitivities to pyrazole and its derivatives. Class III alcohol dehydrogenases have anodic electrophoretic mobility at neutral to alkaline pH, and they are further distinguished by their insensitivity to pyrazole and poor use of ethanol and other short-chain alcohols as substrate. Long-chain alcohols like 1-octanol are, in contrast, used well as substrates by these enzymes (Vallee and Bazzone, 1983). The class III alcohol dehydrogenases occur in mammalian tissues much more widely than the enzymes from the other classes (Julià et al., 1987b). The class III enzymes have been purified from human liver (Wagner et al., 1984) and brain (Beisswenger et al., 1985), and from the livers of horse (Kaiser et al., 1989), rat (Julià et al., 1987b) and mouse (Algar et al., 1983) but the physiological significance of these enzymes has remained obscure although their involvement in the oxidation of long-chain fatty alcohols or ω-hydroxy fatty acids has been postulated (Giri et al., 1989).

Formaldehyde dehydrogenase (EC 1.2.1.1) catalyzes the NAD-dependent reversible reaction in which *S*-hydroxymethylglutathione, nonenzymically formed from formaldehyde and reduced glutathione, is oxidized into the thiolester, *S*-formylglutathione (Uotila and Koivusalo, 1974a) :

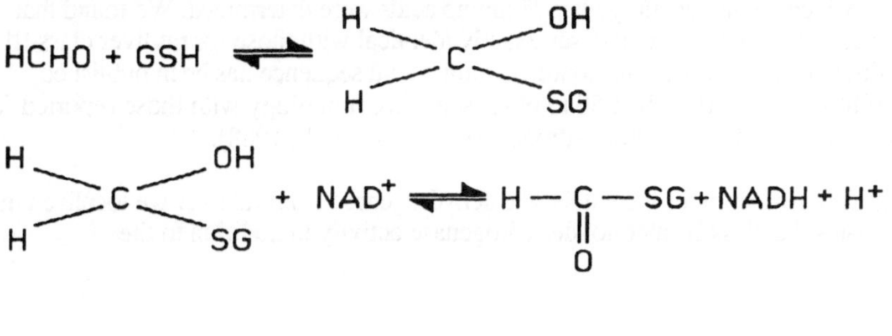

The enzyme is ubiquitously present in all animal tissues studied, and is also present in plant tissues, yeasts, *Escherichia coli* and in some methylotrophic bacteria (Uotila and Koivusalo, 1989). The animal tissues also generally contain a specific enzyme, *S* -formylglutathione hydrolase, which catalyzes the hydrolysis of the thiol ester product of formaldehyde dehydrogenase into formate and glutathione (Uotila and Koivusalo, 1974b). The hydrolase is also present in yeasts, some plant sources and *E. coli* (Uotila, 1989). Both formaldehyde dehydrogenase (Uotila and Koivusalo, 1974a) and *S* -formylglutathione hydrolase (Uotila and Koivusalo, 1974b) have been completely purified and characterized from human liver and they have later also been isolated from a number of other sources (Uotila and Koivusalo, 1989; Uotila, 1989). Formaldehyde dehydrogenase has been found to be specific for formaldehyde as substrate since acetaldehyde and other aldehydes tested are completely inactive (Uotila and Koivusalo, 1989), and for GSH which cannot be replaced with other thiols (Koivusalo and Uotila, 1974). This indicates extreme specificity of the enzyme for *S* -hydroxymethylglutathione, the true substrate of the enzyme (Uotila and Mannervik, 1979).

Here we present evidence that formaldehyde dehydrogenase can under specific conditions also catalyze the oxidation of several long-chain alcohols in the presence of NAD, and the reduction of long-chain aldehydes in the presence of NADH. This and several other lines of evidence suggest that class III alcohol dehydrogenases and formaldehyde dehydrogenase are identical enzymes (see also Koivusalo et al., 1989).

METHODS

Formaldehyde dehydrogenase was isolated from the livers of female rats of the Wistar strain by a procedure which involves affinity chromatography on 5'-AMP-Sepharose and chromatofocusing in the pH gradient 8-5 (Koivusalo et al., 1982; Uotila and Koivusalo, 1983), and yields an electrophoretically homogeneous enzyme preparation. The standard assay mixture for formaldehyde dehydrogenase activity contained 0.1 M sodium pyrophosphate pH 8.0, 1 mM formaldehyde, 1 mM GSH, 1.2 mM NAD and enzyme (forward reaction) or 0.1 M sodium acetate pH 5.7, 1.0 mM *S* -formylglutathione, 0.2 mM NADH and the enzyme (reverse reaction). The standard assay for class III alcohol dehydrogenase activity contained 0.1 M NaOH-glycine pH 9.6, 1 mM 1-octanol, 1.2 mM NAD and the enzyme (forward reaction) or 0.1 M sodium phosphate pH 6.0, 1 mM 1-octanal, 0.2 mM NADH and the enzyme (reverse reaction). The reduction of NAD or the oxidation of NADH was monitored at 340 nm and 25 °C on a Shimadzu UV-240 spectrophotometer. The zinc content of the purified formaldehyde dehydrogenase was assayed by atomic absorption spectrophotometry on a Perkin-Elmer 1100B instrument. The other methods used were similar to those described in a previous communication (Koivusalo et al., 1989).

RESULTS AND DISCUSSION

The amino acid sequences for eight tryptic peptides from the purified formaldehyde dehydrogenase which contained altogether 78 amino acids were determined. We found that the sequences determined were in all cases exactly identical with those for rat liver class III alcohol dehydrogenase, for which the complete amino acid sequence has been published (Julià et al., 1988). The peptides had 50 to 60 % sequence homology with those reported for rat liver class I and class II alcohol dehydrogenases (Kaiser et al., 1988).

During the purification of formaldehyde dehydrogenase from rat liver we localized in the column eluates the class III alcohol dehydrogenase activity in addition to the formaldehyde dehydrogenase. Both of these activities were always eluted in identical positions. When purified formaldehyde dehydrogenase was fractionated by isoelectric focusing on polyacrylamide gel, the same 2-3 main enzyme activity bands were always obtained when the gel was at the completion of the focusing stained either for formaldehyde dehydrogenase (Uotila and Koivusalo, 1987) or for alcohol dehydrogenase activity with 1-octanol and NAD as the substrates.

Purified formaldehyde dehydrogenase was, in agreement with earlier studies (Uotila and Koivusalo, 1989), specific for the oxidation of formaldehyde in the presence of GSH and NAD, and the oxidation of formaldehyde was specifically dependent on GSH. However, the enzyme could at high pH values catalyze the NAD-dependent but GSH-independent oxidation of several long-chain alcohols. Of the alcohols tested at 1 mM concentration, 12-hydroxydodecanoate gave the highest rate which was, at pH 9.6, 48 % of the rate which formaldehyde and GSH gave in the standard assay at pH 8.0. 1-Dodecanol, 1-decanol and 1-nonanol all gave at pH 9.6 30 %, and 1-octanol gave 27 %, 1-heptanol 18 %, and 1-hexanol 7 % of the rate with formaldehyde and GSH at pH 8.0. Butanol and the shorter-chain alcohols did not react at 1 mM concentrations, but even ethanol showed some reactivity when very high alcohol concentrations were used. The pH optimum for the oxidation of the alcohols was at 11.0-11.5, which is much higher than the pH optimum for the oxidation of *S*-hydroxymethylglutathione which was at 8.0-8.5. At pH values below 8.0 the alcohols did not react any more, in contrast to the good reactivity of *S*-hydroxymethylglutathione. The apparent K_m values were 0.09 mM for 12-hydroxydodecanoate and 0.50 mM for 1-octanol at pH 9.6. The apparent K_m value for *S*-hydroxymethylglutathione is below 0.001 mM at pH 8.0 (Uotila and Koivusalo, 1983). It can be calculated that in the k_{cat}/K_m ratio *S*-hydroxymethylglutathione is more than 190-fold more effective substrate than 12-hydroxydodecanoate, and more than 1800-fold more effective than 1-octanol, even when the alcohol dehydrogenase activity is determined at pH 9.6 instead of the more physiological pH 8.0.

In the reverse reaction formaldehyde dehydrogenase was, in accord with earlier results (Uotila and Koivusalo, 1989), among glutathione thiol esters specific for *S*-formylglutathione but could use as the coenzyme either NADH or NADPH. The pH optimum for this reaction was at 5.5-6.0. Only NADH was used above pH 7. We found that the enzyme could also catalyze the reduction of a number of long-chain aliphatic and some aromatic aldehydes to the corresponding alcohols. Either NADH or NADPH could again be used as the coenzyme. Among the aldehydes tested, 1-octanal gave the highest rate. 1-Nonanal gave 95 %, 1-heptanal 42 %, 1-hexanal 40 %, 1-pentanal 20 %, 1-butanal 5 %, 3-nitrobenzaldehyde 35 % and 4-nitrobenzaldehyde 20 % of the rate with 1-octanal (all aldehydes at 1 mM concentration, 0.2 mM NADH, pH=6.0). The pH optimum with 1 mM 1-octanal as the substrate was at 5.0-5.5 with NADPH and at 5.0-6.5 with NADH as the coenzyme. In the oxidation of the alcohols at pH 8.0-11.0, no activity was detected when NAD was replaced with NADP.

Alcohol dehydrogenases are zinc-containing enzymes, and the class III enzymes have also been reported to contain 2 zinc atoms per subunit of the dimeric enzyme (Vallee and Bazzone, 1983; Kaiser et al., 1989). In accord with this we found that the purified formaldehyde dehydrogenase contained 3.7 mol zinc per mol enzyme. We have reported earlier that formaldehyde dehydrogenase from human liver is not sensitive to metal-binding agents, e.g. 10 mM EDTA and 3 mM 1,10-phenanthroline were not inhibitory (Uotila and Koivusalo, 1974a). In this work we found that EDTA up to 30 mM had no effect on rat liver formaldehyde dehydrogenase, but the enzyme was inhibited slightly by 1,10-phenanthroline at 3 mM (10-15 % inhibition) and more clearly by the latter at 10 mM (40

% inhibition). The inhibitory effect was instantaneous. The inhibition was reversed by dilution but not by the addition of zinc to the enzyme. Two non-chelating analogs of 1,10-phenanthroline, 1,7- and 4,7-phenanthroline were more strong inhibitors of formaldehyde dehydrogenase than the chelating compound. Both 1,7- and 4,7-phenanthroline inhibited 30-35 % of the formaldehyde dehydrogenase activity at 3 mM inhibitor concentration, and 50-55 % inhibition was obtained at 10 mM inhibitor concentration. Thus the role of zinc for the enzyme is unclear at present.

We attempted to find out whether the oxidations of *S*-hydroxymethylglutathione and the alcohol substrates by formaldehyde dehydrogenase are mutually competitive. Experiments with varying 12-hydroxydodecanoate concentrations and constant concentrations of formaldehyde and GSH, in which NADH production was monitored at 340 nm, gave in 1/v vs. 1/[S] plots linear curves with varying slopes but with a common intercept on the y axis. This indicates substrate competition between *S*-hydroxymethylglutathione and 12-hydroxydodecanoate. Similarly, 12-hydroxydodecanoate inhibited competitively the production of *S*-formylglutathione from formaldehyde and GSH (assayed specifically at 240 nm according to Uotila and Koivusalo, 1974a).

In conclusion, all of our results suggest that formaldehyde dehydrogenase and class III alcohol dehydrogenase are identical enzymes in rat liver. The experiments on substrate competition indicate that the same active site on the enzyme is responsible for the oxidation of both *S*-hydroxymethylglutathione and of the alcohols. The literature data on formaldehyde dehydrogenase and class III alcohol dehydrogenases also support their identity. Thus both enzymes are reported to be dimers with the subunit molecular weight of 40 000 (Uotila and Koivusalo, 1974a; Uotila and Koivusalo, 1989; Vallee and Bazzone, 1983; Wagner et al., 1984; Julià et al., 1987a,b; Kaiser et al., 1988), and the gene for both enzymes has been located to chromosome 4q21-25 (Hiroshige et al., 1985; Carlock et al., 1985).

ACKNOWLEDGEMENTS

The research has been supported by a grant from the University of Helsinki. Mrs. Eija Haasanen and Miss Anne Hyvönen provided skilful technical assistance.

REFERENCES

Algar, E. M., Seeley, T-L., and Holmes, R. S., 1983, Purification and molecular properties of mouse alcohol dehydrogenase isozymes, **Eur.J.Biochem.**, 137: 139-147.

Beisswenger, T. B., Holmquist, B., and Vallee, B. L., 1985, χ-ADH is the sole alcohol dehydrogenase isozyme of mammalian brains: Implications and inferences, **Proc.Natl.Acad.Sci.USA**, 82: 8369-8373.

Carlock, L., Hiroshige, S., Wasmuth, J., and Smith, M., 1985, Assignment of the ADH5 gene coding for class III ADH to human chromosome 4: 4q21-4q25, **Cytogenet.Cell Genet.**, 40: 598.

Giri, P. R., Linnoila, M., O'Neill, J. B., and Goldman, D., 1989, Distribution and possible metabolic role of class III alcohol dehydrogenase in the human brain, **Brain Res.**, 481: 131-141.

Hiroshige, S., Carlock, L., Wasmuth, J., and Smith, M., 1985, Regional assignment of human formaldehyde dehydrogenase (FDH) to the region 4q21-4q25, **Cytogenet.Cell Genet.**, 40: 651-652.

Julià, P. , Boleda, M. D. , Farrés, J. , and Parés, X. , 1987a, Mammalian alcohol dehydrogenase: characteristics of class III isoenzymes, **Alcohol & Alcoholism,** Suppl. 1: 169-173.
Julià, P. , Farrés, J. , and Parés, X. , 1987b, Characterization of three isoenzymes of rat alcohol dehydrogenase. Tissue distribution and physical and enzymatic properties, **Eur.J.Biochem.**, 162: 179-189.
Julià, P. , Parés, X. , and Jörnvall, H. , 1988, Rat liver alcohol dehydrogenase of class III. Primary structure, functional consequences and relationships to other alcohol dehydrogenases, **Eur.J.Biochem.** , 172: 73-83.
Kaiser, R. , Holmquist, B. , Hempel, J. , Vallee, B. L. , and Jörnvall, H. , 1988, Class III human liver alcohol dehydrogenase: a novel structural type equidistantly related to the class I and class II enzymes, **Biochemistry** 27: 1132-1140.
Kaiser, R. , Holmquist, B. , Vallee, B. L. , and Jörnvall, H. , 1989, Characteristics of mammalian class III alcohol dehydrogenases, an enzyme less variable than the traditional liver enzyme of class I, **Biochemistry** 28: 8432-8438.
Koivusalo, M. , Baumann, M. , and Uotila, L. , 1989, Evidence for the identity of glutathione-dependent formaldehyde dehydrogenase and class III alcohol dehydrogenase, **FEBS Lett.** , 257: 105-109.
Koivusalo, M. , Koivula, T. , and Uotila, L. , 1982, Oxidation of formaldehyde by nicotinamide nucleotide-dependent dehydrogenases, **in** "Enzymology of Carbonyl Metabolism: Aldehyde Dehydrogenase and Aldo/Keto Reductase", H. Weiner and B. Vermuth, eds., pp. 155-168, Alan R. Liss, Inc. , New York.
Koivusalo, M. , and Uotila, L. , 1974, Enzymic method for the quantitative determination of reduced glutathione, **Anal.Biochem.**, 59: 34-45.
Uotila, L. , 1989, Glutathione thiol esterases, **in**: "Coenzymes and Cofactors, vol. III, Glutathione. Chemical, Biochemical and Medical Aspects", part A, D. Dolphin, R. Poulson and O. Avramovic, eds., pp. 767-804, John Wiley & Sons, Inc., NewYork.
Uotila, L. , and Koivusalo, M. , 1974a, Formaldehyde dehydrogenase from human liver. Purification, properties and evidence for the formation of glutathione thiol esters by the enzyme, **J.Biol.Chem.** , 249: 7653-7663.
Uotila, L. , and Koivusalo, M. , 1974b, Purification and properties of *S*-formylglutathione hydrolase from human liver, **J.Biol.Chem.** , 249: 7664-7672.
Uotila, L. , and Koivusalo, M. , 1983, Formaldehyde dehydrogenase, **in**: "Functions of Glutathione. Biochemical, Physiological, Toxicological and Clinical Aspects", A. Larsson, S. Orrenius, A. Holmgren, and B. Mannervik, eds. , pp. 175-186, Raven Press, New York.
Uotila, L. , and Koivusalo, M. , 1987, Multiple forms of formaldehyde dehydrogenase in human red blood cells, **Human Heredity** , 37: 102-106.
Uotila, L. , and Koivusalo, M. , 1989, Glutathione-dependent oxidoreductases: Formaldehyde dehydrogenase, **in** "Coenzymes and Cofactors, vol. III, Glutathione. Chemical, Biochemical and Medical Aspects", part A, D. Dolphin, R. Poulson, and O. Avramovic, eds. , pp. 517-551, John Wiley & Sons, Inc. , New York.
Uotila, L. , and Mannervik, B. , 1979, A steady-state kinetic model for formaldehyde dehydrogenase from human liver. A mechanism involving NAD^+ and the hemimercaptal adduct of glutathione and formaldehyde as substrates and free glutathione as an allosteric activator of the enzyme, **Biochem.J.**, 177, 869-878.
Vallee, B. L. , and Bazzone, T. J. , 1983, Isozymes of human liver alcohol dehydrogenase, **in**: "Isozymes: Current Topics in Biological and Medical Research", vol. 8, M. C. Rattazi, J. C. Scandalios, and G. S. Whitt, eds. , pp. 219-244, Alan R. Liss, Inc. , New York.
Wagner, F. W. , Parés, X. , Holmquist, B. , and Vallee, B. L. , 1984, Physical and enzymatic properties of a class III isozyme of human liver alcohol dehydrogenase: χ-ADH, **Biochemistry**, 23: 2193-2199.

GENETIC POLYMORPHISMS OF ALCOHOL METABOLIZING ENZYMES AND THEIR SIGNIFICANCE FOR ALCOHOL-RELATED PROBLEMS

A. Yoshida

Department of Biochemical Genetics
Beckman Research Institute of the City of Hope
Duarte, California, U.S.A.

INTRODUCTION

Alcohol dehydrogenase (ADH) and aldehyde dehydrogenase (ALDH) isozymes not only play a role in alcohol detoxification, but also participate in metabolism of neurotransmitters, fatty acids and glutathione, and in clearance of biogenic toxic substances. Genetic abnormalities, and consequent alteration of activity and kinetic properties of any one of the ADH and ALDH isozymes would potentially affect one's sensitivity to alcohol and vulnerability in developing alcohol-related problems.

This paper summarizes: a) kinetic properties and possible physiological roles of various ADH and ALDH isozymes; and b) our current knowledge of genetic variations and polymorphisms of these isozymes, and discuss the relationships between these polymorphisms and alcohol-related problems.

KINETIC PROPERTIES AND PHYSIOLOGICAL ROLES

Alcohol Dehydrogenase

ADH isozymes of human and other mammals are grouped in three distinctive "classes," based on their enzymatic characteristics and structural similarities (Table 1). The human class I ADH isozymes include homo- and hetero-dimers consisting of three types of subunits (α, β and γ), which are controlled by three non-allelic genes $\underline{ADH_1}$, $\underline{ADH_2}$ and $\underline{ADH_3}$. The human class II ADH is a homo-dimer of π subunit which is controlled by the $\underline{ADH_4}$ gene, and the class III ADH is a homo-dimer of χ subunit controlled by the $\underline{ADH_5}$ gene.

The class I ADH isozymes are most abundant in the liver, and exist also in other tissues including the intestine, lung and kidney, but are not detectable in the brain, skin fibroblasts and red blood cells. The class II ADH is primarily found in the liver. A similar isozyme(s) is also detected in the gastrointestinal tract. The class III ADH is found in all

Alcoholism: A Molecular Perspective
Edited by T.N. Palmer, Plenum Press, New York, 1991

Table 1

ADH Isozymes

Class	Gene	Subunit	Substrate	Biological Role
I	ADH_1	α	Short aliphatic alcohol	Ethanol detoxification
	ADH_2	ß	Dopamine, Norepinephrine 3ß-Hydroxy-5ß-steroid	Metabolism of neurotransmitters Fatty acid metabolism
	ADH_3	γ	1,6-Hydroxyhexadecanoic acid 4-Hydroxy-nonenal	Steroid metabolism Clearance of hydroxy-nonenal
II	ADH_4	π	Ethanol (Km ca 100 mM) Long-chain aliphatic alcohols Aromatic alcohols Serotonin, Norepinephrine	Ethanol detoxification Metabolism of neurotransmitters
III	ADH_5	χ	Long-chain alcohols S-formyl-glutathione	Glutathione metabolism
Testis/Retina ADH			Retinol oxidation	Activation of vitamin A

tissues examined, including the brain, testis and red blood cells (ref. in Yoshida et al., 1990).

The class I ADHs exhibit a high activity for ethanol oxidation, and play a major role in ethanol detoxification. The isozymes also oxidize dopamine and norepinephrine (Mårdh and Vallee, 1986), and ADH consisting of γ subunit exhibits 3ß-hydroxy-5ß-steroid dehydrogenase activity, and testosterone was found to be an allosteric regulator (Mårdh et al., 1986). The class I ADHs can efficiently oxidize 1,6-hydroxyhexa-decanoic acid, hence the enzymes may take part in fatty acid metabolism (Wagner et al., 1983). The rat liver class I ADH, and presumably also human ADHs, can detoxify the endogenously generated toxic aldehyde, 4-hydroxy-nonenal (Esterbauer et al., 1985). Ethanol ingested would compete with the endogenous substrates and disturb the physiological detoxification, resulting in tissue damage.

The class II ADH, which has a higher Km for ethanol than the class I ADHs, may participate in ethanol oxidation, when tissue ethanol level exceeds 20 mM (Li et al., 1977). The class II ADH also oxidizes serotonin and norepinephrine (Consalvi et al., 1986).

The molecular identity of rat class III ADH and glutathione-dependent formaldehyde dehydrogenase (EC 1.2.1.1) was recently demonstrated (Koivusalo et al., 1989). The human class III ADH may also have similar activity and play a role in glutathione metabolism.

The sex-dependent ADH existing in stomach mucous membrane, which may play a role in ethanol oxidation in the male stomach (Frezza et al., 1990), may or may not be identical to the liver class II ($\pi\pi$) isozyme. Testis-specific ADH and retina-specific ADH isozyme(s) are found in rodent (Julià et al., 1987; Keung, 1988). The corresponding isozyme(s) probably exist also in humans, and play a role in activation of vitamin A. The kinetic properties and physiological role of the newly discovered $\underline{ADH_6}$ remain to be elucidated.

Aldehyde Dehydrogenase

At least nine ALDH isozymes have been distinguished based on the separation by physico-chemical methods, tissue and subcellular distributions, and enzymatic properties

Table 2

Human ALDH Isozymes

Isozyme	Major Tissues	Subcellular localization	Major Substrate, Km (µM) and pH	Subunit M.W.	Subunit/active enzyme
$ALDH_1$	Liver	C	acetaldehyde, 22 µM at pH 7.5	54,000	4
$ALDH_2$	Liver	M	acetaldehyde, 3.5 µM at pH 7.5	54,000	4
$ALDH_3$	Liver, Stomach	C	heptaldehyde, 11 µM at pH 8.5	54,000	2
$ALDH_4$	Liver	M	glutamic-γ-semialdehyde, 100 µM at pH 7.0	70,600	2
γ-ABDH	Liver	N	γ-aminobutylaldehyde,	54,000	4
Major			13.8 µM at pH 7.4		
Minor			8.0 µM at pH 7.4		
$ALDH_\chi$	Liver, Testis	M	unknown	54,000	unknown
$ALDH_{2a}$ (disulfiram sensitive)	Brain	M	acetaldehyde, 1 µM at pH 9.0; dopaldehyde, 1 µM at pH 9.0	54,000	4
$ALDH_{2a}$ (disulfiram sensitive)	Brain	M	acetaldehyde, 1 µM at pH 9.0; dopaldehyde, 0.5 µM at pH 9.0	54,000	4
Saliva ALDH	Saliva	C	acetaldehyde, 106 µM at pH 8.0	48,000	4

(Table 2). The cytosolic liver $ALDH_1$ and the mitochondrial liver $ALDH_2$, which exhibit high activity for oxidation of acetaldehyde, are considered to play a major role in acetaldehyde detoxification in the liver. $ALDH_3$ exhibits optimal activity for oxidation of benzaldehyde but it can also oxidize long-chain aliphatic aldehydes. $ALDH_4$ has a broad substrate specificity and oxidizes short chain aliphatic aldehydes; but it is most active with glutamic-γ-semialdehyde as substrate. An unique ALDH (γ-ALDH) has low Km for acetaldehyde, but it is most active for oxidation of γ-aminobutyraldehyde at the physiological pH (ref. in Yoshida et al., 1990).

The brain $ALDH_1$ corresponds to, and is most likely identical to, the liver $ALDH_1$. Human brain contains ALDH isozymes which metabolize neurotransmitters. The two brain mitochondrial ALDH isozymes ($ALDH_{2a}$ and $ALDH_{2b}$) not only oxidize acetaldehyde but also DOPAL (3,4-dihydroxyphenyl-acetaldehyde) efficiently (Ryzlak and Pietruszko, 1987, 1989). Cerebellum, corpus striatum and pons show a high activity for DOPAL oxidation (Hafer et al., 1987). Other ALDH isozymes, which play a role in the metabolism of neurotransmitters, probably exist in human brain. Human saliva contains an ALDH isozyme which can oxidize acetaldehyde (Harada et al., 1989). The biochemical and immunological data suggest that the saliva ALDH is different from liver $ALDH_1$ and $ALDH_2$.

The gene and cDNA for a new ALDH isozyme ($ALDH_\chi$) was recently cloned (Hsu et al., 1989). The $\underline{ALDH_\chi}$ gene is mainly expressed in the testis. Kinetic properties and physiological role of this new ALDH isozyme remain to be elucidated.

GENETIC POLYMORPHISMS

Alcohol Dehydrogenase

The $\underline{ADH_1}$ locus is monomorphic, but the $\underline{ADH_2}$ and $\underline{ADH_3}$ loci are polymorphic. Three common alleles, $\underline{ADH_2}^1$ for β_1 subunit (wild type), $\underline{ADH_2}^2$ for β_2 subunit (common in Orientals), and $\underline{ADH_2}^3$ for β_3 subunit (common in Blacks), have been recognized in the $\underline{ADH_2}$ locus. Two common alleles, i.e. $\underline{ADH_3}^1$ for γ_1 subunit (wild type) and $\underline{ADH_3}^2$ for γ_2 subunit, are found in Caucasians, but not in Orientals. The molecular differences at both the protein level and the gene level, of these wild type and common variant type alleles have been elucidated (Table 3). The common variants $\beta_2\beta_2$ and $\beta_3\beta_3$ enzyme exhibit much higher catalytic activity (about 100-fold) for ethanol oxidation than the usual $\beta_1\beta_1$ enzyme. The activity of $\gamma_1\gamma_1$ enzyme is about 2 times higher than that of the $\gamma_2\gamma_2$ enzyme.

The $\underline{ADH_4}$ locus may be polymorphic, unless the discrepancies observed in the amino acid sequence data of the π subunit and the nucleotide sequence of cDNA (position 312 is Arg in cDNA but Lys in the protein, and position 303 is Val in cDNA but Ile in protein) are due to sequencing errors (Table 3).

The $\underline{ADH_5}$ locus (for χ subunit) may also be polymorphic. A cDNA for the χ subunit can encode 373 amino acid residues matched with the amino acid sequence data, while another cDNA clone differed by one residue (Tyr→Asp at 166), and could encode an additional 19 amino acid residues starting from an upstream initiation codon. It remains to be further confirmed whether or not polymorphism exists in the $\underline{ADH_5}$ locus, and post-translational cleavage of the NH_2-terminal occurred in the χ subunit.

Table 3

Polymorphisms and Mutation Sites of ADH Isozymes

Locus	Common Alleles	Subunit	Mutation Site	Gene Frequency
ADH_2	ADH_2^1	β_1		Wild type, >90% in Caucasians
	ADH_2^2	β_2	Arg→His at 47	Common in Orientals (~70%)
	ADH_2^3	β_3	Arg→Cys at 369	Common in Blacks (~16%)
ADH_3	ADH_3^1	γ_1		Wild type, >90% in Orientals
	ADH_3^2	γ_2	Arg→Glu at 271 Ile→Val at 349	Common in Caucasians (~40%)
ADH_4	ADH_4^1	π_1		
	ADH_4^2	π_2	Arg→Lys at 312 Val→Leu at 303	Unknown

Restriction fragment length polymorphisms (RFLP) have been observed in the ADH loci, i.e. PvuII polymorphism at the ADH cluster locus, RsaI and MaeIII polymorphisms at the ADH_2 locus, XbaI, MspI and StuI polymorphisms at the ADH_3 locus. SacI polymorphism was found at the ADH_5 locus. RFLP of the ADH_4 locus has not been reported (ref. in Yoshida et al., 1990).

Aldehyde Dehydrogenase

Approximately 50% of Orientals lack $ALDH_2$ activity in their livers. The deficiency is due to the genomic mutation in the $ALDH_2$ locus resulting in catalytically inactive protein with amino acid substitution Glu→Lys at the 487th position from the NH_2-terminal (Table 4). The frequency of variant $ALDH_2^2$ gene is 20~35% in various Oriental populations (ref. in Yoshida et al., 1990).

In contrast to the $ALDH_2$, common $ALDH_1$ variants were not found. A variant with severely diminished catalytic activity but strong immunological cross-reactivity, was first found in one out of ten Japanese livers (Yoshida, 1983), and subsequently apparently the same variant was found in two out of 60 Japanese livers. Another $ALDH_1$ variant

Table 4

Polymorphisms and Mutation Sites of ALDH Isozymes

Locus	Common Alleles	Mutation Site	Gene Frequency
$\underline{ALDH_2}$	$\underline{ALDH_2}^1$		Wild type, ~100% in Caucasians
	$\underline{ALDH_2}^2$	Glu→Lys at 487	Common in Orientals (~30%)
$\underline{ALDH_{3b}}$	$\underline{ALDH_{3b}}^1$		Wild type
	$\underline{ALDH_{3b}}^2$	Unknown	Common in Chinese
$\underline{ALDH_x}$	$\underline{ALDH_x}^1$		Wild type
	$\underline{ALDH_x}^2$	Val→Ala at 69 Arg→Leu at 90	>10%

associated with different electrophoretic mobility was found in two Oriental subjects (Eckey et al., 1986). Recently two $ALDH_1$ variants with diminished activity were found in Caucasian alcohol-flushers (Yoshida et al., 1989).

Polymorphism at the $\underline{ALDH_3}$ locus was proposed based on the observed multiplicity of $ALDH_3$ bands in Chinese stomach samples. In the model proposed, $\underline{ALDH_{3a}}$ gene is expressed in the liver as well as in the stomach and lung, while $\underline{ALDH_{3b}}$ gene is not expressed in the liver, and two common alleles $\underline{ALDH_{3b}}^1$ and $\underline{ALDH_{3b}}^2$ exist in the $\underline{ALDH_{3b}}$ locus (Yin et al., 1988). The frequencies of $\underline{ALDH_{3b}}^1$ and $\underline{ALDH_{3b}}^2$ were estimated to be 0.14 and 0.84, respectively, in Chinese.

The newly discovered $\underline{ALDH_\chi}$ locus is polymorphic, i.e. two types of alleles -C-C-T- and -T-T-G-, commonly exist in the $\underline{ALDH_\chi}$ locus (Hsu et al., 1989; Yoshida et al., 1990). The nucleotide differences result in amino acid substitutions at two positions, i.e. Val→Ala at position 69 and Arg→Leu at position 90 of $ALDH_\chi$ isozyme. The two types of $ALDH_\chi$ enzyme produced by the two types of alleles may have different kinetic properties.

Salivary ALDH deficiency was found in about 30% of Japanese (Harada et al., 1989). It remains to be confirmed whether or not the deficiency really has a genetic origin.

In the study of RFLP, *TaqI* polymorphism at the $\underline{ALDH_1}$ locus was found in Orientals but not in Caucasians. *MspI* polymorphism at the $\underline{ALDH_1}$ locus was found in both Caucasians and Orientals, but the frequencies are different. *MspI* type of the $\underline{ALDH_2}$ locus is polymorphic in both populations and the frequencies are the same (Yoshida and Chen, 1989).

POLYMORPHISMS AND ALCOHOL-RELATED PROBLEMS

As reviewed in the preceding sections, the nature (i.e. nucleotide base changes) of polymorphisms of three ADH loci ($\underline{ADH_2}$, $\underline{ADH_3}$ and $\underline{ADH_4}$) and two $\underline{ALDH}$ loci ($\underline{ALDH_2}$ and $\underline{ALDH_\chi}$) have been elucidated. Based on this information, one can devise the method for genotype determination using a pair of allele specific oligonucleotide probes (Hsu et al., 1987; Ikuta et al., 1988).

Since the common Oriental $\beta_2\beta_2$ ADH is far more active than the usual $\beta_1\beta_1$ ADH, and inversely the Oriental atypical $ALDH_2^2$ enzyme has no (or severely diminished) activity, the relationships of genotypes of these two loci to the alcohol sensitivity and alcoholic liver diseases were first examined.

In 49 unrelated control Japanese individuals examined, the frequency of atypical $\underline{ADH_2}^2$ allele is 0.71, and that of usual $\underline{ADH_2}^1$ is 0.29; and the frequencies of atypical $\underline{ALDH_2}^2$ allele is 0.35 and that of the usual $\underline{ALDH_2}^1$ is 0.65. None of the Caucasians examined had the atypical $\underline{ALDH_2}^2$ gene. The frequencies of genotypes determined were compatible to the Hardy-Weinberg expectation, indicating the reliability of the genotyping method.

Since about 50% of Japanese lack $ALDH_2$ activity, the data imply that the heterozygous $\underline{ALDH_2}^1/\underline{ALDH_2}^2$ subjects, as well as the homozygous atypical $\underline{ALDH_2}^2/\underline{ALDH_2}^2$ subjects, lack liver $ALDH_2$ activity, i.e. the atypical $\underline{ALDH_2}^2$ gene is dominant in expression of enzyme activity (Shibuya and Yoshida, 1988a). The notion was supported by other investigators.

Genotypes of $\underline{ADH_2}$ and $\underline{ALDH_2}$ loci in Japanese alcohol-flushers, non-flushers, and patients with alcoholic liver diseases are summarized in Table 5. No difference was found between the subjects with alcoholic liver diseases and controls in the frequency of the atypical $\underline{ADH_2}^2$. The frequency of the $\underline{ADH_2}^2$ was found to be higher in alcohol-flushers than in non-flushers, but the statistical significance was not established in the sample size analyzed. More cases have to be examined to settle the correlation between the $\underline{ADH_2}$ genotypes and alcohol-flushing.

The genotype analysis indicates that all Japanese alcohol-flushers are either heterozygous $\underline{ALDH_2}^1/\underline{ALDH_2}^2$ or homozygous atypical $\underline{ALDH_2}^2/\underline{ALDH_2}^2$, and conversely, most Japanese with alcoholic liver diseases are homozygous usual $\underline{ALDH_2}^1/\underline{ALDH_2}^1$. We concluded that Japanese with genotype $\underline{ALDH_2}^1/\underline{ALDH_2}^2$ or $\underline{ALDH_2}^2/\underline{ALDH_2}^2$ are alcohol sensitive, and consequently they are at low risk in developing alcoholic liver diseases and

Table 5. Genotypes of $\underline{ADH_2}$ and $\underline{ALDH_2}$ Loci in Japanese.

$\underline{ADH_2}$ Locus

	Genotypes			Gene Frequency	
	1-1	1-2	2-2	$\underline{ALDH_2}^1$	$\underline{ALDH_2}^2$
Control (n=49)	4	20	25	0.29	0.71
Patients with Alcoholic Liver Disease (n-23)	3	6	14	0.26	0.74
Alcohol Flusher (n=8)	0	1	7	0.06	0.94
Non-flusher (n=6)	0	3	3	0.25	0.75

$\underline{ALDH_2}$ Locus

	Genotypes			Gene Frequency	
	1-1	1-2	2-2	$\underline{ALDH_2}^1$	$\underline{ALDH_2}^2$
Control (n=49)	21	22	6	0.65	0.35
Patients with Alcoholic Liver Disease (n=23)	20	3	0	0.93	0.07
Alcohol Flusher (n=9)	0	7	2	0.39	0.61
Non-flusher (n=6)	5	1	0	0.92	0.08

Difference in frequencies of the $\underline{ALDH_2}^1$ and $\underline{ALDH_2}^2$ between the controls and alcoholic liver patients is statistically significant ($P<0.005$). Difference in frequencies of the $\underline{ALDH_2}^1$ and $\underline{ALDH_2}^2$ between the alcohol flushers and non-flushers is statistically significant ($P<0.01$). The data are from Shibuya and Yoshida (1988b) and Shibuya et al (1989).

other alcohol-related problems (Shibuya and Yoshida, 1988b). The recent finding that most Japanese alcoholics are homozygous usual $\underline{ALDH_2}^1$/$\underline{ALDH_2}^1$ further confirms this notion (Harada, 1990).

In other words, the atypical $\underline{ALDH_2}^2$ gene is a major "alcohol-rejecting" genetic factor in Orientals. It should be pointed out that a small number of patients (3 out of a total of 23) are heterozygous $\underline{ALDH_2}^1$/$\underline{ALDH_2}^2$ (Table 5). Other investigators (Harada, 1990) confirmed the same situation in alcoholic patients. These few exceptions (about 5~10% of patients suffering from alcoholic diseases) cannot be due to errors in genotype determination, since the method used is highly reliable and the typing has been customarily re-checked in cases of such peculiarity. The heterozygous $\underline{ALDH_2}^1$/$\underline{ALDH_2}^2$ alcoholic patients must be associated with extremely strong "alcohol-seeking" tendency, which overcomes the effect of the "alcohol-rejecting" $\underline{ALDH_2}^2$ gene. It is important to study the nature of this "alcohol-seeking" tendency, which may be related to a major "alcohol-seeking" gene.

The frequency of the atypical $\underline{ALDH_2}^2$ allele was found to be high also in other Orientals, i.e. 0.15 in South Koreans and 0.20 in Chinese (Singh et al., 1989). Atypical subjects, either homozygous or heterozygous at the $\underline{ALDH_2}$ locus in these populations are likely to be alcohol sensitive and less vulnerable in developing alcoholic liver diseases.

The $\underline{ALDH_2}$ gene is expressed in early embryos (Yoshida et al., 1990), hence the atypical $\underline{ALDH_2}^2$ gene, which is a preventive factor against alcohol-related diseases in adult life could be a risk factor in developing the fetal alcohol syndrome. The possibility can be tested by genotyping affected individuals.

A high frequency (50~70%) of alcohol-flushing was observed in a North American Indian tribe (Wolff, 1973). However, unlike Orientals, American Indians have higher frequencies of the alcohol-related problems. North American Indians showed a low (or no) frequency of $\underline{ALDH_2}$ deficiency, while approximately 40% of South American Indians (including Mapuche Indians) were reported to be $ALDH_2$ deficient (Goedde et al., 1986). Recently, $ALDH_2$ genotypes of 38 Mapuche Indians were determined, and it was found that none of them has the atypical $\underline{ALDH_2}^2$ gene (O'Dowd et al., 1990). Should about 40% of the Mapuche Indians lack $ALDH_2$ activity, another defective $\underline{ALDH_2}$ gene, which is different from the Oriental $\underline{ALDH_2}^2$, might exist at a high frequency in this population.

Approximately 5~8% of Caucasians exhibit alcohol-flushing, and, in addition, about 6~10% are reported to be chlorpropamide-induced alcohol-flushers. The characteristic seems to be a dominantly inherited trait. By examining the properties of the $ALDH_1$ isozyme existing in blood samples from Caucasian alcohol-flushers, thus far two types of $ALDH_1$ variants have been found. One variant ($ALDH_1$ Harrow) had very low activity (~15% of control), and the other variant ($ALDH_1$ Columbo) had moderately reduced activity and altered kinetic properties (Yoshida et al., 1989). The enzyme deficiency and flushing character are inherited in relatives of the propositus. The frequencies of these two variants among Caucasians have not yet been determined.

The strong positive correlation between the Oriental atypical $\underline{ALDH_2}^2$ gene with alcohol sensitivity and the negative correlation with development of alcoholic liver diseases have been proven by genotyping of the $\underline{ALDH_2}$ locus using allele specific oligonucleotide probes as described above. Similar genetic studies can be readily carried out in the $\underline{ADH_2}^3$ gene (for β_3 subunit) which is common in Blacks, the $\underline{ADH_3}^2$ (for γ_2 subunit) which is common in Caucasians, the $\underline{ADH_4}$ locus which may be polymorphic, the $\underline{ALDH_{3b}}$ locus which may be polymorphic in Chinese and other populations, and the polymorphic $\underline{ALDH_\chi}$ locus. As reviewed above, ADH and ALDH isozymes not only participate in alcohol detoxification, but also play other physiological roles. An association of RFLP at the dopamine D receptor locus with alcoholism was recently reported (Blum et al., 1990). Genetic abnormalities and polymorphisms of ADH and ALDH isozymes involved in the metabolism of neurotransmitters may affect alcohol-seeking behavior. Recent technological developments in molecular genetics allow us to identify mutation sites and determine the genotypes of these isozymes. The significance of genetic variations of these loci in alcohol-related problems may be elucidated through this approach.

ACKNOWLEDGEMENTS

This work was supported by the U.S. Public Health Service Grant AA05763.

REFERENCES

Blum, K., Noble, E. P., Sheridan, P. J., Montgomery, A., Ritchie, T., Jagadeeswaran, P., Nogami, H., Briggs, A. H., and Cohn, J. B., 1990, Allelic association of human dopamine D2 receptor gene in alcoholism, **JAMA**, 263:2055.

Consalvi, V., Mårdh, G., and Vallee, B. L., 1986, Human alcohol dehydrogenases and serotonin metabolism, **Biochem.Biophys.Res.Comm.,** 139:1009.

Eckey, R., Agarwal, D. P., Saha, N., and Goedde, H. W., 1986, Detection and partial characterization of a variant of cytosolic aldehyde dehydrogenase isozyme, **Hum.Genet.** 72:95.

Esterbauer, H., Zollner, H., and Lang, J., 1985, Metabolism of the lipid peroxidation product 4-hydroxynonenal by isolated hepatocytes and liver cytosolic fractions, **Biochem.J.,** 228:363.

Frezza, M., di Padova, C., Possato, G., Terpin, M., Baraona, E., and Lieber, C. S., 1990, High blood alcohol levels in women: The role of decreased gastric alcohol dehydrogenase activity and first-pass metabolism, **New Eng.J.Med.,** 322:95.

Goedde, H. W., Agarwal, D. P., Harada, S., Rothhammer, F., Whittaker, J.O., and Lisker, R., 1986, Aldehyde dehydrogenase polymorphism in North American, South American and Mexican Indian populations, **Am.J.Hum.Genet.** 38:395.

Hafer, G., Agarwal, D. P., and Goedde, H. W., 1987, Human brain aldehyde dehydrogenase: Activity with DOPAL and isozyme distribution, **Alcohol**, 4:413.

Harada, S., 1990, Genetic polymorphism of aldehyde dehydrogenase and its physiological significance to alcohol metabolism, **Prog.Clin.Biol.Res.,** 344:289.

Harada, S., Muramatsu, T., and Agarwal, D. P., and Goedde, H. W., 1989, Polymorphism of aldehyde dehydrogenase in human saliva, **Prog.Clin.Biol. Res.,** 190:133.

Hsu, L. C., Bendel, R. E., and Yoshida, A., 1987, Direct detection of usual and atypical alleles on human aldehyde dehydrogenase-2 ($ALDH_2$) locus, **Am.J. Hum.Genet.,** 41:1987

Hsu, L. C., Chang, W-C., and Yoshida, A., 1989, Cloning of a new human aldehyde dehydrogenase gene and comparison with liver cytosolic $ALDH_1$ and mitochondrial $ALDH_2$ genes, **Am.J.Hum.Genet.,** 45:A196.

Ikuta, T., Shibuya, A., and Yoshida, A., 1988, Direct determination of usual (Caucasian type) and atypical (Oriental type) alleles of class I human alcohol dehydrogenase-2 locus, **Bioch.Genet.,** 26:519.

Julià, P., Farrés, J., and Parés, X., 1987, Characterization of three isozymes of rat alcohol dehydrogenase: Tissue distribution and physical and enzymatic properties, **Eur.J.Bioch.,** 162:179.

Keung, W-M., 1988, A genuine organ specific alcohol dehydrogenase from hamster testes: Isolation, characterization and developmental changes, **Bioch.Biophys.Res.Comm.,** 156:38.

Koivusalo, M., Baumann, M., and Votila, L., 1989, Evidence for the identity of glutathione-dependent formaldehyde dehydrogenase and class III alcohol dehydrogenase, **FEBS Lett.,** 257:105.

Li, T. K., Bosron, W. F., Defeldecker, W. P., Lange, L. G., and Vallee, B. L., 1977, Isolation of π-alcohol dehydrogenase of human liver: Is it a determinant of alcoholism?, **Proc.Natl.Acad.Sci.U.S.A.,** 74:4378.

Mårdh, G., and Vallee, B. L., 1986, Human class I alcohol dehydrogenase catalyze the interconversion of alcohols and aldehydes in the metabolism of dopamine, **Biochem.,** 25:7279.

Mårdh, G., Falchuck, K. H., Auld, D. S., and Vallee, B. L., 1986, Testosterone allosterically regulates ethanol oxidation by homo- and heterodimeric γ-subunit-containing isozymes of human alcohol dehydrogenase, **Proc.Natl.Acad.Sci.U.S.A.,** 83:2836.

O'Dowd, B. F., Rothhammer, F., and Israel, Y., 1990, Genotyping of mitochondrial aldehyde dehydrogenase locus of native South American Indians, **Alcoholism Clin.Exp.Res.,** in press.

Ryzlak, M. T., and Pietruszko, R., 1987, Purification and characterization of aldehyde dehydrogenase from human brain, **Arch.Bioch.Biophys.,** 255:409.

Ryzlak, M. T., and Pietruszko, R., 1989, Human brain glyceraldehyde-3-phosphate dehydrogenase, succinic semialdehyde dehydrogenase and aldehyde dehydrogenase isozymes: Substrate specificity and sensitivity to disulfiram, **Alcoholism:Clin.Exp.Res.,** 13:755.

Shibuya, A., and Yoshida, A., 1988a, Frequency of the atypical aldehyde dehydrogenase-2 gene ($ALDH_2^2$) in Japanese and Caucasians, **Am.J.Hum.Genet.,** 43:741.

Shibuya, A., and Yoshida, A., 1988b, Genotypes of alcohol metabolizing enzymes in Japanese with alcoholic liver disease: A strong association of the usual Caucasian type aldehyde dehydrogenase gene ($\underline{ALDH_2}^1$) with the disease, **Am.J.Hum.Genet.,** 34:744.

Shibuya, A., Yasunami, M., and Yoshida, A., 1989, Genotypes of alcohol dehydrogenase and aldehyde dehydrogenase loci in Japanese alcohol flushers and non-flushers, **Hum.Genet.,** 82:14.

Singh, S., Fritz, G., Fang, B., Harada, S., Paik, Y. K., Eckey, R., Agarwal, D. P., and Goedde, H. W., 1989, Inheritance of mitochondrial aldehyde dehydrogenase: Genotyping in Chinese, Japanese and South Korean families reveals dominance of the mutant allele, **Hum.Genet.,** 83:119.

Wagner, F. W., Burger, A. R., and Vallee, B. L., 1983, Kinetic properties of human liver alcohol dehydrogenase: Oxidation of alcohols by class I isozymes, **Biochem.,** 22:1857.

Wolff, P. H., 1973, Vasomotor sensitivity to alcohol in diverse Mongoloid populations, **Am.J.Hum.Genet.,** 25:193.

Yin, S-J., Cheng, T-C., Chang, C-P., Chen, Y-J., Chao, Y-C., Tang, H-S., Chang, T-M., and Wu, C-W., 1988, Human stomach alcohol and aldehyde dehydrogenase (ALDH): A genetic model proposed for ALDH III isozymes, **Bioch.Genet.,** 26:345.

Yoshida, A., 1983, A possible structural variant of human cytosolic aldehyde dehydrogenase with diminished activity, **Am.J.Hum.Genet.,** 35:1115.

Yoshida, A., and Chen, S-H., 1989, Restriction fragment length polymorphism of human aldehyde dehydrogenase 1 and aldehyde dehydrogenase 2 loci, **Hum. Genet.,** 83:204.

Yoshida, A., Hsu, L. C., and Yasunami, M., 1990, Genetics of human alcohol-metabolizing enzymes, **in**: "Progress in Nucleic Acid Research and Molecular Biology," W. Cohn and K. Moldave, eds., Academic Press, Orlando, in press.

Yoshida, A., Shibuya, A., Davé, V., Nakayama, M., and Hayashi, A., 1990, Developmental changes of aldehyde dehydrogenase isozymes in human livers: Mitochondrial $ALDH_2$ isozyme is expressed in fetal livers, **Experientia,** in press.

Yoshida, A., Ward, R. J., and Peters, T. J., 1989, Cytosolic aldehyde dehydrogenase ($ALDH_1$) variants found in alcohol flushers, **Ann.Hum.Genet.,** 53:1.

BIOCHEMICAL AND GENETIC STUDIES IN $ALDH_1$-DEFICIENT SUBJECTS

Roberta Ward[1], Andrew Macpherson[1], Margaret Warren-Perry[1], Vibha Dave[2], Lily Hsu[2], Akira Yoshida[2] and Timothy J. Peters[1]

[1]Department of Clinical Biochemistry
King's College School of Medicine & Dentistry
Bessemer Road, London, UK
[2]Molecular Genetics
Beckman Research Institute
Duarte, California, USA

At least nine isoenzymes of aldehyde dehydrogenase (E.C.1.2.1.3.) have been identified in man based on their enzymatic characteristics and physiochemical properties (Yoshida *et al.*, 1990). Liver mitochondrial aldehyde dehydrogenase ($ALDH_2$) probably plays a major role in acetaldehyde metabolism as its low k_m (1-2μM) is compatible with the circulating acetaldehyde levels after alcohol ingestion in both normal subjects and alcohol abusers (Arthur et al.,1984: Schumate et al., 1967; Kelding et al., 1983; Mezey and Tubon, 1971; Peters et al.,1987). The $ALDH_2$ gene is 44 kbp in length and contains at least 12 exons which encode 517 amino acid residues (Hsu *et al.*, 1988). These precisely match the reported protein sequence of $ALDH_2$ (Hempel *et al.*, 1985). A point mutation (C→A) leads to the formation of an enzymatically inactive subunit in exon 12 of the genetic sequence (Yoshida *et al.* , 1984). This occurs in over 50% of Orientals (Goedde *et al.*, 1979) and such individuals suffer an alcohol-flush reaction after ingestion of small amounts of ethanol. This is attributable to high circulating acetaldehyde concentrations (in excess of 15 μmol) and causes clinical symptoms such as increase in blood pressure and pulse rate (Wolffe, 1973).

By contrast it remains unclear as to the role of cytosolic aldehyde dehydrogenase ($ALDH_1$) in acetaldehyde metabolism since it's high k_m for acetaldehyde [30-100 μM] (Tipton, Henehan & Harrington, 1989) would preclude a major role. However it may have a secondary function in limiting acetaldehyde concentration since the intensity of the flush response to ethanol in the Orientals may be lower in individuals with higher erythrocyte $ALDH_1$ activity (Inoue, Fukunaga & Yamasawa, 1980). The $ALDH_1$ gene is 53 kbp in length and has 13 exons, encoding 501 amino acid residues (Hsu *et al.*, 1989). This also matches the previously determined protein sequence (Hempel *et al.*, 1984). As yet no alteration in its genetic or protein sequences has been identified.

Flushing in Caucasians after alcohol ingestion may be caused by a defect in acetaldehyde metabolism and is thought to have a prevalence of between 5-10% in this

Alcoholism: A Molecular Perspective
Edited by T.N. Palmer, Plenum Press, New York, 1991

population (Agarwal and Goedde, 1989). The aetiology is not known. No alteration in either alcohol dehydrogenases or $ALDH_2$ activities in such Caucasian subjects have been identified (Goedde *et al.*, 1979; Teng, 1981).

In the present studies, we have investigated the activity of cytosolic $ALDH_1$ in Caucasian subjects with a history of flushing after small alcohol ingestion, whether there was a familial pre-disposition to alcohol flushing in these subjects and assayed blood acetaldehyde levels and physical signs in these Caucasian flushers after a small dose of alcohol. In a preliminary study we have also investigated the incidence of flushing in a normal Caucasian population by questionnaire.

RESULTS

The red cell activity of $ALDH_1$ and $ALDH_1$ properties were determined in blood samples from 12 unrelated individuals with the alcohol flushing. The results (Yoshida *et al.*, 1989) in essence showed that activity of $ALDH_1$ in 3 subjects was severely reduced (<20%) while another had low utilisation of deamino NAD and ethanoamide NAD as co-enzymes. The remaining 8 samples showed little difference in $ALDH_1$ activity or properties from the control samples. Further blood samples were collected from the families of each of subjects with reduced $ALDH_1$ activity. There was a high prevalence of reduced $ALDH_1$ activity in most members of each respective family, and this coincided with the occurrence of alcohol-induced flushing. A typical pedigree of $ALDH_1$ deficiency is shown in Figure 1.

A test dose of alcohol (0.4g/kg body weight) was administered to probands from two of the pedigrees and the levels of alcohol (Bucher & Redetzki, 1951) and acetaldehyde (Rideout *et al.*, 1986) determined at 15, 30, 45, 60, and 90 mins post-ingestion . Urinary catecholamines and dopamine were assayed in a urine sample collected at the conclusion of the study. The results were compared with Caucasian subjects with normal $ALDH_1$ activity after drinking an equivalent amount of alcohol. No differences in acetaldehyde levels were detected; a similar level of acetaldehyde 1.0 ± 0.2 μmol/l v 0.9 ± 0.3 μmol/l was found in both patient groups. No increases in the urinary excretion of either catecholamines or dopamine were detectable in the urine of either subject. In one $ALDH_1$-deficient subject, there was no change in blood pressure or pulse rate taken at 10-minute intervals for 1 hr after ingestion of alcohol despite the intense alcohol flush reaction.

Since alcohol has a vasodilator effect aldehyde-mediated flushing has been difficult to distinguish from the normal pharmacological effect of ethanol in Caucasians unless flushing responses are measured by earlobe densitometer (Wolffe, 1972). In contrast, questionnaire surveys can still identify possible subjects with the reaction, although this will be somewhat subjective. A survey of 30 healthy medical students by questionnaire showed a high predominance of flushing (37 %) in female students.

DISCUSSION

The biochemical results from the Caucasian flushers after alcohol are noteworthy because of the intense flush reaction despite the low circulating acetaldehyde levels. This was confirmed by the lack of change in both blood pressure and pulse rate. It is reported that the prime role for $ALDH_1$ activity is in the metabolism of the aldehydes derived from a number of physiologically important amines including catecholamines (Haroda *et al.*, 1982;

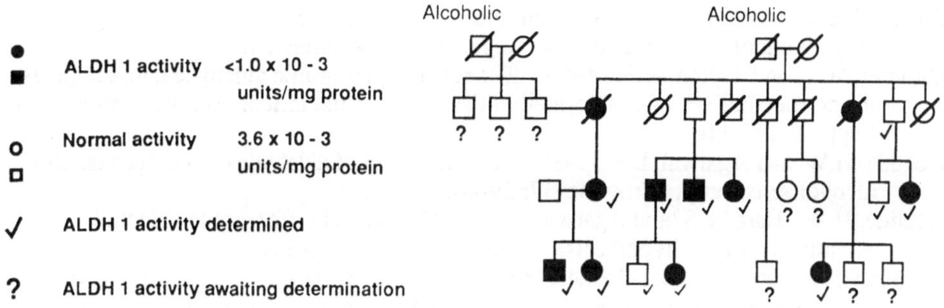

Fig.1

Helander & Tottmar, 1986) and histamine (Henehan *et al.*, 1985; Gitomer & Tipton, 1983). Even though various drugs and alcohol itself are reported to reduce $ALDH_1$ activity, these drugs also affect $ALDH_2$ activity such that it is difficult to assess the actual involvement of these two isoenzymes in acetaldehyde metabolism. A variety of drugs will initiate the flush reaction after alcohol ingestion. Disulfiram will predominantly inhibit $ALDH_1$ and to a lesser degree $ALDH_2$ activity (Helander & Tottmar, 1988) in erythrocytes and leucocytes respectively and also hepatic mitochondrial (Yourick & Faiman, 1989) and cytosolic (Jenkins & Peters, 1983) ALDH. Therefore, the cause of the flush reaction in non-abstaining alcoholics maybe attributable to high circulating levels of acetaldehyde. The intense alcohol-induced facial flushing in diabetics to receiving chlorpropamide therapy is caused by raised levels of acetaldehyde which are similar to that measurable in Orientals (Barnett *et al.*, 1981).

It is clear that further studies of the DNA of these subjects using molecular biological techniques, i.e. amplification of the exons of $ALDH_1$ using synthetic oligonucleotide primers adjacent to the intron-exon junction followed by dideoxynucleotide sequencing will identify whether mutations are causing the inactivation of the $ALDH_1$ enzyme.

CONCLUSIONS

Flushing in Caucasians is not uncommon and may be associated with reduced erythrocyte activity. It is, like flushing in Orientals, aversive, at least in females, to continued alcohol consumption. The pathogenesis of the flushing reaction is currently unknown and may reflect changes in biogenic aldehyde rather than aldehyde metabolism in the subjects.

REFERENCES

Agarwal, D.P. and Goedde, H.W., 1989, Human aldehyde dehydrogenases: Their role in alcoholism. **Alcohol** 6: 517-523.

Arthur, M.J.P., Lee, A. and Wright, R., 1984, Sex differences in the metabolism of ethanol and acetaldehyde in normal subjects. **Clin. Sci.**, 67: 397-401.

Barnett, A.H., Gonzalez-Auvert, C., Pyke, D.A., Saunders, J.B., Williams, R., Dickenson, C.J. and Rawlins, M.D., 1981, Blood concentrations of acetaldehyde during chlorpropamide-alcohol flush. **Br. Med. J.**, 283: 939-941.

Bucher,T. and Redetzki, H., 1951, Eine spezifische photometrische bestimmung von athylalkonhol auf fermentativen wege. **Klin. Wochenschr.**, 29: 615.

Gitomer, W.L. and Tipton, K.F., 1983, The role of cytoplasmic aldehyde dehydrogenase in the metabolism of N-tele-methylhistamine. **Pharmacol Biochem Behav.**, 18 Supp. 1: 113-116.

Goedde, H.W. and Agarwal, D.P., 1987, Polymorphism of aldehyde dehydrogenase and alcohol sensitivity. **Enzyme** 37: 29-44.

Goedde, H.W., Harada, S. and Agarwal, D.P., 1979, Racial differences in alcohol sensitivity: a new hypothesis. **Hum. Genet.**, 5: 331-334.

Haroda,S., Agarwal, D.P. and Goedde, H.W., 1982, Human aldehyde dehydrogenase: 3,4-dihydroxy-phenyl-acetaldehyde metabolizing isoenzymes. **In**: 'Enzymology of Carbonyl Metabolism', Vol. 1. Ed. H. Weiner & B. Wermuth Publ. A.R.Liss New York, USA pp 147-153.

Helander, A. and Tottmar, S.O.C., 1988, Effects of disulfiram, cyanamide and 1-aminocyclopropanol on the aldehyde dehydrogenase activity in human erythrocyte and leucocytes. **Pharmacol Toxic.,** 63: 262-265.

Helander, A. and Tottmar, S.O.C., 1986, Cellular distribution and properties of human blood aldehyde dehydrogenase. **Alcohol. Clin. Exp. Res.,** 10: 71-76.

Hempel, J., von Bahr-Lindstrom, H. and Jörnvall, H., 1984, Aldehyde dehydrogenase from human liver: Primary structure of the cytoplasmic isoenzyme. **Eur. J. Biochem.**, 141: 21-35.

Hempel, J., Kaiser, R. and Jornvall H., 1985, Mitochondrial aldehyde dehydrogenase from human liver. **Eur J Biochem.**, 153: 13-28.

Henehan, G.T.M., Ward, K., Kennedy, N.P., Weir, D.G. and Tipton, K.F., 1985, Subcellular distribution of aldehyde dehydrogenase activities in human liver. **Alcohol** 2: 107-110.

Hsu, L., Bendel, R.E. and Yoshida, A., 1988, Genomic structure of the human mitochondrial aldehyde dehydrogenase gene. **Genomics** 2: 57-65.

Hsu, L.C., Chang, W-C. and Yoshida, A., 1989, Genomic structure of the human cytosolic aldehyde dehydrogenase gene. **Genomics** 5 857-865.

Inoue, K., Fukunaga, M. and Yamasawa, K., 1980. Correlation between human erythrocyte aldehyde dehydrogenase activity and sensitivity to alcohol. **Pharmacol Biochem Behav.**, 13: 295-298.

Jenkins,W.J. and Peters, T.J., 1983, Subcellular localization of aldehyde dehydrogenase in human liver. **Cell Bioch Function.,** 1: 37-40.

Kelding, S., Christensen, N.J., Damgaard, S.E., Dejgard, A., Iverson, H.L., Jacobsen, A., Johansen, S., Lundquist, F., Rubinstein, E. and Winkler, 1983, Ethanol metabolism in heavy drinkers after massive and moderate alcohol intake. **Biochem Pharmacol.**, 32: 3097-3102.

Mezey, E. and Tobon, F., 1971, Rates of ethanol clearance and activities of the ethanol-oxidising enzymes in chronic alcoholic patients. **Gastroenterology** 61: 707-715.

Peters, T.J., Ward, R.J., Rideout, J. and Lim, C.K., 1987, Blood acetaldehyde and ethanol levels in alcoholism , **in**: 'Genetics and Alcoholism', Ed Goedde, H.W. & Agarwal, D.P.. Publ. A.Liss New York USA. pp 215-230.

Rideout, J.M., Lim, C.K. and Peters, T.J., 1986, Assay of blood acetaldehyde by HPLC with fluorescence detection of its 2-diphenylacetyl-1,3-indiandione-1-azine derivative. **Clin Chem Acta.,** 161: 29-35.

Shumate, R.P., Crowther, R.F. and Zarafshan, M., 1967, A study of the metabolism rates of alcohol in the human body. **J Forensic Med.**, 14: 83-100.

Teng, Y.S., 1981, Human liver aldehyde dehydrogenase in Chinese and Asiatic Indians : gene deletion and its possible implications in alcohol metabolism. **Biochem. Genet.** 19: 107-114.

Tipton, K.F., Henehan, G.T.M. and Harrington, M.C., 1989, Cellular and intracellular distribution of aldehyde dehydrogenases. **in**: 'Human Metabolism of Alcohol', Vol 2. ed K.E.Crow & R.D.Batt. Publ. CRC.Press Boca Raton. pp 105-116.

Wolffe P.H., 1972, Ethnic differences in alcohol sensitivity. **Science** 175: 449-450.

Wolffe, P.H., 1973, Vasomotor sensitivity to alcohol in diverse mongoloid populations. **Amer J Hum Genet.**, 25: 193-199.

Yoshida, A., Huang, I-Y. and Ikawa H., 1984, Molecular abnormality of an inactive aldehyde dehydrogenase variant commonly found in Orientals **Proc Natl Acad Sci USA.**, 81: 258-261.

Yoshida A., Dave, V., Ward, R.J. and Peters, T.J., 1989, Cytosolic aldehyde dehydrogenase ($ALDH_1$) variants found in alcohol flushers. **Ann.Hum. Genet.**, 53: 1-7.

Yoshida, A., 1990, Genetic polymorphisms of alcohol metabolizing enzymes and their significance for alcohol-related problems, **in**: 'Alcoholism: A Molecular Perspective', ed. T.N.Palmer, Publ Plenum U.K. pp 118-128.

Yourick, J.J. and Fairman, J.D., 1989, Comparative aspects of disulfiram and its metabolites in the disulfiram-ethanol reaction in rats. **Biochem Pharmacol.**, 38: 413-421.

ETIOLOGY OF SUBGROUPS IN CHRONIC ALCOHOLISM AND DIFFERENT MECHANISMS IN TRANSMITTER SYSTEMS

O.M. Lesch, H. Walter, W. Bonte*, J. Grunberger, M. Musalek, and R. Sprung**

Psychiatric Univ. Clinic, Vienna Wahringer Gurtel 18-20
1090 Wien, Austria
Anton Proksch Institute, Kalksburg, Vienna, Austria
Institute for Forensic Medicine, Düsseldorf, FRG*
Institute for Forensic Medicine, Gottingen, FRG**

INTRODUCTION

The present state of the art in alcoholism research still suffers from the fundamental problem of delineating reliable boundaries between alcohol use, alcohol abuse and alcohol addiction, which includes the distinction between problem drinking and alcoholism (Edwards et al., 1977; Glatt, 1982; Heimann et al., 1989; Jellinek, 1960; Keller, 1976; Lesch, 1985, Lesch et al., 1991; Morey and Blashfield, 1981; Nimmerrichter et al., 1989; Pattison and Kaufman, 1982; Polich et al., 1980; Rodgerson et al., 1980; Schuckit, 1979; Edwards et al., 1977; Sobell, 1981; Uhl, 1983; Vaillant, 1983). The continuously ongoing changes in diagnostic criteria demonstrate quite well these difficulties (Cernovsky, 1985; Ewing, 1984; Fitzgerald et al., 1971; Horn et al., 1983; Keller, 1976; Koehler and Sass, 1984; Lachar et al., 1976; Marlatt, 1976; McLennan et al., 1980; Rohsenow, 1982; Selzer, 1971; Spitzer, 1987; Wanberg and Horn, 1983; Wetzler, 1988; Wittchen et al., 1989). The heterogenity of this illness is undoubted (Babor and Lauerman, 1986: Berner et al., 1986; Besson et al., 1981; Button, 1956; Buydens-Branchey et al., 1989; Cloninger et al., 1981; Conley and Prioleau, 1983; Donovan et al., 1986; Gilligan et al., 1987; Grünberger et al., 1989; Johnson et al., 1986; Knorring et al., 1985; Lesch, 1985; Lesch et al., 1988a,b,c; Martin et al., 1988; Norström and Berglund, 1987; Pattison and Kaufman, 1982; Schuckit, 1979; Schuckit and Irwin, 1989; Thurstin and Alfano, 1988; Victor and Banker, 1978; Winokur et al., 1971). Between 1850 and 1941 thirty-nine different typologies of alcoholism have been proposed and many further attempts have been added since (Babor et al., 1986; Buydens-Branchey, 1989; Cloninger, 1987). Most of these typologies are cross-sectional classifications reflecting the researchers interests rather than the different aspects of the illness (Bean et al., 1981; Nerviano and Gross, 1983). For this reason different types of subgroups exist which are oriented psychologically, socially or biologically. The results of all attempts to subgroup chronic alcoholism still remain unsatisfactory (Alfano et al., 1987; Alterman and Tarter, 1986; Donovan et al., 1986; Litman et al., 1979; Zivich, 1981).

Alcoholism: A Molecular Perspective
Edited by T.N. Palmer, Plenum Press, New York, 1991

Basically we find the following main problems in alcoholism research:

1. The cause of the illness "chronic alcoholism" is chronic intoxication with ethanol and congener alcohols.
2. This intoxication affects different persons with different somatic and psychic conditions.
3. The longer the intoxication lasts and the more pre-damages present, the more marked the alcohol-related disabilities will be. This damage can cover the causal basic disturbances.
4. The heterogenity of chronic alcoholic patients lies in different causalities and different disabilities as a consequence.

These circumstances are mirrored in etiological considerations as well as in therapeutical ones. Also the different importance attributed to predictors clearly delineates the dilemma of alcoholism research (Blake, 1967; Blaney et al., 1975; Conley and Prioleau, 1983; Edwards et al., 1988; Feuerlein et al., 1980; Gilles et al., 1974; Hagnell et al., 1986; Kolb et al., 1976; Polich et al., 1980; Sobell and Sobell, 1976; Voegtlin and Broz, 1949). Some authors favour gender as predictive (Curlee, 1971; Dahlgren, 1975; Davis, 1957; Fox, 1976; Glatt, 1982; Horn et al., 1983; Ruggles et al., 1975), some stress the importance of age (Cloninger et al., 1981). Other authors proved that these factors are no predictors for long-term prognosis (Blake, 1967; Blaney et al., 1975; Feurlein et al., 1980; Gilles et al., 1974; Kolb et al., 1976; Lesch, 1985; Lesch et al., 1989; McLellan et al., 1980; Ogborne, 1978).

All these different results are caused not only by the different ways diagnosis is set up (DSM-III, DSM-IIIR, ICD-9, ICD-10: see Lesch, 1985; Koehler and Sass, 1984; Wittchen et al., 1989) but also by selection criteria (investigation dates during the patients progress to chronic alcohol addiction; different stages of the disease at the time when a patient enters the study [early stage or end of the disease]; place of admission [patients admitted to general psychiatric wards or specialised institutions]; mode of data collection [different inventories under use] (Johnson et al., 1986; Lesch, 1985; Lesch et al., 1989; Rodgerson et al., 1980; Winokur et al., 1971).

It is well known from research in the area of effective psychoses that endogenomorphic depressive symptom constellations may be quite similar if viewed cross-sectionally (Berner, 1982; Jaspers, 1973; Wittchen et al., 1989). In the long-term course genetic as well as therapeutic differences between bipolar and unipolar types of this illness become apparent (Angst, 1973).

Dissatisfaction with the results of the research discussed above motivated us to develop 4 subgroups of chronic alcoholic patients by means of prospective long-term follow-up observation (Lesch et al., 1988a)). Our results show that these 4 subgroups will meet more specifically defined data for basic as well as therapy-orientated research. Methodological claims for long-term follow-up studies in chronic alcoholism were taken into regard (e.g. selection, diagnosis [basic data], follow-up time, follow-up rate, more than 6 controls a year and personal interview: Berner et al., 1986; Lesch, 1985; Lesch et al., 1988a,b,c).

DIAGNOSIS OF SUBTYPES IN CHRONIC ALCOHOLICS

Our case material consisted of 444 chronic alcoholic patients, who live in one region of Austria and who had been admitted for hospital treatment of chronic alcoholism within a period of 3 years. The above mentioned region is divided into sub-regions with different socio-cultural backgrounds (f.i. wine production differs largely between region A and D), which influence the frequency rates of admission, but hold no predictive potential for the further course of the diseased and therefore do not influence the results of this study.

On admission 95.4% of patients already suffered from liver damage, and 54.8% from an organic brain syndrome. These two factors clearly indicate that these patients had reached a severe degree of their illness. Patients were observed for a mean of 5.33 years, 23.2% (101 patients) died within the observation period. Data collection was made by personal interview only. After contact in hospital and in the domestic environment symptoms were semiquantitatively recorded at three dates: (l) admission, (2) 12-15 months after admission and (3) at least 48 months after admission time. For patients who died during the follow-up period the last investigation before death was recorded as third investigation date. In addition to this time-dependent evaluation, we also affected a "time-unrelated" observation of the individual patient, followed by attribution to individual course types. In view of the fact that the "loss of control, total abstinence concept" is undisputed in Austria, the influence of this dogma could not be excluded. All patients except 1.8% (n=8) were followed up, which represents a small drop out rate. We refrained from establishing a control group. Therefore this investigation cannot be interpreted as a therapy study, although therapeutic factors are described. Diagnosis was made according to the WHO criteria of 1982 for chronic alcoholism (National Council on Alcoholism (US): Criteria for the Diagnosis of Alcoholism, 1972) : dosage increase, loss of control, psychic and somatic withdrawal syndrome. At the individual (time dependent) rating dates, additional criteria were used: drinking behaviour, psychopathology, psychic, somatic and social status. Attribution was carried out semi-quantitatively. In addition to the investigation on determined rating dates we performed a time-unrelated evaluation over the whole investigation period (the patient was usually seen every second month; most patients were contacted more than 6 times a year): Before the study was started 4 types of long-term course were defined, according to clinical experience and results of pertinent literature (Glatt, 1982; Heimann et al., 1989; Keller, 1976; Polich et al., 1980; Schuckit, 1979; Sobell, 1981; Vaillant, 1983). Exact long-term course evaluation of these patients made a categorization into 4 different types possible.

Definitions of the Long-term Courses

TYPE 1: (18.5%) an illness course without social or psychological conspicuousness, whereby the therapeutic goal of total abstinence is possible.

TYPE 2: (25.6%) an illness course with no psychic, somatic or social conspicuousness, though from time to time alcohol was consumed (more than 42 months abstinent within 48 months)

TYPE 3: (31.7%) a fluctuating illness course with psychic, somatic and social conspicuousness and frequent fluctuation in these dimensions as a predominant factor.

TYPE 4: (24.2%) an illness course with physic and social deterioration and serious somatic damage unrelated to drinking.

Our study profited from the fact that patients from admission time onwards were attributed to the different types of illness course under use of the abovementioned prospective observation.

Three additional cross-sectional investigations during follow-up time clearly demonstrated the superiority of "time-unrelated" course observation to that of "time-dependent" one. Anamnestic data of the patients were correlated with the 4 types of long-term illness course (time unrelated evaluation) and together with the symptom constellations of the three

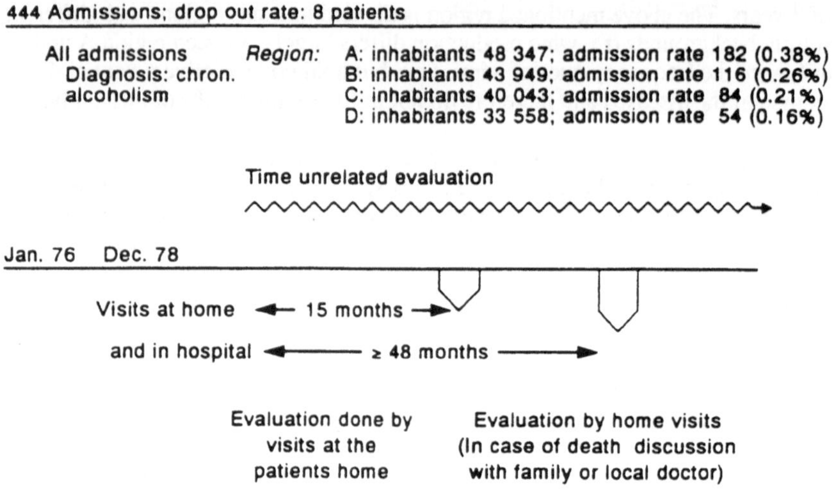

Figure 1. Follow-up study 2 - 7 years (m = 5.33 years).

cross-sectional investigation dates (i.e.time-dependent evaluation). By this method we found "marker symptoms", that enabled us to diagnose the 4 types also cross-sectionally (Lesch, 1985; Lesch et al., 1988a). Statistical methods and a comparison of our typology to other typologies (Jelinek, 1960 and others) has already been published (Lesch et al., 1991).

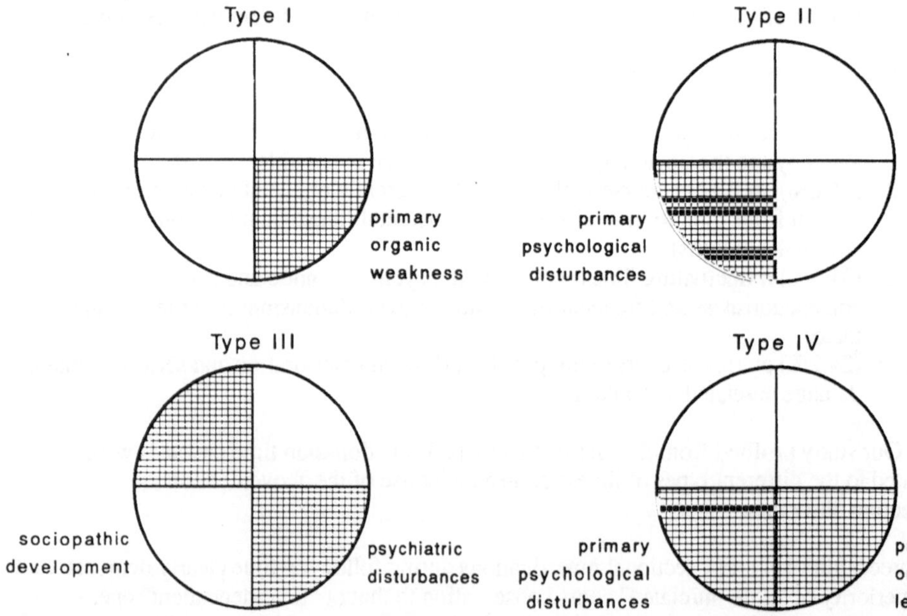

Figure 2. The four subtypes according to Lesch (1985).

Diagnosis of the Four Types

Symptom constellations differed so considerably between the 4 types of illness course that these types can be viewed as qualitatively different diseases (Lesch, 1985; Lesch et al., 1988a,b, c; Lesch et al., 1989)

TYPE 1: "ALCOHOL INTAKE BECAUSE OF BIOLOGICAL CRAVING". In this subgroup no pre-alcoholic characteristics can be found. These patients may have a certain organic weakness on one hand and effects of alcohol consumption may lead to corresponding symptoms on the other. Patients may be socially adapted for a long time and become eventually conspicuous because of somatic and/or psychiatric withdrawal symptoms (e.g. meta-alcoholic psychosis and/or withdrawal seizures).

TYPE 2: "ALCOHOL INTAKE BECAUSE OF PSYCHOLOGICAL CRAVING". In this group clear pre-alcoholic characteristics were found, whereby the relationship to parents and the patient's personal development may represent main factors. Passive behaviour as well as partnership characteristics have to be understood as symptoms rather than as causal features. Alcohol intake as an additional factor represents an attempt of "self-treatment" and can be seen as an aggravating factor soon leading to admission.

TYPE 3: "ALCOHOL INTAKE FOR SELF TREATMENT OF PSYCHIATRIC DISORDERS". A positive family history of alcoholism, together with social conspicuousness and combined with psychiatric disorders (manic depressive illness, mood- and drive- disturbances characterize this type and lead to its fluctuating course

TYPE 4: "ALCOHOL INTAKE BECAUSE OF AN UNCRITICAL POSITION TOWARDS THE ALCOHOL PERMISSIVE CLIMATE IN A CULTURE. BRAIN DAMAGE BEFORE ONSET OF DRINKING." These patients have typical pre-alcoholic characteristics, above all disturbances of effective climate in the parental home, often minimal brain damage (neonatal origin) and behaviour disturbances in childhood (f.i. enuresis nocturna). The patients themselves treat symptoms of their development and somatic disturbances by alcohol intake. This aspect of pre-alcoholic brain damage may be contributory to this special course and should be considered in all studies investigating alcohol-related disabilities (e.g. brain atrophy as well as severeness of seizures, following neonatal brain damage) (Lesch, 1985; Lesch et al., 1988a,b,c).

Summarizing it should be emphasized that all 4 types of chronic alcoholics have to be viewed as being qualitatively different. As can be seen in Figure 3 alcohol intake arises for different reasons in the 4 types (against rebound phenomena, for anxiolytic effects, sedation, or as an unspecific antidepressant). These different effects are based on different transmitter systems.

Operationalisation of Types (See Lesch et al., 1991)

In the next step we developed an inventory for cross-sectional diagnosis of the 4 subtypes. With the reduction of the diagnostic categorization of the 4 types into a decision tree, it became evident, that only a part of all chronic alcoholic patients could be clearly diagnosed as belonging to one of these 4 types. The other patients represent "mixed types".

By a visual check of protocols - if individual symptoms are weighted in most patients - a clinical categorization can be achieved. For this reason our decision tree will have to be further ameliorated.

The hierarchical character of our typology recalls Jaspers' "Schichtregel", whereby only chronic alcohol intoxication and its rebound phenomena constitute a modifying element. The later a patient enters the study the more important intoxication and its rebound phenomena become. Viewed as another possible influencing element drinking behaviour (start of drinking, amount, type of alcohol consumed etc.) did not render any additional information concerning the subtypes of chronic illness course.

We have used our typology in all further studies and found group-specific results for biological as well as neurophysiological parameters. Studies on different therapeutic strategies, as our typology demands as a logical consequence, are presently being performed.

TREATMENT STRATEGIES IN THE 4 SUBTYPES AND THE ROLE OF PSYCHOPHARMACEUTIC AGENTS

	REASONS FOR THE USE OF ALCOHOL	BIOCHEMICAL PROCESSES	THERAPY
TYPE 1 (18%)	"BIOLOGICAL" CRAVING	FORMALDEHYDE +DOPAMINE= = TIQs = MORPHINE EFFECT	BLOCKING OF TIQs MORPHINE ANTAGONISTS ACETYLHOMOTAURINAT-CA (AOTAL)
TYPE 2 (25%)	"PSYCHOLOGICAL" CRAVING (ANTI-ANXIETY) (SEDATION)	GABAergic MECHANISMS	PSYCHOTHERAPY NO PSYCHOPHARMACOLOGY TRANQUILIZERS
TYPE 3 (33%)	"SELF-TREATMENT" OF DEPRESSIVE SYNDROMES	SEROTONERGIC, ADRENERGIC, CHOLINERGIC MECHANISMS	LITHIUM PROPHYLAXIS, ANTIDEPRESSANTS e.g. SEROTONERGIC SUBSTANCES, f.e. RITANSERIN
TYPE 4 (24%)	ORGANIC BRAIN SYNDROME (+SEVERE SOMATIC DISTURBANCES) LOSS OF CRITICISM ALSO TOWARDS OWN DRINKING BEHAVIOUR	LOSS OF INHIBITING PROCESSES	LONG-TERM IN-PATIENT TREATMENT SOMATIC THERAPY NEUROLEPTICS

Figure 3. Alcoholics consume alcohol to treat their psychic symptoms. The typical pharmaceutical profile of alcohol consumption explains the compulsory intake in our subtypes of patients.

We do not have final data yet, because we are of the opinion the effectiveness of a therapeutic strategy in chronic alcoholic patients can only be evaluated after a period of two years (Conley et al., 1983; Glatt, 1982; Koehler, 1984; Polich et al., 1980; Rounsaville et al., 1987; Sobell, 1981; Winokur et al., 1971).

BIOLOGICAL ASPECTS

As Bonte (1987) emphasized all alcoholic beverages contain not only ethanol but also "congener alcohols" like methanol. This has been neglected in several experimental studies and ethanol effects were regarded as "alcohol" effects (Bacopoulos et al., 1978; Begleiter and Platz, 1972; Crossley et al., 1986; Martin et al., 1986; Naranjo et al., 1986; Myers, 1989; Schuckit and Gold, 1988; Sonntag anmd Boyd, 19895).

In an investigation with 63 alcoholics, diagnosed along DSM-III criteria, who were admitted in an alcoholized state to our inpatient unit for treatment and who were compared to 21 healthy controls who underwent drinking experiment, we were able to demonstrate that methanol metabolism showed significant differences between alcoholics and healthy controls (Barz et al., 1988,1989; Bonte et al., 1989). The healthy volunteers showed a long-lasting methanol plateau as long as the ethanol levels remained above 0.4 to 0.5. Alcoholics, however, exhibited a time-dependent exponential decrease in blood methanol levels without any correlation to the actual ethanol concentration. Therefore we conclude that in chronic alcoholic patients different mechanisms governing methanol synthesis and methanol elimination exist, whereby methanol elimination follows a different pathway. (Blum, 1980; Bonte, 1987; Bonte et al., 1987; Bonte and Sprung, 1987; Collins et al., 1973; Sprung and Bronte, 1988; Sprung et al., 199: compare von Wartburg, 1975;1985 120). Methanol elimination via breath and sweat can be neglected, because 95% of consumed alcohol - ethanol as well as methanol - is metabolised. 63 chronic alcoholic patients (DSM-III), who were admitted in an alcoholized state, exhibited an average ethanol concentration of 1.7 promille. Each two hours from admission time onwards blood methanol levels were measured. The rate of decrease in the methanol concentration showed marked differences between individual patients. If the elimination constant rate is related to the typology significant relationships can be found (Figure 4). Type 1 and type 4 especially seem to be of special interest for further research in this field. In type 1 the methanol elimination constant rate per hour is the highest of all types. It could be assumed that higher concentrations of formaldehyde may be the result. Formaldehyde could be "the" addiction promotive substance and provoke craving as well as withdrawal syndromes.

In type 4 biological deterioration is already present before the onset of drinking. Very low methanol elimination constant rates correlate to this subgroup of chronic alcoholic patients (Lesch et al., 1988b; Sprung et al., 1991). In this type peripheral damage (e.g.polyneuropathy) correlate directly with the ethanol and methanol concentrations. Figure 5 shows the significant correlation between polyneuropathy and low methanol elimination constant rates.

NEUROPHYSIOLOGICAL ASPECTS (static and dynamic pupillometry)

In recent years static dynamic light-provoked pupillometry have achieved increasing importance in biological and psychopharmacological research, specially for measurement of activity-levels and of the autonomous reaction patterns (Grünberger et al., 1984, 1985, 1986, 1987a,b, 1988,1990) . We used this method to compare the pupillometry data of 117

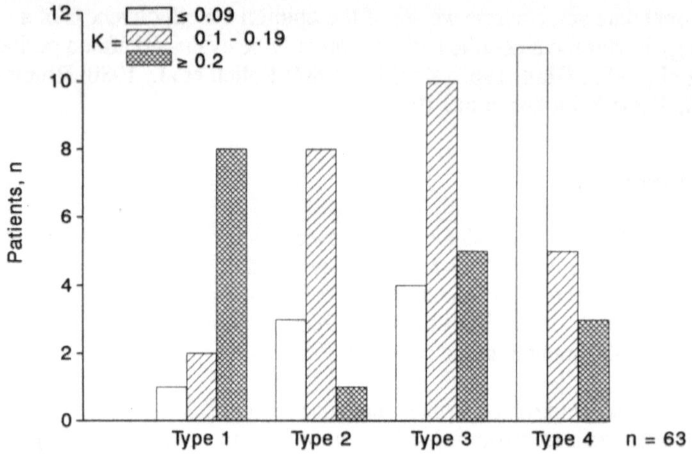

Figure 4. Typology according to Lesch (types 1 - 4) and methanol elimination rates. Metabolism of methanol to formaldehyde could be the toxic agent that causes "biological" craving.

chronic alcoholic patients to that of 107 normal controls (Grünberger et al., 1989) (Figure 6). When classified according to our typology (type 1-4) we found that patients belonging to type 1 were the most similar to normal controls, whereas the other 3 types differed in various parameters. In type 2 and type 3 our results suggest that psychopathological symptoms, as f.i. anxiety, could play an important role (Grünberger et al., 1989).

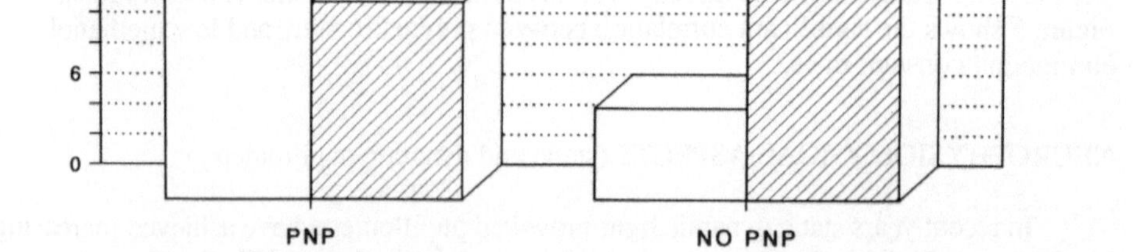

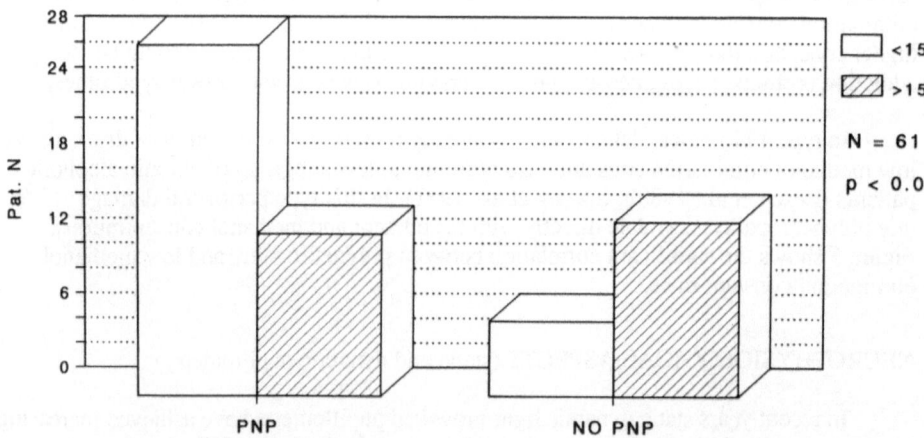

Figure 5. The metabolism of methanol is significantly different in the 4 types of alcoholics. A single symptom such as polyneuropathycorrelates with the methanol elimination constant rate (<15 or >15).

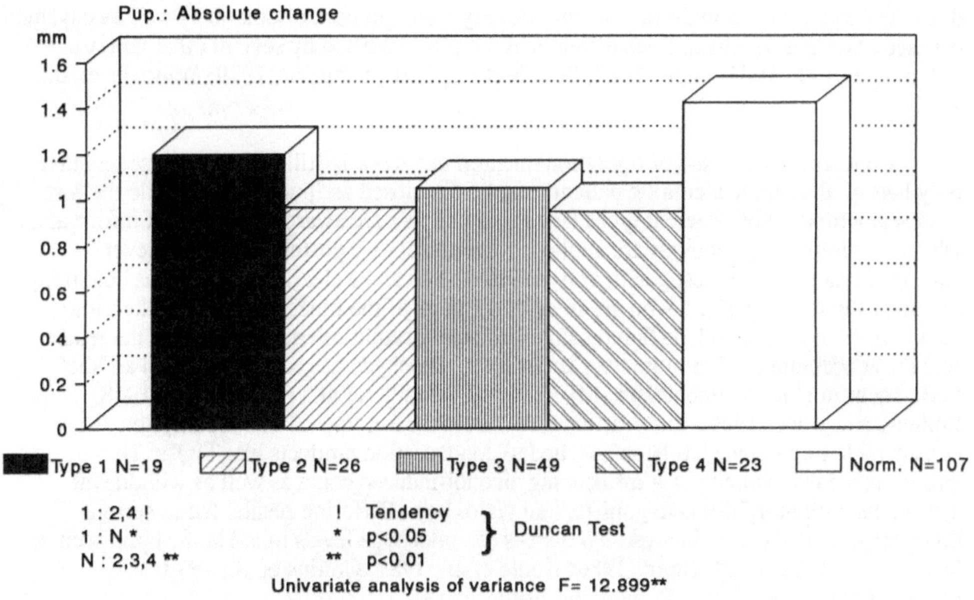

Figure 6. Static and dynamic pupillometry reflects the autonomic nervous system. One important parameter is the absolute change in the pupil, which is determined by cholinergic effects. This figure shows the greatest changes, which reflect the most severe disturbances in the cholinergic system, in the 4 types of alcoholic patients.

DISCUSSION

Nearly all authors (Berner et al., 1986; Buydens-Branchey et al., 1989; Cloninger et al., 1981; Glatt, 1982; Heimann et al., 1989; Knorring et al. 1985; Lesch, 1985; Martin and Eckardt, 1985; Nordström and Berglund, 1987; Rounsaville et al., 1987; Schuckit, 1979) agree on the necessity of dividing chronic alcoholic patients into different subgroups. In our opinion drinking behaviour has always been over-estimated especially in the Anglo-American literature - (Babor and Lauerman, 1986; Button, 1986; Donovan et al., 1986; Glatt, 1982; Jellinek, 1960; Keller, 1976; Pattison and Kaufman, 1982; Vaillant, 1983) and our studies have emphasized this assumption . We are by contrast of the opinion that this overestimation of drinking behaviour has hampered scientific alcoholism research for the last 30 years (Glatt,, 1982; Lesch, 1985).

Our results clearly show, that different alcohol effects (stimulation, sedation, etc) largely depend on the patient's basic disturbances underlying drinking behaviour:

Type 1 submits to alcohol permissive surroundings with genetic factors as a hypothesized etiology.

Type 2 gains anxiolytic effects. We hypothesize in this type a psychogenic etiology.

Type 3 experiences sedating and antidepressive effects. As etiological factors we hypothesize psychiatric conspicuousness (as e.g. manic-depressive illness).

Type 4 experiences more sedating effects, whereby a lowered critical attitude towards drinking behaviour is a main factor. For etiology we hypothesize somatic as well as psychic damage prior to alcoholism. Such differences are also indicated by several other authors (Cloninger et al., 1981; Glatt, 1982; Polich et al., 1980; Schuckit, 1979; Winokur et al., 1971).

Concerning the issue of operationalisation our work is still going on. It seems that only half of all chronic alcoholic patients can be diagnosed as "pure types", while the rest represent a mixture of different types (see Figure 4). We are aware of the loss of information whenever simplifying methods are applied. Nevertheless we postulate that whenever etiological factors are investigated, only patients who can be clearly diagnosed as belonging to one of the four types and who are in an early stage of the illness should be taken into a study . If mixed types with different alcohol-related disabilities are investigated, the result will differ accordingly. For more than ten years acetaldehyde has been considered as "the" toxic agent in chronic alcoholism (Myers, 1989; Naranjo et al., 1986; Ritson, 1968; Sonntag and Boyd, 1989; von Wartburg, 1975,1985). Relying on this assumption considerable importance has been attached to condensation products like TIQ' s. This circumstance led to attempts at influencing alcohol-induced coma as well as withdrawal symptoms with morphine-antagonists, like Naloxone . Differing results led to serious doubts regarding the acetaldehyde-hypothesis (acetaldehyde levels found in the brain seem to be too low) (Barz, 1988; Blum, 1980; Bonte et al., 1987; Collins et al., 1973; NImmerrichter et al., 1989; Sprung and Bonte, 1988) .

Bonte, Sprung and other authors (Bonte, 1987; Bonte et al., 1989; Sprung and Bonte, 1988) have focussed interest on methanol content in alcoholic beverages and on its metabolism to produce formaldehyde . Our present results indicate that - especially in type 1 - formaldehyde may constitute an important etiological factor.

The different results obtained in neurophysiological alcoholism research (Begleiter and Platz, 1972; Pollock et al., 1986), as our results, show could be ascribed to the heterogenity of basic data and our studies appear to confirm this assumption. Our data on activation as well as on autonomous regulation (Grünberger et al., 1989) differ significantly in the 4 different types. As far as we know the international literature does not contain information on similar investigations. EEG studies, f.i. those of Ehlers, Wall and Schuckit, point in the same direction (Ehlers et al., 1989; Lishman, 1981).

CONCLUSION

Our results as well as pertinent literature emphasize the necessity to subgroup patients with the diagnosis "chronic alcoholism" for basic research as well as for the appliance of special therapeutic strategies. The formation of subgroups should consider the underlying symptomatology (Cloninger et al., 1981; Lesch, 1985). This type of diagnostic approach has already been applied by Viennese researchers to patients with delusional disorders. By this approach it could be shown that the structure of delusions is important for the therapeutic handling of the patient, but that the pharmacological approach depends on the basic disturbance underlying the delusional phenomena (Berner et al., 1986; Musalek, 1989; Walter and Berner, 1989). Only a clear separation between different neuronal mechanisms of biologically caused craving (Figure 3) and the exclusion of psychologically- and socially-caused alcoholism syndromes can lead to detection of different biological mechanisms. Our typology represents a contribution to stimulate research in this direction. Only when the varying basic mechanisms are detected, specialized therapeutic programs can be developed, as f.i. identification of a chemical substance capable of blocking specific mechanisms responsible for craving in the individual subtypes.

REFERENCES

Alfano, A.M., Nerviano, V.J. and Thurstin, A.H., 1987, An MMPI based clinical typology for inpatient alcoholic males. Part I: derivation and interpretation. **J.Clin.Psychol.** 43 (4): 431-437.

Alterman, A.I. and Tarter, R.E., 1986, An examination of selected typologies: Hyperactivity, familial and antisocial alcoholism. **Recent Dev Alcohol**, 4: 169-189.

Angst, J., 1973, The course of monopolar depression and bipolar psychosis. **Psychiatria, Neurologia, Neurochirurgia**, 76 (6):489.

Babor, T.F. and Lauerman, R.J., 1986, Classification and forms of inebriety: Historical antecedents of alcoholic typologies. **Recent Dev Alcohol**, 4, p:113144 (1986).

Bacopoulos, N.G., Bhatnagar, R.K. and Van Orden, L.S., 1978, The effects of subhypnotic doses of ethanol on regional catecholamine turnover. J. **Pharmacol. Exp. Ther.** 204:1.

Barz, J., Sprung, R., Fruedenstein, P., Bonte, W., Nimmerrichter, A., Lesch, O.M. and Jakob, B., 1988, Investigation on methanol kinetics in alcoholics. **Blutalkohol** 25: 163.

Barz, J., Freudenstein, P., Bonte, W., Lesch, O.M., Sprung, R. and Nimmerrichter, A., 1989, Besonderheiten des Methanolstoffwechsels bei chronischen Alkoholikern. **Wiener Zeitschrift f. Suchtforschung**, Jg.ll; 4: 25-28.

Bean, M.H., Khantzian, E.J., Mack, J.E., Vaillant, G. and Zinberg, N.E., 1981, **in**: 'Dynamic Approaches to the Understanding and Treatment of Alcoholism'. Eds.: M.H. Bean and N.E. Zinberg. New York, Free Press (Macmillan).

Begleiter, H. and Platz, A., 1972, The effects of alcohol on the CNS in humans. **In**: B.Kissin and H.Begleiter (Eds.)-'The Biology of Alcoholism,. Vol.2, pp: 293-343, Plenum Press, New York.

Berner, P., 1982, Psychiatrische Systematik. 3. Aufl. Huher Verlag. Bern.

Berner, P., 1986a, Psychosis. New aspects of classification and prognosis coming from the Vienna Research Group. **Psychopathology**, 19: 16-29.

Berner, P., Lesch, O.M. and Walter,H., 1986b, Alcohol and Depression. **Psychopathology** 19, suppl.2: 177-183.

Besson, J.A.O., Hutchinson, J.M.S., Mallard, J.R. and Ashcroft, G.W., 1981, Nuclear magnetic resonance observations in alcoholic cerebral disorder and the role of vasopression. **Lancet** 2: 923-924.

Blake, B.G., A follow-up of alcoholics treated by behaviour therapy. **Behaviour Research and Therapy**, 5: 89-94.

Blaney, R., Radford, I.S. and MacKenzie, G., 1975, A Belfast Study of the Prediction of Outcome in the Treatment of Alcoholism. **Brit.J. of Add.** 70: 41-50.

Blum, K., 1980, Alcohol and opiates: A review of common mechanisms. **In**: Y.Lacasse, N.Levy, L. Manzo (Eds.), Proceedings of the Italian Society of Toxicology. pp: 71-90. Pergamon Press, Oxford.

Bonhoeffer, K., 1912, Die Psychosen im Gefolge von akuten Infektionskrankheiten. Allgemeinerkrankungen und inneren Erkrankungen. **In**: G. Aschaffenburg (Hrsg.) Handbuch der Psychiatrie. Deuticke, Leipzig, Wien.

Bonte, W., 1987, Begleitstoffe alkoholischer Getranke, Verlag Max Schmidt-Romhild, Lubeck.

Bonte, W. and Sprung, R., 1987, Methanol as a biological marker of chronic alcoholism (1987). **In**: W. Bonte (Ed.), 'Congener alcohols and their medicolegal significance'. Stockholm, Dalctraf, 154.

Bonte, W., Lesch, O.M. and Sprung, R., 1987, Alcoholism equates methanolism? **Canad Soc Forens Sci J.**,20:74.

Bonte, W., Sprung, R. and Lesch, O.M., 1989, Neue biochemische Oberlegungen zurSuchtentwicklung. Wiener Zeitschrift f. Suchtforschung,Jg.ll: 15-18.

Button, A.D., 1956, The Genesis and Development of Alcoholism, an empirically based Schema, **Quart.J. Stud.Alc**,17: 671-675.

Buydens-Branchey, L., Branchey, M.H. and Noumair, D., 1989, Age of Alcoholism Onset.I.Relationship to Psychopathology. **Arch. Gen Psychiatry**, 46:225-230.

Buydens-Branchey, M.H., Noumair, D and Lieber, C.H., 1989, Age of Alcoholism Onset.II. Relationship to Susceptibility to Serotonin Precursor Availability. **Arch. Gen. Psychiatry**, 46: 231-236.

Cernovsky, Z.Z.,1985, MacAndrew alcoholism scale and repression: detection of false negatives. **Psychol Rep.**, 7: 191-194.

Cloninger, C.R., 1987, Neurogenic adaptive mechanisms in alcoholism. **Science**, 236: 410-416.

Cloninger, C.R., Bohman, M. and Sigvardsson, S., 1981, Inheritance of alcohol abuse cross-fostering analysis of adopted men. **Arch Gen Psychiat**. 42: 1043-1049.

Collins, A.C., Cashaw, J.L. and Davis, V.E., 1973, Dopamine-derived tetrahydroisoquinoline alkaloids - Inhibitors of neuroamine metabolism. **Biochem Pharmacol**., 22: 2337.

Conley, J.J. and Prioleau, L.A., 1983, Personality typology of men and women alcoholics in relation to etiology and prognosis. **J.Stud.Alc**.,44: 996-1010.

Crossley, I.R., Neuberger, J., Davis, M., Williams, R. and Eddleston, A.L.W.F., 1986, Ethanol metabolism in the generation of new antigenic determinants on liver cells. **Gut** 27: 186-189.

Curlee, J., 1971, Sex differences in patient attitudes toward alcoholism treatment. **Quart.Journ. of Stud. on Alc**., 32: 643-650.

Dahlgren, L., 1975, Special problems in female alcoholism. **Brit.J. Addict**., 70 (suppl. 1): 18-24.

Davis, H.G., 1957, Variables associated with recovery in male and female alcoholics following hospitalisation. Doct.Diss., Texas. Technol. College.

Donovan, D.M., Kivlahan, D.R. and Walker, R.D., 1986, Alcoholic subtypes based on multiple assessment domains: validation against treatment outcome. **In**: M.Galander (Ed.) Recent Development in Alcoholism, Chapt.9,pp: 207-222, Plenum Press, New York.

Edwards, G., 1984, Drinking in longitudinal perspective: career and natural history. **Brit.J. of Addiction**, 79: 175-183.

Edwards, G., Gross, M.M., Keller, M., Moser, J. and Room, R., 1977a, Alcohol Related Disabilities, World Health Organisation Offset Publ., No.32; Geneva.

Edwards, G., Orford, J., Egert, S., Guthrie, S., Hawker, A., Hensman, C., Mitcheson, M., Oppenheimer, E. and Taylor, C., 1977b, Alcoholism: A controlled Trial of "Treatment" and "Advice", **J.Stud.Alc**., pp: 1004-1031.

Edwards, G., Duckitt, A., Oppenheimer, E., Sheehan, M. and Taylor, C., 1983, What happens to alcoholics? **Lancet**, 2: 269-271.

Edwards, G., Brown, D., Duckitt, A., Oppenheimer, E., Sheehan, M. and Taylor, C., 1987, Outcome of alcoholism: The structure of patient attribution as to what causes change. **Brit.J.of Addiction**, 82: 533-545.

Edwards, G., Brown, D., Oppenheimer, E., Sheehan, M., Taylor, D. and Duckitt, A., 1988, Long term outcome for patients with drinking problems: the search for predictors. **Brit.J.of Addiction** 83: 917-927.

Ehlers, G., Wall, T.L. and Schuckit, M.A. , 1989, EEG spectral characteristics following ethanol administration in young men. **Electroencephalography and Clin. Neurophysiol**. 73: 179-187.

Ewing, J.A., 1984, Detecting alcoholism, the Cage questionnaire. **JAMA** 252: 1905-1907.

Feuerlein, W., Ringer, C., Küfner, H. and Antons, K., 1980, Diagnosis of alcoholism. The Munich Alcoholism Test (MALT), **In**: M.,Gallanter (Ed), 'Currents in Alcoholism', Vol.7.

Fitzgerald, B.J., Paewark, R.A. and Clark, R., 1971, Four-year Follow-up of Alcoholics at a Rural State Hospital. **Quart.J.** 636-642.

Fox, R., 1976, A multidisciplinary approach to the treatment of alcoholism.**Am.J. Of Psychiatry** 23: 769-778.

Gilles, M., Laverty, S.G. Smart, R.G. and Aharan, C.H., 1974, Outcome; in treated alcoholics, patient and treatment characteristics in a one year follow-up study. **Journal of Alc.**9: 125-134.

Gilligan, S., Reich, T., Cloningrer, C.R., 1987, Etiologic heterogeneity in alcoholism. **Genet Epidemiol.**, 4: 395-414.

Glatt, M., 1982, Alcoholism, care and welfare. Hodder & Stoughton, Sevenoaks.

Grünberger, J., Linzmayer, L. and Saletu, B., 1984, Zur Methodologie der Pupillenmessung. **Psychiatr. Clin.** 3: 157.

Grünberger, J., Linzmayer, L., Gattmann, P. and Saletu, B., 1985, Computerassistierte statische und lichtevozierte dynamische Pupillometrie bei psychosomatischen Patienten. **Wiener Klin.Wschr.** 20: 775.

Grünberger, J., Linzmayer, L., Cepko, H. and Saletu, B., 1986, Pupillometrie im psychopharmakologischen Experiment. **Arzeim.Forschung.** l: 141.

Grünberger, J., Linzmayer, L., Cepko, H. and Saletu, B., 1987a, Lichevozierte dynamische Pupillometrie zur Differenzierung psychotroper Substanzen. **Arzneim.Forsch.** 3: 357.

Grünberger, J., Linzmayer, L., Witek, R. and Saletu, B.,: Faktorenanalytische Untersuchungen und Reliabilitatsbestimmung der statischen und dynamischen Pupillometrie. **Wiener Med. Wschr.** 78; pp: 135-139 (1987b).

Grünberger, J., Linzmayer, L. and Feuerlein, W., 1988, Psychische Veranderungen beim Alkoholmißbrauch. **Internist** 29: 307-313.

Grünberger, J., Lesch, O.M. and Linzmayer, L., 1989, Bestimmung von vier Alkoholikertypen mit Hilfe der statischen und lichtevozierten dynamischen Pupillometrie. **Wiener Zeitschrift fur Suchtforschung**, Jg.ll; 4: 29-34 .

Grünberger, J., Linzmayer, L., Saletu, B. and Lesch, O.M., Klinische und psychophysiologische Diagnostik bei ambulanten Alkoholikern: statische und lichtevozierte Pupillometrie. **Wiener Zeitschr. für Suchtforschung**, in press.

Hagnell, O., Lanke, J., Rosman, B. and Öhman, R., 1986, Predictors of alcoholism in the Lundy study. **Eur.Archs Psychiatr.Neurol.Sci**, 235: 187-191.

Heimann, H., Mayer, K. and Schied, W., 1953, (Hrsg.): Der chronische Alkoholismus. Gustav Fischer Verlag, Stuttgart, New York.

Hoff, E.C. and McKeown, C.E., 1953, An Evolution of the use of tetraethylthiuram disulfide in the treatment of 560 cases alcohol addiction, **Am. J. of Psychiatry** 109: 670-673.

Horn, J.L., Wanberg, K.W. and Foster, F.M., 1983, The Alcohol Use Inventory. Baltimore. Psych. Systems.

Jaspers, K., 1973, Allgemeine Psychopathologie. 9. Auflg., Springer Verlag,Berlin, Heidelberg, New York.

Jellinek, E.M., 1960, The disease concept of alcoholism. Hillhouse Press,New Brunswick.

Johnson, J.L., Adinoff, B., Bisserbe, J.C., Martin, P.R., Rio, D., Rohrbaugh, J.W., Zubovic, E. and Eckardt, J., 1986, Assessment of alcoholism-related organic brain syndromes with positron emission tomography. **Alcoholism: Clinical and Experimental Research** 10: 237-240.

Keller, M., 1976, The disease concept of alcoholism revisited. **J.Stud.Alc.** 37: 1694-1717.

Knorring, L. von, Palm, U. and Andersson, H.E., 1985, Relationship between treatment outcome and subtype of alcoholism in men. **J.Stud.Alc** 46: 388-391.

Koehler, K. and Sass, H., 1984, Diagnostisches und Statistisches Manual Psychischer Storungen, DMS-III. Beltz Verlag, Weinheim (1984) .
Kolb, D., Gunderson, E. and Buvcky, S., 1976, Prognostic indicators for black and white alcoholics in the U.S. navy. **J.Stud. Alc.**, 377: 9890.
Lachar, D., Berman, W. and Grisell, J.L., et al, 1976, The Mac Andrew alcoholism scale as a general measure of substance misuse.**J.Stud. Alc.** 37: 1609-1615.
Lesch, O.M., 1981, Chronischer Alkoholismus. Typen und ihr Verlauf. Eine Langzeitstudie. Georg Thieme Verlag, Stuttgart, New York.
Lesch, O.M., Dietzel, M., Musalek, M., Walter, H. and Zeiler, K., 1988a, The course of alcoholism. Long-term prognosis in different types. **Forensic Science Int.**, 36: 121-138.
Lesch, O.M., Bonte, W. and Grünberger, J., 1988b, Eine Typologie des chronischen Alkoholismus - neue Basisdaten fur Forschung und Therapie. **In**: Ladewig (Ed), 'Drogen und Alkohol', pp: 119-134, ISPA Press, Lausanne.
Lesch,O.M., Walter, H., Mader, R., Musalek, M. and Zeiler, K., 1988c, Chronic alcoholism in relation to attempted or effected suicide. A long-term study. **Psychiatr. & Psychobiol.**, 3: 181-188.
Lesch, O.M., Dietzel, M., Musalek, M., Walter, H. and Zeiler, K., 1989, Therapiekonzepte und Therapieziele im Lichte langfristiger Katamnesen: **In**: Schied, Heimann und Mayer (Hrsg.) : 'Der chronische Alkoholismus'. Gustav Fischer Verlag, Stuttgart, New York.
Lesch, O.M., Kefer, J., Lentner, S., Mader, R., Marx, B., Musalek, M., Nimmer-richter, A., Preinsberger, H., Puchinger, H., Rustembevovic, A. and Walter, H., Diagnosis of Chronic Alcoholism - Classificatory Problems. **Psychopathology** 733, , in press.
Lishman, W.A., 1981, Cerebral disorder in alcoholism: syndromes of impairment.**Brain** 104: 1-20.
Litman, G.K., Eiser, J.R., Rawson, N.S.B. and Oppenheim, A.N., 1979, Towards a typology of relapse: differences in relapse and coping behaviours between alcoholic ralapsers and survivors. **Behaviour Research and Therapy**, 17: 89-94.
Marlatt, G.A., 1976, The drinking profile: a questionnaire for the behavioural assessment of alcoholism. **In**: E.J. Mash, L.G. Terdal (Eds), 'Behaviour Therapy Assessment'. Springer, New York.
Martin, P.R. and Eckardt, M.J., 1985, Pharmacological interventions in chronic organic brain syndromes associated with alcoholism. **In**: C.A. Naranjo and E.M.Sellers (Eds.), 'Research Advances in New Psychopharmacological Treatments of Alcoholism'. pp: 257-272. Elsevier/Biomedical Press.Amsterdam.
Martin, P.R., Adinoff, B., Weingartner, H., Mukherjee, A.B. and Eckardt, J., 1986, Alcoholic chronic brain disease: Nosology and pathophysiologic mechanisms. **Prog. Neuro-Psychopharmacol. & Biol. Psychiat.**, 10: 147-164.
Martin, P.R., Eckardt, M.J. and Linnoila, M., 1988, Treatment of chronic organic mental disorders associated with alcoholism. IV. Clinical pharmacology. **In**: Treatment of Chronic Mental Disorders, pp 329-350.
McLellan, A.T., Luborsky, L and Obrien, C.P. et al., 1980, An improved evaluation instrument for substance abuse patients: The Addiction Severity Index: reliability and validity in three centers. **J. Nerv Ment Dis** 168: 26.33.
Morley, L.C. and Blashfield, R.K., 1981, The empirical classification of alcoholism a review. **J.Stud.Alcohol** 42: 925-937.
Musalek, M.:, 1989, Les indicateurs psychopathologiques et biologiques pour la therapeutique des delires chroniques. **Psychologie Medicale** 21: 1355-1359.
Myers, R.D., 1989, Neuroanatomical circuitry underlying alcohol drinking revealed by anatomical mapping of THP reactive sites in rat brain. **American College of Neuropsychopharmacology**, 28th Annual Meeting, Dec. 10th - 15th; p: 47.

Naranjo, C.A., Sellers, E.M. and Lawrin, M., 1986, Modulation of ethanol intake by serotonin uptake inhibitors. **J.Clin. Psychiatry** 47: 16-22

Nationalj Council jon Alcoholism (US), 1972, Criteria for the Diagnosis of Alcoholism, **Annals of Int. Med** 77: 249-258.

Nerviano, V.J. and Gross, H.W., 1983, Personality types of alcoholics on objective inventories: a review. **J.Stud.Alcohol** 44: 837-851.

Nimmerrichter, A., Grohs-Kellner, G. and Lesch, O.M., 1989, Ein Modell fur organisch bedingte psychische Storungen chronischer Alkoholiker. **Wiener Zeitschr.f. Suchtforschung**, Jg.ll; 4: 3-13.

Nordström, G. and Berglund, M., 1987, Type 1 and type 2 alcoholics (Cloninger & Bohman) have different patterns of successful long-term adjustment. **Brit. J. Addict** 82: 761-769.

Ogborne, A.C., 1978, Patient characteristic as predictors of treatment outcomes for alcohol and drug abusers. **In**: Israel et al (Eds.) Research advances in alcohol and drug problems, Vol.4, Plenum Press, New York.

Pattison, E.M. and Kaufman, F., 1982, The alcoholism syndrome: Definitions and models. **In**: E.M.Pattison & F. Kaufman (Eds), 'Encyclopedic Handbook of Alcoholism', Chap. 1, pp. 3-30. Gardner Press, New York.

Polich, M.J., Armor, D.J. and Braiker, H.B., 1980, The Course of Alcoholism, Four Years after Treatment. St.Monica, The Rand Corp.

Pollock, V.E., Teadsale,T.W., Gabrielli, W.F. and Knop, J., 1986, Subjective and objective measures of response to alcohol among young men at risk for alcoholism. **J.Stud.Alcohol** 47: 297-304.

Ritson, B., 1968, The prognosis of alcohol addicts treated by a specialised unit. **Brit.J. of Psychiatry** 114: 1019-1029.

Rodgerson, M.J., Wiseman, B.L. and Ford, W.E., 1980, Rate of Alcoholism diagnoses in community mental health centers, The Effect of the presence of an alcoholism treatment program. **In**: M.Gallanter (Ed), 'Currents in Alcoholism'. Grune/Stratton, pp: 149-159.

RohsenowD.J., 1982, The Alcohol Use Inventory as a predictor of drinking by male heavy social drinkers. **Addict Behav** 7: 387-395.

Rounsaville, B.J., Dolinsky, Z.S., Babor, T.F. and Meyer, R.E., 1987, Psychopathology as a predictor of treatment outcome in alcoholics. **Arch Gen Psychiat** 44: 505-513.

Ruggles, W.L., Armor, D.J., Polich, J.M., Mothershead, A. and Stephen, M., 1975, A follow-up study of clients at selected alcoholism treatment centers funded by NIAA, Menlo Park, Calif., Stanford Research Instr.

Schuckit, M., 1979, Alcoholism and effective disorders. Diagnostic confusion. **In**: D.W. Goodwin, C.K. Erickson (Eds.), 'Alcoholism and effective disorders'. SP Med. Books, New York, London.

Schuckit, M. et al., 1969, Alcoholism I. Two types of alcoholism in women. **Arch.Gen.Psychiatry**, 20: 301-306.

Schuckit, M. and Gold, E.O., 1988, A simultaneous evaluation of multiple markers of ethanol/placebo challenges in sons of alcoholics and controls. **Arch.Gen. Psychiatry** 45: 211-216.

Schuckit, M. and Irwin, M., 1989, An analysis of the cinical relevance of Type 1 and Type 2 alcoholics. **Brit.J.Addict.** 84: 869-876.

Selzer, M.L., 1971, The Michigan Alcoholism Screening Test: the quest for a new diagnostic instrument. **Am J Psychiatry** 127 : 1653-1658.

Sobell, M.B., 1981, The nature of alcohol problems. **Prog.Neuropsychopharmacol.** 5: 475-481.

Sobell, M.B. and Sobell, L.C., 1976, Second year treatment outcome of alcoholics treated by individualized behaviour therapy. **Results.Behav.Res.Ther.**, 14: 195-215.

Sonntag, W.E. and Boyd, R.L., 1987, Diminished insulin-like growth factor 1 levels after chronic ethanol: Relationship to pulsatile growth hormone release. **Alcoholism: Clinical and Experimental Research** 13: 3-7.

Spitzer, M., 1987, Gegenuberstellung der Diagnosekategorien von IDC-9 und DSM-III. **Fortschr.Neurol.Psychiat.** 55: 16-30.

Sprung, R. and Bonte, W., 1988, Die Bedeutung der sogenannten Begleitalkohole fur Suchtentwicklung und Spatschaden des chronischen Alkoholismus. **Wiener Zeitschr.f.Suchtf.**, 11:19-24.

Sprung, R., Bonte, W. and Lesch, O.M., 1991, Methanol - Ein bisher verkannter Bestandteil aller alkoholischen Getranke. Eine neue biochemische Annaherung an das Problem des chronischen Alkoholismus. **Wiener klin. Wochenschrift** pp: 282-288, in press.

Thurstin, A.H. and Alfano, A.M., 1988, The Association of alcoholic subtype with treatment outcome: An 18-month follow-up. **Int.J. Addiction**, 23: 321-330.

Uhl, A., 1983, Die Probleme der Vagheit und Mehrdeutigkeit des Begriffes "Alkoholismus". **In**: R.Mader (Ed), 'Alkohol- und Drogenabhangigkeit. Neue Ergebnisse aus Theorie und Praxi'. pp: 211-226 Verlag Bruder Hollinek, Wien.

Vaillant, G.E., 1983, The natural history of alcoholism: causes, patterns and paths to recovery. Harvard Univ.Press, Cambridge.

VIictor, M. and Banker, B.Q., 1978, Alcohol and dementia. **In**: R.Katzmann, R.D. Terry,K.L. Bick (Eds), 'Alzheimer's Disease, Senile Dementia and Related Disorders'.New York, Raven Press, Vol.7.

Voegtlin, W.L. and Broz, W.R., 1949, The conditioned reflex treatment of chronic alcoholism. An analysis of 3125 admissions over a period of ten and a half years. **Annals of Int.Med.**, 30: 580-597.

Walter, H. and Berner, P., 1989, Symptoms affectis chez les delirants chroniques. **Psychologie Medicale** 21: 1351-1352.

Wanberg, K.W. and Horn, J.L., 1983, Assessment of alcohol use with multidimensional concepts and measures. **Am Psychol** 38: 1055-1069.

Wartburg, J.P. von, 1975, Biochemie der Alkoholintoxikation und des Alkoholismus.**In**: W.Steinbrecher und H. Solms (Hrsg.), Sucht und Mißbrauch. pp: II/78-II/85; Georg Thieme Verlag, 2. Aufl.

Wartburg, J.P. von, 1985, Genetische Suchtdisposition: mogliche biochemische Mechanismen. **In**: W.Keup (Hrsg.), 'Biologie der Sucht'. Springer Verlag, Berlin, Heidelberg, New York, Tokyo.

Wetzler, S., 1988, Assessment of alcoholism and substance abuse. **In**: Scott Wetzler (ed.), 'Measuring Mental Illness: Psychometric Assessment for Clinicians'. Chapt.8.; American Psychiatric Press,Inc., Washington, DC, pp: 163181.

Winokur, G., Rimmer, J. and Reich, T., 1971, Alcoholism IV: Is there more than one type of alcoholism? **Brit.J.Psychiatry** 118: 525-531.

Wittchen, H.U., Sass, H., Zaudig, M. and Koehler, K., 1989, Diagnostisches und statistisches Manual psychischer Störungen. DSM-III-R Revision, Beltz Verlag, Weinheim.

Zivich, M., 1981, Alcoholic subtypes and treatment effectiveness. **J. Consult. Clin.Psychol.** 49: 72-80.

GENETIC LINKAGE ANALYSIS: SIMPLE SOLUTIONS FOR COMPLEX TRAITS?

Luis A. Giuffra

Department of Human Genetics
Yale School of Medicine
New Haven, CT 06510 USA

Many human disorders are either partially or wholly determined by genetic factors. The identification of these genetic defects at the DNA level is a major goal with important implications. Such identification could greatly increase our understanding of diseases whose pathophysiology and clinical course are currently unknown. In turn, this could lead to logical improvements in the way the disorder is treated. Furthermore, antenatal diagnosis and the identification of carriers could provide the opportunity for more accurate genetic counseling.

In disorders where the biochemical abnormality is known, identification of the DNA defect is relatively straightforward: from the amino acid sequence of both the normal and abnormal gene product, the corresponding DNA sequences can be deduced. However,in many heritable disorders little or nothing is known about the biochemical abnormality. In these conditions, identification of the genetic defect needs to be approached in a different way.

For many years, geneticists have attempted to take advantage of the familial aggregation of disorders in their search for defective genes. Simply stated, in a large family where many members are affected by a particular genetic abnormality, all those affected share the same piece of chromosome containing the defective gene. This chromosome segment has been inherited from a single copy introduced into the pedigree by an ancestor suffering from the disease. All those affected in the family carry an identical copy of this segment which not only contains the abnormal gene but also a sizable amount of other surrounding DNA. Therefore, if a piece of the surrounding DNA can be identified, then the chromosomal location of the genetic defect is immediately known despite complete absence of information about the gene product. Once a precise location is achieved, molecular biological techniques can be applied to ultimately clone and sequence the genetic defect underlying the disease.

Two pieces of DNA that tend to be inherited together are said to be linked. During meiosis, homologous chromosomes exchange whole regions of DNA in a process known as crossing over or genetic recombination. The closer two pieces of DNA are to each other, the less likely it is that a meiotic crossover will separate them. Tightly linked pieces of DNA are almost always inherited together. If we can follow one of them through a pedigree we

Alcoholism: A Molecular Perspective
Edited by T.N. Palmer, Plenum Press, New York, 1991

can assume that the other has been inherited with it. The goal of linkage analysis is to identify a piece of DNA, called a marker, which lies so close to a defective gene that they tend to be co-inherited. Since the chromosomal location of the marker is usually known beforehand, localization of the defective gene is made, paving the way for its ultimate characterization.

LOD SCORES AND RFLPs

In order to perform a genetic linkage analysis, families where the disease is segregating need to be collected. DNA from both affected and unaffected family members is first extracted and typed for genetic markers. Next, the pattern of segregation of each genetic marker is compared to the pattern of segregation of the disease in the family. A statistic is then constructed that compares, for each marker in turn, the likelihood that the pattern of segregation of the marker and the disease gene occurred because the two are linked, ie, show some degree of cosegregation, with the likelihood that it arose because they are unlinked, ie, segregate independently of each other. In other words, this statistic compares the odds of linkage versus non-linkage as explanations for the observed segregation of the data. These odds are usually presented as their logarithm to the base 10, hence the name, LOD score (Logarithm of the ODds).

Until a few years ago, the possibilities of using the LOD score method to map genetic diseases was severely limited by the small number of markers available. These markers were usually protein variants or polymorphisms such as the ABO and Rh blood types and included the HLA system on chromosome 6, a highly polymorphic genetic locus. Our ability to map heritable disorders is now much improved by the rapid accumulation of new DNA markers scattered throughout the human genome (Watkins, 1988). These markers often occur in regions of DNA that do not code for any gene. Mutations occurring within these regions are therefore biologically silent and have no adverse effects. Their occurrence, however, has been exploited advantageously because they can alter the specific recognition sequence of restriction enzymes. These enzymes cut DNA into various lengths. A mutation can either destroy a pre-existing restriction enzyme site or generate a new one, thereby creating variants or polymorphisms in the length of DNA cut. These variants are termed RFLPs (restriction fragment length polymorphisms) and segregate within families. These RFLPs seem to be located everywhere in the genomes of all species studied so far, and are invaluable tools for gene mapping as they can be tracked through pedigrees They have allowed geneticists to map a large number of genetic disorders for whom no gene product was known. For some of these diseases, such as Duchenne's muscular dystrophy, their mapping has ultimately led to the identification of the primary defect (Kunkel et al., 1986).

SIMPLE AND COMPLEX TRAITS

A common characteristic of all disorders successfully mapped to date, is that their mode of inheritance was well known before the linkage analysis was performed. All followed typical mendelian patterns of segregation, either dominant (such as Huntington's disease and polycystic kidney disease) or recessive (such as cystic fibrosis and Friedreich's ataxia). In others, the lack of transmission of the disease from father to son placed the defect on the X chromosome and linkage was used to achieve a more detailed localization.

Another characteristic shared by these diseases is that they tend to occur rarely in the population. Strong selection against them usually exists, with decreased fertility among those affected reducing the presence of the defective gene in the population.

On the other hand, there is another group of conditions which also demonstrate familial aggregation, suggesting a genetic origin. These include insulin-dependent diabetes mellitus (IDDM), hypertension, schizophrenia and many others. These conditions differ from the mapped disorders in two ways. Firstly, they are commonly found in all populations, affecting large numbers of individuals, often in spite of strong selection forces acting against them. For example, despite a large decrease in fertility among schizophrenics, schizophrenia affects nearly 1% of all populations studied. Secondly, their mode of inheritance is not known, or, at least, is known not to follow mendelian laws. Although largely genetic in origin, the environment plays an important role in their etiology (monozygotic twins are never 100% concordant, as is usually the case in mendelian disorders). Furthermore, careful study of the pattern of familial aggregation suggests that the genetic component of some of these diseases is due to the synergistic interaction of several unlinked genes, probably located on different chromosomes. Since no single gene produces the disease by itself, its passage through many generations can occurr without causing deleterious effects. This may explain why some of these disorders are so common, and why their incidence is not naturally declining through selection processes.

Disorders that are known to run in families, whose expression seem to depend at least partly on the environment, and whose genetic component does not follow mendelian laws, are called complex. No simple mechanism can be invoked to group them; even within one disorder, the factors causing the disease in one person could be different to the factors causing it in a close relative.

LINKAGE ANALYSIS IN PSYCHIATRIC DISORDERS: TYPE I AND TYPE II ERRORS

Does linkage analysis have a role in the study of these genetically complex traits? The familial aggregation observed in these disorders, and the relative robustness of the LOD score method (Clerget-Darpeaux et al., 1986) prompted researchers to adopt this approach in an attempt to map genes underlying some of the more common complex traits. Major psychiatric disorders, because of both their high prevalence and social cost, have been among the first examined using RFLPs and LOD scores (Giuffra and Kidd, 1989). However, controversial results have been obtained for both schizophrenia with chromosome 5 markers and bipolar affective disorder with chromosome 11 markers.

The original report of linkage between markers on the short arm of chromosome 11 and bipolar affective disorder in the Old Order Amish (Egeland et al., 1987) was published simultaneously with the results from two independent groups who where unable to replicate this finding (Detera-Wadleigh et al., 1987; Hodgkinson et al., 1987). Initially this non-replication was thought to reflect the uniqueness of the Amish population, whose isolation and limited gene pool made them an ideal group for genetic research. However, the addition of new family members and revision in the disease status of several of the original members (from unaffected to affected) resulted in the positive LOD score largely disappearing, and the original claim for linkage has been recently withdrawn (Kelsoe et al., 1990).

Similarly, linkage between schizophrenia and markers on the proximal long arm of chromosome 5 has been reported (Sherrington et al., 1988) although several other groups have been unable to replicate this finding (reviewed in McGuffin et al., 1990). Inability to replicate should not be confused with genetic heterogeneity. The former is the most likely

Table 1. Multipoint linkage analysis is performed to calculate LOD scores across a genetic map formed by three markers (MK 5, MK 6 and MK 7) and containing a recessive gene between two of them (marked by an asterisk between MK 6 and MK 7; between markers, genetic distances shown are recombination fractions). The gene is underlying the trait in some of the families only (for a description of the dataset, see Elston et al, 1989). Using high penetrances (p) and/or low allele frequencies (f) generate very negative LOD scores all across the map (Model I: p=0.95 and f=0.001; Model II: p=0.0 and f=0.001). A more conservative analysis with moderate penetrance and high gene frequency, however, generate very positive LOD scores (Model III: p=0.50 and f=0.30). These results show the strong sensitivity of the analysis to the parameters used.

MODEL	LOD SCORES ACROSS THE GENETIC MAP				
	MK5---------0.5 ----------- MK6------0.1 ------ * ------0.2 MK7				
I	-76.07	-52.08	-130.32	-64.97	-117.62
II	-34.21	-27.58	-57.47	-36.64	-54.57
III	8.02	9.53	9.10	9.17	6.71

explanation for these opposing results, while the latter requires solid and reproducible evidence of genetic linkage to two or more different genetic locations. Tuberous Sclerosis is a good example of genetic heterogeneity, with its two autosomal forms, one on chromosome 9 and the other on chromosome 11 (Jansen et al., 1990).

Explanations for these conflicting results are beginning to emerge. Careful analysis of the LOD score approach to complex traits reveals that the method suffers from a significant risk of generating both type I and type II errors.

Type I errors (falsely concluding linkage when linkage does not exist) are generated when the traditional LOD score of 3 is used as a threshold for establishing linkage. This value is appropriate for disorders with an established mode of inheritance, when families are collected sequentially and when the analysis is stopped after reaching a LOD score of either 3 (supporting linkage) or of -2 (rejecting linkage). Application of the LOD score method to the same dataset used iteratively with several different genetic models introduces the problem of multiple comparisons. Two groups have demonstrated recently the effects of multiple comparisons and have shown that multiple testings can generate high LOD scores more easily than if one single genetic model is used in the analysis (Clerget-Darpeaux et al., 1990; Weeks et al., 1990).

Type II errors (wrongly rejecting linkage when linkage does exist) can arise when the mode of inheritance is misspecified. Because in complex traits the mode of inheritance is unknown, misspecifications are very likely. It can be shown that misspecifications are particularly damaging when multipoint linkage analysis is used (Risch and Giuffra, 1990).

Table I shows the effect on LOD score calculations of changing the genetic parameters for a trait with unknown mode of inheritance. A simulated dataset of a genetically heterogeneous condition was used, whose true mode of inheritance was not known when performing the linkage analysis (Elston et al., 1989). Multipoint analysis (testing for linkage between a disease locus and a set of linked markers simultaneously) using the LOD score method was performed. The results shown correspond to the inner segments of the map, where the defective gene causing the disease in a subset of the families is located. When stringent parameters are used, very negative LOD scores are readily obtained despite testing the correct mode of inheritance (in this case, recessive). These results demonstrate that multipoint linkage analysis, as opposed to pairwise analysis, is very susceptible to misspecification of the parameters used to define the mode of inheritance. Furthermore, in view of these type II errors, "candidate genes" for a complex trait should not be excluded on the basis of negative LOD scores, particularly in multipoint analysis, because these LOD scores can arise merely from misspecifications.

In summary, the LOD score method of linkage analysis has not yet been successfully applied to complex human traits, although researchers have been able to learn more about its strengths and weaknesses when used in non-mendelian situations. A careful evaluation of the current methods used in genetic analysis should help us design new strategies to study these complex disorders. The massive amount of information that is accumulating on the human genome and its polymorphisms promises to be our best tool in attempting to dissect out the genetic component of complex traits.

ACKNOWLEDGEMENTS

Dr. Chris Dudley kindly revised and commented on drafts of this manuscript. Supported by a training fellowship from the John D. and Catherine T. MacArthur Foundation Mental Health Research Network I .

REFERENCES

Clerget-Darpeaux, F., Bonaiti-Pellie, C. and Hochez, J., 1986, Effects of misspecifying genetic parameters in LOD score analysis. **Biometrics** 42:393-399.

Clerget-Darpeaux, F., Babron, M-C. and Bonaiti-Pellie, C., 1990, Assessing the effect of multiple linkage tests in complex diseases. **Genetic Epidemiology** 7(4):245-253.

Detera-Wadleigh, S., Berrettini, W.H., Goldin, L., Boorman, D., Anderson, S. and Gershon, E., 1987. Close linkage of c-Harvey-ras-1 and the insulin gene to affective disorders is ruled out in three North American pedigrees. **Nature** 325:806-808.

Egeland, J.A., Gerhard, D.S., Pauls, D.L., Sussex, J., Kidd, K.K., Allen, C.R., Hostetter, A.M. and Housman, D.E., 1987, Bipolar affective disorder linked to DNA markers on chromosome 11. **Nature** 325:783-787.

Elston, R.C., MacCluer, J., Hodge, S., Spence, M.A. and King, R.H., 1989, Genetic Analysis Workshop 6: Linkage analysis based on affected pedigree members. **Progress in Clinical and Biological Research** 329:93-103.

Giuffra, L.A. and Kidd, K.K., 1989, Linkage Analysis in Psychiatry. **International Review of Psychiatry** 1:231-242.

Hodgkinson, S., Sherrington, S.R., Gurling, H., Marchbanks, R., Reeders, S., Mallet, J., McInnins, M., Petursson, H. and Brynjolfsson, J., 1987, Molecular genetic evidence for heterogeneity in manic depression. **Nature** 325:805-806.

Jansen, L.A., Sandkuyl, L.A., Merkens, E.C., Maat-Kievit, J.A., Sampson, J.R., Fleury,

P., Hennekam, R.C.M., Grosveld, G.C., Lindhout, D. and Halley, D.J.J., 1990, Genetic Heterogeneity in Tuberous Sclerosis. **Genomics** 8:237-242.

Kelsoe, J.R., Ginns, E.I., Egeland, J.A., Gerhard, D.S., Goldstein, A.M., Bale, S.J., Pauls, D.L., Long, R.T., Kidd, K.K., Conte, G., Housman, D.E. and Paul, S.M., 1990, Re-evaluation of the linkage relationship between chromosome 11p loci and the gene for bipolar affective disorder in the Old Order Amish. **Nature** 342:238-243.

Kunkel, L.M. & co-authors, 1986, Analysis of deletions in DNA from patients with Becker and Duchenne muscular dystrophy. **Nature** 322:73-77.

McGuffin, P., Sargeant, M., Hett, G., Tidmarsh, S., Whatley, S. and Marchbanks, R.M., 1990, Exclusion of a schizophrenia susceptibility gene from the chromosome 5q11-q13 region: new data and a reanalysis of previous reports. **American Journal of Human Genetics** 47(3):524-535.

Risch, N. and Giuffra, L.A., 1990, Multipoint linkage analysis of genetically complex traits. **American Journal of Human Genetics** 47 (3):A197.

Sherrington, R, Brynjolfsson, J., Petursson, H., Potter, M., Dudleston, K., Barraclough, K., Wasmuth, J., Dobbs, M. and Gurling, H., 1988, Localization of a susceptibility locus for schizophrenia on chromosome 5. **Nature** 336:164-167.

Watkins, P.C., 1988. Restriction fragment length polymorphism (RFLP): applications in human chromosome mapping and genetic disease research. **Biotechniques** 6:310-320.

Weeks, D.E., Lehner, T., Squires-Wheeler, T.E., Kauffman, C. and Ott, J., 1990, Measuring the inflation of the LOD score due to its maximization over model parameter values in human linkage analysis. **Genetic Epidemiology** 7(4):237-243.

THE CHANGING VIEW OF ETHANOL'S ACTIONS: FROM GENERALITIES TO SPECIFICS*

Boris Tabakoff and Paula L. Hoffman

Dept. of Pharmacology
Univ. of Colorado Hlth. Sci. Ctr.
Denver, CO 80262, USA
and Section on Receptor Mechanisms
Lab. of Physiologic and Pharmacologic Studies
NIAAA
Bethesda, MD 20852, USA

The development of an addiction on ethanol involves a number of phenomena arising from the actions of ethanol on the central nervous system (CNS). These phenomena include ethanol- induced intoxication, reinforcement, the development of functional tolerance and physical dependence and ethanol-induced brain damage. One of the more popular hypotheses regarding the cellular and molecular mechanisms by which ethanol produces its effect on the CNS proposes that ethanol acts by disordering (fluidizing) the lipids of neuronal membranes (Meyer and Gottlieb, 1926; Chin and Goldstein, 1977). The disorder of the structural matrix of the neuronal membrane is thought to secondarily affect the activity of the functional proteins (receptors, ionophores, enzymes) residing in this (lipid) matrix. The "lipid hypothesis" of ethanol's actions has recently been challenged, however, because relatively high concentrations of ethanol are necessary to produce lipid perturbations of the magnitude produced by even normal variations in body temperature (Tabakoff, Hoffman, and MacLaughlin, 1988). In addition, recent research has provided evidence of a selectivity of ethanol's effects on various neuronal membrane proteins which cannot be explained simply by invoking an ethanol-induced perturbation of membrane lipid structure. Some of the most ethanol sensitive membrane proteins are the receptor-gated ion channels. The burgeoning knowledge of the physiologic roles of these channels and the responses of these proteins to the acute and chronic effects of ethanol suggests that particular systems may play a major role in the expression of the various physiologic and behavioral effects of ethanol. This overview will attempt to summarize the evidence that certain receptor-gated ion channels may be considered as specific "receptive systems" for ethanol in the CNS. We will also provide evidence on how certain neurohormones can modify the actions of ethanol within the CNS.

The effects of ethanol on the N-methyl-D-aspartate (NMDA) receptor have been avidly investigated over the last two years. The NMDA receptor is activated *in vivo* by the

Edited by T.N. Palmer, Plenum Press, New York, 1991

neurotransmitter amino acid glutamate. Glutamate is the major excitatory neurotransmitter in brain, and there are at least three other subtypes of glutamate receptor, the kainate, quisqualate and 4-aminophosphonobutyrate receptors, named for their selective interactions with these respective ligands (Collingridge and Lester, 1989). The NMDA receptor is coupled to an ion channel that, when open, is permeable to calcium, as well as monovalent cations. The channel can be blocked by magnesium, which is cleared from the channel as the cell is depolarized. Thus, responses to NMDA or glutamate at the NMDA receptor depend on the depolarization state of the cell. There is also a binding site within the channel for the dissociative anesthetics, phencyclidine (PCP) and ketamine, and the compound MK-801 binds to this site as well. All of these agents act as noncompetitive inhibitors of NMDA actions. The NMDA receptor is also subject to "allosteric" regulation by various substances. Glycine, acting at a strychnine-insensitive site, has been suggested to be required for the action of NMDA, i.e., glycine and NMDA or glutamate act as co-agonists at the NMDA receptor. Zinc ion and polyamines also influence NMDA responses (Collingridge and Lester, 1989).

The NMDA receptor has been implicated in learning processes (long-term potentiation), in neuronal development and in brain damage (Harris et al., 1984: Kleinschmidt et al., 1987). It also plays a key role in the generation of epileptiform seizures (Dingledine et al., 1986). These functions of the NMDA receptor sparked an interest in evaluating the effects of ethanol on responses to NMDA, since one could postulate a role for the NMDA receptor system in the cognitive impairment produced by ethanol, the psychological deficits associated with the fetal alcohol syndrome and in the occurrence of ethanol withdrawal seizures.

Initial biochemical studies in cultured cerebellar granule cells demonstrated ethanol to be a potent and selective inhibitor of the NMDA receptor/ionophore function (Hoffman et al., 1989). Ethanol concentrations as low as 10 mM (46mg%) produced significant inhibition of this system while ethanol was shown to be much less potent at inhibiting responses to agonists acting at other subtypes of glutamate receptors. The initial biochemical studies were quickly confirmed by electrophysiologic studies using patch-clamp recording of NMDA-stimulated ion currents (Lovinger et al., 1989) and other biochemical studies monitoring NMDA-stimulated release of neurotransmitters in brain tissue (Gothert and Fink, 1989; Woodward and Gonzales, 1990). More recently the mechanism of action of ethanol has been evaluated and the evidence to date suggests that ethanol may affect the NMDA-glycine interaction that is necessary for channel opening (Rabe and Tabakoff, in press). Thus, in both biochemical and electrophysiological studies, high concentrations of glycine can reduce the inhibition caused by ethanol. These data suggest that ethanol, far from being a non-specific drug, may interact quite selectively with particular sites on the NMDA receptor complex, and the sensitivity of this system to low concentrations of ethanol makes it a prime candidate for mediating cognitive and/or intoxicating effects of ethanol.

Certain of the chronic effects of ethanol, particularly the development of physical dependence as exemplified by ethanol withdrawal hyperexcitability of the CNS, seem to be related to ethanol's actions on the NMDA receptor. There is evidence that a compensatory up-regulation of NMDA receptors occurs during the chronic presence of ethanol in the CNS. After chronic ethanol ingestion by mice, binding of the NMDA antagonist, MK-801, was increased in brain. MK-801, as already noted, labels ion channels associated with the NMDA receptor. MK-801 binding was increased in several brain areas, and the time course of changes in hippocampal MK-801 binding paralleled that for susceptibility to ethanol withdrawal seizures (Gulya et al., in press). Furthermore, treatment of animals with NMDA receptor antagonists reduced ethanol withdrawal seizure severity, while NMDA, at a dose that was not convulsant in control animals, exacerbated ethanol withdrawal seizures (Grant et al., 1990).

While the NMDA receptor may mediate the ability of ethanol to produce acute cognitive impairment and aspects of physical dependence, other receptors and receptor-channel complexes may be postulated to contribute to the subjective, and possibly reinforcing effects of ethanol. Very recently, it has been found that low concentrations of ethanol (20-25 mM) can **enhance** electrophysiologic responses to serotonin (5-HT) acting at the 5-HT_3 receptor subtype, both in NCB-20 cells and in mammalian neurons in culture (Lovinger, in press). This action of ethanol may have important consequences *in vivo*. In a discriminative stimulus paradigm, animals can be trained to recognize when they receive ethanol, and to respond on a particular lever that they have learned to associate with ethanol administration in order to receive food reinforcement. When rats and pigeons that had been trained to recognize ethanol were given 5-HT_3 receptor antagonists along with the ethanol, their responding on the lever associated with ethanol was completely blocked (Grant and Barrett, in press). That is, under the influence of the 5-HT_3 antagonists, the animals no longer recognized that they had received ethanol, although blood ethanol levels and physiologic signs of intoxication were not affected. These data indicate that the action of ethanol at the 5-HT_3 receptor is important in mediating subjective effects of ethanol. Given previous results indicating that alterations in synaptic concentrations of serotonin can influence ethanol intake (e.g. Naranjo et al., 1987), these findings are consistent with a role for the 5-HT_3 receptor in the reinforcing properties of ethanol.

The mesolimbic dopaminergic system has also been considered to be important in mediating the reinforcing properties of ethanol and other drugs (Wise and Bozarth, 1981). Although there is little evidence for a direct effect of ethanol on the function of dopamine receptors, it appears that ethanol may alter the activity of dopaminergic neurons and enhance dopamine release. For example, concentrations of ethanol as low as 40 mM produced a significant increase in the firing rate of dopaminergic neurons in slices taken from the ventral tegmental area of rat brain (Brodie et al., 1990). These neurons project to the frontal cortex, including the nucleus accumbens, and some recent studies using *in vivo* microdialysis have indicated that ethanol increased the release of dopamine in the nucleus accumbens (Wozniak et al., 1990), an area thought to be crucial for drug and ethanol reinforcement (Wise and Bozarth, 1981). Interestingly, the action of ethanol on dopamine release, like ethanol's discriminative stimulus properties, is reduced in the presence of a selective 5-HT_3 receptor antagonist (Wozniak et al., 1990). These results suggest that an interaction between dopaminergic and serotonergic systems in the nucleus accumbens could be an important factor in the subjective and/or reinforcing effects of ethanol.

Another receptor-gated ion channel which is affected by low to moderate concentrations of ethanol is the GABA/benzodiazepine-Cl^- channel complex and it has been speculated that γ-aminobutyric acid (GABA), the major inhibitory neurotransmitter in brain, might mediate the CNS depressant effects of ethanol. The literature regarding the effects of ethanol on GABA *turnover* is conflicting, in part because of the difficulty in distinguishing metabolic and neurotransmitter "pools" of GABA (see Hoffman and Tabakoff, 1985). However, recent research has revealed that concentrations of ethanol in the range of 10-40mM can enhance GABA-stimulated $^{36}Cl^-$ flux measured in synaptoneurosomal preparations of brain (Suzdak et al., 1986; Allan and Harris, 1987). This action of ethanol was observed to be particularly evident in preparations of cerebellar tissue taken from mice selectively bred to be sensitive to the hypnotic effect of ethanol ("long-sleep", LS mice) (Allan and Harris, 1986). Similarly, when mRNA from brains of LS mice was expressed in *Xenopus* oocytes, ethanol potentiation of GABA-induced Cl^- currents was observed, while ethanol actually *inhibited* these responses when mRNA from SS ("short sleep") mouse brain was expressed (Wafford et al., 1990). In contrast to the biochemical findings, ethanol-induced potentiation of GABA-induced responses has not been consistently demonstrated in electrophysiologic experiments (e.g. Siggins et al., 1987). Recent work, however, in which ion currents were measured in cultured rat cerebral cortical and hippocampal neurons using a whole cell patch-clamp technique, has shown that ethanol can enhance the GABA response but only in certain neurons (Aguayo, 1990). An explanation for this finding may lie in the molecular heterogeneity of the $GABA_A$ receptor. Molecular biological experiments have shown that the $GABA_A$ receptor consists of several subunits (α, β, and γ have been identified, and there may also be δ and ξ subunits), and that these subunits exist in a number of isoforms (for example, at least six α subunits have been cloned) (Levitan et al., 1988). It is not yet clear how these subunits are assembled *in vivo*. However, the various subunits are differentially distributed among brain areas, (Garrett et al., 1990), and expression of cloned subunits in varying combinations gives rise to $GABA_A$ receptors with different pharmacological properties. Recently, for example, it has been reported that the α_6 subunit of the $GABA_A$ receptor, which is found in cerebellar granule cells, is necessary in order to observe binding of the benzodiazepine partial inverse agonist, Ro 15-4513 (Luddens et al., 1990). The recombinant receptors that contain this subunit bind Ro 15-4513 and the GABA agonist muscimol, but not other benzodiazepines. Ro 15-4513 can antagonize certain actions of ethanol (Suzdak et al., 1986; Hoffman et al., 1987). Thus, it is possible that the molecular heterogeneity of the $GABA_A$ receptors, including the anatomical distribution of the α_6 subunit, may account for varying sensitivity to the effects of ethanol among neurons. Genetic regulation of the subunit composition of these receptors could also contribute to the differential sensitivity to ethanol in the selected lines of mice. The knowledge of $GABA_A$ receptor structure gained from molecular biological analysis promises to significantly advance our understanding of the site and mechanism of action of ethanol at this receptor.

It is likely that enhancement of $GABA_A$ responses could contribute to the anxiolytic as well as the incoordinating and hypnotic actions of ethanol. In addition, with chronic ethanol treatment of animals, there is a reduced response of chloride flux to GABA, and a reduced potentiation of the GABA response by ethanol (Allan and Harris, 1987; Morrow et al., 1988). This decreased response to ethanol is observed in animals that are functionally tolerant to ethanol, and may in fact contribute to some aspects of ethanol tolerance (e.g., tolerance to the hypnotic effect of ethanol).

Although the GABA systems of brain may be the "intrinsic" systems which are actually responsible for expression of certain aspects of tolerance, there are a number of other systems (e.g.: noradrenergic, serotonergic) that can modulate the development of alcohol tolerance (Hoffman and Tabakoff, 1989). Tolerance to ethanol is also powerfully modulated by at least one other factor in brain, the neuropeptide arginine vasopressin (AVP). This peptide has been known for some time to reduce the rate of loss of functional tolerance over time after ethanol withdrawal (Hoffman et al., 1978). Recent work has provided evidence for a mechanism of action of vasopressin in brain that may be applicable not only to its effects on tolerance, but also to effects on other neuroadaptive processes, such as learning or memory (Hoffman, 1987). It was demonstrated that the ability of vasopressin to maintain ethanol tolerance was mediated by its interaction with the V_1 subtype of vasopressin receptor in the brain (Szabo et al., 1988). The effects of vasopressin on memory are also mediated through V_1 receptors (see Hoffman, 1987). Receptor binding and autoradiographic studies have demonstrated a high density of V_1 receptors in the lateral septum, with a lower concentration of receptors in other limbic areas, including the hippocampus (Ishizawa et al., 1990). Activation of the V_1 receptors by vasopressin results in stimulation of phospholipase C activity and polyphosphoinositide metabolism resulting in increased levels of intracellular calcium (Shewey and Dorsa, 1988).

It has also recently been found that vasopressin, acting at V_1 receptors in brain, stimulates the expression of the proto-oncogene c-*fos* (Rathna Giri et al., 1990). Intracerebrovenicular injection of AVP or selective V_1 agonists in mice led to a large increase in c-*fos* mRNA in both septum and hippocampus. The c-*fos* gene, which is the cellular homolog of a viral gene, v-*fos*, is normally involved in growth and development (Curran, 1988). The stimulation of its expression by neurotransmitters led to the suggestion that, in the mature brain, c-*fos* could play a role in synaptic plasticity, possibly through induction of morphological changes (Greenberg et al., 1986). The product of the c-*fos* gene is a nuclear protein that, in combination with the product of another proto-oncogene, c-*jun*, can bind to particular transcription activation elements on DNA, leading to changes in expression of other as-yet undetermined proteins. The ability of vasopressin to increase c-*fos* expression therefore provides a means by which the short-lived peptide signal can be transformed into a long-term change in CNS function, such as tolerance or memory. Structure-activity studies with several neuropeptides support the postulate that activation of *septal* c-*fos* expression by vasopressin could be a key factor in the effects of vasopressin on ethanol tolerance (Rathna Giri et al., 1990). Interestingly, the administration of vasopressin doses that produced significant effects on the expression of ethanol tolerance had little effect on the expression of ethanol physical dependence. The overt signs and symptoms of withdrawal hyperexcitability were not modified by administration of vasopressin during the period of chronic consumption of ethanol supporting the earlier contentions that ethanol tolerance and ethanol physical dependence do not arise from identical neuroadaptive processes (Hoffman and Tabakoff, 1989).

The recently accumulated data on the acute and chronic effects of ethanol transforms the perception of ethanol's pharmacology from consideration of ethanol's actions as non-specific "solvent" effects, to a view that ethanol's actions are mediated by specific neuronal elements which are selectively affected by ethanol. Knowledge of the structure of the receptors, and ion channels and the action of peptides that are affected by ethanol, or influence ethanol responses, is constantly growing, and the advances made by molecular biology should greatly enhance our understanding of genetic influences on responses to ethanol, as well as the mechanism of action of ethanol at the molecular level. This understanding in turn will promote the development of medications to treat alcohol related problems, the development of intervention strategies to arrest the progression of the addictive process, and the generation of means for early identification of individuals susceptible to the negative consequences of alcohol's actions. In all, both alcoholism prevention and treatment shall benefit from current findings on the effects of ethanol on brain function.

*Portions of this review were presented at and submitted to the 5th Congress of the International Society for Biomedical Research on Alcoholism.

REFERENCES

Aguayo, L. G., 1990, Ethanol potentiates the $GABA_A$-activated Cl^- current in mouse hippocampal and cortical neurons, **Eur. J. Pharmacol.**, 187:127-130.

Allan, A. M. and Harris, R. A., 1986, Gamma-aminobutyric acid and alcohol actions: neurochemical studies of long sleep and short sleep mice, **Life Sci.**, 39:2005-2015.

Allan, A. M. and Harris, R. A., 1987, Acute and chronic ethanol treatments alter GABA receptor-operated chloride channels, **Pharm. Biochem. Behav.**, 27:665-670.

Brodie, M. S., Shefner, S. A. and Dunwiddie, T. V., 1990, Ethanol increases the firing rate of dopamine neurons of the rat ventral tegmental area *in vitro*, **Brain Res.**, 508:65-69.

Chin, J. H. and Goldstein, D. B., 1977, Drug tolerance in biomembranes: a spin label study of the effects of ethanol, **Science**, 196:684-685.

Collingridge, G. L. and Lester, R. A. J., 1989, Excitatory amino acid receptors in the vertebrate central nervous system, **Pharmacol. Rev.**, 41:143-210.

Curran, T., 1988, The *fos* oncogene, **in**: 'The Oncogene Handbook', E. P. Reddy, A. M. Skalka and T. Curran, eds., Elsevier Science Publishers B.V. (Biomedical Division).

Dingledine, R., Hynes, M. A. and Kinge, G. L., 1986, Involvement of N-methyl-D-aspartate receptors in epileptiform bursting in the rat hippocampal slice, **J. Physiol.**, 380:175-189.

Garrett, K. M., Saito, N., Duman, R. S., Abel, M. S., Ashton, R. A., Fujimori, S., Beer, B., Tallman, J. F., Vitek, M. P. and Blume, A. J., 1990, Differential expression of γ-aminobutyric $acid_A$ receptor subunits, **Mol. Pharmacol.**, 37:652-657.

Gothert, M. and Fink, K., 1989, Inhibition of N-methyl-D-aspartate (NMDA)- and L-glutamate-induced noradrenaline and acetylcholine release in the rat brain by ethanol, **Arch. Pharmacol.**, 340:515-521.

Grant, K. A., Valverius, P., Hudspith, M. and Tabakoff, B., 1990, Ethanol withdrawal seizures and the NMDA receptor complex, **Eur. J. Pharmacol.**, 176:289-296.

Grant, K. A. and Barrett, J. E., Blockade of the discriminative stimulus effects of ethanol with 5-HT_3 receptor antagonists, **Psychopharmacol.**, in press 1991.

Greenberg, M. E., Ziff, E. B. and Greene, L. A., 1986, Stimulation of neuronal acetylcholine receptors induces rapid gene transcription, **Science**, 234:80-83.

Gulya, K., Grant, K. A., Valverius, P., Hoffman, P. L. and Tabakoff, B., Brain regional specificity and time course of changes in the NMDA receptor-ionophore complex during ethanol withdrawal, **Brain Res.**, in press.

Harris, E. W., Ganong, A. H. and Cotman, C. W., 1984, Long-term potentiation in the hippocampus involves activation of N-methyl-D-aspartate receptors, **Brain Res.**, 323:132-137.

Hoffman, P. L., Ritzmann, R. H., Walter, R. and Tabakoff, B., 1978, Arginine vasopressin maintains ethanol tolerance, **Nature**, 276:614-616.

Hoffman, P. L. and Tabakoff, B., 1985, Ethanol's action on brain biochemistry, **in**: 'Alcohol and the Brain: Chronic Effects', R. E. Tarter, D. H. van Thiel and K. L. Edwards, eds., Academic Press, New York.

Hoffman, P. L., 1987, Central nervous system effects of neurohypophyseal peptides, **in**: 'The Peptides', C. W. Smith, ed.

Hoffman, P. L., Tabakoff, B., Szabo, G., Suzdak, P. D. and Paul, S. M., 1987, Effect of an imidazobenzodiazepine, Ro15-4513, on the incoordination and hypothermia produced by ethanol and pentobarbital, **Life Sci.**, 41:611-619.

Hoffman, P. L. and Tabakoff, B., 1989, Mechanisms of alcohol tolerance, **Alc. Alcoholism**, 24:251-252.

Hoffman, P. L., Rabe, C. S., Moses, F. and Tabakoff, B., 1989, N-methyl-D-aspartate receptors and ethanol: inhibition of calcium flux and cyclic GMP production, **J. Neurochem.**, 52:1937-1940.

Ishizawa, T., Tabakoff, B., Mefford, I. N. and Hoffman, P. L., 1990, Reduction of arginine vasopressin binding sites in mouse lateral septum by treatment with 6-hydroxydopamine, **Brain Res.**, 507:189-194.

Kleinschmidt, A., Bear, M. F. and Singer, W., 1987, Blockade of NMDA receptors disrupts experience-dependent plasticity of kitten striate cortex, **Science**, 238:355-358.

Levitan, E. S., Schofield, P. R., Burt, D. R., Rhee, L. M., Wisden, W., Kohler, M., Fujita, N., Rodriguez, H. F., Stephenson, A., Darlison, M. G., Barnard, E. A. and Seeburg, P. H., 1988, Structural and functional basis for GABA-A receptor heterogeneity, **Nature**, 335:76-79.

Lovinger, D. M., White, G. and Weight, F. F., 1989, Ethanol inhibits NMDA-activated ion current in hippocampal neurons, **Science**, 243:1721-1724.

Lovinger, D. M., Ethanol potentiates ion current mediated by 5-HT_3 receptors on neuroblastoma cells and isolated neurons, **Proc. 5th Congress Intl. Soc. Res. Alcoholism**, in press.

Luddens, H., Pritchett, D. B., Kohler, M., Killisch, I., Keinanen, K., Monyer, H., Sprengel, R. and Seeburg, P. H., 1990, Cerebellar $GABA_A$ receptor selective for a behavioral alcohol antagonist, **Nature**, 346:648-651.

Meyer, H. H. and Gottlieb, R., 1926, Theory of narcosis, **in**: Experimental Pharmacology as a Basis for Therapeutics (V.E. Henderson, trans.), Lippincott, Philadelphia, PA.

Morrow, A. L., Suzdak,. P. D., Karanian, J. W. and Paul, S. M., 1988, Chronic ethanol administration alters γ-aminobutyric acid, pentobarbital and ethanol-mediated $^{36}Cl^-$ uptake in cerebral cortical synaptoneurosomes, **J. Pharmacol. Exp. Ther.**, 246:158-164.

Naranjo, C. A., Sellers, E. M., Sullivan, J. T., Woodley, D. V., Kadlec, K. and Sykora, K., 1987, The serotonin uptake inhibitor citalopram attenuates ethanol intake, **Clin. Pharmacol. Ther.**, 41:266-274.

Rabe, C. S. and Tabakoff, B., Glycine site directed agonists reverse ethanol's actions at the NMDA receptor, **Mol. Pharmacol.**, in press.

Rathna Giri, P., Dave, J. R., Tabakoff, B. and Hoffman, P. L., 1990, Arginine vasopressin induces the expression of c-*fos* in the mouse septum and hippocampus, **Mol. Brain Res.**, 7:131-137.

Shewey, L. M. and Dorsa, D. M., 1988, V_1-type vasopressin receptors in rat brain septum: binding characteristics and effects on inositol phospholipid metabolism, **J. Neurosci.**, 8:1671-1677.

Siggins, G. R., Pittman, Q. J. and French, C. D., 1987, Effects of ethanol on CA1 and CA3 pyramidal cells in the hippocampal slice preparation: an intracellular study, **Brain Res.**, 414:22-34.

Suzdak, P. D., Glowa, J. R., Crawley, J. N., Schwartz, R. D., Skolnick, P. and Paul, S. M., 1986, A selective imidazobenzodiazepine antagonist of ethanol in the rat, **Science**, 234:1243-1247.

Szabo, G., Tabakoff, B. and Hoffman, P. L., 1988, Receptors with V_1 characteristics mediate the maintenance of ethanol tolerance by vasopressin, **J. Pharmacol. Exp. Ther.**, 247:536-541.

Tabakoff, B., Hoffman, P. L. and McLaughlin, A., 1988, Is ethanol a discriminating substance?, **Seminars Liver Dis.**, 8:26-35.

Wafford, K. A., Burnett, D. M., Dunwiddie, T. V. and Harris, R. A., 1990, Genetic differences in the ethanol sensitivity of $GABA_A$ receptors expressed in *Xenopus* oocytes, **Science**, 249:291-293.
Wise, R. A. and Bozarth, M. A., 1981, Brain substrates for reinforcement and drug self-administration, **Prog. Neuropsychopharmacol.**, 5:467-474.
Woodward, J. J. and Gonzales, R. A., 1990, Ethanol inhibition of N-methyl-D-aspartate-stimulated endogenous dopamine release from rat striatal slices: reversal by glycine, **J. Neurochem.**, 54:712-715.
Wozniak, K. M., Pert, A. and Linnoila, M., 1990, Antagonism of 5-HT_3 receptors attenuates the effects of ethanol on extracellular dopamine, **Eur. J. Pharmacol.**, 187:287-289.

THE NEUROCHEMISTRY OF ETHANOL TOLERANCE

Boris Tabakoff and Paula L. Hoffman

Department of Pharmacology
Univ. of Colorado Hlth. Sci. Ctr.
Denver, CO 80262 USA
and Division of Intramural Clinical and Biological Research
National Institute on Alcohol Abuse and Alcoholism
Rockville, MD 20852 USA

Tolerance to ethanol (alcohol) or any other drug is defined as an acquired resistance to the physiological and/or behavioral effects of the drug. Another empirical definition of tolerance is that a given dose of drug has less effect in an individual on his/her later exposures to the drug than on the first exposure. The phenomenon of tolerance, as well as the mechanisms underlying its development, can be complex (Tabakoff et al., 1982), ranging from changes in drug disposition or metabolism ("metabolic tolerance"), to alterations in the function of cells including the neurons of the central nervous system which result in drug resistance. The cellular adaptations to resist a drug's actions are part of the phenomenon of "functional" tolerance and the processes that contribute to functional tolerance will be the focus of this discussion.

Tolerance represents one example of the ability of an individual to adapt to stimuli in the external environment. Another example of such a neuroadaptive process is learning, and many of the mechanisms that have been proposed to underlie learning may also be applicable to tolerance. In fact, as will be described later, learning seems to play a role in the development of some forms of tolerance (Tabakoff et al., 1982).

The importance of tolerance, with respect to the actions of ethanol, may lie in its postulated effect on ethanol intake. Ethanol has both reinforcing and aversive properties, and it has been hypothesized that development of tolerance to the aversive effects of ethanol could result in enhancement of its reinforcing effects (Cappell and LeBlanc, 1979), thus promoting ethanol intake. It is only relatively recently that experimental data have been obtained to support this view. Rats that have been selectively bred to "prefer" ethanol (i.e., to drink relatively large amounts of ethanol) were found to rapidly develop a high degree of tolerance to the sedative effects of ethanol, and to maintain this tolerance for a long period, compared to rats that were bred to avoid ethanol (Waller, et al., 1983). It has also been reported that prior ingestion of ethanol is necessary in order to demonstrate a reinforcing effect of ethanol in rats (Numan, 1981). In addition to possibly promoting ethanol intake, the development of tolerance to the (aversive) effects of high doses of ethanol could also, by reducing the incapacitating effects of the drug, *allow* an individual to ingest large amounts

Alcoholism: A Molecular Perspective
Edited by T.N. Palmer, Plenum Press, New York, 1991

of ethanol, leading to physical dependence and organ damage. One way to mitigate these consequences of ethanol ingestion is to modulate tolerance acquisition or loss, and an understanding of the neurochemical changes that are responsible for tolerance is essential to such efforts.

A construct that we have found useful for analyzing the biochemical determinants of tolerance is that of intrinsic and extrinsic neuronal systems (Tabakoff and Hoffman, 1989). This framework has also been applied to concepts of learning and memory (Squire and Davis, 1981). Intrinsic systems are those that encode tolerance within themselves, presumably by changes in cellular structure or function. These intrinsic systems would be the ones that contribute directly to the physiological process or behavior which demonstrates tolerance to ethanol, and might be different for each such behavior that is studied. Extrinsic systems are those which modulate the development, expression or maintenance of tolerance, but which do not encode tolerance within themselves. Thus, extrinsic systems could be expected to affect tolerance to a number of effects of ethanol.

Two neuronal systems that appear to have the characteristics of extrinsic systems, with regard to ethanol tolerance, are the noradrenergic and serotonergic systems. Treatment of mice or rats with drugs that deplete serotonin in the brain delayed the development of tolerance to the hypnotic, hypothermic and motor-impairing effects of ethanol (Lê et al., 1981; Lê et al., 1981; Melchior and Tabakoff, 1981). In rats, the specific pathway involved connects the median raphe nucleus to the hippocampus (Lê et al., 1981). In other experiments, depletion of brain norepinephrine in mice *blocked* the development of tolerance to the hypnotic and hypothermic effects of ethanol (Tabakoff and Ritzmann, 1977). In rats, it was necessary to deplete both serotonin *and* norepinephrine in order to completely block tolerance to the hypnotic effect of ethanol (serotonin depletion alone only delayed tolerance development) (Lê et al., 1981), although tolerance to the hypothermic effect of ethanol in rats was blocked by depletion of norepinephrine alone (Trzaskowska et al., 1986). Thus, while interactions among neuronal pathways may differ among species, and are most likely subject to genetic influence, the noradrenergic and serotonergic systems seem to modify the development (or loss (Lê et al., 1981)) of tolerance to a number of effects of ethanol. It should be noted that the animals that were subject to neurotransmitter depletion ingested as much ethanol as the controls (Tabakoff and Ritzmann, 1977), indicating that the presence of ethanol in the brain is necessary but not sufficient to produce tolerance; both ethanol and the activity of certain neuronal pathways are necessary. Ethanol can influence the activity of noradrenergic and serotonergic systems, as well as other neurotransmitter systems (Hoffman and Tabakoff, 1985), and this influence is no doubt important for the development of tolerance.

Recently there has been some indication of the postsynaptic mechanisms that may be involved in the modulation of ethanol tolerance by norepinephrine. When brain norepinephrine was depleted in mice, the blockade of tolerance to the hypnotic effect of ethanol could be overcome by treating the mice with forskolin, an activator of adenylate cyclase, during chronic ethanol exposure (Szabo et al., 1988). These data can be interpreted to mean that, in the presence of ethanol, the interaction of norepinephrine with ß-adrenergic receptors, leading to activation of adenylate cyclase, and presumably accumulation of cyclic AMP, is necessary for the development of tolerance. It has been found that ethanol, at concentrations that can occur *in vivo*, enhances the activation of adenylate cyclase by ß-adrenergic agonists in brain preparations (Saito et al., 1985). Thus, stimulation of cyclic AMP accumulation during ethanol ingestion may be important in order for tolerance to develop. Increases in this "second messenger" activate cellular protein kinases, and resulting changes in protein phosphorylation can have many consequences. It needs to be noted that these consequences are of a modulatory nature in the development of tolerance (extrinsic), since the destruction of noradrenergic systems after tolerance has developed has little or no effect on the expression of tolerance.

One modulatory consequence may be an increase in activity and/or an "up-regulation" of L-type voltage-gated calcium channels, the protein subunits of which (in peripheral tissue) can be phosphorylated by a cyclic AMP-dependent mechanism (Hosey and Lazdunski, 1988). The L-type calcium channels can then be considered intrinsic systems since acute ethanol treatment inhibits channel function and an "up-regulation" of L-type voltage-gated calcium channels in brain during chronic ethanol ingestion would produce a resistance to ethanol's actions (Dolin and Little, 1989). Therefore, one could propose that activation of ß-adrenergic receptors and increased production of cyclic AMP, which would be enhanced by ethanol in the brain, would promote the up-regulation of L-type calcium channels. This up-regulation would, in turn, contribute to the development of ethanol tolerance. On the other hand, the noradrenergic system does not act in isolation, and interactions among the extrinsic and intrinsic systems must define the characteristics of tolerance to various effects of ethanol.

Another possible extrinsic system, and one which also seems to be involved in learning and memory (Hoffman, 1987), consists of neurons synthesizing the neuropeptide, arginine vasopressin (AVP). A well-characterized effect of this peptide is to maintain ethanol tolerance once it has been acquired; thus, AVP seems to prevent, or reduce the rate of loss of, the change(s) in CNS function that has already occurred during chronic ethanol treatment, and that contributes to ethanol tolerance (Hoffman et al., 1979; Lê et al., 1982). Adding a layer of complexity to the control of tolerance are the observations that depletion of brain norepinephrine in mice (Hoffman et al., 1983), or serotonin (specifically the dorsal serotonergic afferent pathways to the hippocampus) in rats (Speisky and Kalant, 1985), blocked the ability of vasopressin to maintain tolerance to the hypnotic or motor-incoordinating effects of ethanol indicating functional interactions between noradrenergic, serotonergic and vasopressin systems in brain.

Some progress has been made in elucidating the mechanism by which vasopressin can influence tolerance. The peptide acts at CNS receptors that have the characteristics of the V_1 subtype of vasopressin receptor (Jard, 1983) when it maintains tolerance (Szabo et al., 1988). Some of these receptors appear to be localized presynaptically on catecholaminergic terminals (Ishizawa et al., 1990), suggesting that vasopressin-induced changes in turnover or release of norepinephrine (Tanaka et al., 1977) could influence ethanol tolerance. Evidence for a postsynaptic mechanism of action (which could be either direct or mediated by neurotransmitter release) is provided by the recent finding that vasopressin, administered *in vivo*, can stimulate the expression of the proto-oncogene c-*fos* in mouse hippocampus and septum (Rathna Giri et al., 1990). This proto-oncogene is normally involved in growth and development (Curran, 1988), and its expression in the mature CNS is thought to be associated with synaptic plasticity (Greenberg et al., 1986). The c-*fos* product is a DNA-binding protein that can alter the transcription of other genes (Greenberg et al., 1986), and therefore increased c-*fos* expression is a means by which the action of a short-lived peptide like vasopressin could influence longer-term changes in CNS function, such as tolerance or memory.

The extrinsic systems discussed so far affect the development or maintenance (loss) of ethanol tolerance, and seem to modulate tolerance to a number of ethanol's behavioral effects. In addition to the L-type, voltage sensitive calcium channels, there have also been several other proposals for *intrinsic* systems involved in ethanol tolerance (Tabakoff and

Hoffman, 1989). One might postulate that as with the L-channels such intrinsic neurochemical systems would be sensitive to perturbation by ethanol, and that "tolerance", or resistance to the effects of ethanol, would be manifest in these systems in an animal that shows behavioral tolerance to ethanol. While there have been several instances in which cellular resistance to ethanol's effects at the biochemical level has been observed in ethanol-tolerant animals (Hoffman and Tabakoff, 1985), few of these biochemical systems are initially sensitive to pharmacologically-relevant concentrations of ethanol. The function of the gamma-aminobutyric acid (GABA) receptor-gated chloride channel is, however, potentiated by very low concentrations of ethanol *in vitro* (Allan and Harris, 1987; Suzdak et al., 1986). GABA is the major inhibitory neurotransmitter in brain, and its actions at the $GABA_A$ receptor (coupled to the ion channel) are also modulated by benzodiazepines and barbiturates (Skolnick and Paul, 1982). These characteristics of the receptor suggest the possibility of its participation in the sedative-hypnotic, as well as the anxiolytic, effects of ethanol. Ethanol stimulation of GABA-induced chloride ion flux has been measured in brain preparations of mice selectively bred to be sensitive (LS) or resistant (SS) to the hypnotic effect of ethanol, as measured by duration of loss of righting reflex. Ethanol enhanced the effect of GABA in cerebellar preparations from LS, but not SS mice (Allan and Harris, 1986). These data are consistent with a role for GABA in the sedative-hypnotic effect of ethanol, and, since the differential effect was observed in cerebellum, but not hippocampus, could also indicate a role in ethanol-induced motor incoordination. Similarly, a number of investigators have found that GABA agonists increase, and GABA antagonists decrease, the motor-impairing effects of ethanol (Hakkinen and Kulonen, 1976; Frye and Breese, 1982).

After chronic treatment of mice with ethanol, such that the mice were behaviorally tolerant, the ability of a GABA agonist to stimulate chloride flux in a mouse cerebellar preparation was unchanged, but ethanol potentiation of the GABA response was attenuated; i.e., the system became "tolerant" to ethanol (Allan and Harris, 1987). Although some investigators have also found a reduced response to GABA in ethanol-tolerant animals (Morrow et al., 1988), binding studies have not demonstrated any change in GABA receptors *per se* (Liljequist et al., 1986; Rastogi et al., 1986). The reduced GABA-potentiating effect of ethanol is consistently observed in tolerant animals, and could well underlie the expression of tolerance to the sedative-hypnotic, anxiolytic and/or motor-impairing effects of ethanol. Investigations of the time course for appearance and disappearance of tolerance to these effects, and comparison to the time course of changes in the response of the GABA receptor system, would be one means to further characterize the GABA receptor-coupled channel as an intrinsic system for ethanol tolerance.

The role of the $GABA_A$ receptor-gated ion channel in the acute and chronic effects of ethanol has recently come under more scrutiny because of rapid advances in determining the molecular structure of this receptor. Molecular biological studies have revealed that the receptor consists of up to five distinct polypeptide units, and that multiple forms of these subunits exist (Dingledine et al., 1990). The actual *in vivo* composition of the receptor is not yet known, but this heterogeneity at the molecular level, which includes differences in brain regional distribution of various forms of the subunits (Garrett et al., 1990), may contribute to the differential sensitivity of the receptor to ethanol among brain areas (Allan and Harris, 1986) and the differential development of tolerance. Very recent work, for example, has shown that reconstituted receptors that contain a particular form of the α subunit of the $GABA_A$ receptor (the "α_6" subunit), which occurs in cerebellar granule cells, bind GABA agonists and the benzodiazepine inverse agonist, Ro15-4513, but not other benzodiazepines (Luddens et al., 1990). Ro15-4513 is a compound that antagonizes some actions of ethanol (e.g., Suzdak et al., 1986; Hoffman et al., 1987), implying that the presence of the α_6 subunit may influence the response of the receptor to ethanol. Furthermore, cloning of the $GABA_A$ receptor has allowed the discovery that the mRNA for some subunits of this receptor is reduced in certain brain areas of animals treated chronically with ethanol (Morrow et al., 1990). The functional significance of this change remains to be determined, particularly since no comparable differences in ligand binding have been reported, but these new data provide further evidence that changes in the $GABA_A$ receptor complex may be involved in ethanol tolerance.

This review has concentrated on only one form of tolerance, i.e. chronic functional tolerance, but there are several forms of tolerance, and the relationships among them have not been completely elucidated at present. For example, tolerance can occur over the course of a single exposure to ethanol (acute tolerance), and presumably both acute and chronic forms of tolerance can be either metabolic (pharmacokinetic) or functional (pharmacodynamic; cellular). In addition, "conditioned" or "environment-dependent" forms of tolerance have been described. In this type of tolerance, the individual learns to associate the environmental cues with the effect of ethanol, and develops a conditioned compensatory response that counteracts ethanol's effect (i.e. produces tolerance). However, the conditioned compensatory response only occurs in the ethanol-associated environment, so that tolerance is not observed in a unique environment, regardless of the individual's previous exposure to ethanol. This type of tolerance obviously has an important learned component. "Behaviorally-augmented" tolerance has also been described, in which the rate of development of tolerance is enhanced when an individual must perform a task under the influence of ethanol.

It is not clear whether all forms of functional tolerance depend on the same underlying neurochemical mechanisms. "Environment-dependent" tolerance was shown to be influenced by noradrenergic and vasopressin systems in a manner similar to environment-independent tolerance (Melchior and Tabakoff, 1981; Melchior et al., 1983), and acute tolerance may be related to chronic tolerance (Tabakoff et al., 1982), but more work needs to be done in order to characterize these relationships, as well as the neurochemistry that influences each form of tolerance. It may be hoped that an understanding of these phenomena will eventually allow for the development of pharmacotherapies that will alter tolerance, and thereby perhaps ameliorate some aspects of ethanol abuse.

REFERENCES

Allan, A. M. and Harris, R. A., 1986, Gamma-aminobutyric acid and alcohol actions: Neurochemical studies of long sleep and short sleep mice, **Life Sci.**, 39:2005-2015.

Allan, A. M. and Harris, R. A., 1987, Acute and chronic ethanol treatments alter GABA receptor operated chloride channels, **Pharm. Biochem. Behav.**, 27:665-670.

Cappell, H. and LeBlanc, A. E., 1979, Tolerance to, and physical dependence on, ethanol: Why do we study them? **Drug Alc. Dependence**, 4:15-31.

Curran, T., 1988, The *fos* oncogene, **in**: The Oncogene Handbook, E.P. Reddy, A.M. Skalka and T. Curran, eds., Elsevier Sciences Publishers B.V., Amsterdam.

Dingledine, R., Myers, S. J. and Nicholas, R. A., 1990, Molecular biology of mammalian amino acid receptors, **FASEB J.**, 4:2636-2645.

Dolin, S. J. and Little, H. J., 1989, Are changes in neuronal calcium channels involved in ethanol tolerance? **J. Pharmacol. Exp. Ther.**, 250:985-991.

Frye, G. D. and Breese, G. R., 1982, GABAergic modulation of ethanol-induced motor impairment, **J. Pharmacol. Exp. Ther.**, 223:750-756.

Garrett, K. M., Saito, N., Duman, R. S., Abel, M. S., Ashton, R. A., Fujimori, S., Beer, B., Tallman, J. F., Vitek, M. P. and Blume, A. J., 1990, Differential expression of γ-aminobutyric acid$_A$ receptor subunits, **Mol. Pharmacol.**, 37:652-657.
Greenberg, M. E., Ziff, E. B. and Greene, L. A., 1986, Stimulation of neuronal acetylcholine receptor induces rapid gene transcription, **Science**, 234:80-83.
Hakkinen, J. M. and Kulonen E., 1976, Ethanol intoxication and gamma-aminobutyric acid, **J.Neurochem.**, 27:631-633.
Hoffman, P. L., Ritzmann, R. F., Walter, R. and Tabakoff, B., 1979, Arginine vasopressin maintains ethanol tolerance, **Nature**, 276:614-616.
Hoffman, P. L., Melchior, C. L. and Tabakoff, B., 1983, Vasopressin maintenance of ethanol tolerance requires intact brain noradrenergic systems, **Life Sci.**, 32:1065-1071.
Hoffman, P. L. and Tabakoff, B., 1985, Ethanol's action on brain biochemistry, **in**: Alcohol and the Brain: Chronic Effects, R.E. Tarter and D.H. van Thiel, eds., Plenum Medical Book Company, New York.
Hoffman, P. L., 1987, Central nervous system effects of neurohypophyseal peptides, **in**: The Peptides, C.W. Smith, ed., Academic Press, New York.
Hoffman, P. L., Tabakoff, B., Szabó, G., Suzdak, P. D. and Paul, S. M., 1987, Effect of an imidazobenzodiazepine, Ro15-4513, on the incoordination and hypothermia produced by ethanol and pentobarbital, **Life Sci.**, 41:611-619.
Hosey, M. M. and Lazdunski, M., 1988, Calcium channels: Molecular pharmacology, structure and regulation, **J. Membrane Biol.**, 104:81-105.
Ishizawa, H., Tabakoff, B., Mefford, I. N. and Hoffman, P. L., 1990, Reduction of arginine vasopressin binding sites in mouse lateral septum by treatment with 6-hydroxydopamine, **Brain Res.**, 507:189-194.
Jard, S., 1983, Vasopressin isoreceptors in mammals: Relation to cyclic AMP-dependent and cyclic AMP-independent transduction mechanisms, **Current Topics Membr. Transp.**, 18:255-285.
Lê, A. D., Khanna, J. M., Kalant, H. and LeBlanc, A. E., 1981, Effect of modification of brain serotonin (5-HT), norepinephrine (NE) and dopamine (DA) on ethanol tolerance, **Psychopharmacol.**, 75:231-235.
Lê, A. D., Khanna, J. M., Kalant, H. and LeBlanc, A. E., 1981, The effect of lesions in the dorsal, median and magnus raphe nuclei on the development of tolerance to ethanol, **J. Pharmacol. Exp. Ther.**, 218:525-529.
Lê, A.D., Kalant, H. and Khanna, J.M. (1982) Interaction between des-glycinami^9-[Arg^8]vasopressin and serotonin on ethanol tolerance. **Eur. J. Pharmacol.**, 80:337-345.
Liljequist, S., Culp, S. and Tabakoff, B., 1986, Effect of ethanol on the binding of [^{35}S]-*t*-butylbicyclophosphorothionate to mouse brain membranes, **Life Sci.**, 38:1931-1939.
Lüddens, H., Pritchett, D. B., Köhler, M., Killisch, I., Keinänen, K., Monyer, H., Sprengel, R. and Seeburg, P. H., 1990, Cerebellar GABA$_A$ receptor selective for a behavioral alcohol antagonist, **Nature**, 346:648-651.
Melchior, C. L. and Tabakoff, B., 1981, Modification of environmentally cued tolerance to ethanol in mice, **J. Pharmacol. Exp. Ther.**, 219:175-180.
Melchior, C. L., Hoffman, P. L. and Tabakoff, B., 1983, Influencing environment-dependent tolerance to ethanol, **in**: Ethanol Tolerance and Dependence: Endocrinological Aspects (National Research Monograph No. 13), T.J. Cicero, ed., U.S. Government Printing Office, Washington, D.C.
Morrow, A. L., Suzdak, P. D., Karanian, J. W. and Paul, S. M., 1988, Chronic ethanol administration alters γ-aminobutyric acid, pentobarbital and ethanol-mediated $^{36}Cl^-$ uptake in cerebral cortical synaptoneurosomes, **J. Pharmacol. Exp. Ther.**, 246:158-164.

Morrow, A. L., Montpied, P., Lingford-Hughes, A. and Paul, S. M., 1990, Chronic ethanol and pentobarbital administration in the rat: Effects on $GABA_A$ receptor function and expression in brain, **Alcohol**, 7:237-244.
Numan, R., 1981, Multiple exposures to ethanol facilitates intravenous self-administration of ethanol by rats, **Pharm. Biochem. Behav.**, 15:101-108.
Rastogi, S. K., Thyagarajan, R., Clothier, J. and Ticku, M. K., 1986, Effect of chronic treatment of ethanol on benzodiazepine and picrotoxin sites on the GABA receptor complex in regions of the brain of the rat, **Neuropharmacol.**, 26:1179-1184.
Rathna Giri, P., Dave, J. R., Tabakoff, B. and Hoffman, P. L., 1990, Arginine vasopressin induces the expression of c-*fos* in the mouse septum and hippocampus, **Mol. Brain Res.**, 7:131-137.
Saito, T., Lee, J. M. and Tabakoff, B., 1985, Ethanol's effects on cortical adenylate cyclase activity, **J. Neurochem.**, 44:1037-1044.
Skolnick, P. and Paul, S. M., 1982, Benzodiazepine receptors in the central nervous system, **Int. Rev. Neurobiol.**, 23:103-140.
Speisky, M. B. and Kalant, H., 1985, Site of interaction, of serotonin and desglycinamide-arginine-vasopressin in maintenance of ethanol tolerance, **Brain Res.**, 326:281-290.
Squire, L. R. and Davis, H. P., 1981, The pharmacology of memory: A neurobiological perspective, **Ann. Rev. Pharmacol. Toxicol.**, 21:323-356.
Suzdak, P. D., Schwartz, R. D., Skolnick, P. and Paul, S. M., 1986, Ethanol stimulates γ-aminobutyric acid receptor mediated chloride transport in rat brain synaptoneurosomes, **Proc. Natl. Acad. Sci. USA**, 83:4071-4075.
Suzdak, P. D., Glowa, J. R., Crawley, J. N., Schwartz, R. D., Skolnick, P. and Paul, S. M., 1986, A selective imidazobenzodiazepine antagonist of ethanol in the rat, **Science**, 234:1243-1247.
Szabó, G., Hoffman, P. L. and Tabakoff, B., 1988, Forskolin promotes the development of ethanol tolerance in 6-hydroxydopamine-treated mice, **Life Sci.**, 42:615-621.
Szabó, G., Tabakoff, B. and Hoffman, P. L., 1988, Receptors with V_1 characteristics mediate the maintenance of ethanol tolerance by vasopressin, **J. Pharmacol. Exp. Ther.**, 247:536-541.
Tabakoff, B. and Ritzmann, R. F., 1977, The effects of 6-hydroxydopamine on tolerance to and dependence on ethanol, **J. Pharmacol. Exp. Ther.**, 203:319-331.
Tabakoff, B., Melchior, C. L. and Hoffman, P. L., 1982, Commentary on ethanol tolerance, **Alcoholism: Clin. Exp. Res.**, 6:252-259.
Tabakoff, B. and Hoffman, P. L., 1989, Adaptive responses to ethanol in the central nervous system, **in:** Alcoholism: Biomedical and Genetic Aspects, H.W. Goedde and D.P. Agarawal, eds., Pergamon Press, New York.
Tanaka, M., de Kloet, E. R., de Wied, D. and Versteeg, D. H. G., 1977, $Arginine^8$-vasopressin affects catecholamine metabolism in specific brain nuclei, **Life Sci.**, 20:1799-1808.
Trzaskowska, E., Pucilowski, O., Dyr, W., Kostowski, W. and Hauptmann, M., 1986, Suppression of ethanol tolerance and dependence in rats treated with DSP-4, a noradrenergic neurotoxin, **Drug Alc. Dependence**, 18:349-353.
Waller, M. B., McBride, W. J., Lumeng, L. and Li, T-K., 1983, Initial sensitivity and acute tolerance to ethanol in the P and NP lines of rats, **Pharm. Biochem. Behav.**, 19:683-686.

THE ROLE OF CALCIUM CHANNELS IN ETHANOL DEPENDENCE

H. J. Little[1] and J. M. Littleton[2]

[1]Pharmacology Department
The Medical School
Bristol, BS8 1TD, United Kingdom
[2]Division of Biomedical Science
King's College
London, WC2R 2LS, United Kingdom

CALCIUM AND NEURONAL FUNCTION

Calcium channels in neurones are divided into those activated by neurotransmitter-receptor interactions, and voltage-operated channels. The conductance of the latter increases as the membrane potential decreases. Calcium ions also act as second messengers and are involved in many neuronal functions, including postsynaptic responses and neurotransmitter release. Calcium in the cytosol interacts with calmodulin and other calcium-sensitive proteins.

Voltage-dependent channels have been found to be of several different subtypes. In cultured cells of neuronal origin, Tsien and collegues have demonstrated three voltage-dependent conductances (Nowycky et al., 1985). The "T"-subtype were activated at low voltages, and opened transiently. These have been suggested to be involved in bursting behaviour of neurones (Hernandez-Cruz & Pape, 1989). The "L"-subtype was activated by strong depolarisation and the conductance change is long lasting. Dihydropyridine compounds were found to be selective blocking agents for the "L" channels in cultured cells (see below). "N"-channels were also activated by strong depolarisations, but were shorter acting. They are thought to be involved in neurotransmitter release from neurones (Hirning et al., 1988).

Calcium channels with activation and inactivation properties similar to those found in cultured cells have been described in mammalian tissues. However, the selectivity of the dihydropyridine compounds for the "L" subtype of channel is not evident in normal mammalian neurones. In hypothalamic and hippocampal cells, dihydropyridines blocked channels with the voltage dependence and inactivation characteristics of "T" channels as well as those with the characteristics of the "L" type (Akaike et al., 1989; Takahashi et al., 1989; 1990). Whichever subtype is involved, the high affinity binding sites for dihydropyridines, found in the central nervous system, are considered to represent voltage-sensitive calcium channels (Miller, 1987).

Ethanol depresses excitable cells by a variety of mechanisms. It has been found to decrease neuronal calcium conductances at concentrations that are achieved in the body during its behavioural effects. There is some evidence to suggest that chloride conductance mechanisms, activated by the inhibitory neurotransmitter GABA, are potentiated by acute ethanol, although this has not been found in all cases ((Davidoff, 1973; Gruol, 1982; Carlen et al., 1982; Gage and Robertson, 1985; Siggins et al., 1987). Potentiation by ethanol of the inhibitory effect of GABA would produce a reduction in depolarisation-induced calcium entry through voltage-operated calcium channels. Ion movement through calcium-dependent potassium channels has also been suggested to be increased by concentrations of ethanol relevant to its actions in vivo (Carlen et al., 1982) but this has been disputed (Siggins et al., 1987). Ethanol also appears to have similar actions on nonneuronal tissue. It inhibits both smooth and cardiac muscle contractility by mechanisms that probably involve inhibition of voltage-activated calcium channels (Mayer et al., 1980).

We have considered how excitable cells might respond to the inhibitory actions of ethanol in an attempt to overcome the inhibition. The multiplicity of functions of calcium in neuronal excitability offers many possibilities for adaptation. We have carried out an investigation of the effect of ethanol on calcium channels, using cell cultures of neuronal origin, mammalian tissues maintained in vitro, and whole animal studies our working hypothesis was that ethanol might evoke an adaptive increase in voltage-operated calcium channels in the membranes of excitable cells. This adaptation would be designed by the cell to restore excitability, but might also have implications for cell survival under certain conditions. The rest of this chapter will discuss the results of experiments designed to test this hypothesis.

The Effects of Ethanol on Neurotransmitter Release

Bovine adrenal chromaffin cells represent undifferentiated neurones and can be used in primary culture as models in which to study stimulus-secretion coupling mechanisms. Catecholamine secretion can be induced either by nicotinic receptor stimulation or by generalised membrane depolarisation with potassium. In both cases the final stimulus is entry of calcium through voltage-operated calcium channels, but the types of channel involved probably differ, in that carbachol-stimulated release is relatively insensitive to dihydropyridine calcium channel inhibitors, whereas potassium-stimulated release is sensitive to these drugs. In addition, the stimulation by carbachol of catecholamine release appears to be limited by GABA, which is co-released with the catecholamines (Brennan et al., 1989).

Carbachol-induced release was very much more sensitive to inhibition by ethanol than was potassium-stimulated release (Harper and Littleton, 1990). This suggests that the major effect of ethanol on voltage-activated calcium channels is on those that are dihydropyridine-insensitive. This result is consistent with an action of ethanol on GABA receptors in the membrane of the chromaffin cells. Experiments in our laboratory using GABA receptor antagonists and the imidazodiazepine, Ro 15 4513, suggest that this is the case (see below).

Growth of adrenal-derived cells in culture in ethanol (200 mM for 6 days) resulted in a marked increase in the number of [^{3}H]-dihydropyridine binding sites on the cell membrane, with no change in binding affinity (Messing et al., 1986). ^{45}Ca flux studies showed that this increase in binding sites represented a functional increase in calcium channels. Therefore, the presence of ethanol induced an increase in the type of calcium channel least affected by the acute presence of ethanol. This has an obvious logic as an adaptive mechanism, since the presence of ethanol would be less inhibitory to calcium entry on depolarisation, and less inhibitory to catecholamine release. The resulting tolerance to ethanol can be shown readily in adrenal cell cultures (Harper et al., 1990).

On removal of ethanol the increased number of dihydropyridine-sensitive calcium channels would be predicted to cause an increase in catecholamine release in response to any depolarising stimulus. When this was tested, the results were complex. Adrenal cell cultures grown in the presence of ethanol and then stimulated with potassium did release a greater fraction of stored catecholamines than control cells. However, when a similar experiment was carried out using carbachol-induced stimulation, the cells grown in ethanol released a smaller fraction of stored catecholamine than control cells. Since carbachol-induced release was relatively insensitive to dihydropyridine calcium channel inhibitors, this suggests that the number of dihydropyridine-*insensitive* calcium channels may actually be reduced in ethanol-grown cultures. There is as yet no specific ligand for these calcium channels, so this interpretation cannot be convincingly challenged.

The above changes in calcium channels in adrenal chromaffin cells show the characteristics of an adaptive mechanism that could underlie ethanol tolerance and physical dependence. Experiments on the mechanism by which the dihydropyridine-sensitive type of calcium channel is "upregulated" in response to ethanol suggest that synthesis of new channel proteins is required (Harper et al., 1989). It is likely that the gene that regulates synthesis of the dihydropyridine-sensitive calcium channel is "switched on" as a result of the inhibitory action of ethanol on neuronal excitability, and that this process restores relatively normal activity in the continuing presence of ethanol. Experiments on the intracellular mechanisms that lead to this increase in calcium channel gene expression suggest the involvement of the inositol lipid-protein kinase C system (Brennan and Littleton, 1990).

Ethanol and Phospholipid Metabolism

There is much evidence that ethanol affects the membranes of neuronal cells. Changes in lipid fluidity have been described (Goldstein, 1986), but it has been suggested that these are insufficient to produce the neuronal effects of ethanol. The actions of ethanol may be exerted via the lipid/protein interface or directly on the hydrophobic areas of proteins and at present it is not possible to distinguish between these two possibilities. Our results show that changes in metabolism of membrane phospholipids and the important biochemical processes associated with the neuronal membrane are seen after either acute or chronic ethanol treatment.

When added in vitro at 50 mM, ethanol decreased the activity of phospholipase A_2, the enzyme that catalyses the breakdown of membrane phospholipids (Hudspith et al, 1985). Ethanol, at 100 mM, inhibited the basal activity of phospholipase C, the enzyme that transforms inositol phospholipids (phosphatidylinositol 4,5-biphosphate, PIP_2, and phosphatidyl inositol, PI) into inositol triphosphate (IP_3) and diacylglycerol. The activity of these enzymes is dependent on calcium.

Our results also suggested that protein kinase C was involved in the adaptations to chronic ethanol treatment. Battaini et al. (1989), found that chronic ethanol treatment decreased the number of binding sites for phorbol esters in rat cerebral cortex and hippocampus, but not in cerebellar or hypothalamic tissues. Phorbol esters are known to stimulate protein kinase C activity, in the same manner as diacylglycerol. Chronic ethanol treatment decreased protein kinase activity, measured by the incorporation of ^{32}P into the protein histone type III S, in cytosolic and particulate fractions of hippocampus and cortex.

These results show that ethanol not only has important effects on calcium entry through voltage-operated channels, but it changes the membrane and intracellular actions of calcium and its target proteins.

ETHANOL AND DIHYDROPYRIDINE-SENSITIVE CALCIUM CHANNELS IN THE INTACT ANIMAL

Although the function of dihydropyridine-sensitive binding sites in normal neuronal function is not well understood, there is increasing evidence that they do have a role in excitability and may be involved in neuronal plasticity. Dihydropyridine antagonists, such as nifedipine, are used for their effects on the cardiovascular system, but have recently been shown to have actions on the central nervous system (Ramkumar and El-Fakahany, 1986; Raeburn and Gonzales, 1988). They appear to have a selective action on drug-induced withdrawal syndromes, as in addition to their protective effect on the ethanol withdrawal (see below), they have been reported to decrease morphine withdrawal (Bongianni et al., 1986) and nitrous oxide withdrawal (Dolin and Little, 1989a).

Changes in Dihydropyridine-Sensitive Binding Sites in the CNS After Chronic Ethanol Administration

As found in the cell culture studies, the number of dihydropyridine binding sites in the CNS was increased after chronic ethanol treatment (Dolin et al., 1987; Dolin & Little, 1989b). We have demonstrated increases in binding after various methods of administration of ethanol, including inhalation of ethanol by rats (10 - 20g/kg/day) for seven days, intraperitoneal injection of ethanol in rats at 2g/kg, once daily, for ten days, and administration of ethanol in the drinking fluid to mice (11 - 14g/kg/day) for 12 weeks. All of these chronic treatment schedules caused tolerance to ethanol and a mild withdrawal syndrome (convulsive behaviour on handling), with the exception of the intraperitoneal schedule that caused tolerance, but not withdrawal signs (Littleton et al., 1990; Whittington & Little, 1988; Dolin & Little, 1989b). Wu et al. (1987) found a temporal correlation between the increase in binding sites and the development of ethanol tolerance. These authors also reported a correlation between the time course of this increase in dihydropyridine binding site number and that of the development of ethanol tolerance.

One report, by Lucchi et al. (1985), found a more complex pattern of changes in dihydropyridine binding. These workers studied striatal tissue and showed that calcium-independent binding was increased, and calcium-dependent binding decreased, after administration of ethanol to rats at approximately 20 g/kg/day for 25 days. A later communication from this laboratory (Bergamaschi et al., 1987) found an increase in dihydropyridine binding B_{max} values in cerebral cortex on acute administration of ethanol and a decrease in B_{max} after chronic treatment. The authors suggested that the increased binding described above may be associated with the hyperexcitability state of withdrawal.

Binding in human tissues has so far not been found to be significantly increased in alcoholics (Marks et al., 1989b; Kril et al., 1989), although the former authors, studying small samples, found that the binding site number was increased by 26%. Details of any pathological changes or cell loss were not reported. This is important given the possibility that an increase in voltage-operated calcium channels may itself be cytotoxic (see later). It has been shown that the characteristics of dihydropyridine binding are altered by postmortem delays (Quirion et al., 1988); these authors cautioned against extrapolation from postmortem results on dihydropyridine binding to physiological conditions. It is also notable that results from postmortem studies on human alcoholic brains using ligands for a variety of receptors have not been found to parallel those from animal studies, probably because of the difficulties in controlling the conditions and "treatment" in human measurements.

No experimental studies have yet reported the effects of chronic ethanol treatment on dihydropyridine binding measured in vivo. This is an important aspect, as dihydropyridine calcium channel antagonists bind preferentially to the inactivated form of the channel complex (Bean, 1984). Binding to homogenate preparations measures the interaction in a totally depolarised membrane, in which channels will be largely inactivated, and thus may not reflect the situation in the normal physiological state. This was clearly demonstrated by the results of Rius et al. (1987), who showed that a single dose (3g/kg, i.p.) of ethanol, in vivo, caused a short duration increase in the B_{max} for dihydropyridine binding, 40 min after ethanol administration, followed by an increase in affinity that peaked at 8h, returning to normal by 36h. Acute addition of ethanol in vitro has been shown by several authors not to affect dihydropyridine binding (Greenberg et al., 1984; Harris et al., 1985; Rius et al., 1987).

Increased Effects of Dihydropyridines After Chronic Ethanol Treatment

A change in the number of voltage-sensitive calcium channels is of interest only if it is accompanied by functional changes. That the increase in dihydropyridine binding is of functional importance was demonstrated by Dolin et al. (1987) and Littleton et al. (1988) using mammalian tissue. The in vitro effects of the dihydropyridine calcium channel activator, Bay K 8644, on neurotransmitter release were increased in cortical tissue from rats treated chronically with ethanol by inhalation for seven days. Chronic ethanol administration also increased the sensitivity of inositol phospholipid breakdown to the action of dihydropyridines, illustrated in Figure 1 (Dolin et al., 1987; Hudspith et al. 1987). The stimulation of inositol phospholipid turnover by carbachol and noradrenaline was also increased. The increase in dihydropyridine binding sites appears, therefore, to reflect an increase in functional proteins, though whether these are voltage-sensitive calcium channels can be determined only by intracellular electrophysiological recording. These experiments are in progress.

Studies Using Cell Cultures on the Mechanism of the Changes in Dihydropyridine Binding.

We have investigated the mechanism of the increase in dihydropyridine binding, using cultured bovine chromaffin cells (Brennan & Littleton, 1990). Pertussis toxin, that prevents the coupling of G-proteins, mimicked the upregulation of binding sites. When the calcium concentration in the growth medium was increased, the upregulation due to ethanol was prevented. These results suggested that inhibition of membrane calcium flux, and a G-protein, are involved in the response of dihydropyridine binding sites to chronic ethanol. Prolonged action of a phorbol ester causes inhibition of protein kinase and this was also found to produce up-regulation of dihydropyridine binding sites in these cells. The fact that this up-regulation was not additive with that due to ethanol suggested that protein kinase-C is involved in the ethanol-induced alteration of dihydropyridine-sensitive calcium channels.

The change in dihydropyridine binding sites was prevented by the protein synthesis inhibitor, cycloheximide, and by the mRNA synthesis inhibitor, anisomycin, suggesting that it was the result of synthesis of new channel protein and was under genetic regulation

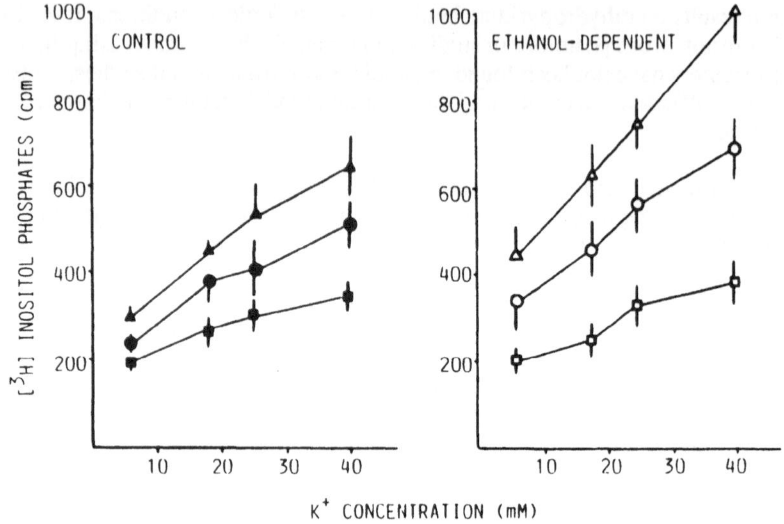

Figure 1. The effects of chronic ethanol treatment on the actions of Bay K 8644 on inositol phospholipid turnover, measured over a range of potassium concentrations. Preparations of cortical tissue made after chronic ethanol treatment (on right) showed higher basal and depolarisation-induced accumulation of [^{3}H]-inositol phosphates compared with control tissues (on left). (Reproduced with permission from Dolin et al., 1987)

(Harper et al., 1989). Marks et al. (1989a) demonstrated that the increase occurred in intact PC12 cells, in which the second messenger systems are functional.

Evidence for a causal relationship between increases in central dihydropyridine binding and the ethanol withdrawal syndrome was provided by a study of the binding changes in withdrawal-resistant and withdrawal-susceptible lines of mice. These strain lines have been selectively bred for many generations and show large differences in the severity of the ethanol withdrawal syndrome. The increase in dihydropyridine binding site number after chronic ethanol administration was greater in the withdrawal-susceptible line than in the withdrawal-resistant line (Brennan et al, 1990).

ACUTE INTERACTIONS BETWEEN ETHANOL AND DIHYDROPYRIDINES IN VIVO

Calcium channel antagonists potentiate the acute, in vivo, pharmacological effects of ethanol (Isaacson et al., 1985; Dolin & Little, 1986; Pucilowski et al., 1989). The calcium channel activator, Bay K 8644, had the opposite action (Dolin et al., 1988a). Calcium channel antagonists possess some sedative properties, at high doses, so the interaction with ethanol may have been partly additive. However, they do not produce general anaesthesia, even at extremely high doses (Dolin & Little, 1986). The interaction could also have been due to effects of ethanol on the dihydropyridine receptor. Rius et al. (1987), found that ethanol, given acutely, increased the affinity of the dihydropyridine binding sites in rat brain, even though ethanol had no effect on dihydropyridine binding measured in vitro (Greenberg and Cooper, 1984).

The Acute Effects of Dihydropyridines in Ethanol Withdrawal

Dihydropyridine calcium channel antagonists decrease ethanol withdrawal hyperexcitability. This effect was seen when a mild withdrawal syndrome was measured by convulsive behaviour and tremor on handling (Littleton et al., 1990) and in severe withdrawal measured by the incidence of spontaneous and audiogenic convulsions (Little et al., 1986). The action of dihydropyridine antagonists in the above studies was a selective effect. These compounds have weak anticonvulsant actions against some other forms of convulsions, particularly those produced by pentylenetetrazol, but not against others, such as N-methyl-d-aspartate or electroshock seizures (Hoffmeister et al., 1982; Dolin et al., 1988b). We have repeatedly found that they have no protective action against convulsions induced by bicuculline.

The dose of PN 200-110 effective against the ethanol withdrawal syndrome (Littleton et al., 1990) was chosen as being the same as that reported by Supervilai & Karobath (1984) to be needed for displacement of radiolabelled dihydropyridine from the central nervous system in vivo. Brain concentrations of nitrendipine and nimodipine, after the doses effective against the ethanol withdrawal syndrome were in the low micromolar/high nanomolar range (Dolin, 1988b; Dolin & Little, 1989b). This order of concentration is the same as that required to produce effects on brain tissues in vitro (see above).

The doses of the dihydropyridines required for central nervous system actions are higher than those needed to affect the cardiovascular system, although the binding affinities of neuronal and smooth muscle preparations, measured on homogenates, are similar. This difference is not due to problems in entering the brain, as the compounds penetrate the CNS well. The apparent discrepancy is due to the influence of membrane potential on the binding affinity. Dihydropyridine calcium channel antagonists have a higher affinity for the inactivated state of the channel than for the open or closed states (Bean, 1984). At normal membrane potentials, in neuronal tissue, few channels are activated, therefore few would be in the inactivated state. In homogenate preparations the tissues are totally depolarised, and binding affinities measured in this way do not reflect the affinity in the physiological situation. Studies on binding in vivo, such as that by Supervilai & Karobath (1984) provide a better indication of the affinities in functioning neurones.

The stereospecificity of the isomers of the dihydropyridine calcium channel antagonist, PN 200-110, in protecting against the ethanol withdrawal syndrome, was the same as that demonstrated on calcium channel conductance (Littleton et al., 1990). The stereospecificity, and the correlation with the dose required to displace a radiolabelled dihydropyridine from central binding sites (Littleton et al., 1990), showed that the protective effect on the withdrawal was probably mediated by neuronal voltage-sensitive calcium channels, rather than a nonspecific action.

Nitrendipine, given acutely, was not found to be effective in preventing the anxiogenic aspects of the ethanol withdrawal syndrome (File et al., 1989). The dose used was the same as that which prevented the tremor and convulsions in other studies (Littleton et al., 1990), so the results suggest that anxiety during ethanol withdrawal may be caused by different biochemical changes than the convulsive behaviour. Of more interest are the effects of chronic administration of nitrendipine (see below) on ethanol withdrawal-induced anxiogenic behaviour, but these have not yet been investigated.

Bay K 8644 is a dihydropyridine calcium channel activator. It increases the opening of the channels, thereby having the opposite effect to that of the antagonists. The (-) form possesses this action, but the (+) isomer is a calcium channel antagonist (Franckowiak et al., 1985; O'Neill and Bolger, 1988). The racemic form of Bay K 8644 prevented the protective action of the calcium channel antagonists in ethanol withdrawal, but did not itself potentiate the withdrawal syndrome (Littleton et al., 1990). The (-) isomer, however, increased the convulsive behaviour seen in mice on handling during ethanol withdrawal (unpublished results).

A calcium channel blocking agent has been found to protect against the alcohol withdrawal syndrome in humans (Koppi et al., 1987), and clinical trials of the dihydropyridines are in progress.

MODULATION OF ADAPTIVE CHANGES BY CHRONIC ADMINISTRATION OF DIHYDROPYRIDINES

A causal relationship between a biochemical change and a behavioural syndrome is very difficult to demonstrate. The production of a biochemical change by chronic ethanol treatment does not mean that such a change is the cause of the behavioural manifestations of dependence, such as tolerance and the withdrawal syndrome. Prevention of both the biochemical change and the tolerance and withdrawal, however, provides good evidence for a causal relationship.

Prevention of Ethanol Tolerance

Chronic dihydropyridine treatment, given during the ethanol intake prevented the development of alcohol tolerance, in animal models of this condition (Wu et al., 1987; Little & Dolin, 1987; Dolin & Little, 1989b). When nifedipine or nitrendipine were given continuously with the ethanol, significantly less tolerance developed to the ataxic action of ethanol than when rodents received the same amount of ethanol given alone. The brain concentrations of ethanol were unaltered by the dihydropyridine treatment, so the effect on tolerance was not due to changes in ethanol pharmacokinetics. At the time the tolerance was measured, the central nitrendipine concentrations were an order of magnitude lower than those that are necessary to have an acute effect on the actions of ethanol.

The results therefore showed that concurrent, chronic, administration of nitrendipine prevented the development of ethanol tolerance, even though the acute effect of this compound was to increase the behavioural actions of ethanol.

Prevention of the Ethanol Withdrawal Syndrome

Chronic administration of a dihydropyridine calcium channel antagonist, for two weeks or more, also decreased the ethanol withdrawal syndrome, even though the dihydropyridine treatment was stopped 24h or 48h before the measurement of ethanol withdrawal severity (Whittington & Little, 1988). Again, when we measured the central dihydropyridine concentrations and found that they were found to be too low at the time of the ethanol withdrawal tests to have had an acute action on the withdrawal behaviour. The effect on withdrawal was not seen when only two days' dihydropyridine treatment was given. This results suggested that the chronic administration of a dihydropyridine calcium channel antagonist produced some form of adaptive response that prevented or reversed the adaptations to ethanol.

PREVENTION OF THE UPREGULATION OF DIHYDROPYRIDINE BINDING SITES BY CHRONIC ADMINISTRATION OF DIHYDROPYRIDINES

Concurrent dihydropyridine administration also prevented the increase in the number of dihydropyridine sensitive calcium channels caused by chronic ethanol treatment (Dolin et al., 1988c; Dolin & Little, 1989b). Panza et al. (1985) reported that chronic administration of a dihydropyridine calcium channel antagonist caused downregulation of dihydropyridine receptors in mouse brain. An effect such as this would counteract the upregulation caused by chronic ethanol administration. Prevention of the increase in dihydropyridine binding caused by chronic ethanol treatment was also seen when when bovine adrenal chromaffin cells were grown in ethanol and a dihydropyridine (Brennan et al., 1989).

ELECTROPHYSIOLOGICAL STUDIES

Definition of the Ethanol Withdrawal Syndrome In Vitro and the Effects of Dihydropyridines

Evidence that the changes observed in vivo after these treatments were due to changes in neuronal excitability was obtained using hippocampal slices (Whittington & Little, 1989b; 1990). Extracellular recordings from isolated slices showed changes in field potentials immediately following withdrawal from in vivo chronic ethanol treatment. The changes included decreased thresholds for elicitation of population spikes and increases in paired pulse potentiation. The individual changes followed different time courses during the recording period, suggesting that a series of alterations takes place in this tissue on withdrawal from ethanol.

The dihydropyridine calcium channel antagonist, PN 200-110, stereospecifically blocked all the signs of hyperexcitability in the hippocampal field potentials seen after the ethanol treatment (Figure 4, Whittington & Little; 1989a). This was a selective effect, as this compound did not alter the hyperexcitability in hippocampal slices from untreated mice produced by addition of the GABA antagonist, bicuculline.

The Effects of Chronic Dihydropyridine Treatment on the Electrophysiological Manifestations of the Ethanol Withdrawal Syndrome

When nitrendipine was included in the drinking fluid during the chronic ethanol treatment, all the signs of hyperexcitability in the hippocampal slices were reduced (Whittington and Little, 1989b). The effects of such concurrent treatment in the electrophysiological studies was therefore the same as that found in the behavioural measurements of withdrawal.

ADAPTIVE RESPONSES IN CALCIUM CHANNELS OF NONNEURONAL TISSUE

As discussed previously, the functional role of dihydropyridine-sensitive calcium channels in neurones is not well understood. However, in other excitable cells, such as smooth muscle as well as cardiac muscle and conducting tissue, the dihydropyridine calcium antagonist drugs have potent inhibitory effects on contraction or conduction, arguing an important functional role for these channels in these cells. If an upregulation of dihydropyridine-sensitive calcium channels were to prove a generalised adaptive response to ethanol of excitable cells, then functional consequences in smooth muscle and heart should be more easy to demonstrate than in brain.

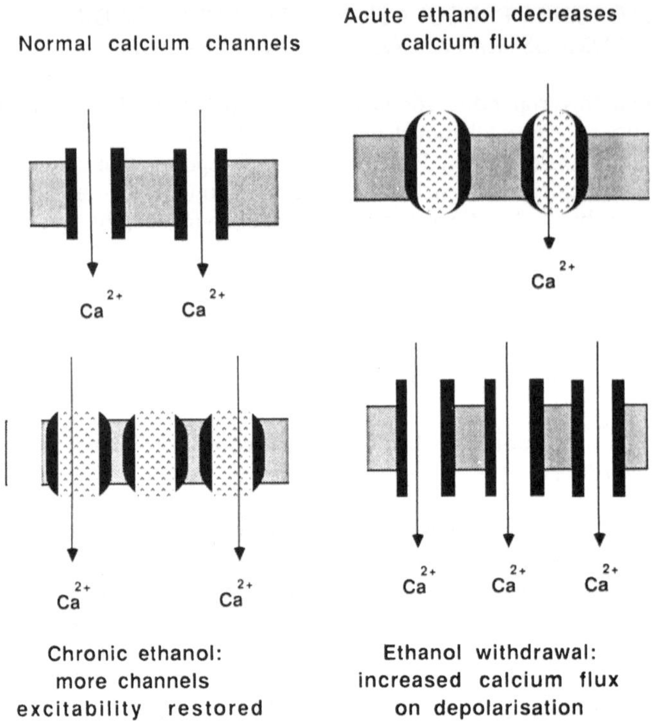

Figure 2. Schematic diagram of the hypothesis that an increase in neuronal calcium channels is involved in ethanol dependence

In membranes from vasa deferentia the number of dihydropyridine binding sites was increased about 80% by chronic ethanol treatment, and this correlates well with functional alterations in smooth muscle contraction and inositol lipid breakdown (Guppy and Littleton, 1990; Hudspith et al., 1987). We believe, in this preparation, that increases in dihydropyridine-sensitive calcium channels occur in both the smooth muscle cells and in the intrinsic autonomic nerves. This, however, is difficult to prove experimentally. The end result is that smooth muscle in general may become hyperexcitable to neuronal influences after chronic administration of ethanol.

In cardiac membranes, there was an increase of about 150% in the number of dihydropyridine binding sites, with no change in binding affinity, after chronic exposure of rats to ethanol (Guppy and Littleton, 1990). As yet we do not know whether this occurs on cardiac muscle membranes or conducting tissue (although we suspect both). However, we have found no direct evidence of functional alterations in force of contraction, or of rhythm, in isolated hearts from ethanol-treated animals that could be attributed to these changes in dihydropyridine-sensitive calcium channels. Long-term ethanol administration to man is known to produce dysfunction in cardiac rhythm consistent with such a change (Regan et al., 1977). We have, however, observed that isolated hearts from ethanol-dependent rats were more susceptible to cellular damage during periods of anoxia than are controls. This increase in damage could be reversed by perfusion with dihydropyridine calcium channel antagonists (Guppy and Littleton, unpublished results). This provides clear evidence for a pathological, rather than a physiological, role of an increase in the number of dihydropyridine-sensitive calcium channels in the heart.

CONCLUSIONS

These findings suggest that wherever excitable tissues are depressed by ethanol (possibly by a final common path of inhibition of calcium entry) the cellular response is to up-regulate dihydropyridine-sensitive calcium channels (see Figure 2). This provides an adaptive response restoring normal excitability in the presence of ethanol, but it could have severe disadvantages for the cells that have responded in this way. In the CNS, the removal of ethanol while neurones are in the "calcium-channel upregulated" state, could expose latent hyperexcitablity and cause the physical withdrawal syndrome. Hyperexcitability and excess calcium entry could in addition increase the possibility of cytotoxicity in these cells.

The effects of chronic dihydropyridine administration described here are the first example of a pharmacological treatment preventing both the adaptive changes in the behavioural actions of ethanol (tolerance and the withdrawal syndrome) and a related biochemical change (the increase in number of dihydropyridine binding sites). Consistent results have been obtained from behavioural, receptor binding and electrophysiological studies on this effect (Wu et al., 1987, Dolin et al., 1987; Dolin & Little, 1989b; Whittington & Little, 1988; Whittington & Little, 1989a,b,c). The evidence from all these types of study suggest that chronic administration of a dihydropyridine calcium channel antagonist with ethanol, during long-term ethanol intake, prevents the manifestations of ethanol dependence.

REFERENCES

Akaike, N., Kostyuk, P.G. and Osipchuk, Y.V., 1989, Dihydropyridine sensitive low-threshold calcium channels in isolated rat hypothalamic neurones., **J. Physiol.**, 412:181

Battaini, F., Del Vasco, R., Govoni, S. and Trabucci, M., 1989, Chronic alcohol intake modifies phorbol ester binding in selected rat brain areas., **Alcohol**, 6:169

Bean, B.P., 1984, Nitrendipine block of cardiac calcium channels: high affinity binding to the inactivated state., **Proc. Natl. Acad. Sci.,** 81:6388

Bergamaschi, A., Govoni, S., and Trabucchi, M., 1987, Alcohol and calcium interactions at the CNS level: an in vivo study., **Pharmacol. Res. Commun.**, 19:975

Bongianni, F., Carla, V., Moroni, F. and Pellegrini-Giampietro, D.E., 1986, Calcium channel inhibitors suppress the morphine-withdrawal syndrome in rats., **Br. J. Pharmacol.,** 88:561

Brennan, C.H., and Littleton, J.M., 1990, Second messengers involved in genetic regulation of the number of calcium channels in bovine adrenal chromaffin cells in culture., **Neuropharmacology**, 29:689

Brennan, C.H., Lewis, A. and Littleton, J.M., 1989, Membrane receptors involved in upregulation of calcium channels in bovine adrenal chromaffin cells, chronically exposed to ethanol., **Neuropharmacology**, 28:1303

Brennan, C.H., Crabbe, J. and Littleton, J.M., 1990 Genetic regulation of dihydropyridine-sensitive calcium channels may determine susceptibility to alcohol physical dependence. **Neuropharmacology**, 29:429

Carlen, P.L., Gurevich, N. and Durand, D., 1982, Ethanol in low doses augments calcium-mediated mechanisms measured intracellularly in hippocampal neurones., **Science**, 215:306

Davidoff, R.A., 1973, Alcohol and presynaptic inhibition in an isolated spinal cord preparation., **Arch. Neurol.,** 28:60

Dolin, S.J. and Little, H.J., 1986, Augmentation by calcium channel antagonists of general anaesthetic potency in mice., **Br. J. Pharmacol.**, 88:909
Dolin, S.J. and Little, H.J., 1989a, Effects of the calcium antagonist, nitrendipine, on N_2O anaesthesia, tolerance and physical dependence., **Anaesthesiology**, 70:91
Dolin, S.J. and Little, H.J., 1989b, Are changes in neuronal calcium channels involved in ethanol tolerance?, **J. Pharmacol. Exp. Ther.**, 250:985
Dolin, S.J., Little, H.J., Hudspith, M., Pagonis, C. and Littleton, J., 1987, Increased dihydropyridine sensitive calcium channels in rat brain may underly ethanol physical dependence., **Neuropharmacology**, 26:275
Dolin, S.J., Halsey, M.J., and Little, H.J., 1988a, Effects of the calcium channel agonist, BAY K 8644, on a general anaesthetic potency in mice., **Br. J. Pharmacol.**, 94:413
Dolin, S.J., Hunter, A.B., Halsey, M.J., and Little, H.J., 1988b Anticonvulsant profile of the dihydropyridine calcium channel antagonists, nitrendipine and nimodipine., **Eur. J. Pharmacol.**, 152:19
Dolin, S.J., Patch, T., Siarey, R.J., Whittington, M.A. and Little, H.J., 1988c, Evidence for the involvement of dihydropyridine-sensitive calcium channels in ethanol dependence., **Br. J. Pharmacol.**, 95:877P
Dolin, S.J., 1988, The effects of calcium antagonists on sedative drug action in rodents: anaesthesia, physical dependence and tolerance., PhD Thesis, University of London
File, S.E., Baldwin, H.A. and Hitchcott, P.K., 1989, Flumazenil but not nitrendipine reverses the increased anxiety during ethanol withdrawal in the rat., **Psychopharmacology**, 98:262
Franckowiak, G., Bechem, M., Schramm, M. and Thomas, G. 1985, The optical isomers of the 1,4-dihydropyridine BAY K 8644 show opposite effects on calcium channels., **Eur. J. Pharmacol.**, 114: 223
Gage, P.W. and Robertson, B., 1985, Prolongation of inhibitory postsynaptic currents by pentobarbitone, halothane and ketamine in CA1 pyramidal cells in rat hippocampus., **Br. J. Pharmacol.**, 85:675
Greenberg, D.A. and Cooper, E.C., 1984, Effect of ethanol on [^{3}H]-nitrendipine binding to calcium channels in brain membranes., **Alcoholism Clin. Exp. Res.**, 8:568
Greenberg, D.A., Cooper, E.C., Gordon, A. and Diamond, I., 1984, Ethanol and the gamma-aminobutyric acid-benzodiazepine receptor complex., **J. Neurochem.**, :1062
Gruol, D.L., 1982, Ethanol alters synaptic activity in cultured spinal neurones., **Brain Res.**, 243:25
Guppy, L.J. and Littleton, J.M., 1989, Increased [^{3}H]-dihydropyridine binding in brain, heart and smooth muscle of ethanol-dependent rats. **Br. J. Pharmacol.**, 92:662P
Harper, J.C., Brennan, C.H. and Littleton, J.M., 1989, Genetic upregulation of calcium channels in a cellular model of ethanol dependence., **Neuropharmacology**, 28:1299
Harper, J. C. and Littleton, J. M. 1990, Alcohol tolerance in bovine adrenal chromaffin cells. **Alc. Clin. Exp. Res.**, 14: 508
Harris, R.A., Jones, S.B., Bruno, P. and Bylund, D.B., 1985, Effects of dihydropyridine derivatives and anticonvulsant drugs on [^{3}H]-nitrendipine binding and calcium and sodium fluxes in brain., **Biochem. Pharmacol.**, 34:2187
Hernandez-Cruz, A. and Pape, H. C., 1989, Indentification of two calcium currents in acutely dissociated neurones from the rat lateral geniculate nucleus. **J. Neurophys.**, 61, 1270-1283
Hirning, L.D., Fox, A.P., McCleskey, E.W., Olivera, R.M., Thayer, S.A., Miller, R.J. and Tsien, R.W., 1988, Dominant role of N-type Ca^{2+} channels in evoked release of norepinephrine from sympathetic neurones., **Science**, 239:57

Hoffmeister, F., Benz, U., Heise, A., Krause, H.P. and Neuser, V., 1982, Behavioural effects of nimodipine in animals., **Arzneim-Forsch.**, 32:347

Hudspith, M., John, G.R., Nhamburo, P.T. and Littleton, J.M., 1985, Effect of ethanol in vitro and in vivo on Ca^{2+} activated metabolism of membrane phospholipids in rat synaptosomal and brain slice preparations., **Alcohol**, 2:133

Hudspith, M.J., Brennan, C.H., Charles, S., and Littleton, J.M., 1987, Dihydropyridine-sensitive calcium channels and inositol phospholipid metabolism in ethanol physical dependence., **Ann. N.Y. Acad. Sci.**, 492:156

Isaacson, R.L., Molina, J.C., Draski, L.J. and Johnston, J.E., 1985, Nimodipine interactions with other drugs, 1. Ethanol., **Life Sci.**, 36:2195

Koppi, S., Eberhardt, G., Haller, R. and Konig, P., 1987, Calcium channel blocking agent in the treatment of acute alcohol withdrawal. Caroverine versus meprobamate in a randomized double blind study., **Neuropsychobiol.**, 17:49

Kril, J.J., Gundlach, A.L., Dodd, P.R., Johnston, G.A.R. and Harper, C.G., 1989, Cortical dihydropyridine binding sites are unaltered in human alcoholic brain., **Ann. Neurol.**, 26:395

Little, H.J. and Dolin, S.J., 1987, Lack of tolerance to ethanol after concurrent administration of nitrendipine., **Br. J. Pharmacol.**, 92:606P

Little, H.J., Dolin, S.J. and Halsey, M.J., 1986, Calcium channel antagonists decrease the ethanol withdrawal syndrome., **Life Sci.**, 39:2059

Littleton, J., Harper, J., Hudspith, M., Pagonis, Dolin, S.J. and Little, H.J., 1988, Adaptation in neural Ca^{2+} channels may cause alcohol physical dependence, **in**: The Psychopharmacology of Addiction., ed., M. Lader, pp 61-72

Littleton, J.M., Little, H.J. and Whittington, M.A., 1990, Effects of dihydropyridine calcium channel antagonists in ethanol withdrawal; doses required, stereospecificity and actions of BAY K 8644, **Psychopharmacology**, 100:387

Lucchi, L., Govoni, S., Battaini, F., Pasinetti, G. and Trabucci, M., 1985, Ethanol administration in vivo alters calcium ions control in rat striatum., **Brain Res.**, 332:376

Marks, S.S., Watson, D.L., Carpenter, C.L., Messing, R.O. and Greenberg, D.A., 1989, Comparative effects of chronic exposure to ethanol and calcium channel antagonists on calcium channel antagonist receptors in cultured neural (PC12) cells., **J. Neurochem.**, 53:168
Marks, S.S., Watson, D.L., Carpenter, C.L., Messing, R.O. and Greenberg, D.A., 1989, Comparative effects of chronic exposure to ethanol and calcium channel antagonists on calcium channel antagonist receptors in cultured neural (PC12) cells., **J. Neurochem.**, 53:168
Mayer, J.M., Khanna, J.M., Kalant, H. and Spero, L., 1980, Cross tolerance between ethanol and morphine in the isolated guinea-pig ileum myenteric plexus preparation., **Eur. J. Pharmacol.**, 63:223

Mayer, J.M., Khanna, J.M. and Kalant, H., 1980, A role for calcium in the acute and chronic actions of ethanol in vitro., **Eur. J. Pharmacol.**, 68:223

Messing, R.O., Carpenter, C.L., Diamond, I. and Greenberg, D.A., 1986, Ethanol regulates calcium channels in clonal neural cells., **Proc. Natl. Acad. Sci.**, 83:6213

Miller, R.J., 1987, Multiple calcium channels and neuronal function., **Science**, 235:46

Nowycky, M.C., Fox, A. and Tsien, R.W., 1985, Three types of neuronal calcium channel with different calcium agonist sensitivity., **Nature**, 316:440

O'Neill, S.K. and Bolger, G.T., 1988, Enantiomer selectivity and the development of tolerance to the behavioural effects of the calcium channel activator, Bay K 8644., **Brain Res. Bull.**, 21: 865

Panza, G., Grebb, J.A., Sanna, E., Wright, A.G. and Hanbauer, L., 1985, Evidence for down regulation of [^{3}H]-nitrendipine recognition sites in mouse brain after long term treatment with nifedipine or verapamil., **Neuropharmacology**, 24:1113

Pucilowski, O., Krzascik, P., Trzaskowska, E., and Kostowski, W., 1989, Different effect of diltiazem and nifedipine on some central actions of ethanol in the rat., **Alcohol**, 6:165

Quirion, R., Lal, S., Olivier, A., Robitaille, Y., Vassavan Nair, N.P., Ford, R.M. and Stratford, J.G., 1988, Calcium channel binding sites in human brain., **Ann. N.Y. Acad. Sci.**, 522:203

Raeburn, D. and Gonzales, R.A., 1988, CNS disorders and calcium antagonists., **Trends Pharmacol. Sci.**, 9:117

Ramkumar, V. and El-Fakahany, E.E., 1986, The current status of the dihydropyridine calcium channel antagonist binding sites in the brain., **Trends Pharmacol. Sci.**, May:171

Regan, T.J., Ettinger, P.O., Haider, B., Ahmed, S.S., Oldewurtel, H.A., and Lyons, M.M., 1977, The role of ethanol in cardiac disease., **Ann. Rev. Med.,** 28:393

Reynolds, I.J., Wagner, J.A., Snyder, S.A., Thayer, S.A., Olivera, B.B. and Miller, R.J., 1986, Brain voltage-sensitive calcium channel subtypes: differentiation by omega-conotoxin fractions GVIA., **Proc. Natl. Acad. Sci.**, 83:8804

Rius, R.A., Bergamaschi, S., DiFonso, F., Govoni, S., Trabucci, M. and Rossi, F., 1987, Acute ethanol effect on calcium antagonist binding in rat brain., **Brain Res.,** 402:359

Siggins, G.R., Pittman, Q.J. and French, E.D., 1987, Effects of ethanol on CA1 and CA3 pyramidal cells in the hippocampal slice preparation: an intracellular study., **Brain Res.,** 414:22

Supervilai, P., and Karobath, M., 1984, The interaction of [^{3}H]-PY 108-068 and of [^{3}H]-PN 200-110 with calcium channel binding sites in the rat brain., **J. Neural Transmission**, 60:149

Takahashi, K., Wakamori, M., and Akaike, N., 1989, Hippocampal CA1 pyramidal cells of rats have four voltage-dependent calcium conductances., **Neurosci. Lett.,** 104:229

Takahashi, K. and Akaike, N. 1990 Nicergoline inhibits T-type calcium channels in rat isolated hippocampal CA1 pyramidal neurones. **Br. J. Pharmacol.,** 100:705

Whittington, M.A. and Little, H.J., 1988, Nitrendipine prevents the ethanol withdrawal syndrome, when administered chronically with ethanol prior to withdrawal., **Br. J. Pharmacol.,** 94:885P

Whittington, M.A. and Little, H.J., 1989a, Stereospecificity of the actions of PN 200-110 on the changes due to ethanol withdrawal in hippocampal field potentials., **Br. J. Pharmacol.,** 98:923P

Whittington, M.A. and Little, H.J., 1989b, Chronic ethanol increases excitability of hippocampal cell bodies; effects decreased by chronic nitrendipine treatment., **Br. J. Pharmacol.,** 97:435P

Whittington, M.A. and Little, H.J., 1990, Patterns of changes in field potentials in the isolated hippocampal slice on withdrawal from chronic ethanol treatment of mice in vivo., **Brain Res.**

Wu, P.H., Fan, T. and Naranjo, C.A., 1987, Nifedipine delays the acquisition of ethanol tolerance, **Eur. J. Pharmacol.,** 139: 236

STRUCTURAL CHANGES IN THE HIPPOCAMPAL FORMATION AND FRONTAL CORTEX AFTER LONG-TERM ALCOHOL CONSUMPTION AND WITHDRAWAL IN THE RAT

A. Cadete-Leite

Department of Anatomy
Porto Medical School
Alameda Hernani Monteiro
4200 Porto, PORTUGAL

Alcohol-induced neurocytological changes in humans were first reported by Courville (1955). However, such changes needed to be cautiously interpreted because other factors, as vitamin deficiency (Yew et al., 1981) and malnutrition (Victor and Adams, 1961; Paula-Barbosa et al., 1989), commonly accompany chronic alcohol consumption (CAC) and are known also to affect the neuronal structure. Thus, the pioneer results obtained by Riley and Walker (1978) were of paramount importance because they described marked neuronal loss in the rodent hippocampal formation after 5 months of alcohol intake using a nutritional-controlled model.

In our laboratory, over the last 12 years, by using an animal model of alcohol consumption in which the periods of ethanol intake were extended as far as 18 months, we also contributed to provide a comprehensive picture of the significant alterations in the structural organization of the CNS. The most striking findings were the massive degenerative changes: neuronal death (Tavares and Paula-Barbosa, 1982; Tavares et al., 1987a), impoverishment of dendritic trees (Tavares et al., 1983a), loss of synaptic contacts (Tavares et al., 1987b), formation of neuritic plaques (Paula-Barbosa and Tavares, 1984) and increased accumulation of lipofuscin (Tavares and Paula-Barbosa, 1983; Tavares et al., 1985). In spite of these deteriorative changes, there was evidence of remodeling activity which included synaptic plasticity (Tavares and Paula-Barbosa, 1984; Tavares et al., 1986a,b, 1987b) and lengthening of dendritic spines (Tavares et al., 1983b).

The effects of CAC upon the cerebellum were the first to be studied because its structural arrangement is simple. Subsequently we extended the investigations to the study of the hippocampal formation (Borges et al., 1986; Paula-Barbosa et al., 1986) and the prefrontal cortex (Cadete-Leite et al., 1986). The reasons for the selection of hippocampal formation (HF) were: (1) its role in cognitive deficits including those observed in humans (Butters et al., 1977) and animals consuming alcohol (Walker and Hunter, 1978), (2) its relatively simple and highly characteristic structure (Andersen, 1975), and (3) its vulnerability to ethanol aggression (Walker et al., 1980, 1981). In relation to the prelimbic (PL) area of the prefrontal medial cortex (PFCm), the reasons for its study were: (1) the PFCm of the rodents is considered to be the equivalent of the human frontal cortex (Kolb,

Edited by T.N. Palmer, Plenum Press, New York, 1991

1984), (2) it is implicated in cognitive tasks (Kolb, 1984), (3) it possesses well-defined neuronal layers (Krettek and Price, 1977; Van Eden and Uylings, 1985), (4) it is connected to other limbic areas (Ferino et al., 1987), and (5) there are neuropsychological studies which demonstrate its involvement after CAC (Walsh, 1983).

More recently, we have also studied the effects of withdrawal after CAC. The animals were alcohol-fed according to our model and it was possible to gain an insight into the effects of the prolonged alcohol abstinence and into the controversy concerning the reversibility of the alcohol-induced deteriorative changes (Carlen et al., 1978, 1986; Walker et al., 1980, 1981; Phillips and Cragg, 1983, 1984; McMullen et al., 1984; Phillips, 1985; King et al., 1988; Durand et al., 1989).

In this investigation we used 2-month-old male Sprague-Dawley rats separated into 8 groups of 6 animals individually caged and treated as follows: (1) animals alcohol-fed for 6, 12 and 18 months with a 20% (v/v) aqueous ethanol solution as the only available liquid source; food and fluid intake were measured every other day and the amounts consumed were calculated, (2) respective pair-fed controls; these animals were given the same amount of food and liquid in a solution where sucrose replaced isocalorically the ethanol, (3) alcohol-fed for 6 (Wl) and 12 (W2) months and then switched to standard laboratory pellet food and water for 6 months - withdrawal groups (Wl and W2). In the different groups all liquids were supplemented with vitamins and minerals.

At the end of each experimental period the animals were anaesthetized with ether and transcardially perfused with a solution of 1% glutaraldehyde and 1% paraformaldehyde in 0.12 M phosphate buffer at pH 7.2 (Palay and Chan-Palay, 1974). The brains were removed and after 2 hr in the perfusion solution they were blocked into left and right halves. The right frontal pole and HF were processed as for electron microscopy, whereas the left ones were post-fixed in the solution used for the perfusion for a week in the case of the HF and for 2 weeks in the case of frontal pole. After these periods of post-fixation, the material was then Golgi-impregnated following the modifications introduced by Eckenhoff and Rakic (1984) to the Stensaas method (1967). For electron microscopy, the HF were horizontally sectioned and the tissue samples were taken from the midtemporal part; the slices of the frontal pole were coronally cut. In both cases, sections were osmicated, block stained in uranyl acetate, dehydrated in ethanols and epon embedded. Three blocks of the HF and two blocks of the PFCm were taken per animal and from each block 4 sets of semithin sections, separated at least 40 μm, were obtained and stained with toluidine blue for layers thickness measurements and cell counts. From each block of PFCm ultrathin sections were obtained from the layer III of the PL area. From each block of the HF ribbons of 8~10 ultrathin sectinns were collected.

The thickness of the hippocampal granular and CA3 pyramidal cell layers and of the PL cortical layers I-III was calculated after drawing the respective semithin sections with the aid of a camera lucida. The number of the different neurons (granular and CA3 pyramidal cells of the HF and PL cortical neurons) per unit volume of the respective layer (Nv - numerical density) was estimated also in the semithin sections using the disector method (Sterio, 1984; Braendgaard and Gundersen, 1986). From each set of sections a mean of 6-8 disectors was made, totaling an average of 18-24 disectors per animal (see Cadete-Leite et al., 1988c, 1990a for details).

In Golgi-impregnated material, granule cell dendritic trees were sampled from both blades of the midtemporal part of the dentate gyrus granule cell layer and pooled per animal in a single group. In the PL cortex only the basal dendrites of layer III pyramidal cells were selected for quantification. An average of 15 granule cells and 15 pyramidal cells per animal were studied. Two types of quantitative evaluation were performed:

(1) The three-dimensional branching pattern was analysed with a semiautomatic dendrite-measuring system developed at the Netherlands Institute for Brain Research (Overdijk et al., 1978; Uylings et al., 1986), in cells selected according to a set of criteria similar to those described by De Ruiter and Uylings (1987). From each cell, the following parameters were studied: projected area of the cell soma (somatic surface), total number of dendrites per cell, total number of segments per cell, total dendritic length, length of the individual segments and radial distance from the cell soma to terminal tips. The dendritic segments were ordered centrifugally and the intermediate segments were grouped according to their degree (see Uylings et al., 1975, 1986 and Cadete-Leite et al., 1988c for details).

(2) The dendritic spines of the granule cells were counted separately on dendritic branches located in the inner and outer thirds of the molecular layer and those of the PL layer III pyramidal cells on a terminal and first order intermediate segments. The counts were performed along 20 μm of the dendritic branch (Cadete-Leite et al., 1988c, 1990b).

The number of asymmetrical synapses established between the mossy fiber (MF) terminals and the complex spines on the proximal part of the CA3 pyramidal dendrites per unit volume of stratum lucidum (Nv - numerical density) was calculated also applying the disector method. Corresponding zones of the neuropil-containing profiles of MF were photographed in adjacent sections at a primary magnification of X6000 and the analysed EM photographs were printed at a final magnification of X18,000. Per animal, 30-36 disectors were obtained. With the aid of a MOP-Videoplan, the fraction of plasmalemma occupied by synaptic contacts was obtained by measurements of the perimeter of the MF profiles and of the length of the postsynaptic densities (Cadete-Leite et al., 1989a). The number of synapses per unit volume of PL layer III neuropil was calculated using a discrete unfolding procedure (Weibel, 1979) according to the modifications suggested by Goldsmith (1967) and Cruz-Orive (1978), as well as the corrections for missed profiles and finite thickness (Weibel, 1979). The estimation of the volumetric densities (Vv) of synaptic terminals and glial processes in the PL layer III neuropil was performed in the same EM micrographs. The fraction of neuropil occupied by these structures (Vv) was determined according to classical stereological techniques (Weibel, 1979), applied with the aid of a MOP-Videoplan system (Cadete-Leite et al., 1990a).

Statistical analysis of paired-group data was performed using the two-tailed nonparametric Mann-Whitney U-Test. Differences were considered significant when the p-value was less than 0.05.

MORPHOLOGICAL CHANGES INDUCED BY CHRONIC ALCOHOL CONSUMPTION

Qualitative Observations

In the HF, it was possible to recognize that the number of granule and pyramidal cells was larger in the control rats than in the experimental animals. The morphology of the granule cells dendritic trees and its spines from the alcohol-fed animals were identical to those from controls. At EM level, we could find degenerating neurites in alcohol-fed rats. Although scanty, they were observed more frequently than in respective age-matched controls, although signs of degeneration could also be seen in the 18-month control group. The degenerated dendrites often appeared as dystrophic profiles markedly distended by accumulation of dense and lamellar bodies and lipofuscin granules. Normal synaptic contacts were identified between these profiles and terminals. Degenerating MF could be seen dispersed in the neuropil. They appeared as electron dense profiles with clumped synaptic vesicles in a dense matrix. Glial processes engulfing the degenerated terminals were commonly seen (Cadete-Leite et al., 1989a).

The light microscopic observation of the semithin sections from the PL cortex displayed the normal cytoarchitectonic organization of this area and did not reveal apparent differences among controls and alcohol-fed animals. No differences were also detected in the morphology of dendritic trees. At the ultrastructural level the neuropil of PL layer III revealed few changes in alcohol-treated animals when compared with respective controls. Scattered degenerated terminals were seen after 6 months of alcohol feeding. Additionally, degenerated dendritic and spine profiles could also be found, although fewer in number. Abundant glial processes were seen surrounding both the degenerated terminals and dendritic profiles. Degenerated neurites were numerous as well (Cadete-Leite et al., 1990b).

Quantitative Analysis

Layers thickness and cell counts. In the dentate gyrus, the width of the granule cell layer was significantly thinner in all groups of ethanol-exposed rats, both in their medial and lateral blades. The Nv of the granule cells for 6-, 12- and 18-month alcohol-fed rats was significantly decreased (35%-40%) compared to their respective controls. It is known that among the entire range of alterations that are found in the brains of alcohol-treated animals, cell loss is one of the most striking. In the HF of rodents alcohol fed for 20 weeks, Walker et al. (1980, 1981) showed that the number of dentate granule cells per section is reduced by approximately 20%, despite the short period of alcohol feeding. In early work from our laboratory it was shown that the rate of cerebellar cell loss depends upon the length of alcohol-exposure (Tavares and Paula-Barbosa, 1982; Tavares et al., 1987a). The present data demonstrate that a reduction of about 40% of granule cells per unit volume occurs after 6 months of alcohol feeding. This loss is connected with a significant reduction in the width of the granule layer thickness, and remains significant throughout the experiment (Cadete-Leite et al., 1988c). The thickness of CA3 pyramidal layer showed a significant decrease in the 6-, 12- and 18-month alcohol-fed groups when compared to controls. The numerical density of CA3 pyramidal cells was reduced in all alcohol-fed groups when compared with respective age-matched controls. In the Walker et al. (1980, 1981) studies it was shown that the number of both CAl and CA2-4 pyramidal cells was reduced, whereas the results of Lescaudron and Verna (1985), in a similar experimental model, point to a reduction of CAl neurons. Moreover, Phillips and Cragg (1983) were unable to detect differences in the number of CAl pyramidal cells after 4 months of alcohol feeding. We found a significant loss of CA3 pyramidal cells, reaching 30% after 6 months of alcohol feeding, along with a reduction of the pyramidal cell layer thickness (Cadete-Leite et al., 1989a).

In the PL cortex, the thickness of layers I-III was significantly reduced in alcohol-fed groups, ranging from 8% to 16%. The numerical density of PL layer III neurons was also significantly reduced in 12- and 18-month ethanol-treated groups (28% and 26%, respectively) when compared with controls (Cadete-Leite et al., 1990a).

Three dimensional dendritic analysis. Previous data from experimental studies concerning the alcohol effects on the CNS of adult rodents have shown that dendritic arborizations are markedly altered (Riley, 1977; Riley and Walker, 1978; Pentney, 1982; Tavares et al., 1983a,b; Durand et al., 1989; Pentney and Quackenbush, 1990). In granule cells, the three dimensional analysis of the dendritic arborizations showed that the somatic surface was significantly increased in 6- and 18-month alcohol-fed groups when compared to controls. There were no significant differences in the number of dendrites per cell, although in alcohol-fed animals a significant increase in the total dendritic length was detected. No pair-wise difference was found in the number of dendritic segments per neuron. The radial distance of the terminal segments was significantly increased after 6 months of alcohol intake. The density of dendritic spines was not significantly different between alcohol-fed animals and controls in both the outer and inner thirds of granule cell dendritic trees (Cadete-Leite et al., 1988c). These findings might indicate that the surviving neurons take over additional areas for new synaptic connections with the input fibers present. This means that, as previously reported in HF and cerebellum under the same circumstances (Tavares et al., 1983b; McMullen et al., 1984; Pentney and Quackenbush, 1990), the expected reduction of the dendritic network is partially compensated by the proliferation of the remaining granule cells dendrites. Dendritic plasticity was quantitatively shown in granule cells dendrites after 5 months of alcohol feeding using a shifted linear analysis (Durand et al., 1989). However, Riley and Walker (1978) were unable to detect it using less subtle methods. Plastic alterations of dendritic spines following CAC have been already described. In our laboratory, an elongation in Purkinje cells spines was reported (Tavares et al., 1983b), whereas King et al. (1988) found a decrease in spine density in CAl pyramidal cells and an increase in granule cells after 5 months of alcohol ingestion. Lescaudron et al. (1989) were able to report a reduction ranging 25% to 33% on CAl pyramidal cell dendritic spines after 6.5 and 9.5 months of alcohol intake. Our results do not show significant differences in the density of granule cell spines.

The dendrites of the pyramidal neurons from the PL cortex appear to be more resistant to alcohol treatment than those of the cerebellum (Riley, 1977; Pentney, 1982; Tavares et al., 1983a; Pentney and Quackenbush, 1990) and hippocampal formation (Riley, 1977; Walker et al., 1981; McMullen et al., 1984; Cadete-Leite et al., 1988c; Durand et al., 1989). In fact, the dendritic parameters which were altered in the granule cells remained either unchanged or displaying minor abnormalities (Cadete-Leite et al., 1990b). This reveals the presence of differential responses among different areas of the CNS to alcohol aggression.

Quantitative ultrastructural analysis. In the neuropil of the stratum lucidum, no significant differences were found in the numerical density of the MF-CA3 synapses when 6- and 12-month alcohol-fed groups were compared with their respective controls. Conversely, after 18 months of experiment there was a significant reduction in the Nv of the MF-CA3 synapses. The percentage of MF plasmalemma occupied by postsynaptic densities was significantly increased in the alcohol-fed groups when compared with respective controls (Cadete-Leite et al., 1989a). The significant reduction (23%) of the MF-CA3 synapses found following 18 months of alcohol consumption is probably related to the degenerated MF and CA3 dendrites screened throughout the stratum lucidum of alcohol-fed animals. Yet, it must be emphasized that a synaptic decrease should be expected to occur precociously, as marked granule and pyramidal cells reductions are already present after 6 and 12 months of alcohol exposure. Fewer synapses are expected in all alcohol-fed groups than in the respective controls. The maintenance of the number of synapses at control levels in 6- and 12-month alcohol-fed groups suggest that new synapses must have been formed in order to compensate for the synapse loss due to neuronal degeneration. Once cell loss increases (18 months), it is likely that this plastic process collapses and leads to a decline in the density of synapses. The increase in the synaptic extension, which in physiological terms is known to cause a net increase in the primary region of synaptic activation (Desmond and Levy, 1986), represents an additional alternative way for structural compensation in order to preserve relatively constant the total amount of postsynaptic contact area (Cadete-Leite et al., 1989a).

In the neuropil of the PL layer III, a significant increase (14%) of the numerical density of synapses in 18-month alcohol-fed group was detected when compared with the respective control group. This finding does not have a straight forward interpretation since it could be taken as a relative increase in face of the reduction of synapse number observed in the age-matched animals. As already stated, similar synaptic changes were previously demonstrated after CAC in adult rats (Tavares and Paula-Barbosa, 1984; Tavares et al., 1987b). The evaluation of the synapse-to-neuron ratio showed that an increase can be observed in all groups of alcohol-treated rats, which reaches significant values after 12 months (36%) and 18 months (62%) of alcohol exposure. The assumption that a plastic response is ongoing is thus fully corroborated (Cadete-Leite et al., 1990a). As concerns the evaluation of volumetric density (Vv) of PL synaptic terminals, the data did not reveal any significant differences when alcohol-fed groups were compared with respective controls. On the contrary, the fraction of the neuropil occupied by the glial processes was increased (25% to 41%) in ethanol-treated animals (Cadete-Leite et al., 1990a).

MORPHOLOGICAL CHANGES AFTER WITHDRAWAL FROM ETHANOL

Qualitative Observations

Light microscopy of the semithin sections from the HF of withdrawn animals did not reveal differences between alcohol-fed rats and controls concerning the morphological characteristics of the dentate granule and CA3 pyramidal cells. In the Golgi-impregnated material from withdrawn animals, the morphology of granule cell dendritic trees and its spines was also identical among all groups studied. At the ultrastructural level, there was evidence, in the hippocampal stratum lucidum, of an increased degenerative activity including both axons and dendrites (Cadete-Leite et al., 1989a,b).

With respect to PL cortex, the analysis either of the semithin sections or the Golgi-impregnated sections did not show qualitative differences in the cytoarchitectonic organization of this cortical area, whatever the group studied. In the PL layer III neuropil, the presence of perforated synapses with more extensive synaptic appositions appearing in patches dispersed throughout the neuropil (Cadete-Leite et al., 1986) was more frequently observed in withdrawn than in alcohol-fed animals. Apart from this finding, no other qualitative morphological changes were detected between withdrawn and alcohol-treated rats (Cadete-Leite et al., 1990a,b).

Quantitative Analysis

Layers thickness and neuronal counts. The thickness of the granule cell layer was significantly reduced in withdrawal groups when compared with the respective age-matched controls, both in their medial and lateral blades. The values found matched those obtained in the alcohol-fed groups. The Nv of the granule cells was lower than in alcohol-fed animals, which indicates that withdrawal from alcohol does not impede the ongoing cell death (Cadete-Leite et al., 1988b). This observation is consistent with the numerical reduction observed in the CAl pyramidal and cerebellar Purkinje cells after shorter periods of alcohol feeding and withdrawal (Phillips and Cragg, 1983, 1984). In the withdrawal groups (Wl and W2) the CA3 pyramidal cell layer thickness was also significantly reduced when compared with age-matched control groups. Withdrawal groups also showed a significant decrease of the Nv of the CA3 pyramidal cells when compared to respective age-matched control groups. Although the values of the numerical density of the CA3 pyramidal cells were reduced in both Wl and W2 groups when compared with age-matched alcohol-treated animals, the differences were not significant. As in the case of dentate granule cells, withdrawal after CAC does not impede the progressive loss of CA3 pyramidal cells (Cadete-Leite et al., 1989a).

Significant differences were also detected in the thickness of PL cortical layers I-III when withdrawal groups were compared with age-matched controls. The percentage of reduction was 19% when the values of Wl group were compared with 12-month controls, and 15% when W2 group was compared with its age-matched control group. Besides, Wl group showed a significantly lower numerical density of the PL neurons (20%) when compared with 12-month control group; the Nv of W2 neurons was also reduced (33%) when compared with the 18-month control group and this reduction was still more marked than the mean value found in 18-month alcohol-fed group (10%) (Cadete-Leite et al., 1990a). This increased cell loss after the end of alcohol intake supports the above data and the results reported by different authors in other CNS areas after ethanol withdrawal (Phillips and Cragg, 1983, 1984; Phillips, 1985). This finding is rather unexpected. Recent studies indicate that it is conceivable that withdrawal from alcohol involves a complex sequence of modifications which cause further neuronal changes leading to alterations in the functioning of the CNS (Begleiter and Porjesz, 1977; Begleiter et al., 1980).

Three dimensional dendritic analysis. The three dimensional analysis of the granule cells dendritic trees from withdrawal group showed similar results to those of age-matched (18 months) control group. Previously we described an increase in the dendritic length in alcohol-fed rats. Thus, withdrawn animals display a global reduction of their dendritic arborizations when compared with the group alcohol-treated for 18 months, as it can be seen from the somatic surface, total number of dendrites, total dendritic length and the radial distance of dendritic terminal tips, which are significantly reduced. No significant differences were found in the density of dendritic spines among the three groups, although a slight increase was found in alcohol-fed and withdrawal groups as compared with age-matched control (Cadete-Leite et al., 1989b); these results do not fit with those reported by King et al. (1988) and Lescaudron et al. (1989) in hippocampal granule and CAl pyramidal cells which found that dendritic spine number returned to normal values after withdrawal from ethanol. As previously mentioned for alcohol-fed animals, an increase in the total dendritic length following neuronal death indicates the existence of plastic compensatory mechanisms. The non-existence of such mechanisms in withdrawn animals shows that the substrate available for synaptic connections with afferents and for interaction with the surviving neighbour neurons is reduced.

In the prefrontal cortex, comparisons between the withdrawal and 18-month control groups revealed a significant decrease in the number of basal dendrites, number of dendritic segments and total dendritic length in the withdrawal group. These differences were also detected when the withdrawal group was compared with the group of rats alcohol-fed for 18 months. This impoverishment also indicates a reversal in the previous growth trend observed in control and alcohol-exposed groups. The eventual effects of the regression are even more dramatic because, despite the maintenance of the spine density at control levels in all groups, there was a marked reduction in the number of dendritic spines per cell as a consequence of the reduction of total dendritic lengths (Cadete-Leite et al., 1990b).

Quantitative ultrastructural analysis. In the neuropil of the stratum lucidum, a significant reduction in the Nv of MF-CA3 synapses was found when the W2 group was compared with 18-month control group. No differences were detected when the Wl group was compared with age-matched controls. Significant increases in the percentage of MF plasmalemma occupied by post-synaptic densities were found when Wl and W2 groups were compared with respective age-matched controls (Cadete-Leite et al., 1989a). These data allow us to conclude that the reduction of the numerical density of MF-CA3 synapses verified after 18 months of CAC remained unchanged after withdrawal. In addition, the increase found on the fraction of MF plasmalemma occupied by active zones after alcohol consuming was enhanced by ethanol suppression.

Although Wl and W2 groups showed a general increase of the numerical density of synapses in the neuropil of PL layer III, significant differences were only found between W2 and respective age-matched control groups. The magnitude of differences observed in the values of synapse-to-neuron ratio between alcohol-fed and respective control groups was even enhanced in the withdrawal groups when compared with respective age-matched controls. Besides, we detect a significant increase in this ratio when comparison was made between W2 and 18-month alcohol-fed groups. The results obtained after the evaluation of the volumetric density of the terminals revealed no changes in withdrawal groups. The Vv of glial processes, which was increased after 12 months of alcohol intake, remained unchanged in the withdrawal groups (Cadete-Leite et al., 1990a).

SUMMARY OF THE RESULTS

The data obtained in this investigation allow us to conclude that:

From a Qualitative Point of View

1. The analysis of Golgi-impregnated material of HF and PL cortex did not reveal alterations of dendritic organization when alcohol-fed and withdrawal groups were compared with their respective controls;
2. Conversely, the semithin sections showed a reduction of the number of dentate granule cells and CA3 pyramidal cells in alcohol-treated and withdrawn animals;
3. The ultrastructural analysis showed evidence of axonal and dendritic degeneration in parallel with signs of regrowing activity in alcohol-fed and withdrawn animals.

From a Quantitative Point of View

In the hippocampal formation there was:

1. Reduction in the thickness of the dentate granular and CA3 pyramidal layers in alcohol-fed and withdrawal groups;
2. Reduction in the numerical density of granule cells, which worsened after withdrawal;
3. Reduction in the numerical density of CA3 pyramidal cells, which remained unchanged after ethanol suppression;
4. Increase in the total length of granule cell dendritic arborizations due to the increase of the intermediate and terminal segments after alcohol exposure; a reduction of these parameters was detected after withdrawal;
5. No change in the density of granule cell dendritic spines throughout the experiment in alcohol-fed and withdrawal groups;
6. Reduction in the numerical density of synapses between mossy fiber terminals and CA3 pyramidal cell dendrites after 18 months of consuming alcohol, which remained unchanged after withdrawal;
7. Increase in the fraction of the mossy fiber terminals plasmalemma occupied by postsynaptic densities, which was enhanced after ethanol suppression.

In the prelimbic cortex, there was:

1. Reduction in the thickness of layers I-III, which remained unchanged after withdrawal from the ethanol;
2. Reduction in the numerical density of layer III pyramidal cells, which worsened after withdrawal;
3. No change in the metric parameters of the pyramidal cell basal dendrites in alcohol-fed animals, followed by a decrease of these parameters after withdrawal;
4. No changes in the density of basal dendritic spines in alcohol-fed and withdrawal groups;
5. No changes in the volumetric density of layer III terminals in alcohol-fed and withdrawal groups;
6. Increase in the volumetric density of glial processes after 12 months of ethanol exposure, which remained unchanged after alcohol suppression;
7. Increase in the numerical density of layer III synapses following 18 months of alcohol intake, more evident after withdrawal;
8. Increase in the synapse-to-neuron ratio after 12 months of alcohol feeding, more evident after ethanol suppression.

CONCLUSIONS

1. The structural organization of the hippocampal formation and of prelimbic cortex is markedly affected after chronic alcohol consumption.

2. The characteristics and the degree of involvement of the areas studied are different. The magnitude of changes was also correlated with the length of the alcohol treatment.

3. Marked plastic potentialities in the hippocampal formation and prefrontal cortex were detected, although it is not possible to infer or deduce their functional implications.

4. Withdrawal from alcohol consumption does not interfere with the structural changes occuring in the hippocampal formation and prefrontal cortex.

5. On the contrary, the degenerative processes observed during withdrawal were more prominent than those seen after alcohol consumption, as happened with the increase in cell loss. Besides, plastic dendritic changes were no longer seen.

6. Synaptic plasticity was more evident after withdrawal. However, it did not appear to be enough to compensate for the degeneration observed.

ACKNOWLEDGEMENTS

This study was supported by grants from Stiftung Volkswagenwerk, Instituto Nacional de Investigação Científica (CMEUP) and Junta Nacional de Investigação Científica e Tecnológica.

REFERENCES

Andersen, P., 1975, Organization of hippocampal neurons and their interconnections, **in**: "The Hippocampus," Vol. 1, Structure and Development, R.L. Isaacson and K.H. Pribram, eds., pp. 155-175, Plenum Press, New York, London.

Begleiter, H. and Porjesz, B., 1977, Persistence of brain hyperexcitability following chronic alcohol exposure in rats, **Adv.Exp.Med.Biol.**, 85B:209.

Begleiter, H., De Noble, V. and Porjesz, B., 1980, Protracted brain dysfunction after alcohol withdrawal in monkey, **Adv.Exp.Med.Biol.**, 126:237.

Borges, M. M., Paula-Barbosa, M. M. and Volk, B., 1986, Chronic alcohol consumption induces lipofuscin deposition in the rat hippocampus, **Neurobiol.Aging**, 7:347.

Braendgaard, H. and Gundersen, H. J. G., 1986, The impact of recent stereological advances on quantitative studies of the nervous system, **J.Neurosci.Meth.**, 18:39.

Butters, N., Cermak, L. S., Montgomery, K. and Adinolfi, A., 1977, Some comparisons of the memory and visuoperceptive deficits of chronic alcoholics, and patients with Korsakoff's disease, **Alcoholism: Clin.Exp.Res.**, 1:73.

Cadete-Leite, A., Tavares, M. A., Paula-Barbosa, M. M. and Gray, E. G., 1986, "Perforated" synapses in frontal cortex of chronic alcohol-fed rats, **J.Submicrosc.Cytol.**, 18:495.

Cadete-Leite, A., Alves, M. C. and Tavares, M. A., 1988a, Lysosomal abnormalities in the pyramidal cells of the rat medial prefrontal cortex after chronic alcohol consumption and withdrawal, **J.Submicrosc.Cytol. Pathol.**, 20:115.

Cadete-Leite, A., Tavares, M. A. and Paula-Barbosa, M. M., 1988b, Alcohol withdrawal does not impede hippocampal granule cell progressive loss in chronic alcohol-fed rats, **Neurosci.Lett.**, 86:45.

Cadete-Leite, A., Tavares, M. A., Uylings, H. B. M. and Paula-Barbosa, M.M., 1988c, Granule cell loss and dendritic regrowing in the hippocampal dentate gyrus of the rat after chronic alcohol consumption, **Brain Res.**, 473:1.

Cadete-Leite, A., Tavares, M. A., Pacheco, M. M., Volk, B. and Paula-Barbosa, M. M., 1989a, Hippocampal mossy fiber-CA3 synapses after chronic alcohol consumption and withdrawal, **Alcohol**, 6:303.

Cadete-Leite, A., Tavares, M. A., Alves, M. C., Uylings, H. B. M. and Paula-Barbosa, M. M., 1989b, Metric analysis of hippocampal granule cell dendritic trees after alcohol withdrawal in rats, **Alcoholism: Clin.Exp.Res.**, 13:837.

Cadete-Leite, A., Alves, M. C., Tavares, M. A. and Paula-Barbosa, M. M., 1990a, Effects of chronic alcohol intake and withdrawal on the prefrontal neurons and synapses, **Alcohol**, 7:145.

Cadete-Leite, A., Alves, M. C., Paula-Barbosa, M. M., Uylings, H. B. M. and Tavares, M. A., 1990b, Quantitative analysis of basal dendrites of prefrontal pyramidal cells after chronic alcohol consumption and withdrawal in the adult rat, **Alcohol Alcoholism**, 25:467.

Carlen, P. L., Wortzman, G., Holgate, R. C., Wilkinson, D. A. and Rankin, J.G., 1978, Reversible cerebral atrophy in recently abstinent chronic alcoholics measured by computed tomography scans, **Science**, 200:1076.

Carlen, P. L., Peen, R. D., Fornazzari, L., Bennett, J., Wilkinson, D. A. and Wortzman, G., 1986, Computerized tomographic scan assessment of alcoholic brain damage and its potential reversibility, **Alcoholism: Clin.Exp.Res.**, 10:226.

Courville, C. B., 1955, "Effects of Alcohol on the Nervous System of Man", San Lucas Press, Los Angeles.

Cruz-Orive, L. M., 1978, Particle size-shape distributions: the general spheroid problem. II. Stochastic model and practical guide. **J.Microsc.**, 112:153.

De Ruiter, J. P. and Uylings, H. B. M., 1987, Morphometric and dendritic analysis of fascia dentata granule cells in human aging and senile dementia, **Brain Res.**, 402:217.

Desmond, N. L. and Levy, W. B., 1986, Changes in the numerical density of synaptic contacts with long-term potentiation in the hippocampal dentate gyrus, **J.Comp.Neurol.**, 253:466.

Durand, D., Saint-Cyr, J. A., Gurevich, N. and Carlen, P. L., 1989, Ethanol-induced dendritic alterations in hippocampal granule cells, **Brain Res.**, 477:373.

Eckenhoff, M. F. and Rakic, P., 1984, Radial organization of the hippocampal dentate gyrus: a Golgi, ultrastructural, and immunocytochemical analysis in the developing rhesus Monkey, **J.Comp.Neurol.**, 223:1.

Ferino, F., Thierry, A. M. and Glowinski, J., 1987, Anatomical and electrophysiological evidence for a direct projection from Ammon's horn to the medial prefrontal cortex in the rat, **Exp.Brain Res.**, 65:421.

Goldsmith, P. L., 1967, The calculation of true particle size distribution from the sizes observed in a thin slice. **Br.J.Appl.Physiol.**, 18:813.

King, M. A., Hunter, B. E. and Walker, D. W., 1988, Alterations and recovery of dendritic spine density in rat hippocampus following long-term ethanol ingestion, **Brain Res.**, 459:381.

Kolb, B., 1984, Functions of the frontal cortex of the rat: a comparative review, **Brain Res.Rev.**, 8:65.

Krettek, J. E. and Price; J. L., 1977, The cortical projections of the mediodorsal nucleus and adjacent thalamic nuclei in the rat, **J.Comp. Neurol.**, 171:157.

Lescaudron, L. and Verna, A., 1985, Effects of chronic ethanol consumption on pyramidal neurons of the mouse dorsal and ventral hippocampus: a quantitative histological analysis, **Exp.Brain Res.**, 58:362.

Lescaudron, L., Jaffard, R. and Verna, A., 1989, Modifications in number and morphology of dendritic spines resulting from chronic ethanol consumption and withdrawal: a Golgi study in the mouse anterior and posterior hippocampus, **Exp.Neurol.**, 106:156.

McMullen, P. A., Saint-Cyr, J. A. and Carlen, P. L., 1984, Morphological alterations in rat CAl hippocampal pyramidal cell dendrites resulting from chronic ethanol consumption and withdrawal, **J.Comp.Neurol.**, 225:111.

Overdijk, J., Uylings, H. B. M., Kuypers, K. and Kamstra, A. W., 1978, An economical semi-automatic system for measuring cellular tree structures in three dimensions, with special emphasis on Golgi-impregnated sections, **J.Microsc.**, 114:271.

Palay, S. L. and Chan-Palay, V., 1974, "Cerebellar Cortex. Cytology and Organization," Springer, Berlin.

Paula-Barbosa, M. M. and Tavares, M. A., 1984, Neuritic plaque-like structures in the rat cerebellum following prolonged alcohol consumption, **Experientia**, 40:110.

Paula-Barbosa, M. M., Borges, M. M., Cadete-Leite, A. and Tavares, M. A., 1986, Giant multivesicular bodies in the rat hippocampal pyramidal cells after chronic alcohol consumption, **Neurosci.Lett.**, 64:345.

Paula-Barbosa, M. M., Andrade, J. P., Castedo, J. L., Azevedo, F. P., Camões, I., Volk, B. and Tavares, M. A., 1989, Cell loss in the cerebellum and hippocampal formation of adult rats after long-term low-protein diet, **Exp.Neurol.**, 103:186.

Pentney, R. J., 1982, Quantitative analysis of ethanol effects on Purkinje cell dendritic tree, **Brain Res.**, 249:397.

Pentney, R. J. and Quackenbush, L. J., 1990, Dendritic hypertrophy in Purkinje neurons of old Fisher 344 rats after long term ethanol treatment, **Alcoholism: Clin. Exp.Res.**, 14:878.

Phillips, S. C. 1985, Qualitative and quantitative changes of mouse cerebellar synapses after chronic alcohol consumption and withdrawal, **Exp.Neurol.**, 88:748.

Phillips, S. C. and Cragg, B. G., 1983, Chronic consumption of alcohol by adult mice: effect on hippocampal cells and synapses, **Exp.Neurol.**, 80:218.

Phillips, S. C. and Cragg, B. G., 1984, Alcohol withdrawal causes a loss of cerebellar Purkinje cells in mice, **J.Stud.Alcohol**, 45:475.

Riley, J. N., 1977, Alterations in dendritic morphology following chronic alcohol consumption: a Golgi analysis, PhD. Dissertation, University of Florida.

Riley, J. N. and Walker, D. W., 1978, Morphological alterations in hippocampus after long-term alcohol consumption in mice, **Science**, 201:646.

Stensaas, L. J., 1967, The development of hippocampal and dorsolateral pallial regions of the cerebral hemisphere in fetal rabbits. I. Fifteen millimeter stage, spongioblast morphology, **J.Comp.Neurol.**, 129:59.

Sterio, D. C., 1984, The unbiased estimation of number and sizes of arbitrary particles using the disector, **J.Microsc.**, 134:127.

Tavares, M. A. and Paula-Barbosa, M. M., 1982, Alcohol-induced granule cell loss in the cerebellar cortex of the adult rat, **Exp.Neurol.**, 78:574.

Tavares, M. A. and Paula-Barbosa, M. M., 1983, Lipofuscin granules in Purkinje cells after long-term alcohol consumption in rats, **Alcoholism: Clin.Exp.Res.**, 7:302.

Tavares, M. A. and Paula-Barbosa, M. M., 1984, Remodeling of the cerebellar glomeruli after long-term alcohol consumption in the adult rat, **Brain Res.**, 309:217.

Tavares, M. A., Paula-Barbosa, M. M. and Gray, E. G., 1983a, A morphometric Golgi analysis of the Purkinje cell dendritic tree after long-term alcohol consumption in the adult rat, **J.Neurocytol.**, 12:939.

Tavares, M. A., Paula-Barbosa, M. M. and Gray, E. G., 1983b, Dendritic spine plasticity and chronic alcoholism in rats, **Neurosci.Lett.**, 42:235.

Tavares, M. A., Paula-Barbosa, M. M., Barroca, H. and Volk, B., 1985, Lipofuscin granules in cerebellar interneurons after long-term alcohol consumption in the adult rat, **Anat.Embryol.**, 171:61.

Tavares, M. A., Paula-Barbosa, M. M. and Cadete-Leite, A., 1986a, Morphological evidence of climbing fiber plasticity after long-term alcohol intake, **Neurobehav. Toxicol.Teratol.**, 8:481.

Tavares, M. A., Paula-Barbosa, M. M. and Volk, B., 1986b, Chronic alcohol consumption induces plastic changes in granule cell synaptic boutons of the rat cerebellar cortex, **J.Submicrosc.Cytol.**, 18:725.

Tavares, M. A., Paula-Barbosa, M. M. and Cadete-Leite, A., 1987a, Chronic alcohol consumption reduces the cortical layer volumes and the number of neurons of the rat cerebellar cortex, **Alcoholism: Clin.Exp.Res.**,11:315.

Tavares, M. A., Paula-Barbosa, M. M. and Verwer, R. W. H., 1987b, Synapses of the cerebellar cortex molecular layer after chronic alcohol consumption, **Alcohol**, 4:109.

Uylings, H. B. M., Smit, G. J. and Veltman, W. A. M., 1975, Ordering methods in quantitative analysis of branching structures of dendritic trees, **in**: "Advances in Neurology," Vol. 12, G.W. Kreutzberg, ed., pp. 247-254, Raven Press, New York.

Uylings, H. B. M., Ruiz-Marcos, A. and Van Pelt, J., 1986, The metric analysis of three-dimensional dendritic tree patterns: a methodological review, **J.Neurosci.Meth.**, 18:127.

Van Eden, C. G. and Uylings, H. B. M., 1985, Cytoarchitectonic development of the prefrontal cortex in the rat, **J.Comp. Neurol.**, 241:253.

Victor, M. and Adams, R. D., 1961, On the etiology of the alcoholic neurologic diseases with special reference to the role of nutrition, **Am.J.Clin.Nutr.**, 9:379

Walker, D. W. and Hunter, B. E., 1978, Short-term memory impairment following chronic alcohol consumption in rats, **Neuropsychologia**, 16:545.

Walker, D. W., Barnes, D. E., Zornetzer, S. F., Hunter, B. E. and Kubanis, P., 1980, Neuronal loss in hippocampus induced by prolonged ethanol consumption in rats, **Science**, 209:711.

Walker, D. W., Hunter, B. E. and Abraham, W. C., 1981, Neuroanatomical and functional deficits subsequent to chronic ethanol administration in animals, **Alcoholism: Clin.Exp.Res.**, 5:267.

Walsh, K., 1983, Alcohol related brain damage: an hypothesis, **Aust.Alcohol/ Drug Rev.**, 2:84.

Weibel, E. R., 1979, "Stereological Methods: Practical Methods for Biological Morphometry," Vol. 1, Academic Press, London.

Yew, M. L. S., Moore, S. and Biesele, M. M., 1981, Effects of chronic "moderate" alcohol consumption on vitamins A and C status of male Sprague-Dawley rats, **Nutr.Rep.Intern.**, 23:427.

EFFECTS OF ETHANOL ON THE RAT MEDIAL SEPTUM NUCLEUS : AN IN VIVO STUDY

P. Verbanck [1,2], J. Scuvée [2], I. Giesbers [2],
C. Kornreich [1] and A. Dresse [2]

Laboratoire de Psychologie Médicale - Unité de Recherche sur la Biologie des Dépendances
Université Libre de Bruxelles
Campus Brugmann. Pl. Van Gehuchten, 4
1020 Brussels, BELGIUM
Laboratoire de Pharmacologie - Université Liège
BELGIUM

INTRODUCTION

Among basal forebrain structures, the medial septum and diagonal band complex (MSDB) is interesting because it provides the major source of extrinsic cholinergic fibers to the hippocampus (Lewis and Shute, 1967; Mellgren and Srebo, 1973; Amaral and Kurz, 1985). Physiological and pharmacological studies have provided evidence that this septo-hippocampal projection plays a major role in the rhythmic slow activity (theta rhythm) of the hippocampal formation (Bland, 1986). Furthermore, cholinergic septal neurons seem to be necessary for the achievement of learning tasks by the hippocampus (Ikegami et al., 1989).

Recently, Givens and Breese (1990) have reported that ethanol, a substance known to impair memory and learning during acute intoxication, inhibits the firing activity of medial septum neurons. However, they observed an important heterogeneity in the neuronal response to ethanol.

Several studies have provided evidence that only part of the septo-hippocampal projection is cholinergic (Lynch et al., 1978; Baisden et al., 1984, Amaral and Kurz, 1985), and that neurons using gamma-aminobutyric acid (GABA) as a neurotransmitter form a second major group of cells in the septal region (Köhler and Chan-Palay, 1983; Panula et al., 1984; Brashear et al., 1986 and Leranth and Frotscher, 1989). Similarly, at least two types of cells have been identified electrophysiologically and pharmacologically in MSDB (Lamour et al., 1984 and Griffith and Matthews, 1986).

In preliminary experiments (Scuvée et al, unpublished data), we have also observed two types of MSDB neurons: the first type has short duration potentials (SP) (2.01 ± 0.08

msec) and high firing rate , the second long duration potentials (LP) (3.34 ± 0.10 msec) and low firing rate. LP cells are typically sensitive to acetylcholine, arecoline and atropine and are presumed to be cholinergic.

We present here the results of a study exploring the relation between the sensitivity of MSDB neurons to ethanol and their electrophysiological characteristics.

MATERIAL AND METHODS

All experiments were performed on male Wistar rats (200-300 g). The animals were anesthetized by intraperitoneal chloral hydrate (400 mg/kg) and then set in a stereotaxic David Kopff apparatus.

The activity of medial septum neurons was estimated by extracellular recordings of the spontaneous firing rate of single neurons using extracellular glass microelectrodes filled with 2M NaCl saturated with Fast Green which had an impedance between 5 and 15 MegOhms . The glass microelectrodes were positioned according to Paxinos and Watson coordinates. Action potentials were amplified and displayed on a Tektronix oscilloscope (Model 2430). The firing was counted for each 10 seconds after rejecting the artefacts by an electronic home-made voltage-gating window discriminator. The mean firing was then calculated for each minute. Two solutions of ethanol were prepared in 0.9 % w/v NaCl: 2g/ml/kg (50% ethanol) and 40 mg/ml/kg (1% ethanol) and infused by an intravenous jugular catheter at a rate of 0.1 ml/min. Control animals were perfused with 0.9% NaCl.

After recording, Fast Green was iontophoretically ejected through the top of the electrode, the brain removed and fixed in 4% formaldehyde for histological control.

The percentages of inhibition or activation (limited at 100%) by comparison to the base line were calculated in relation to the total dose perfused. The effect of ethanol and NaCl were evaluated by comparing the mean number of spikes/10 sec calculated during the last minute of perfusion with the firing activity during the control period. Statistical analysis of the results was performed using Student's t test.

RESULTS

A total of 50 neurons were recorded, 25 SP cells and 25 LP cells.

As shown in Figure 1, 8 SP neurons received 1% ethanol, 9 50% ethanol and 8 control cells 0.9% NaCl. No significant modification of firing was present with 1% ethanol or 0.9% NaCl. A dose-dependent inhibition of firing was observed for 50% ethanol, resulting in a 40% mean inhibition for a total dose of 2g/kg/ ethanol.

In Figure 2 are summarized the recordings from the 25 LP neurons. As in the SP experiments, 8 neurons received 1% ethanol, 9 50% ethanol and 8 0.9% NaCl. The neurons were not altered by either 1 % ethanol or 0.9% NaCl, but a marked dose-dependent inhibitory effect was present for 50% ethanol, with a 80% mean inhibition for a total dose of 2g/kg ethanol. Furthermore, in 4 of the 9 which received 50% ethanol, a transient but marked activation was present at the begin of perfusion and preceded the inhibition.

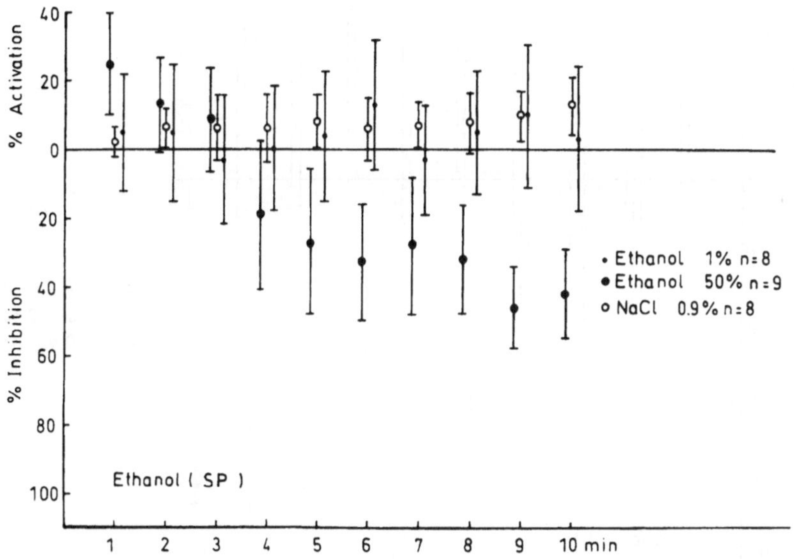

Figure 1. Summary of the recordings in 25 SP cells. The data are presented as mean ± SEM of the relative modification of neuronal firing during perfusion of 1% ethanol (n=8), 50% ethanol (n=9) and 0.9% NaCl (n=8). A dose-dependent inhibition is present only for 50% ethanol.

DISCUSSION

In this study, the acute effect of ethanol on rat MSDB neurons was investigated by an in vivo electrophysiological technique. In all the neurons, amounts of ethanol that correlate with behavioral intoxicating effects in the rat inhibit the spontaneous firing activity. Polysynaptic mechanisms could be involved in the effects of ethanol on the medial septum. However, these changes could not be attributed to an indirect action of ethanol on afferents from the lateral septum, because ethanol did not alter neural activity of cells in the lateral septum (Givens and Breese, 1990). The heterogeneity of responses reported by Givens and Breese seems to be due to the presence of at least 2 types of neurons in MSDB. The first type of neurons (LP cells) are sensitive to acetylcholine, arecoline and atropine and are probably cholinergic neurons. The second type of neurons (SP cells) are presumed to be GABAergic.

We have observed that LP cells are much more sensitive to ethanol than SP cells. This is in accordance with previous observations about the influence of acute ethanol on cholinergic and GABAergic transmission in the central nervous system. Indeed, the influence of acute ethanol on brain GABA metabolism has not been clearly delineated (Sutton and Simmons, 1973; Rawat, 1974; Hakkinen and Kulonen, 1979 and Wixon and Hunt, 1980). On the contrary, cholinergic neurotransmitter system was reported to be particularly sensitive to the effects of ethanol (Carmichael and Israel, 1975). Acute ethanol exposure has been found to decrease spontaneous and electrically stimulated acetylcholine release in vivo. This decreased release is accompanied by an increase in brain acetylcholine levels (Kalant et al., 1967; Erickson and Graham, 1973; Hunt and Dalton, 1976 and Sinclair and Lo, 1976).

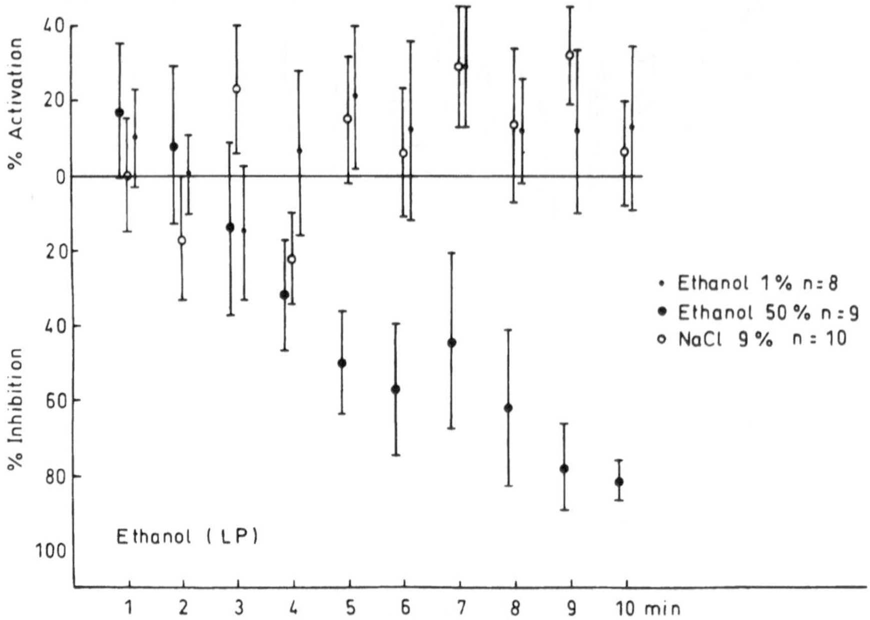

Figure 2. Summary of the recordings in 25 LP cells. The data are presented as mean ± SEM of the relative modification of neuronal firing during perfusion of 1% ethanol (n=8), 50% ethanol (n=9) and 0.9% NaCl (n=8). A dose-dependent inhibition is present only for 50% ethanol.

Furthermore, we observed a biphasic effect of ethanol on some LP neurons. Such a biphasic effect was already reported in other neural structures, like striatum (Schoener, 1984), hippocampus (Grupp, 1980) and monoaminergic neurons (Mereu and Gessa, 1985 and Verbanck et al., 1990).

CONCLUSIONS

The heterogeneity of the response of MSDB neurons to ethanol seems to be due to the fact that there are at least two types of neurons in this structure. They can be differentiated by their electrophysiological characteristics. The type of neurons that are presumed to be cholinergic are particularly sensitive to ethanol. This could be related to the alteration of memory and learning abilities during ethanol intoxication.

REFERENCES

Amaral, D.G. and Kurz, J., 1985, An analysis of the origins of the cholinergic and non-cholinergic septal projections to the hippocampal formation of the rat. **J Comp Neurol** 240: 37-59

Baisden, R.H., Woodruff, M.L. and Hoover, D.B., 1984, Cholinergic and non-cholinergic septo-hippocampal projections : A double-labeled horseradish peroxidase-acetylcholinesterase study in the rabbit. **Brain Res** 290: 146-51

Bland, B.H., 1986, The physiology and pharmacology of hippocampal formation theta rhythms. **Prog Brain Res** 26: 1-54

Brashear, H.R., Zaborszky, L. and Heimer, L., 1986, Distribution of GABA-ergic and cholinergic neurons in the rat diagonal band. **Neuroscience** 17: 439-51

Carmichael, F.J. and Israel, Y.,1975, Effects of ethanol on neurotransmitter release by rat brain cortical slices. **J Pharmacol Exp Ther** 193: 824-34

Erickson, C.K. and Graham, D.T., 1973, Alteration of cortical and reticular acetylcholine release by ethanol in vivo. **J Pharmacol Exp Ther** 185: 583-93

Givens, B.S. and Breese, G.R., 1990, Electrophysiological evidence that ethanol alters function of medial septal area without affecting lateral septal function. **J Pharmacol Exp Ther** 253: 95-103

Griffith, W.H. and Matthews, 1986, Electrophysiology of AchE-positive neurons in basal forebrain slices. **Neurosci Lett** 71: 169-74

Grupp, L.A.,1980, Biphasic action of ethanol on single units of the dorsal hippocampus and the relationship to the cortical EEG. **Psychopharmacology** 70: 95-103

Hakkinen, H.M. and Kulonen, E., 1979, Ethanol intoxication and the activities of glutamate decarboxylase and gamma-aminobutyric aminotransferase in rat brain. **J Neurochem** 33: 943-6

Hunt, W.A. and Dalton, T.K., 1976, Regional brain acetylcholine levels in rats acutely treated with ethanol or rendered ethanol dependent. **Brain Res** 109: 628-31

Ikegami, S., Nihonmatsu, T., Hatanaka, W., Takei, N. and Kawamura, H., 1989, Transplantation of septal cholinergic neurons improves memory impairments of spatial learning in rats treated with AF64A. **Brain Res** 496:321-6

Kalant, H., Israel, Y. and Mahon, M.A., 1967, The effect of ethanol on acetylcholine synthesis, release and degradation in brain . **Can J Physiol Pharmacol** 45: 172-6

Köhler, C. and Chan-Palay, V., 1983, Distribution of gamma aminobutyric acid containing neurons and terminal in the septal area. **Anat Embryol** 167: 53-75

Lamour, Y., Dutar, P. and Jobert, A., 1984, Septo-Hippocampal and Other Medial Septum-Diagonal Band Neurons: Electrophysiological and Pharmacological Properties. **Brain Res** 309: 227-39

Leranth, C. and Frotscher, M., 1989, Organization of the Septal Region in the Rat Brain: Cholinergic-GABAergic Interconnections and the Termination of Hippocampo-Septal Fibers. **J Comp Neurol** 289: 304-14

Lewis, P.R. and Shute, C.C.D., 1967, The cholinergic limbic system: Projection to hippocampal formation medial cortex, nuclei of ascending cholinergic reticular system and the subfornical organ and supraoptic crest. **Brain** 90: 521-40

Lynch, G., Rose, G. and Gall, C., 1978, Anatomical and functional aspects of the septo-hippocampal projections. In: 'Functions of the Septo-Hippocampal System'. Ciba Foundation Symp. 58. Amsterdam: **Elsevier**, pp. 5-24

Mellgren, S.I. and Srebo, B.,1973, Changes in acetylcholinesterase and distribution of degenerating fibres in the hippocampal region after septal lesions in the rat. **Brain Res** 52: 19-36

Mereu, G. and Gessa, G.K., 1985, Low doses of ethanol inhibit the firing of neurons in the susbtantia nigra, pars reticulata: a GABAergic effect? **Brain Res** 360: 325-30

Panula, P., Revuelta, A.V., Cheney, D.L., Wu, J.Y. and Costa, E., 1984, An immunohistochemical study on the location of GABAergic neurons in rat septum. J **Comp Neurol** 222: 69-80

Paxinos, G. and Watson, C., 1982, The Rat Brain in Stereotaxic Coordinates. Academic Press. Sydney

Rawat, A.K., 1974, Brain levels and turnover rates of presumptive neurotransmitters as influenced by administration and withdrawal of ethanol in mice. **J Neurochem** 22: 915-22

Schoener, E.P., 1984, Ethanol effect on striatal neuron activity. **Alcohol Clin Exp Res** 8: 266-8

Sinclair, J.G. and Lo, G.F., 1976, Acute tolerance to ethanol on the release of acetylcholine from the cat cerebral cortex. Can **J Physiol Pharmacol** 56: 668-70

Sutton, J. and Simmonds, M.A., 1973, Effects of acute and chronic ethanol on the gamma-aminobutyric acid system in rat brain. **Biochem Pharmacol** 22: 1685-92

Verbanck, P., Seutin, V., Dresse, A., Scuvée, J., Massotte, L., Giesbers, I. and Kornreich, C., 1990, Electrophysiological effects of ethanol on monoaminergic neurons : An in vivo and in vitro study. **Alcohol Clin Exp Res**, 14:728-35

Wixon, H.N. and Hunt, W.A., 1980, Effect of acute and chronic ethanol treatment on gamma-aminobutyric acid levels and on aminoacetic acid-induced GABA accumulation. **Subst Alcohol Actions Misuse** 1: 481-91

LIVER TRYPTOPHAN PYRROLASE, BRAIN 5-HYDROXYTRYPTAMINE AND ALCOHOL PREFERENCE

Abdulla A.-B. Badawy

South Glamorgan Health Authority
Biomedical Research Laboratory
Whitchurch Hospital
Cardiff CF4 7XB
Wales, U.K.

INTRODUCTION

There is considerable evidence suggesting that a deficiency in the cerebral indolylamine 5-HT (5-hydroxytryptamine or serotonin) may be a key factor in predisposition to alcohol (ethanol) consumption in both man and experimental animals (for a review, see Naranjo et al.,1986). Because of obvious methodological limitations, a central 5-HT deficiency in human alcoholism has not been demonstrated directly, but is inferred through indirect evidence, e.g. decreased availability of circulating tryptophan to the brain (Branchey et al., 1984) and decreased ratios in the urine of the concentration of the major 5-HT metabolite 5-HIAA (5-hydroxyindol-3-ylacetic acid) to that of the metabolites representative of other tryptophan-degradative pathways (namely indol-3-ylacetic acid for the tryptamine pathway and anthranilic acid for the hepatic kynurenine-nicotinic acid pathway) (Thomson & McMillen, 1987) in abstinent chronic alcoholics, and the ability of selective 5-HT uptake blockers to lower the extent of alcohol consumption by non-abstinent alcoholics (Naranjo et al., 1986, 1987). In addition to the ability of these 5-HT uptake blockers and of other modulators of central 5-HT metabolism and/or function to influence ethanol consumption in experimental animals (for references, see Naranjo et al., 1986), there is direct evidence for a central 5-HT deficiency in alcohol-preferring, but not in -non-preferring, strains of mice (Serri & Ely, 1984; Yoshimoto & Komura, 1987) and rats (Murphy et al., 1986, 1987), though with one exception (Korpi et al., 1988).

POSSIBLE MECHANISMS OF THE CENTRAL 5-HT DEFICIENCY IN ALCOHOL PREFERENCE

The central 5-HT deficiency in alcohol preference could be due to an increase in the rate of 5-HT degradation (turnover), a decrease in that of its synthesis or both. As regards degradation, there is no evidence to suggest that the activity of monoamine oxidase type A (the form of the enzyme responsible for 5-HT degradation) is higher in alcohol-preferring mice (Zimmer & Geneser, 1987) or rats (Pispa et al., 1986). Accordingly, it is more likely that the above deficiency is the result of impaired synthesis. Cerebral 5-HT synthesis is

Edited by T.N. Palmer, Plenum Press, New York, 1991

controlled mainly by brain tryptophan (Trp) concentration, because tryptophan hydroxylase, the rate-limiting enzyme of the 5-HT-biosynthetic pathway, is unsaturated with its tryptophan substrate (Fernstrom & Wurtman, 1971; Carlsson & Lindqvist, 1978; Curzon, 1979). It follows therefore that peripheral factors influencing the availability of circulating tryptophan to the brain must play important roles in the control of central 5-HT synthesis. These factors include: (1) tryptophan binding to circulating albumin (Curzon & Knott, 1974); (2) competition between Trp and five other circulating amino acids (Val, Leu, Ile, Phe and Tyr) for the same cerebral uptake mechanism (Fernstrom & Wurtman, 1971; Carlsson & Lindqvist, 1978); (3) activity of the major tryptophan-degrading enzyme, hepatic tryptophan pyrrolase (tryptophan 2,3-dioxygenase, EC 1.13.11.11) (Badawy, 1977).

BRAIN 5-HT SYNTHESIS IN ALCOHOL PREFERENCE IN MICE

As the results in Table 1 show, brain [5-HT] was 23% lower in alcohol-preferring male C57BL mice, compared with the control value in the non-preferring CBA/Ca strain. Concentration of the major 5-HT metabolite 5-HIAA was also significantly lower (by 10%) in C57 mice. A 22% lower brain tryptophan concentration was also observed in C57 mice. Taken together, these findings strongly suggest that, in alcohol-preferring C57BL mice, brain 5-HT synthesis and, hence, turnover are lower in comparison with those in non-preferring CBA/Ca mice, because of a lower concentration of the tryptophan precursor. As these results do not support the possibility of an enhanced cerebral tryptophan degradation, the low brain tryptophan concentration must be caused by a decrease in the availability of the circulating amino acid to the brain.

TRYPTOPHAN AVAILABILITY TO THE BRAIN IN ALCOHOL PREFERENCE IN MICE

As stated in the Introduction, tryptophan availability to the brain is controlled by at least three peripheral factors. One of these, competition between tryptophan and five other circulating amino acids (usually expressed as the ratio of concentration of Trp to the sum of those of its competitors, or [Trp]/[CAA] ratio) was found (Table 1) to be significantly lower (by 19%) in sera of C57 mice, compared with the ratio in the CBA strain. A decrease in this ratio could be due to a decrease in [Trp], an increase in [CAA] or both. An increase in the sum of [CAA] is ruled out by the results in Table 1, which actually show a small (10%), though insignificant, decrease in this sum. The 26% decrease in total serum [Trp] observed in C57 mice is therefore the cause of the lower ratio in this alcohol-preferring strain.

Another peripheral factor influencing circulating tryptophan availability to the brain is the extent of binding of the amino acid to serum albumin. As shown in Table 1, tryptophan binding to albumin, expressed as the percentage free serum tryptophan, was not significantly different between mouse strains, thus excluding increased binding as the cause of the lower availability of circulating tryptophan to the brain in alcohol-preferring C57 mice. The absence of altered albumin binding of circulating tryptophan is because, although total serum [Trp] was lower in C57 mice, there was also a proportionate decrease in that of free (ultrafiltrable) serum Trp (23% and 27% respectively). Such proportionate decreases in free and total serum [Trp] in the absence of altered binding are characteristic of enhancement of liver tryptophan pyrrolase activity leading to increased hepatic degradation of the amino acid and a consequent decrease in its availability in the circulation (see, e.g. Badawy & Evans, 1983). An increased hepatic Trp degradation is suggested by the observed (Table 1) decrease (31%) in liver [Trp] in C57, compared to CBA, mice.

Table 1. Tryptophan, tyrosine and related metabolic parameters in alcohol-preferring C57BL and non-preferring CB/Ca Mice

In Expt. 1, all parameters listed were determined in individual mouse brains, sera and/or livers, whereas those in Expt. 2 were measured in sera pooled from at least six mice per individual determination. In Expt. 1, concentrations of total serum tryptophan and the sum of its five competitors were determined by auto-analyzer, and are therefore expressed in μM, whereas in Expt. 2, both total and free serum [Trp] were determined fluorimetrically and are expressed in μg/ml. Liver tryptophan pyrrolase and tyrosine aminotransferase activities are expressed in μmol of product (kynurenine and 4-hydroxyphenylpyruvate respectively) formed/h per g wet wt. The holoenzyme and total enzyme activities of both enzymes are those obtained respectively in the absence and presence of added cofactor (2μM haematin or 40μM pyridoxal 5'-phosphate respectively). Apo- (tryptophan pyrrolase) activity was obtained by difference. Serum corticosterone is expressed in μg/dl. All other expressions (except the percentage free serum tryptophan and the tryptophan:sum of competitors ratio) are in μg/ml of serum or per g wet wt. of tissue. Values are means ± S.E.M. for each group of six mice, except for liver tryptophan pyrrolase and tyrosine aminotransferase activities (eight mice per group each), and free and total serum tryptophan concentrations (determined fluorimetrically) and the percentage free serum tryptophan (five samples per group, with each sample representing the pooled sera of six mice). Values in C57BL mice were compared with those in CBA/Ca controls and the significance of the differences is indicated as follows: *p < 0.05; **p < 0.01; ***p < 0.001. Abbreviations used: CAA, sum of competing amino acids; 5-HT, 5-hydroxytryptamine; 5-HIAA, 5-hydroxyindol-3-ylacetic acid; TAT, tyrosine aminotransferase; TP, tryptophan pyrrolase; Trp, tryptophan. Experimental and other details have been described in a previous publication (Badawy et al., 1989), from which the data in this Table are reproduced by permission.

Ex.no.	Parameter	CBA/Ca mice	C57BL mice
1.	Brain [5-HT]	0.94 ± 0.024	0.72 ± 0.027***
	Brain [5-HIAA]	0.43 ± 0.019	0.39 ± 0.009*
	Brain [Trp]	1.84 ± 0.09	1.43 ± 0.07**
	Total serum [Trp] (μM)	123.00 ± 8.00	91.00 ± 10.00*
	Sum of serum Trp [CAA] (μM)	954.00 ± 81.00	857.00 ± 24.00
	Serum [Trp]/[CAA] ratio	0.130 ± 0.005	0.105 ± 0.011*
	Liver [Trp]	9.10 ± 0.83	6.24 ± 0.39**
	Liver TP activity:		
	Holoenzyme	0.90 ± 0.05	1.90 ± 0.07***
	Total enzyme	1.80 ± 0.11	4.30 ± 0.18***
	Apoenzyme	0.90 ± 0.07	2.40 ± 0.20***
	Liver TAT activity:		
	Holoenzyme	153.00 ± 15.00	161.00 ± 11.00
	Total enzyme	206.00 ± 13.00	222.00 ± 9.00
2.	Free serum [Trp]	1.82 ± 0.10	1.33 ± 0.12**
	Total serum [Trp]	20.58 ± 1.22	15.82 ± 0.46**
	Free serum Trp (%)	8.84 ± 0.36	8.41 ± 0.71
	Serum [corticosterone]	8.1 ± 1.3	16.4 ± 1.6**

ENHANCEMENT OF LIVER TRYPTOPHAN PYRROLSE ACTIVITY IN ALCOHOL PREFERENCE IN MICE

Liver tryptophan pyrrolase exists in man, rat, mouse and some, but not all, other animal species in at least two forms, the haem-containing active form or holoenzyme and the haem-free inactive apoenzyme (Badawy & Evans, 1976). The activity of the former can be observed by incubating a suitable liver preparation with tryptophan in the absence of an added source of the haem cofactor, such as haematin, whereas that of the apoenzyme is calculated by subtracting the holoenzyme activity from that of the total enzyme, which is the activity observed after incubation in the presence of added haematin. As shown in Table 1, the holoenzyme, total enzyme and apoenzyme activities of tryptophan pyrrolase were 111%, 139% and 167% higher in livers of C57, than CBA, mice. Tryptophan pyrrolase is the first and rate-limiting enzyme of the quantitatively most important of all the tryptophan-degradative routes, namely the hepatic kynurenine-nicotinic acid pathway. Pyrrolase activity is thus a major determinant of tryptophan degradation and, hence, availability in the circulation for various tissues including the brain. The observed higher pyrrolase activity in C57, compared to CBA, mice therefore explains the lower availability of circulating tryptophan to the brain and the consequent defect in cerebral 5-HT synthesis and turnover seen in the former, alcohol-preferring, mouse strain.

MECHANISM OF ENHANCEMENT OF LIVER TRYPTOPHAN PYRROLASE ACTIVITY IN ALCOHOL PREFERENCE IN MICE

Liver tryptophan pyrrolase activity could be enhanced by a variety of mechanisms, notably hormonal induction by glucocorticoids, substrate activation and stabilization by tryptophan and cofactor activation by haem (see Badawy & Evans, 1975). In the latter two mechanisms, the haem saturation of the enzyme is enhanced, such that the rise in the holoenzyme activity is relatively greater than that in the total enzyme and, hence, also the apoenzyme. By contrast, the hormonal induction mechanism does not lead to an increase in the haem saturation of the pyrrolase; in relative terms, the increases in the total enzyme and apoenzyme activities are either similar to, or slightly greater than, that in the holoenzyme, because hormonal induction is associated primarily with increased de novo enzyme synthesis, rather than activation of pre-existing apoenzyme. The increases in activities of the three forms of the enzyme described above in C57 mice are therefore characteristic of the hormonal induction mechanism, and this is further supported by the finding (Table l) that C57 mice had twice as much circulating corticosterone, the pyrrolase-inducing major glucocorticoid in the mouse, as did the CBA strain. It may therefore be concluded that glucocorticoid induction of liver tryptophan pyrrolase activity is the cause of the low brain [5-HT] in alcohol preference in mice.

SPECIFIC GLUCOCORTICOID INDUCTION OF LIVER TRYPTOPHAN PYRROLASE IN ALCOHOL PREFERENCE IN MICE

The results in Table 1 show that there were no significant strain differences in the activity of another glucocorticoid-inducible enzyme, liver tyrosine aminotransferase, despite the higher circulating corticosterone concentration in C57 mice. A possible explanation of this interesting observation is that liver tryptophan pyrrolase may be more sensitive than tyrosine aminotransferase to induction by glucocorticoids in C57 than in other mouse strains. Evidence for this has already been presented (Monroe, 1968). More recently, Hirota et al. (1985) discovered a new glucocorticoid receptor, coined "receptor C", in rat liver cytosolic and nuclear fractions under a variety of stressful conditions, and implicated it specifically in mediating the glucocorticoid induction of tryptophan pyrrolase, but not tyrosine aminotransferase.It would therefore be of interest to find out if alcohol-preferring C57BL mice possess this new receptor either exclusively or in relatively greater abundance than other non-preferring strains, and whether induction and/or expression of the tryptophan pyrrolase message is enhanced under such possible conditions.

CONCLUSIONS AND COMMENTS

The present results have demonstrated that the low brain [5-HT] in alcohol-preferring C57BL mice, compared with the -non-preferring CBA/Ca strain, is due to a defective indolylamine synthesis brought about by a decrease in the availability of circulating tryptophan to the brain secondarily to a higher liver tryptophan pyrrolase activity associated with an increase in circulating corticosterone concentration. It remains to be seen if similar conclusions could also be applied to the 5-HT deficiency in alcohol-preferring rat lines and strains. In view of the important role of brain 5-HT in control of alcohol intake, we suggest that activity or expression of tryptophan pyrrolase and/or their induction by glucocorticoids may be important biological determinants of alcohol consumption and, hence, dependence.

ACKNOWLEDGEMENTS

This work was done in collaboration with various colleagues. I am grateful to Dr P V Taberner (University of Bristol Medical School, Bristol, U.K.), through whom mice were purchased, for additional provision of samples of mouse sera, and the Editorial Manager of the Biochemical Journal for his kind permission to reproduce the data given in Table 1 here.

REFERENCES

Badawy, A. A.-B. 1977, Minireview: Functions and regulation of tryptophan pyrrolase, **Life Sci.**, 21: 755.

Badawy, A. A.-B., and Evans, M., 1975, The regulation of rat liver tryptophan pyrrolase by its cofactor haem - experiments with haematin and 5-amino-laevulinate and comparison with the substrate and hormonal mechanisms, **Biochem. J.,** 150: 511.

Badawy, A. A.-B., and Evans, M., 1976, Animal liver tryptophan pyrrolases -absence of apoenzyme and of hormonal induction mechanism from species sensitive to tryptophan toxicity, **Biochem. J.**, 158: 79.

Badawy, A. A.-B., and Evans, M., 1983, Opposite effects of chronic administration and subsequent withdrawal of drugs of dependence on the metabolism and disposition of endogenous and exogenous tryptophan in the rat, **Alcohol Alcohol.**, 18: 369.

Badawy, A. A.-B., Morgan, C. J., Lane, J., Dhaliwal, K., and Bradley, D. M., 1989, Liver tryptophan pyrrolase: A major determinant of the lower brain 5-hydroxytryptamine concentration in alcohol-preferring C57BL mice, **Biochem. J.**, 264: 597.

Branchey, L., Branchey, M., Shaw, S., and Lieber, C. S., 1984, Depression, suicide, and aggression in alcoholics and their relationship to plasma amino acids, **Psychiatry Res.,** 12: 219.

Carlsson, A., and Lindqvist, M., 1978, Dependence of 5-HT and catecholamine synthesis on concentrations of precursor amino acids in rat brain, **Naunyn-Schmiedeberg's Arch. Pharmacol.**, 303: 157.

Curzon, G., 1979, Relationship between plasma, CSF and brain tryptophan, **J. Neural Transm.**, 15: 81.

Curzon, G., and Knott, P. J., 1974, Effects on plasma and brain tryptophan in the rat of

drugs and hormones that influence the concentration of unesterified fatty acids in the plasma, **Br. J. Pharmacol.**, 50: 197.
Fernstrom, J. D., and Wurtman, R. J., 1971, Brain serotonin content: physiological dependence on plasma tryptophan levels, **Science**, 173: 149.
Hirota, T., Hirota, K., Sanno, Y., and Tanaka, T., 1985, A new glucocorticoid receptor species: Relation to induction of tryptophan dioxygenase by glucocorticoids, **Endocrinology (Baltimore)**, 117: 1788.
Korpi, E. R., Sinclair, J. D., Kaheinen, P., Viitamaa, T., Hellevuo, K., and Kiianmaa, K., 1988, Brain regional and adrenal monoamine concentrations and behavioral responses to stress in alcohol-preferring AA and alcohol avoiding ANA rats, **Alcohol**, 5: 417.
Monroe, C. B. 1968, Induction of tryptophan oxygenase and tyrosine aminotransferase in mice, **Am. J. Physiol.**, 214: 1410.
Murphy, J. M., McBride, W. J., Lumeng, L., and Li, T.-K., 1986, Alcohol preference and regional brain monoamine contents of N/NIH heterogeneous stock rats, **Alcohol Drug Res.**, 7: 33.
Murphy, J. M., McBride, W. J., Lumeng, L., and Li, T.-K., 1987, Contents of monoamines in forebrain regions of alcohol-preferring (P) and -nonpreferring (NP) lines of rats, **Pharmacol. Biochem. Behav.**, 26: 389.
Naranjo, C. A., Sellers, E. M., and Lawrin, M. O., 1986, Modulation of ethanol intake by serotonin uptake inhibitors, **J. Clin. Psychiat.**, 47, (Suppl):16.
Naranjo, C. A., Sellers, E. M., Sullivan, J. T., Woodley, D. V., Kadlec, K., and Sykora, K., 1987, The serotonin uptake inhibitor citalopram attenuates ethanol intake, **Clin. Pharmacol. Ther.**, 41: 266.
Pispa, J. P., Huttunen, M. O., Sarviharju, M., and Ylikahri, R., 1986, Enzymes of catecholamine metabolism in the brains of rat strains differing in their preference for or tolerance of ethanol, **Alcohol Alcohol.**, 21: 181.
Serri, G. A., and Ely, D. L., 1984, A comparative study of aggression related changes in brain serotonin in CBA, C57BL, and DBA mice, **Behav. Brain Res.**,12: 283.
Thomson, S. M., Jr., and McMillen, B. A., 1987, Test for decreased serotonin/tryptophan metabolite ratios in abstinent alcoholics, **Alcohol**, 4: 1.
Yoshimoto, K., and Komura, S., 1987, Re-examination of the relationship between alcohol preference and brain monoamines in inbred strains of mice including senescence-accelerated mice, **Pharmacol. Biochem. Behav.**, 27:317.
Zimner, J., and Geneser, F. A., 1987, Difference in monoamine oxidase B activity between C57 black and albino NMRI mouse strains may explain differential effects of the neurotoxin MPTP, **Neurosci. Lett.**, 78: 253.

ALCOHOL ABUSE AND FUEL HOMEOSTASIS

[1]T. Norman Palmer, [2]Elisabeth B. Cook and [1]Paul G. Drake

[1]Department of Biochemistry
University of Western Australia
Nedlands, Perth, Western Australia 6009, Australia
[2]Department of Clinical Biochemistry
King's College Hospital School of Medicine and Dentistry
London, U.K.

INTRODUCTION

Ethanol has profound effects on whole-body fuel homeostasis and nutrition: Chronic alcohol abuse is recognised to be the primary cause of malnutrition in developed countries (see Lieber, 1988; World et al., 1985; Morgan and Levine, 1988; Halsted and Keen, 1990; Wheeler, 1990; Halsted, 1991). This is in part because fuel homeostasis in man is dependent on complex inter-organ relationships that govern the supply and utilization of energy substrates. The interaction of ethanol with fuel metabolism in any one specific tissue has the capacity to compromise these inter-organ axes in substrate supply and disposal and to have systemic effects on fuel homeostasis. It is therefore necessary to approach the question of alcohol-related malnutrition and impaired fuel metabolism from the perspective of whole-body fuel homeostasis.

ACUTE ALCOHOL ABUSE AND HEPATIC FUEL METABOLISM

The primary pathway of ethanol oxidation is hepatic in location and involves the enzymes alcohol dehydrogenase (EC 1.1.1.1) and aldehyde dehydrogenase (EC 1.2.1.3) (see Lieber, 1991). This pathway oxidizes ethanol to acetate via the intermediate acetaldehyde with the concomitant production of excess reducing equivalents as NADH. The products of this pathway, acetate and excess NADH, are primary mediators of the acute effects of ethanol on fuel homeostasis (Palmer, 1990). Rates of ethanol absorption following even moderate alcohol consumption significantly exceed the hepatic capacity for ethanol oxidation. The implication is that ethanol metabolism is intrinsically inefficient in comparison to the metabolism of primary nutrients like glucose or fat. The homeostatic mechanisms that govern the supply and disposition of lipid-derived fuels are acutely perturbed by ethanol oxidation (reviewed Lieber, 1991). Acetate inhibits triacylglycerol lipolysis in adipose tissue (Crouse et al., 1968) and as a consequence acute ethanol administration in man reduces circulating FFA concentration via suppression of peripheral FFA release (Lieber, 1982). By contrast, in the liver the excess reducing equivalents as NADH generated following ethanol oxidation may inhibit fatty acid oxidation. Fatty acids,

Alcoholism: A Molecular Perspective
Edited by T.N. Palmer, Plenum Press, New York, 1991

which are oxidised via ß-oxidation and the tricarboxylic acid (TCA) cycle), are normally the main oxidative substrate for the liver. This function is supplanted by ethanol following its acute administration since the NADH generated from the metabolism of ethanol suppresses ß-oxidation and TCA cycle flux and replaces NADH derived from fatty acid oxidation as the primary source of reducing equivalents for mitochondrial oxidation. As a corollary, fatty acids (derived from dietary fat or endogenous lipolysis in adipose tissue) may be diverted from oxidation into triacylglycerol synthesis and accumulation and the increased availability of acetyl-CoA may provide precursors for the enhanced rates of hepatic lipogenesis and ketogenesis that may accompany ethanol oxidation. Ethanol oxidation also acutely suppresses hepatic amino acid oxidation, including that of alanine, via its inhibition of TCA cycle flux.

Research over the last few years (see Sugden et al., 1989; Katz et al., 1986; McGarry et al., 1987; Radziuk, 1989a,b) indicates that gluconeogenic flux contributes not only to hepatic glucose production but also to hepatic glycogen repletion in response to dietary carbohydrate intake after prolonged starvation. The term 'the glucose paradox' has been invoked to describe the finding that dietary glucose promotes hepatic glycogen deposition and lipogenesis but not via acting as a direct precursor. It is now recognised that a substantial proportion (up to 70%) of the liver glycogen deposited in response to carbohydrate feeding is synthesised from glucose not via its direct phosphorylation to glucose 6-phosphate but following its conversion to 3-carbon metabolites, mainly lactate. There is controversy concerning the nutritional modulation of this 'indirect' pathway of glyconeogenesis and its contribution to total hepatic glycogen synthesis (Huang and Veech, 1988; Lang et al., 1986; Landau and Wahren, 1988; Rognstad, 1989). The primary site of disposal of a dietary glucose load is extrahepatic and the conversion of dietary glucose to lactate and related 3-carbon metabolites is presumed to occur at a number of sites (McGarry et al., 1987), including the gastrointestinal tract, skeletal muscle, skin (Cox and Palmer, 1988), central nervous system and erythrocytes . Hepatic glycogen synthesis is therefore dependent on glucose-to-lactate conversion mainly at extrahepatic sites (but see Pilkis et al., 1985; Soley et al., 1985a,b) and the liver may have a pivotal function (as a 'lactastat') in lactate disposal in the absorptive state (Cox et al., 1988).

The acute effects of ethanol on hepatic carbohydrate metabolism are mediated primarily via excess NADH generation and involve the inhibition of hepatic gluconeogenic flux from 3-carbon precursors and glucogenic amino acids. In poorly-nourished individuals with depleted hepatic glycogen stores, this inhibition of gluconeogenesis may provoke severe hypoglycaemia (see Hoffman and Goldfrank, 1989). Moreover ethanol reduces hepatic glucose production in diabetic patients (see Lewis and Kendall, 1988). Ethanol oxidation, which is associated with marked hyperlactacidemia and impaired lactate clearance (Kreisberg et al., 1971), acutely compromises hepatic glycogen repletion following glucose re-feeding after starvation in the rat (Cox et al., 1988; Cook et al., 1988). Hepatic glycogen deposition and synthesis (measured either as ^{3}H incorporation into glycogen from $^{3}H_2O$ or ^{14}C incorporation from [^{14}C]bicarbonate) is decreased by approximately 70% at 1h after glucose (2mmol/100g body wt.) re-feeding of 24h-starved rats (Cox et al., 1988; Cook et al., 1988). This impairment in hepatic glycogen repletion is mediated primarily via the inhibition of gluconeogenic flux, although this process may be exacerbated by delayed gastric emptying (Cook et al., 1988), intestinal glucose malabsorption (Chang et al., 1967) and the activation of phospholipase C (see French, 1991). Ethanol may additionally affect the provision of gluconeogenic precursors by its acute inhibitory effects on muscle protein synthesis (Preedy and Peters, 1988a): ethanol has a profound effect on whole-body nitrogen metabolism (Rodrigo et al., 1971; McDonald and Margen, 1976; Bunout et al., 1987; Reinus et al., 1989).

A question posed by the inhibition of hepatic glycogen repletion following carbohydrate re-feeding relates to the fate of lactate carbon in the liver under these circumstances. Does ethanol subvert the normal disposition of dietary carbohydrate via diverting 3-carbon metabolites from hepatic glyco- and gluconeogenesis into lipogenesis? Does this presumed stimulation of lipogenic flux contribute to the abnormal hepatic triacylglycerol accumulation associated with suppression of fatty acid oxidation and tricarboxylic acid-cycle flux? In this context, it is noteworthy that whereas increasing doses of ethanol reduced hepatic glycogen deposition and synthesis in 24h-starved rats re-fed glucose, rates of hepatic lipogenesis were not increased to a statistically significant extent (Table 1). The implication is that acute inhibition by ethanol of glyco- and gluconeogenic flux is not obligatorily coupled to channeling of excess lactate carbon into lipogenesis. By contrast ketone body concentrations were increased up to in excess of 4-fold, consistent with diversion of excess acetyl-CoA into ketogenesis (Table 1). As a corollary, whereas rates of lipogenesis in white and brown adipose tissue were not affected by ethanol pretreatment of glucose re-fed rats, rates of intestinal lipogenesis were increased by up to approximately 50% (results not shown).

ALCOHOL ABUSE AND EXTRAHEPATIC FUEL METABOLISM

Aside from its hepatic effects, ethanol acutely perturbs fuel homeostasis by extrahepatic effects on nutrient absorption, fuel metabolism and endocrine function. There is evidence that ethanol impairs the intestinal absorption of glucose (Cook et al., 1988) and amino acids (reviewed World et al., 1985), in part via its interaction with carrier-mediated

Table 1 Effects of increasing oral doses of ethanol on hepatic glycogen repletion and lipogenesis following glucose re-feeding of 24h-starved rats

Ethanol (0.69ml) or water was administered by intragastric intubation 1h prior to glucose re-feeding (2mmol/100g body wt.) and ^3H_2O administration (5mCi; intraperitoneally) and animals sampled 1h thereafter. Rates of glycogen synthesis and lipogenesis were determined by measurement of ^3H incorporation from ^3H_2O into glycogen or saponifiable fatty acid and are expressed as μg-atoms ^3H/h per g wet wt. Blood ketone bodies (mM: acetoacetate + 3-hydroxybutyrate) were determined in hepatic venous blood. Results are shown as means±S.E.M. and statistically significant effects of ethanol treatment are denoted as a, $p<0.05$: b, $p<0.01$: c, $p<0.001$

Dose of ethanol (mmol/kg body wt.)	Liver glycogen (μmol/g wet wt.)	Glycogen synthesis	Lipogenesis	Blood ketone bodies
0	73.2±4.4	118.5±4.9	2.96±0.20	0.2±0.03
20	49.2±3.1[b]	61.9±2.6[c]	3.46±0.04	0.36±0.03[b]
40	22.3±6.1[c]	26.7±3.8[c]	3.22±0.07	0.88±0.09[c]
60	29.3±5.5[c]	39.2±6.7[c]	3.06±0.04	0.62±0.09[c]
80	14.9±2.7[c]	34.8±5.3[c]	3.06±0.05	0.59±0.10[c]

transport systems. Ethanol oxidation is associated with deranged fuel metabolism: Ethanol oxidation is associated with a profound acute derangement in fuel homeostasis with diminished oxidative metabolism of lipid and protein coupled with insulin resistance, suppression of glucose oxidation and reduced hepatic gluconeogenic flux. Alcohol abuse may be associated with overt diabetes mellitus (Phillips and Safrit, 1971; Feingold and Siperstein, 1983), although it remains controversial whether ethanol pretreatment in normal man is associated with improved (McMonagle and Felig, 1975), diminished (Dornhorst and Ouyang, 1971), or unchanged glucose tolerance (Nikkila and Taslinen, 1975; Singh et al., 1988). Ethanol decreases peripheral glucose utilization (Lockner et al., 1967; Shah, 1988: see also Jorfeldt and Julin-Dannfelt, 1977,1978) and recent research using indirect calorimetry indicates that intravenous ethanol administration prior to glucose infusion in normal man decreases total body fat and protein oxidation, almost completely abolishes the rise in glucose oxidation normally seen in response to glucose infusion, and causes acute insulin resistance which is compensated for by hypersecretion of insulin (Shelmet et al., 1988). The relationship between ethanol oxidation and fuel metabolism is dynamic since dietary carbohydrate may stimulate ethanol oxidation (Rogers et al., 1987).

The Glucose/Fatty Acid Cycle

To gain insight into the mechanism underlying these acute effects of ethanol on extrahepatic fuel metabolism, it is first necessary to examine the mechanisms that govern fuel homeostasis under normal circumstances. The body uses two primary fuels in energy generation. These are glucose and lipid-derived fuels (fatty acids and ketone bodies) and there is a well-established reciprocal relationship between the use of these fuels as oxidative substrates in muscle (Randle et al., 1964, 1981; Randle and Tubbs, 1979; reviewed Sugden et al., 1989). This relationship, commonly termed the glucose/fatty acid cycle, serves to suppress glucose uptake and oxidation in muscle by promoting the preferential utilization of lipid-derived fuels in response to starvation or diets low in available carbohydrate. The maintenance of adequate blood glucose concentrations is a primary homeostatic function because certain tissues including the brain, erythrocytes and skin have an obligate requirement for glucose as a substrate. Glucose metabolism by the brain is oxidative, with the production of CO_2, and imposes a continual drain on the body's glucose stores. Implicit in the concept of the glucose/fatty acid cycle is that restricted carbohydrate availability ('carbohydrate stress') promotes the activation of triacylglycerol lipolysis in adipose tissue and the consequent release of free fatty acids (FFA) into the bloodstream. These FFA and the ketone bodies produced by their partial oxidation in the liver are preferentially oxidized in skeletal and cardiac muscle, their metabolism actively suppressing glucose utilization and oxidation. Since muscle accounts for approximately 40% of total body mass, this suppression markedly reduces whole-body glucose utilization. The mechanism underlying the suppression of glucose metabolism in muscle secondary to the oxidation of lipid-derived fuels is multifactorial and mediated by the co-ordinate inhibition of hexokinase, 6-phosphofructo-1-kinase and the pyruvate dehydrogenase complex (Randle et al., 1964, 1981; Randle and Tubbs, 1979; Sugden et al., 1989).

It is important to recognise that under conditions of dietary carbohydrate restriction there is a mandatory requirement for de novo glucose synthesis to compensate for the glucose carbon irrevocably 'lost' from the body by oxidative metabolism, particularly by the brain. In this context it is important to differentiate between gluconeogenesis via the recycling of the products of incomplete glucose metabolism (Cori and glucose/alanine cycling) and net gluconeogenesis from precursors of non-carbohydrate origin (Palmer et al., 1985), since only the latter can compensate for net glucose carbon loss via oxidation to CO_2. This provides the functional link between carbohydrate and protein metabolism under conditions of dietary carbohydrate restriction since the precursors for this de novo glucose synthesis are the carbon skeletons of glucogenic amino acids derived from the diet or endogenous proteolysis primarily in skeletal muscle. The suppression of peripheral glucose utilization by lipid-derived fuels reduces the body's requirement for gluconeogenesis for the maintenance of glycemia and fatty acids stimulate hepatic gluconeogenic flux.

Acute Alcohol Abuse, Fuel Homeostasis and the Glucose/Fatty Acid Cycle

Ethanol oxidation is associated with deranged lipid metabolism caused by abnormalities in peripheral lipolysis and hepatic fatty acid oxidation. The important question is whether these derangements are associated with abnormal functioning of the glucose/fatty acid cycle via the acute perturbation of the metabolism of glucose and lipid-derived in muscle. Is the function of the glucose/fatty acid cycle blunted or otherwise perturbed by acute ethanol administration? Skeletal muscle contains little, if any, capacity to oxidise ethanol implying that any metabolic derangement secondary to acute ethanol administration must be mediated by ethanol per se or ethanol metabolites derived from the liver. Ethanol may be esterified with long chain fatty acid membrane components in certain organs including pancreas, liver and, heart (Laposata and Lange, 1983) and it cannot be precluded that this process may compromise metabolism. Hepatic ethanol oxidation generates acetate and excess reducing equivalents as NADH, which in turn lead to increased net hepatic lactate output. In theory, acetate and NADH (derived from the oxidation of lactate) may act like fatty acids and ketone bodies to suppress glucose metabolism in muscle. The pyruvate dehydrogenase (PDH) complex is inhibited by protein phosphorylation catalysed by a specific PDH kinase which is activated by excess NADH, whereas acetate may inhibit glycolytic flux and pyruvate oxidation (see Sugden et al., 1989). In man, but not the rat, ethanol may acutely potentiate glucose-mediated insulin secretion (Shelmet et al., 1988). In human volunteers ethanol infusion potentiates glucose-stimulated insulin secretion, decreases basal rates of glucose appearance and disappearance and hyperinsulinemic-euglycemic clamp and other studies indicate a marked decrease in glucose disposal (Shelmet et al., 1988; Shah, 1988). The implication is that ethanol causes acute insulin resistance possibly mediated by a post-receptor defect in insulin signaling but whether this defect is mediated by the ethanol metabolites acetate and NADH remains to be established. The role of the PDH complex in this peripheral insulin resistance is open to question, particularly as *in vitro* studies with isolated rat soleus and extensor digitorum longus (EDL) muscles indicate that neither ethanol (50mM) nor acetaldehyde (250 μM) affect rates of glucose oxidation (measured as $^{14}CO_2$ production from [U-^{14}C]glucose), whereas lactate (approx 5mM) produces marked (>50%: $p < 0.001$) inhibition.

Ethanol and the Starved-to-Fed Transition in Skeletal Muscle

Carbohydrate re-feeding after starvation is associated with the suppression of triacylglycerol lipolysis in adipose tissue and a rapid decline in plasma FFA concentrations. The concept of the glucose/fatty acid cycle predicts that this reduction in fatty acid supply should be associated with the abolition of the constraint imposed by the oxidation of lipid-derived fuels on glucose uptake and metabolism in muscle and to rapid restoration of high rates of glucose uptake and phosphorylation. This is indeed the case: rates of glucose uptake by different muscles after carbohydrate re-feeding as assessed by rates of glycogen deposition are at least equal to rates of glucose phosphorylation in the post-absorptive state (Holness et al., 1988). The increase in glucose uptake by skeletal muscle is, however, not accompanied by parallel increases in pyruvate oxidation or lactate and alanine output (Berger et al., 1976; James et al., 1985; Jackson et al., 1987).

The implication is that the primary fate of glucose in skeletal muscle immediately following carbohydrate re-feeding after starvation is glycogen synthesis. This finding has led to the concept that the glucose/fatty acid cycle, albeit in a modified form, continues to operate following re-feeding after starvation to specifically spare dietary carbohydrate for repletion of the body's glycogen reserves (Sugden et al., 1989). It is proposed that in the first few hours following re-feeding after starvation mechanisms exist to 'uncouple' glucose metabolism in muscle so as to restore rapid rates of glucose uptake, phosphorylation and glycogenesis without the associated reactivation of glycolysis and pyruvate oxidation. The question posed by this postulated mechanism relates to the identity of the factor(s) that mediates the continued inhibition of glycolytic flux and pyruvate oxidation following carbohydrate re-feeding. The mechanism may involve multiple glucose 6-phosphate pools (Dully et al., 1968), reduced fructose 2,6-bisphosphate levels, delayed dephosphorylation of the PDH complex (Holness et al., 1989; see Sugden et al., 1989), and the selective activation of glycogen synthase. Muscle contains significant amounts of endogenous triacylglycerol (see Oscai et al., 1988) and continued intramuscular lipolysis after re-feeding may generate FFA, the in situ oxidation of which may act as a restraint on glycolysis and pyruvate oxidation. It is of interest that chow re-feeding of 40h-starved rats is accompanied not by accretion of intramuscular lipid but by the progressive loss of muscle triacylglycerol, at least in certain muscles (R. Thambirajah, P.G. Drake, D.L. Faulkner and T.N. Palmer., unpublished results).

Although skeletal muscle accounts for approx. 40% of total body mass and is the primary site of glucose disposition following carbohydrate re-feeding after starvation (Drake et al., 1989; Cox et al., 1989), it is not homogeneous in structure and function. Two categories of muscle can be broadly delineated in reference to oxidative metabolism and glucose disposal: (1) oxidative muscles that are constantly working even in the resting state, and (2) non-oxidative muscles (Issad et al., 1987; Holness and Sugden, 1990). Oxidative muscles (viz. diaphragm, postural skeletal muscles) have far higher rates of glucose uptake and phosphorylation than non-oxidative muscles in the fed state but, unlike non-oxidative muscles, respond to starvation by a pronounced suppression of glucose uptake (Issad et al., 1987; Holness and Sugden, 1990). The response of different muscles in the rat to chow re-feeding after 40h-starvation is heterogeneous: rates of glycogen deposition (but see Holness et al., 1988) and glucose uptake and phosphorylation (measured as 2-deoxy[^{3}H]glucose uptake and phosphorylation to 2-deoxy[^{3}H]glucose 6-phosphate: see Ferré et al., 1985; Issad et al., 1987; Drake et al., 1989) in response to re-feeding are significantly higher in oxidative muscles, including diaphragm and soleus, than in non-oxidative muscles (see Figure 1). In the context of the acute modulation by ethanol of peripheral glucose disposal (Shelmet et al., 1988), preliminary results from this laboratory indicate that ethanol administration is associated with the selective suppression of glycogen deposition and 2-deoxy[^{3}H]glucose uptake and phosphorylation in oxidative muscles in response to re-feeding after starvation. As a corollary, it is of interest that ethanol reduces the stimulatory effect of prolonged exercise on glucose uptake and oxidative metabolism by the exercising leg in healthy male volunteers (Juhlin-Dannfelt et al., 1977)

CHRONIC EFFECTS OF ETHANOL ON SKELETAL MUSCLE METABOLISM IN THE RAT

The preceding discussion has focussed on the acute effects of ethanol on fuel homeostasis. The question we would now like to address is the perturbation in fuel homeostasis associated with chronic alcohol abuse and particularly the effects of chronic abuse on muscle metabolism. There is compelling evidence that chronic alcohol abuse is associated with

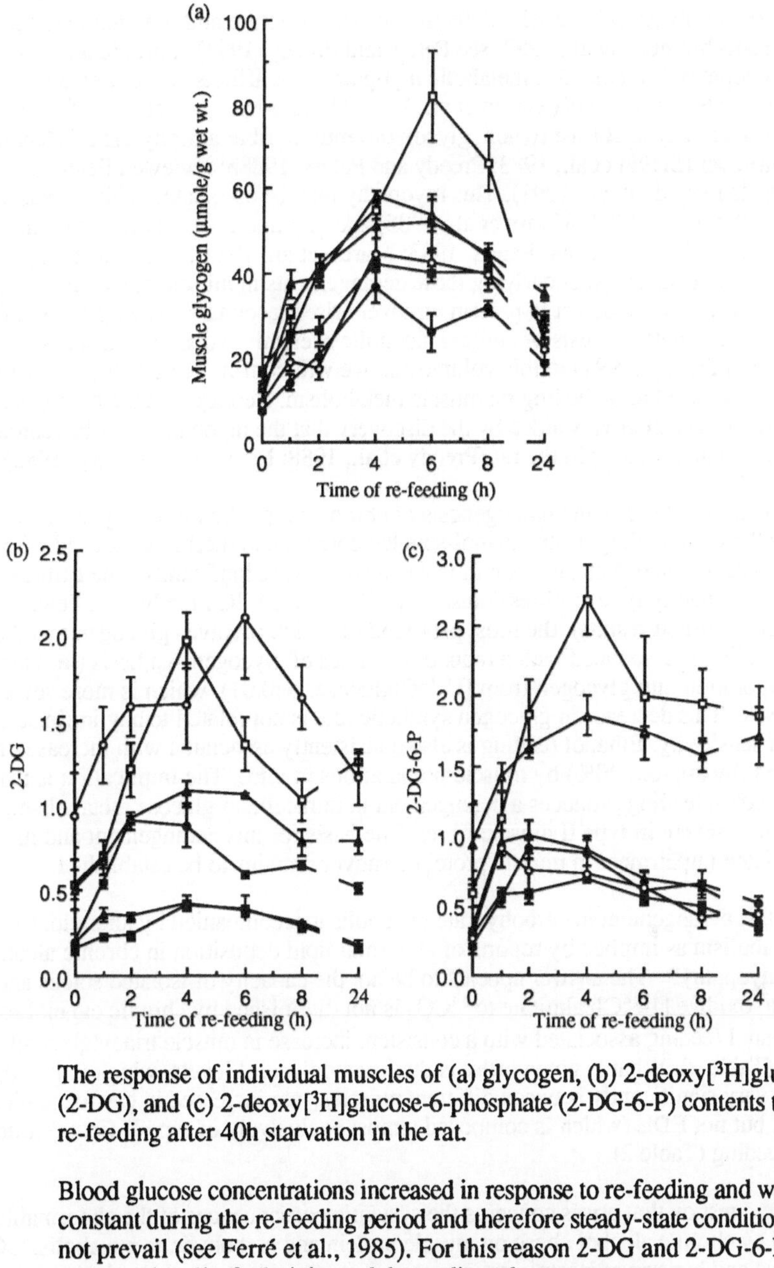

Figure 1 The response of individual muscles of (a) glycogen, (b) 2-deoxy[^{3}H]glucose (2-DG), and (c) 2-deoxy[^{3}H]glucose-6-phosphate (2-DG-6-P) contents to chow re-feeding after 40h starvation in the rat.

Blood glucose concentrations increased in response to re-feeding and were not constant during the re-feeding period and therefore steady-state conditions did not prevail (see Ferré et al., 1985). For this reason 2-DG and 2-DG-6-P are expressed as % of administered dose adjusted to per g wet wt.

The muscles examined were diaphragm (□), EDL (●), gastrocnemius (■), plantaris (▲), soleus (▲), and tibialis anterior (○). Each point shown is the mean (±S.E.M.) of at least 12 determinations.

profound alterations in muscle protein, lipid and carbohydrate metabolism. Chronic alcohol abuse is commonly associated with defective muscle structure and function (Martin et al., 1985; Urbano-Marquez et al., 1989; see Peters and Preedy, 1991). Chronic alcoholic skeletal myopathy is a common metabolic myopathy that afflicts approximately one-half to two-thirds of alcohol abusers (Martin et al., 1985; Urbano-Marquez et al., 1989) and involves selective type II (fast twitch, glycolytic) muscle fibre atrophy via a defect in protein turnover (Slavin et al., 1983; Preedy and Peters, 1988b: reviewed Peters and Preedy, 1991; see also Preedy et al., 1991). This myopathy may be associated with increased lipid deposition (Sunnasy, 1983; Martin et al., 1985) and alterations in carbohydrate metabolism (Perkoff et al., 1966; Ward and Peters, 1983; Martin et al., 1984). The question posed concerns the mechanism(s) underlying these derangements in muscle metabolism and specifically whether the defect in protein turnover arises secondary to compromised fuel metabolism. The pathogenesis of chronic alcoholic skeletal myopathy is discussed at length by Peters and Preedy (1991) in this volume and we will confine our attention here to the effects of chronic ethanol feeding on muscle metabolism. Research on the pathogenesis of this disorder has been greatly aided by the discovery that the myopathy may be reproduced by chronic ethanol feeding in the rat (Preedy et al., 1988; Lieber and DeCarli, 1982).

The active phase in the pathogenesis of chronic alcoholic myopathy in the rat (as applies at 2 weeks feeding on an ethanol-supplemented liquid diet) is associated with abnormalities in glucose metabolism in isolated soleus (predominantly type I fibres) and EDL (predominantly type II fibres) muscles (E.B. Cook., V.R. Preedy, T.J. Peters and T.N. Palmer, unpublished results), the most profound of which involves glycogen metabolism. Ethanol feeding is associated with a reduction in rates of glycogen synthesis (measured as ^{14}C-incorporation into glycogen from [U-^{14}C]glucose: $p<0.01$), which is more severe in EDL muscle. This decrease in glycogen synthetic rate is not related to any intrinsic change in insulin sensitivity. Ethanol feeding is also consistently associated with increased alanine production (by approx. 20%) by muscle preparations in vitro. The implication is that chronic alcohol feeding produces a derangement in intracellular glucose metabolism, which may be more severe in type II muscle fibres. The basis for this derangement and its role in the associated impairment in muscle protein turnover remains to be established.

Is this derangement in carbohydrate metabolism accompanied by alterations in muscle lipid metabolism as implied by reports of abnormal lipid deposition in chronic alcoholic skeletal myopathy? The answer appears to be no: the capacity of isolated soleus and EDL muscles to oxidize [1-^{14}C]palmitate to $^{14}CO_2$ is not diminished by chronic ethanol feeding nor is ethanol feeding associated with a consistent increase in muscle triacylglycerol contents (Table 2). What is noteworthy is that triacylglycerol levels, which unlike those of glycogen vary significantly between different muscles, are increased in gastrocnemius and plantaris, but not EDL (which is composed almost exclusively of type II fibres) following ethanol feeding (Table 2).

The question that arises concerns the systemic consequences of the abnormalities provoked by chronic alcohol abuse on muscle protein and carbohydrate metabolism. Glucose intolerance and hyperinsulinaemia are common features of hepatic cirrhosis, implying that insulin resistance is a feature of this condition (Shankar et al., 1983; Taylor et al., 1985; Cavalto-Perin et al., 1985). The pathogenesis of this lesion is uncertain, and the relative contributions of the liver and peripheral tissues to the metabolic abnormality are unknown, although there is evidence that depressed peripheral glucose utilization may make a more substantial contribution to the insulin resistance than increased splanchnic glucose production (Cavalto-Perin et al., 1985). This may imply that peripheral abnormalities in glucose disposal provoked by the deleterious effects of chronic alcohol abuse on muscle metabolism may be a major factor in the pathogenesis of the condition (Kruszynska et al., 1988).

Table 2 Effects of chronic ethanol feeding on muscle glycogen and triacylglycerol content in the rat

Rats (8 per group) were pair fed for 2 weeks either on a nutritionally complete liquid diet (controls) or on the same diet in which glucose was replaced by isocaloric ethanol . Tissues were dissected out under halothane anasthesia and rapidly frozen in liquid nitrogen until analysis. Results are shown as means±S.E.M. and statistically significant effects associated with ethanol feeding are designated by a, p<0.05, b, p<0.01

Muscle	Glycogen (μmol/g wet wt.)		Triacylglycerol (μmol/g protein)	
	Control	Ethanol-fed	Ethanol-fed	Control
Soleus	22.4±0.6	20.5±1.1	36.9±3.1	30.6±4.9
Gastrocnemius	25.8±1.0	24.3±1.0	15.5±3.4	34.5±5.5[a]
EDL	26.2±1.1	22.5±1.9[a]	6.5±0.8	6.4±0.8
Plantaris	30.5±1.5	28.6±1.3	5.4±0.3	7.5±0.8[b]
Tibialis anterior	32.1±1.6	30.1±1.1	8.2±0.9	9.6±0.7
Quadratus lumborum	51.5±5.1	36.6±1.5[a]	49.2±6.7	54.9±7.1

EDL, extensor digitorum longus

CONCLUSIONS

The purpose of this paper, which has reviewed the acute and chronic effects of ethanol on fuel homeostasis and nutrition, has been to highlight the importance of the integrated whole-body approach in understanding the impact of ethanol on metabolism. There remain several outstanding problems in this field, particularly with regard to the dynamic relationship between lipid and carbohydrate metabolism in muscle and glucose homeostasis.

ACKNOWLEDGEMENTS

This work was supported by the Brewer's Society (London) and the Central Research Fund of the University of London

REFERENCES

Berger, M.I., Hagg, S.A., Goodman, M.N. and Ruderman, N.B., 1976, Glucose metabolism in perfused skeletal muscle. Effects of starvation, diabetes, fatty acids, acetoacetate, insulin and exercise on glucose uptake and disposition. **Biochem. J.**, 158: 191.

Bunout, D., Petermann, M. and Iturriaga, H., 1987. Nitrogen economy in alcoholic patients without liver disease. **Metabolism Clin. Exp.**, 36: 651.

Cavalto-Perin, P. Cassader, M., Bozzo, C., Bruno, A., Nuccio, P., Dall'Omo, A.M., Marucci, M. and Pagano, G., 1985, Mechanism of insulin resistance in human liver cirrhosis. Evidence of a combined receptor and postreceptor defect. **J. Clin. Invest.**, 75: 1659.

Chang, T., Lewis, J. and Glazko, A.J., 1967, Effects of ethanol and other alcohols on the transport of amino-acids and glucose by everted sacs of rat small intestine. **Biochim. Biophys. Acta**, 135: 1000.

Cook, E.B., Preece. J.A., Tobin, S.D.M., Sugden, M.C., Cox, D.J. and Palmer, T.N., 1988, Acute inhibition by ethanol of intestinal absorption of glucose and hepatic glycogen synthesis on glucose re-feeding after starvation in the rat. **Biochem. J.**, 254: 59.

Cox, D.J. and Palmer, T.N., 1988, The role of skin in glucose disposal on re-feeding after starvation. **Biochem. Soc. Trans.** 16: 332.

Cox, D.J., Sugden, M.C. and Palmer, T.N., 1988, Ethanol and glucose disposal in the rat **Biochem. Soc. Trans.**, 16: 247.

Cox, D.J., Sugden, M.C. and Palmer, T.N., 1989, Glucose disposal as glycogen on glucose re-feeding after starvation. **Biochem. Soc. Trans.**, 17: 155.

Crouse, J.R., Gerson, C.D., DeCarli, L.M. and Lieber, C.S., 1968, Role of acetate in the reduction of plasma fatty acids produced by ethanol in man. **J. Lipid Res.**, 9: 509.

Dornhorst, A. and Ouyang, A., 1971, Effect of alcohol on glucose tolerance. **Lancet** 2: 957.

Drake, P.G., Palmer, T.N. and Cox, D.J., 1989, Site of glucose disposal on glucose refeeding after starvation. **Biochem. Soc. Trans.**, 17: 154.

Dully, C.C., Bocek, R.M. and Beatty, C.H., 1969, Presence of two or more glucose-6-phosphate pools in voluntary skeletal muscle and their sensitivity to insulin. **Endocrinology** 84: 855.

Ferré, P., Leturque, A., Burnol, A.-F., Pénicaud, L. and Girard, J., 1985, A method to quantify glucose utilization in vivo in skeletal muscle and white adipose tissue of the anaesthetized rat. **Biochem. J.**, 228: 103.

Feingold, K.R. and Siperstein, M.D., 1983, Normalization of fasting blood glucose levels in insulin-requiring diabetes: The role of ethanol abstention. **Diabetes Care** 6: 186.

French, S.W., 1991, The molecular pathology of alcoholic liver disease: An overview. **In**: 'Alcoholism: A molecular perspective', T.N. Palmer, ed., pp 55-67, Plenum Press, New York, pp.

Halsted, C.H., 1991, Chronic alcoholism, malnutrition , and folate deficiency. **In**: 'Alcoholism: A Molecular Perspective', T.N. Palmer, ed., pp XXX-XXX, Plenum Press, New York and London.

Halsted, C.H. and Keen, C.L., 1990, Alcoholism and micronutrient metabolism and deficiencies. **Eur. J. Gastroent. Hepatol.**, 2: 399.

Hoffman, R.S. and Goldfrank, L.R., 1989, Ethanol-associated metabolic disorders. **Emerg. Med. Clin. North Amer.**, 7: 943.

Holness, M.J. and Sugden, M.C., 1989, Pyruvate dehydrogenase activities during the fed-to-starved transition and on re-feeding after acute and prolonged starvation. **Biochem. J.**, 258: 529.

Holness, M.J. and Sugden, M.C., 1990, Glucose utilization in heart, diaphragm and skeletal muscle during the fed-to-starved transition. **Biochem. J.**, 270: 245.

Holness, M.J., Schuster-Bruce, M.J.I. and Sugden, M.C., 1988, Skeletal-muscle glycogen synthesis during the starved-to-fed transition in the rat. **Biochem. J.**, 254: 855.

Huang, M.T. and Veech, R.L., 1988, Role of the direct and indirect pathways for glycogen synthesis in rat liver in the postprandial period. **J. Clin. Invest.** 81: 872.

Issad, T., Pénicaud, L., Ferré, P., Kandé, J., Baudon, M.-A. and Girard, J., 1987, Effects of fasting on tissue glucose utilization in conscious rats. Major glucose-sparing effect in working muscle. **Biochem. J.**, 246: 241.

Jackson, R.A., Hamling, J.B., Sim., B.M., Hawa, M.I., Blix, P.M. and Nabarro, J.D.N., 1987, Peripheral lactate and oxygen metabolism in man: the influence of oral glucose loading. **Metabolism** 36: 144.

James, D.E., Jenkins, A.B. and Kraegen, E.W., 1985, Heterogeneity of insulin action in individual muscles in vivo: euglycemic clamp studies in rats. **Am. J. Physiol.** 248: E567.

Jorfeldt, L. and Juhlin-Dannfelt, A., 1977, The influence of ethanol on human splanchnic and skeletal muscle metabolism during exercise. **Scand. J. Clin. Invest.**, 37: 609.

Jorfeldt, L. and Juhlin-Dannfelt, A., 1978, The influence of ethanol on human splanchnic and skeletal muscle metabolism during exercise. **Metabolism** 27: 97.

Juhlin-Dannfelt, A., Ahlborg, G., Hagenfeldt, L., Jorfeldt, L. and Felig, P., 1977, Influence of ethanol on splanchnic and skeletal muscle substrate turnover during prolonged exercise in man. **Am. J. Physiol.**, 233: E195.

Katz, J., Kuwajima, M., Foster, D.W. and McGarry, J.D., 1986, The glucose paradox: new perspectives on hepatic carbohydrate metabolism. **Tr. Biochem. Sci.**, 11, 136.

Kreisberg, R.L., Siegel, A.M. and Owen, W.C., 1971, Glucose-lactate interrelationships: Effect of ethanol. **J. Clin. Invest.**, 50: 175.

Kruszynska, Y., Williams, N., Perry, M. and Home, P., 1988, The relationship between insulin sensitivity and skeletal muscle enzyme activities in hepatic cirrhosis. **Hepatol.**, 8: 1615.

Landau, B.R,. and Wahren, J., 1988, Quantification of the pathways followed in hepatic glycogen formation from glucose. **FASEB J.**, 2: 2368.

Lang, C.H., Bagby, G.J., Blakesley, H.L., Johnson, J.L. and Spitzer, J.J., 1986, Plasma glucose concentrations determines direct versus indirect liver glycogen synthesis. **Am. J. Physiol.** 251: E584.

Laposata, E.A. and Lange, L.G., 1986, Presence of nonoxidative ethanol metabolism in human organs commonly damaged by ethanol abuse. **Science** 231: 497.

Lewis, H. and Kendall, M.J., 1988, Alcohol and the treatment of diabetes. **J. Clin. Pharm. Ther.**, 13: 321.

Lieber, C.S., 1982, Ethanol and lipid disorders, including fatty liver, hyperlipidemia and atherosclerosis. **In**: 'Medical Disorders of Alcoholism: Pathogenesis and Treatment', C.S. Lieber, ed., pp 141-177, W.B. Saunders, Philadelphia.

Lieber, C.S., 1988, The influence of alcohol on nutritional status. **Nutr. Rev.**, 46: 241.

Lieber, C.S., 1991, Pathways of ethanol metabolism and related pathology. **In**: 'Alcoholism: A Molecular Perspective', T.N. Palmer, ed., pp 1-25, Plenum Press, New York and London.

Lieber, C.S. and DeCarli, L.M., 1982, The feeding of alcohol in liquid diets: two decades of applications and 1982 update. **Alcohol: Clin. Exp. Res.**, 6: 523.

Lochner, A., Wulff, J. and Madison, L.L., 1967, Ethanol-induced hypoglycemia: the acute effects of ethanol on hepatic glucose output and peripheral glucose utilization in fasted dogs. **Metab. Clin. Exp.**, 16: 1.

Martin, F.C., Levi, A.J., Slavin, G. and Peters, T.J., 1984, Glycogen content and activities of key glycolytic enzymes in muscle biopsies from control subjects and patients with chronic alcoholic skeletal myopathy. **Clin. Sci.**, 66: 69.

Martin, F.C., Ward, K., Slavin, G., Levi, A.J. and Peters, T.J., 1985, Alcoholic skeletal myopathy, a clinical and pathological study. **Quat. J. Med.**, 55: 233.

McDonald, J.T. and Margan, S., 1976, Wine versus ethanol in human nutrition. I. Nitrogen and calorie balance. **Amer. J. Clin. Nutr.**, 29: 1093.

McMonagle, J. and Felig, P., 1975, Effect of ethanol ingestion on glucose tolerance and insulin secretion in normal and diabetic subjects. **Metab. Clin. Exp.**, 24: 625.

McGarry, J.D., Kuwajima, M., Newgard, C.B. and Foster, D.W., 1987, From dietary glucose to liver glycogen: The full circle round, **Annu. Rev. Nutr.**, 7: 51.

Morgan, M.Y. and Levine, J.A., 1988, Alcohol and nutrition. **Proc. Nutr. Soc.**, 47: 85.

Nikkila, E.A. and Taskinen, M.R., 1975, Ethanol-induced alterations in glucose tolerance, postglucose hypoglycemia and insulin secretion in normal, obese, and diabetic subjects. **Diabetes** 24: 933

Oscai, L.B., Gorski, J., Miller, W.C. and Palmer, W.K., 1988, Role of the alkaline TG lipase in regulating intramuscular TG content. **Med. Sci. Sports Exerc.**, 20: 539.

Palmer, T.N., 1990, Fuel homeostasis and alcohol abuse. **Eur. J. Gastroent. Hepatol.**, 2: 406.

Palmer, T.N., Caldecourt, M.A., Snell, K. and Sugden, M.C., 1985, Alanine and inter-organ relationships in branched-chain amino acid and 2-oxo acid metabolism. **Biosci. Rep.**, 5: 1015.

Perkoff, G.T., Hardy, D. and Velez-Garcia, E., 1966, Reversible acute muscular syndrome in chronic alcoholism. **New Engl. J. Med.**, 274: 1277.

Peters, T.J. and Preedy, T.J., 1991, Chronic alcoholic skeletal myopathy: An overview. **In**: 'Alcoholism: A molecular perspective', ed. T.N. Palmer, pp. 301-308, Plenum Press, New York and London.

Phillips, G.B. and Safrit, H.F., 1971, Alcoholic diabetes, **JAMA** 217: 1513.

Pilkis, S.J., Regen, D.M., Claus, T.H. and Cherrington, A.D., 1985, Role of hepatic glycolysis and gluconeogenesis in hepatic glycogen synthesis. **BioEssays** 2: 273.

Preedy, V.R. and Peters, T.J., 1988a, The effect of chronic ethanol ingestion on protein metabolism in Type-I- and Type-II-fibre-rich skeletal muscles of the rat. **Biochem. J.**, 254: 631.

Preedy, V.R. and Peters, T.J., 1988b, Acute effects of ethanol on protein synthesis in different muscles and muscle protein fractions of the rat. **Clin. Sci.**, 74: 461.

Preedy, V.J., Duane, P. and Peters, T.J., 1988, Biological effects of chronic alcohol consumption: a reappraisal of the Lieber-DeCarli liquid-diet model with reference to skeletal muscle. **Alcohol Alcohol.**, 23: 151.

Preedy, V.R., Siddiq, T., Cook, E., Black, D., Palmer, T.N. and Peters, T.J., 1991, Alcohol and protein turnover. **In**: 'Alcoholism: A molecular perspective', ed. T.N. Palmer, pp. 241-255, Plenum Press, New York and London.

Reinus, J.F., Heymsfield, S.B., Wiskind, R., Casper, K. and Galambos, J.T., 1989, Ethanol: relative fuel value and metabolic effects in vivo. **Metabolism**, 38: 125.

Radzuik, J., 1989a, Hepatic glycogen in humans. I. Direct formation after oral or intravenous glucose or after 24 hour fast. **Am. J. Physiol.** 257: E145.

Radzuik, J., 1989b, Hepatic glycogen in humans. I. Gluconeogenic formation after oral or intravenous glucose. **Am. J. Physiol.** 257: E158.

Randle, P.J., 1964, the interrelationship of hormones, fatty acid and glucose in the provision of energy. **Postgrad. Med. J.**, 40: 457.

Randle, P.J., 1981, Molecular mechanisms regulating fuel selection in muscle. **In**: 'Biochemistry of Exercise', J. Poortmans and G. Niset, eds., Vol. IV-A, pp. 13-32, University Park Press, Baltimore.

Randle, P.J. and Tubbs, P.K., 1979, Carbohydrate and fatty acid metabolism. **In**: 'Handbook of Physiology: The Cardiovascular System', R.M. Berne, ed., Vol. 1, pp 805-844, American Physiological Society, Bethesda.

Rodrigo, C., Antezana, C. and Baraona, E., 1971, Fat and nitrogen balances in rats with alcohol-induced fatty liver. **J.Nutr.**, 101: 1307.

Rogers, J., Smith, J., Starmer, G.A. and Whitfield, J.B., 1987, Differing effects of carbohydrate, fat and protein on the rate of ethanol metabolism. **Alcohol Alcohol.**, 22: 345.

Rognstad, R., 1989, Errors in isotopic estimations of hepatic glycogen synthesis and glucose output. **Metabolism** 7: 619.

Shankar, T.P., Solomon, S.S., Duckworth, W.C., Himmelstein, S., Gray, S., Jerkins, T., Bobal, M.A. and Ramamurthy, S.I., 1983, Studies on glucose intolerance in cirrhosis of the liver. **J. Lab. Clin. Med.**, 102: 459.

Shah, J.H., 1988, Alcohol decreases insulin sensitivity in healthy subjects. **Alcohol Alcohol.**, 23: 103.

Shelmet, J.J., Reichard, G.A., Skutches, C.L., Hoeldtke, R.D., Owen, O.O. and Boden, G., 1988, Ethanol causes acute inhibition of carbohydrate, fat, and protein oxidation and insulin resistance, **J. Clin. Invest.**, 81: 1137.

Shelmet, J.J., Reichard, G.A., Skutches, C.L., Hoeldke, R.D., Owen, O.E. and Boden, G., 1988, Ethanol causes acute inhibition of carbohydrate, fat, and protein oxidation and insulin resistance. **J. Clin. Invest.**, 81: 1137.

Singh, S.P., Kumar, Y., Snyder, A.K., Ellyin, F.E. and Gilden, J.L., 1988, Effect of alcohol on glucose tolerance in normal and noninsulin-dependent diabetic subjects. **Alcoholism** 12: 727.

Slavin, G., Martin, F., Ward, K., Levi,, J. and Peters, T.J., 1983, Chronic alcohol excess is associated with selective but reversible injury to type 2B muscle fibres. **J. Clin. Pathol.**, 36: 772.

Soley, M., Chieri, R., Llobera, M. and Herrera, E., 1985a, Glucose infused through the portal vein enhances liver gluconeogenesis and glycogenesis from [3-^{14}C]glucose in the starved rat. **Int. J. Biochem.** 17: 685.

Soley, M., Chieri, R. and Herrera, E., 1985b, Short-term insulin infused through the portal vein enhances liver gluconeogenesis and glycogenesis from [3-^{14}C]glucose in the starved rat. **Int. J. Biochem.** 17: 689.

Sugden, M.C., Holness, M.J. and Palmer, T.N., 1989, Fuel selection and carbon flux during the starved-to-fed transition. **Biochem. J.**, 263: 313.

Sunnasy, D., Cairns, S.R., Martin, F., Slavin, G. and Peters, T.J., 1983, Chronic alcoholic skeletal myopathy: a clinical, histological and biochemical assessment of muscle lipid. **J. Clin. Pathol.**, 36: 778.

Taylor, R., Heine, R.J., Collins, J., James, O.F.W. and Alberti, K.G.M.M., 1985, Insulin action in cirrhosis. **Hepatol.**, 5: 64.

Urbano-Marquez, A., Estruch, A., Navarro-Lopez, F., Grau, J.M., Mont, L. and Rubin, E., 1989, The effects of alcoholism on skeletal and cardiac muscle. **New Engl. J. Med.**, 320: 409.

Ward, K. and Peters, T.J., 1983, Ischaemic lactate response in alcoholism - a reappraisal. Clin. Sci., 65: 21P.

Wheeler, E.F., 1990, The effect of alcohol abuse on energy and nutrient intake. **Europ. J. Gastroent. Hepatol.**, 2: 395.

World., M.J., Ryle, P.R. and Thomson, A.D., 1985, Alcoholic malnutrition and the small intestine. **Alcohol and Alcoholism**, 20: 89.

CHRONIC ALCOHOLISM, MALNUTRITION, AND FOLATE DEFICIENCY

Charles H. Halsted

Division of Clinical Nutrition
Department of Internal Medicine
University of California
Davis, California 95616 USA

MALNUTRITION AND CHRONIC ALCOHOLISM

The association between chronic alcoholism and malnutrition, long a matter of controversy, depends upon the population under study. Thus, surveys of alcoholic patients admitted to US Veterans Administration hospitals indicated a high prevalence of protein-calorie malnutrition, associated with consumption of protein-deficient diets and/or with alcoholic liver disease (Patek et al, 1975; Mendenhall et al., 1984; Mezey et al., 1988). On the other hand, studies of economically self-sufficient chronic alcoholics suggested a low incidence of malnutrition (Neville et al., 1968; Hurt et al., 1979; Morgan and Levine, 1988), often associated with deficiency of a single nutrient (Camilo et al., 1981). Studies of derelict alcoholics admitted to municipal hospitals in the USA demonstrated a high prevalence of vitamin deficiencies, in particular those of folate, thiamin, and pyridoxine (Leevy et al., 1970), and hepatic levels of these vitamins appeared to decrease in proportion to severity of alcoholic liver disease (Baker et al., 1964). More recent studies indicate that vitamin A deficiency is predictable in alcoholic liver disease, as reflected by serum and hepatic levels that decrease according to severity of the histopathology (Leo and Lieber, 1982).

Unlike other drug dependencies, chronic alcohol abuse has far-reaching effects on health and on the functioning of diverse organ systems, including the nervous, gastrointestinal, and hematopoietic systems and the liver. In each target organ, typical diseases have recognizable nutritional bases. Thus, the Wernicke-Korsakoff syndrome of ophthalmoplegia, cerebellar dysfunction, and cerebral degeneration is directly related to thiamin deficiency (Rueler et al., 1985). As detailed below, the chronic diarrhea of binge drinkers is caused in part by folate deficiency. Anemia in chronic alcoholism is caused by single or combined deficiencies of folate, pyridoxine, and iron (Savage and Lindenbaum, 1986). Alcoholic liver disease enhances the likelihood of malnutrition (Patek et al., 1975; Mendenhall et al., 1984; Mezey et al., 1988; Morgan and Levine, 1988) and may be accelerated by deficiencies and/or altered metabolism of trace minerals and vitamins (Halsted and Keen, 1990).

Poor Diet

Although alcohol provides 7 kcal/g, alcoholic beverages are a poor source of protein and are nearly devoid of micronutrients (Darby, 1979). Chronic alcoholic patients are more likely to exhibit signs of protein-calorie malnutrition when the percentage of daily calories as alcohol exceeds 50% (Mezey et al., 1988; Patek et al., 1975). On the other hand, individuals who consume alcohol as 20%-30% of dietary calories may appear well- or even overnourished (Hurt et al., 1979).

Effects of Alcohol Metabolism on Energy Storage and Nutrient Turnover

Although cytosolic alcohol dehydrogenase generates adenosine triphosphate (ATP), the microsomal ethanol oxidizing system (MEOS) operates at a higher K_m (approximately 10 mM or at blood levels of 0.050 g/dl) and generates NADP without subsequent metabolism to ATP (Lieber, 1988). This energy-wasteful aspect of alcohol metabolism may account for the experimental observation that substituting isocaloric amounts of alcohol for dietary carbohydrate resulted in weight loss in normal volunteers (Pirola and Lieber, 1972). In addition, as recently summarized by Lieber (1988), alcohol metabolism has extended effects on metabolism of other major nutrients, including decreased gluconeogenesis, increased synthesis of fatty acids and very low-density lipoproteins, decreased synthesis of visceral proteins, and altered transaminations.

Products of alcohol metabolism and induction of the MEOS have secondary effects on micronutrient metabolism and turnover. Thus, acetaldehyde displaces pyridoxal phosphate, the active form of vitamin B6, from its protein binder, with resultant accelerated degradation to pyridoxic acid, the excreted form of the vitamin (Lumeng, 1978). In vitro, the combination of acetaldehyde and xanthine oxidase in the presence of superoxide results in catalytic destruction of the folate molecule (Shaw et al., 1989); whether this occurs in vivo during drinking has not been tested. Vitamin A depletion in chronic alcoholism can be ascribed to alcohol induction of microsomal enzymes that accelerate conversion of this vitamin to polar metabolites excreted in the bile (Leo and Lieber, 1985).

Intestinal Malabsorption

Following ingestion of an acute intoxicating dose, alcohol levels in the upper intestine reach levels in excess of 2-4 g/dl and subsequently assume equilibrium with serum alcohol levels (Halsted et al., 1973a). These concentrations produce intestinal mucosal injury in experimental animals (Baraona et al., 1974). Ultrastructural mitochondrial changes occur in the small intestine of volunteers fed alcohol (Rubin et al., 1972), and decreased jejunal disaccharidase activity is seen in about one-third of binge drinking alcoholics (Perlow et al., 1977). The chronic diarrhea found in about one-third of binge drinkers results from combinations of altered intestinal motility (Robles et al., 1974), from possible direct mucosal damage by alcohol (Baraona et al., 1974), and from secondary effects of folate deficiency, including cellular enlargement of absorbing enterocytes (Hermos et al., 1972) and decreased absorption of water and electrolytes (Mekhjian and May, 1977; Halsted et al., 1973b). Clinical studies using oral tolerance tests or intestinal perfusion methods have demonstrated that acute or chronic exposure to alcohol induced malabsorption of several water-soluble nutrients including thiamin (Thomson et al., 1970), vitamin B12 (Lindenbaum and Lieber, 1975), folic acid and glucose (Halsted et al., 1973b), and several amino acids (Israel et al., 1969). Steatorrhea, with associated weight loss and risk of fat-soluble vitamin deficiency, occurs in association with alcoholic liver disease (Linscheer, 1970; Roggin et al., 1972) and reversible pancreatic lipase deficiency (Mezey et al., 1970).

Effects of Alcoholic Liver Disease

The likelihood of malnutrition is greatest in patients with established alcoholic liver disease (Patek et al., 1975; Morgan and Levine, 1988; Mendenhall et al., 1984; Mezey et al., 1988). The presence of active liver disease alters energy and protein metabolism as well as micronutrient storage. Decreased visceral proteins, reduced skeletal muscle mass, altered amino acid patterns, and negative nitrogen balance are observed in patients with alcoholic hepatitis (Mendenhall et al., 1984; Soberon et al., 1987; Weber and Reiser, 1982); these observations may relate to increases in the ratio of circulating glucagon to insulin (Kabadi et al., 1985; Marchesini et al., 1981). On the other hand, stable cirrhotics appear to have altered serum amino acid patterns but normal protein turnover kinetics and energy requirements (Morgan et al., 1982; Owen et al., 1983; Weber et al., 1990). Direct measurements in liver biopsies demonstrated reduced hepatic levels of most water-soluble vitamins as well as vitamin A, correlating with the stage of alcoholic liver injury (Baker et al., 1964; Leo and Lieber, 1982).

Although the nearly universal finding of protein malnutrition in alcoholic liver disease (Patek et al., 1975; Mendenhall et al., 1984) suggests a role for dietary protein deficiency in its pathogenesis, epidemiological studies show that amount and duration of alcohol exposure, rather than diet, is the major predictor of cirrhosis (Lelbach, 1975). The alcohol-hepatotoxin argument was strengthened by studies in the baboon model in which controlled, prolonged exposure to alcohol induced cirrhosis despite an adequate diet (Lieber and DeCarli, 1972). On the other hand, the finding of hepatic micronutrient alterations in animal models of alcoholism suggests that nutrition plays a role in the pathogenesis of alcoholic liver disease by modulating free-radical defenses to injury (Keen et al., 1985; Zidenberg-Cherr et al., 1990) and/or by affecting the relationship between hepatic vitamin A stores and transitional cell collagen metabolism (Mak et al., 1984).

Controlled feeding of alcohol in monkey and pig models of alcoholism altered hepatic levels of the trace minerals that are essential cofactors in antioxidant defense. In each model, animals fed alcohol developed lowered hepatic copper and zinc but elevated hepatic manganese levels, with corresponding decreases in CuZn- and increases in Mn-superoxide dismutases (Keen et al., 1985; Zidenberg-Cherr et al., 1990). Because these effects occurred with adequate dietary intake and preceded evidence of hepatic injury, they probably relate to alcohol metabolism and may play a role in hepatic free-radical defense and injury (Halsted and Keen, 1990).

FOLATE DEFICIENCY IN CHRONIC ALCOHOLISM

Normal Requirements, Biochemistry, and Physiology

Folates, a family of water-soluble vitamins, occur in leafy vegetables, nuts, and grains and bound to animal proteins. The biochemical structure of folate or PteGlu is that of a pteridine ring linked to p-aminobenzoate to form the pteroyl moiety (Pte), which is linked to glutamate (Glu). Dietary folate exists mainly as pteroylpolyglutamate ($PteGlu_n$) with up to 7 glutamyl units in gamma-peptide linkage. The pteroyl ring structure is usually reduced and methylated. The recommended dietary allowance of folate for adults is 200 μg, or about 3 μg/kg body weight, with higher requirements during pregnancy and lactation.

Folate coenzymes function in the transfer of single-carbon atoms during reactions essential to the metabolism of amino acids, purine, and pyrimidine. In the form of reduced methyl-PteGlu, folate interacts with vitamin B12 in the methyltransferase reaction in which the methyl group is transferred to homocysteine to produce methionine and reduced PteGlu. The latter compound is required for intracellular polyglutamylation and is converted to reduced 5'10-methylene-PteGlu, the folate coenzyme for thymidylate synthetase in the ultimate synthesis of DNA. Thus, cellular deficiency of folate or vitamin B12 reduces DNA synthesis and alters cell maturation. The tissue effects of folate deficiency are most apparent in systems with rapid cell turnover, particularly the hematopoietic system and the small intestinal surface epithelium. Clinical expression of folate deficiency in these tissues includes megaloblastic anemia and chronic diarrhea.

Defining the etiology of a nutrient deficiency requires understanding normal regulation of its absorption and subsequent metabolism (Table 1). Dietary folates, as a mixture of $PteGlu_n$, are absorbed in the upper small intestine by a two-stage process: hydrolysis followed by transport of the PteGlu derivative. Human and pig intestine contains two hydrolytic enzymes, one on the brush border surface and the other in the intracellular compartment of the intestinal epithelial cell. Data from clinical studies and animal models indicate that brush-border folate hydrolase (BBFH) is required for digestion of dietary folates, whereas the intracellular enzyme probably serves other metabolic functions within the enterocyte.

In both human and pig intestine, BBFH is an exopeptidase with a K_m less than 1 μM, a neutral pH optimum, and a requirement for zinc at the active site. Each enzyme has a subunit molecular weight between 100 and 150 kD (Halsted, 1990). Following hydrolysis, the PteGlu derivative is bound and transported across the brush-border membrane by an active anion-exchange mechanism, followed by intracellular reduction and methylation reactions and subsequent transport across the basal membrane of the enterocyte into the portal venous system (Mason, 1990). Regulation of these complex processes is not well defined. In isolated pig jejunal brush-border vesicles, different kinetics were observed for the hydrolytic and transport reactions (Reisenauer et al., 1986). A recent study suggests that pig intestinal transport and binding proteins are the same protein, with a molecular weight of about 55 kD, much smaller than BBFH (Reisenauer, 1990).

Following intestinal absorption, methylated PteGlu crosses the sinusoidal membrane to the hepatocyte and is then reconverted by folate synthetase to $PteGlu_n$, which is stored or involved in metabolic functions while bound to several folate-dependent enzymes in cytosol and mitochondria (Wagner, 1982). After intracellular hydrolysis to PteGlu, folates are excreted across the bile canalicular membrane and enter the enterohepatic circulation or are released directly to the systemic circulation. Both systemic and biliary folate consist of methylated PteGlu (Pratt and Cooper, 1971), and diversion of biliary secretions lowers circulating folate in the rat (Steinberg et al., 1979). Thus, there appear to be three separate but interrelated folate pools--the systemic and biliary circulation pools containing methylated PteGlu and the liver pool containing various substituted forms of $PteGlu_n$. More than 90% of folate is excreted by the kidneys and less than 5% appears in the feces (Tamura and Halsted, 1983). Renal excretion involves glomerular filtration of methylated PteGlu, followed by reabsorption across the tubular epithelium (Williams and Hueng, 1982). Thus, regulation of folate metabolism involves intestinal hydrolases and hepatic synthetase, several binding or transport proteins, and the requirement to cross several cellular membranes in the intestine, liver, and kidney.

Clinical Folate Deficiency

Folate deficiency is found in patients consuming inadequate diets, in intestinal malabsorption syndromes, in association with drugs like sulfasalazine that block absorption, and in chronic alcoholism (Halsted, 1990). Because the minimum daily folate requirement of 50 μg is relatively large compared to normal hepatic stores of 7-10 mg (Herbert, 1971), 140 days of experimental dietary depletion were required to produce evidence of tissue deficiency (megaloblastic bone marrow) in a human subject (Herbert, 1962).

Incidence of Folate Deficiency in Alcoholism

Chronic alcoholism is the leading cause of folate deficiency in the developed world. Among chronic alcoholics, folate deficiency is more prevalent among the impoverished and among those with liver disease (Halsted and Tamura, 1978). Low serum folate levels were found in 80% of recently drinking alcoholics (Herbert et al., 1963), whereas low levels of red-cell folate, an index of tissue stores, were found in about 40% of poor chronic alcoholics (Hines and Cowan, 1974). In a prospective study of 121 anemic chronic alcoholic patients, 34% demonstrated the megaloblastic bone marrow lesion typical of folate deficiency (Savage and Lindenbaum, 1986), confirming a prior study in which megaloblastic changes were found in 40% of patients (Eichner and Hillman, 1971). In each of the latter two studies, the bone marrow changes of folate deficiency were often accompanied by sideroblastic changes of pyridoxine deficiency and absent iron stores. The greater incidence of low serum folate levels arose because serum folate indicates negative folate balance, whereas low red-cell

Table 1. Folate Metabolism

Dietary Folate ($PteGlu_n$)

Intestine

Hydrolysis of $PteGlu_n$ to PteGlu by brush-border folate hydrolase
Transport of PteGlu into enterocyte
Reduction, methylation and transport to portal vein

Liver

Transport of PteGlu into cell
Synthesis of $PteGlu_n$
Intracellular binding, storage as $PteGlu_n$
Transport of methyl-PteGlu to bile, systemic circulation
Enterohepatic circulation of methyl-PteGlu

Kidney

Glomerular filtration
Tubular reabsorption (binding, transport of methyl-PteGlu)
Urinary excretion of PteGlu and catabolites

folate indicates depletion, and megaloblastic bone marrow indicates clinical deficiency (Herbert, 1990).

Clinical Effects of Folate Deficiency in Alcoholics

Anemia. The most obvious sign of folate deficiency in chronic alcoholism is anemia with megaloblastic bone marrow. Although deficiencies of folate and vitamin B12 produce identical histology, megaloblastic anemia in alcoholism almost always denotes folate deficiency and, depending on the presence or absence of other deficiencies, responds to folic acid therapy (Jandl and Lear, 1956). Vitamin B12 deficiency is rare in alcoholic patients because of the greater storage capacity of this vitamin relative to its daily requirement (Halsted and Keen, 1990). The folate-deficient patient presents with anemia, low serum and red-cell folate levels, macrocytic red cells with hypersegmented polymorphonuclear neutrophils, megaloblastic bone marrow changes, and normal or elevated serum vitamin B12. Elevated serum homocysteine occurs in patients with folate or vitamin B12 deficiency, but elevated serum methylmalonic acid occurs only in patients with vitamin B12 deficiency (Lindenbaum et al., 1988).

Diarrhea and intestinal malabsorption. Folate deficiency also affects the small intestinal mucosa, where the absorbing epithelium is renewed every four days. Case reports of severely folate-deficient alcoholics with megaloblastic anemia describe macrocytosis of absorbing enterocytes that responded to folic acid therapy (Bianchi et al., 1970; Hermos et al., 1972). Experimental folate deficiency in rats results in enterocyte macrocytosis, with tissue folate deficiency and reversal of intestinal function from net fluid absorption to net secretion (Klipstein et al., 1973; Goetsch and Klipstein, 1977). Two prospective clinical studies demonstrated that feeding low-folate diets with alcohol inhibits fluid and electrolyte absorption from the human jejunum (Mekhjian and May, 1977; Halsted et al., 1973b). These studies suggest that the absorbing enterocyte depends on dietary folate. The absorptive defect is correctable within 1 week with folic acid therapy despite continued exposure to alcohol (Halsted et al., 1973b). Thus, the intestinal lesion of alcoholic folate deficiency is analogous to that of tropical sprue, in which the morphologic and functional abnormality responds partially or fully to oral folic acid (Swanson and Thomassen, 1965).

Hepatic regeneration. Folate deficiency may also affect hepatic regeneration in alcoholic liver disease. Poor prognosis in alcoholic hepatitis correlated with decreased in vitro uptake of ^{3}H-thymidine and DNA synthesis in liver slices; both of these activities were stimulated by folic acid (Leevy, 1966). The response to alcoholic liver injury may be mediated by the tissue supply of glutathione and its precursor methionine (Shaw et al., 1981), which depends on folate for de novo synthesis.

ETIOLOGIES OF FOLATE DEFICIENCY IN ALCOHOLISM

The multiple causes of folate deficiency in alcoholism all relate to one or more facets of the requirement, physiology, and biochemistry of the vitamin (Table 2). Furthermore, alcohol appears to have acute effects on folate homeostasis that are distinct from effects seen during chronic exposure to alcohol.

Clinical Observations

Among patients with alcoholic liver disease, hepatic folate concentrations are less than half normal and tend to decrease according to the histologic severity of the lesion

(Baker et al., 1964). Storage was functionally assessed in a prospective study of 2 alcoholic patients without clinical liver disease. A folate-deficient diet induced megaloblastic bone marrow within 5 weeks in one patient and within 10 weeks in the other (Eichner et al., 1971)--less than half the time required to induce deficiency by dietary depletion in a normal subject (Herbert, 1962). Thus, alcoholic patients appear to have lower than normal folate stores and higher than normal risk for rapid development of overt folate deficiency, especially when consuming marginal or inadequate diets (Eichner and Hillman, 1971).

Acute effects of alcohol on folate homeostasis. Serum folate levels fall within 8-10 hours of administration of acute intoxicating doses of alcohol to normal or alcoholic subjects (Eichner and Hillman, 1973). This observation suggests that alcohol rapidly shifts folate from the circulation to another body pool or induces destruction of the vitamin. Alcohol administration also sharply decreases the availability of folate to the bone marrow. A careful study of 3 folate-deficient alcoholic patients demonstrated that the hematopoietic response to oral or parenteral folic acid could be prevented by daily consumption of moderate amounts of alcohol (Sullivan and Herbert, 1964).

Acute and chronic effects of alcohol on intestinal absorption of folate. Two clinical studies demonstrated decreased absorption of ^{3}H-labelled folic acid in recently drinking alcoholic patients. In the first study, serum levels and urinary excretion of ^{3}H after administration of an oral dose of ^{3}H-folic acid were lower than normal in binge drinkers studied within 24 hours of their last drink, whereas acute administration of an intoxicating amount of alcohol had no effect on folic acid absorption in normal control subjects (Halsted et al., 1967). The second study used the intestinal perfusion method to evaluate more directly the effects of alcoholism on folate absorption from the jejunum. Low jejunal mucosal uptake of ^{3}H-folic acid was found in 8 malnourished, folate-deficient alcoholics. Folic acid uptake by the jejunum improved with abstinence and nutritional repletion and was not suppressed when the study was repeated following 2 weeks of alcohol ingestion in the hospital (Halsted et al., 1971).

Another prospective study evaluated the combined and separate effects of diet-induced folate deficiency and prolonged alcohol exposure on ^{3}H-folic acid absorption. Two folate-repleted, alcoholic patient volunteers initially had normal jejunal uptakes of ^{3}H-folic acid,

Table 2. Causes of Folate Deficiency in Alcoholism

Acute Alcohol Intake

Lowers serum levels of folate
Decreases tissue availability of folate

Chronic Alcohol Intake

Decreases tissue stores of folate resulting from
— dietary deficiency
— intestinal malabsorption
— decreased hepatic uptake and/or retention
— increased urinary excretion

glucose, sodium, and water. After 35 days of consuming a folate-deficient diet with 250 g alcohol, both patients developed megaloblastic bone marrow and suppressed jejunal uptake of all four substances. Oral folic acid treatment corrected jejunal uptakes within 1 week despite continued exposure to alcohol. In 2 other patients, jejunal uptakes were unaffected by diet-induced folate deficiency without alcohol or by consumption of alcohol with an adequate diet (Halsted et al., 1973b). These studies suggest that chronic alcohol exposure and folate deficiency synergistically affect the absorptive functions of the jejunal mucosa.

Effects of alcoholism on urinary folate excretion. Two groups demonstrated enhanced urinary folate excretion in human subjects exposed to alcohol. Russell et al. (1983) showed increased folate excretion in 4 of 5 chronic alcoholic patient volunteers after 2-3 weeks of drinking. Others showed that an acute intoxicating dose of alcohol enhanced urinary folate excretion in normal subjects (McMartin et al., 1986).

Animal Models of Alcoholism

Rat, monkey, and pig models have been used to evaluate the pathogenesis of folate deficiency. Animal models provide the opportunity to administer alcohol under controlled conditions and to study tissue effects using invasive techniques. Each phase in the absorption and metabolism of folate has been studied and is summarized as follows.

Intestinal malabsorption. The two-step process of intestinal absorption of dietary folates involves hydrolysis of $PteGlu_n$ followed by uptake and transport of the PteGlu derivative. In miniature pigs, 1 year of feeding alcohol as 50% of calories decreased hydrolysis of $PteGlu_n$ perfused through surgically implanted jejunal tubes and significantly inhibited jejunal BBFH activity in mucosal tissue (Reisenauer et al., 1989; Naughton et al., 1989). However, alcohol feeding had no affect on the lumenal disappearance of PteGlu during jejunal perfusion or on the uptake of the compound as measured in isolated jejunal brush-border vesicles.

In monkeys fed alcohol as 50% of calories for 2 years, liver biopsies demonstrated significant folate depletion in the alcoholic animals (Romero et al., 1981). The absorption of orally administered ^{3}H-PteGlu was assessed by measuring isotope recovery in urine and feces. Urinary isotope recovery from alcohol-fed monkeys was 50% lower than that from control-fed monkeys, and fecal isotope recovery was five-fold higher. Calculated isotope retention in the alcohol-fed group was also significantly lower.

In summary, 2 years of alcohol feeding in monkeys induced folate deficiency in spite of adequate diet and significantly inhibited intestinal absorption of PteGlu. By contrast, 1 year of alcohol feeding in pigs did not induce folate deficiency but did inhibit BBFH activity. Discounting possible interspecies variations, the data suggest that inhibition of BBFH is an early stage of alcohol-induced folate malabsorption, which precedes inhibition of PteGlu transport in the development of folate malabsorption and deficiency.

Hepatobiliary metabolism. Alcohol could affect folate metabolism by blocking folate uptake into the hepatocyte, by inhibiting intracellular metabolism, and/or by accelerating biliary excretion. Acute exposure to alcohol accelerated the uptake of methyl-PteGlu by isolated rat hepatocytes (Horne et al., 1979). Short-term (3-day) exposure of rats to 10% alcohol in drinking water appeared to increase hepatic retention and to decrease biliary excretion of ^{3}H-PteGlu, suggesting "trapping" of folate in the hepatic pool (Hillman et al., 1977). Other studies in the rat showed that feeding alcohol in drinking water or as 30% of calories for longer than 12 weeks did not decrease liver folate levels nor inhibit hepatic synthesis of labelled $PteGlu_n$ from parenterally administered ^{3}H-PteGlu (Keating et al., 1985; Wilkinson and Shane, 1982).

In monkeys made folate deficient by feeding alcohol as 50% of calories for 2 years, the parenteral administration of ^{3}H-PteGlu decreased the hepatic uptake of the label but did not change the pattern of labelled $PteGlu_n$ or other folate metabolites in the liver (Tamura et al., 1981). In a subsequent study, performed after 4 years of feeding, folate turnover was measured after labeling the folate pool by injecting ^{3}H-PteGlu (Tamura and Halsted, 1983). In the alcoholic animals, labelled folate was excreted more rapidly in both urine and feces, resulting in a shorter early-elimination half-life. However, the slow-elimination half-life of the labelled folate was similar in both groups, and urinary appearances of folate metabolites were also similar. Thus, in the monkey model, chronic alcohol feeding depleted liver folate stores by decreasing hepatic uptake and increasing urinary and fecal excretion without changing the metabolism of the vitamin. The increased fecal folate excretion could reflect either increased biliary excretion or decreased intestinal reabsorption of biliary folate. The findings of decreased liver folate levels with increased fecal folate excretion run counter to the hypothesis that alcohol exposure diverts folate from the biliary pool to the liver (Hillman et al., 1977).

Urinary folate excretion. Folate that circulates as methyl-PteGlu is excreted in the urine by glomerular filtration followed by reabsorption across the renal tubular epithelium (Williams and Hueng, 1982). Several studies in rats demonstrated that acute and chronic alcohol administration accelerates urinary folate excretion (McMartin and Collins, 1983; McMartin et al., 1986). Studies in the alcohol-fed monkey confirmed that urinary wastage played a significant role in the rapid-phase turnover of injected ^{3}H-PteGlu (Tamura and Halsted, 1983). The mechanism for accelerated urinary folate excretion following exposure to alcohol is unclear but does not appear to relate to alcohol metabolism (McMartin and Collins, 1983) or to alcoholic diuresis (McMartin et al., 1986; Russell et al., 1983).

MOLECULAR BASIS FOR PATHOPHYSIOLOGY OF FOLATE DEFICIENCY IN ALCOHOLISM

A scheme to explain the pathogenesis of folate deficiency in alcoholism must account for separate observations on the acute and chronic effects of alcohol (Table 2). Summarizing the experimental data, acute exposure to alcohol appears to cause a rapid fall in the serum folate level and to decrease availability of folate to the bone marrow. On the other hand, chronic alcohol exposure appears to inhibit folate transport across a number of biological membranes, resulting in intestinal malabsorption (mucosal brush- border and/or basolateral membranes), decreased hepatic uptake (sinusoidal membranes of the hepatocyte) and increased urinary excretion (tubular brush-border membrane). As a result of these effects, less folate is absorbed and/or retained, and body folate stores decrease, placing the chronic alcoholic at increased risk of overt folate deficiency. Acute and/or chronic exposure to alcohol could cause these outcomes by direct, acute metabolic effects on the folate molecule and by altering the properties of membranes regulating folate homeostasis (Table 3). Altered membrane composition could secondarily affect the functional capacity of membrane-based folate hydrolases and transport proteins. Alternatively, or in addition, chronic alcoholism could alter synthesis and post-translational processing of folate proteins.

Data from animal models associate alcohol metabolism with production of free-radical ions and with alteration of antioxidant defense. Mechanisms that enhance formation of superoxide and hydroxyl radicals include increased activity of microsomal NADPH and xanthine oxidases in the presence of iron, both stimulated by metabolism of alcohol and/or acetaldehyde (Lieber, 1988). Altered defenses to free-radical metabolism during prolonged exposure to alcohol include reduced levels of hepatic glutathione, reflecting in part a relative lack of methionine and cysteine (Shaw et al., 1981), altered levels of Mn- and Cu, Zn-dependent superoxide dismutase (Keen et al., 1985, Zidenberg-Cherr et al., 1990), and reduced alpha-tocopherol (Kawase et al., 1989). The resulting accumulation of superoxide and hydroxyl radicals has two potential effects on folate economy direct catabolism of the folate molecule and altered membrane lipid composition.

Two in vitro studies suggest free-radical-mediated destruction of the folate molecule, either at the pteridine-p-aminobenzoate bond by a xanthine oxidase mediated mechanism (Shaw et al., 1989) or within the pteridine structure in the presence of iron (Taber and Lakshmaiah, 1987). Data from an experimental mouse model suggest that alcohol feeding may increase urinary excretion of folate catabolites (Kelly et al., 1981). More work in this area is required to document quantitatively significant effects on the folate molecule and on the body pool of folate.

Considerable evidence documents the effects of chronic alcohol feeding on membrane lipid composition and fluidity. However, no consistent pattern of effects has been shown, and different membranes appear to respond differently to acute or chronic alcohol administration. Membrane fluidity, a property of the unsaturation of phospholipid fatty acids, increased with chronic alcohol feeding in liver plasma membranes by a vitamin A-mediated process (Kim et al., 1988) but decreased in mitochondrial membranes (Waring et al., 1981). In vitro, alcohol enhanced intestinal membrane fluidity while inhibiting sodium transport (Hunter et al., 1983). However, another study suggested that chronic alcohol feeding exerts opposite effects on fluidity of brush-border and basolateral intestinal membranes (Harris et al., 1987). Chronic exposure to alcohol could alter membrane composition by inhibiting fatty acid desaturase (Cunningham and Spach, 1987; Wang and Reitz, 1983) or by free-radical-generated lipid peroxidation of membranes.

Present experimental evidence does not provide clear relationships among alcohol exposure, membrane composition and fluidity, and transport function. In pigs fed alcohol for 1 year, isolated jejunal brush-border membranes had unchanged fluidity or transport

Table 3. Proposed Pathophysiology of Folate Deficiency in Alcoholism

Acute

Free-radical destruction of folate molecule

Chronic

Altered membrane composition and function, including
— alcohol-induced inhibition of fatty-acid metabolism
— free-radical lipid peroxidation
Altered folate protein synthesis and/or processing

kinetics for PteGlu, whereas acute in vitro exposure to alcohol increased fluidity in membranes from control and alcoholic animals without changing transport (Naughton et al., 1989). Chronic alcohol feeding significantly inhibited BBFH, whereas this and other hydrolases were affected only by acute alcohol exposure in vitro. These studies suggest a selective effect of chronic alcoholism on synthesis and/or post-translational processing of BBFH.

Although the mechanism by which alcohol inhibits synthesis of intestinal enzymes is speculative, it may relate to free-radical effects on intestinal DNA metabolism (Halliwell, 1987) or to metabolic effects of alcohol on post-translational processing (Rothschild et al., 1987). Additional studies are required to determine whether production of BBFH and transport proteins and/or the membranes in which they function can be influenced by more prolonged exposure to alcohol and by modifying dietary constituents that may increase the risk for membrane lipid peroxidation with altered structure and function.

REFERENCES

Baker, H., Frank, O., Ziffer, H., Goldfarb, S., Leevy, C. M., and Sobotka, H., 1964, Effect of hepatic disease on liver B-complex vitamin titers, **Am. J. Clin. Nutr.**, 14:1-6.

Baraona, E., Pirola, R. C. and Lieber, C. S., 1974, Small intestinal damage and changes in cell population produced by ethanol ingestion in the rat, **Gastroenterology**, 66:226-34.

Bianchi, A., Chipman, D. W., Dreskia, A. and Rosensweig, N., 1970, Nutritional folic acid deficiency with megaloblastic changes in the small bowel epithelium, **N. Engl. J. Med.**, 282:859-61.

Camilo, M. D., Morgan, M. Y. and Sherlock, S., 1981, Erythrocyte transketolase activity in alcoholic liver disease, **Scand. J. Gastroenterol.**, 16:273-9.

Cunningham, C. C. and Spach, P. I., 1987, The effect of chronic ethanol consumption on the lipids in liver mitochondria, **Ann. N. Y. Acad. Sci.**, 492:181-91.

Darby, W. J., 1979, The nutrient contributions of fermented beverages, **in**: "Fermented Food Beverages in Nutrition," C. F. Gastineau, W. J. Darby, and T. B. Turner, eds., pp. 61-79, Academic Press, New York.

Eichner, E. R. and Hillman, R. S., 1971, The evolution of anemia in alcoholic patients, **Am. J. Med.**, 50:218-32.

Eichner, E. R. and Hillman, R. S., 1973, Effect of alcohol on serum folate level, **J. Clin. Invest.**, 52:584-91.

Eichner, E. R., Pierce, H. I. and Hillman, R. S., 1971, Folate balance in dietary induced megaloblastic anemia, **N. Engl. J. Med.**, 284:933-8.

Goetsch, C. A. and Klipstein, F. A., 1977, Effect of folate deficiency of the intestinal mucosa on jejunal transport in the rat, **J. Lab. Clin. Med.**, 89:1002-8.

Halliwell, B., 1987, Oxidants and human disease: Some new concepts, **FASEB J.**, 1:358-64.

Halsted, C. H., 1990, Intestinal absorption of dietary folates, **in**: "Folic Acid Metabolism in Health and Disease," M. F. Picciano, E. L. R. Stokstad, and J. F. Gregory, III, eds., pp. 23-45, Wiley-Liss, New York.

Halsted, C. H. and Keen, C. L., 1990, Alcohol and micronutrient metabolism and deficiencies, **Eur. J. Gastroenterol. Hepatol.**, 2:399-405.

Halsted, C. H. and Tamura T., 1978, Folate deficiency in liver disease, **in**: "Problems in Liver Disease," C. S. Davidson, ed., pp. 91-100, Stratton Intercontinental Medical Book, New York.

Halsted, C. H., Griggs, R. C. and Harris, J. W., 1967, The effect of alcoholism on the absorption of folic acid (H^3-PGA) evaluated by plasma levels and urine excretion, **J. Lab. Clin. Med.**, 69:116-31.
Halsted, C. H., Robles, E. A. and Mezey, E., 1971, Decreased jejunal uptake of labelled folic acid (^{3}H-PGA) in alcoholic patients: Roles of alcohol and malnutrition, **N. Engl. J. Med.**, 285:701-6.
Halsted, C. H., Robles, E. A. and Mezey, E., 1973a, Distribution of ethanol in the human gastrointestinal tract, **Am. J. Clin. Nutr.**, 26:831-4.
Halsted, C. H., Robles, E. A. and Mezey, E., 1973b, Intestinal malabsorption in folate-deficient alcoholics, **Gastroenterology**, 64:526-32.
Harris, R. A., Burnett, R., McQuilkin, S., McClard, A. and Simon, F. R., 1987, Effects of ethanol on membrane order: Fluorescent studies, **Ann. N. Y. Acad. Sci.**, 492:125-33.
Herbert, V., 1962, Experimental nutritional folate deficiency in man, **Trans. Assoc. Am. Phys.**, 75:307-20.
Herbert, V., 1971, Predicting nutrient deficiency by formula, **N. Engl. J. Med.**, 284:976-7.
Herbert, V., 1990, Development of human folate deficiency, **in**: "Folic Acid Metabolism in Health and Disease," M. F. Picciano, E. L. R. Stokstad, and J. F. Gregory, III, eds., pp. 195-210, Wiley-Liss, New York.
Herbert, V., Zalusky, R. and Davidson, C. S., 1963, Correlates of folate deficiency with alcoholism and associated macrocytosis, anemia, and liver disease, **Ann. Intern. Med.**, 58:977-88.
Hermos, J. A., Adams, W. H., Lin, Y. K. and Trier, J., 1972, Mucosa of the small intestine in folate deficient alcoholics, **Ann. Intern. Med.**, 76:951-75.
Hillman, R. S., McGuffin, R. and Campbell, C., 1977, Alcohol interference with the folate enterohepatic cycle. **Trans. Assoc. Am. Phys.**, 90:145-56.
Hines, J. D. and Cowan, D. H., 1974, Anemia in alcoholism, **in**: "Drugs and Hematopoietic Reactions," N. V. Dimitrov and J. H. Nodine, eds., pp. 141-53, Grune and Stratton, New York.
Horne, D. W., Briggs, W. T. and Wagner, C., 1979, Studies on the transport mechanism of 5-methyltetrahydrofolic acid in freshly isolated hepatocytes: Effect of ethanol, **Arch. Biochem. Biophys.**, 196:557-65.
Hunter, C. L., Treanor, L. L., Gray, J. B., Halter, S. A., Hoyumpa, A. and Wilson, F. A., 1983, Effects of ethanol in vitro on rat intestinal brush-border membranes, **Biochim. Biophys. Acta**, 732:256-65.
Hurt, R. D., Nelson, R. A., Dickson, E. R., Higgins, J. A. and Morse, R. M., 1979, Nutritional status of alcoholics before and after admission to an alcoholism treatment unit, **in**: "Fermented Food Beverages in Nutrition," C. F. Gastineau, W. J. Darby, and T. B. Turner, eds., pp. 397-408, Academic Press, New York.
Israel, Y., Valenzuela, J. E., Salazar, I. and Ugarte, G., 1969, Alcohol and amino acid transport in the human small intestine, **J. Nutr.**, 98:222-4.
Jandl, J. H., and Lear, A. A., 1956, The metabolism of folic acid in cirrhosis, **Ann. Intern. Med.**, 45:1027-44.
Kabadi, V. M., Eisenstein, A. B. and Konda, J., 1985, Elevated plasma ammonia levels in hepatic cirrhosis: Role of glucagon, **Gastroenterology**, 88:750-6.
Kawase, T., Kato, S. and Lieber, C. S., 1989, Lipid peroxidation and antioxidant defense system in rat liver after chronic ethanol feeding, **Hepatology**, 10:815-21.
Keating, J. N., Weir, D. G. and Scott, J. M., 1985, The effect of ethanol consumption on folate polyglutamate biosynthesis in the rat, **Biochem. Pharmacol.**, 34:1913-6.
Keen, C. L., Tamura, T., Lonnerdal, B., Hurley, L. S. and Halsted, C. H., 1985, Changes in hepatic superoxide dismutase activity in alcoholic monkeys, **Am. J. Clin. Nutr.**, 41:929-32.

Kelly, D., Reed, B., Weir, D. and Scott, J., 1981, Effect of acute and chronic alcohol ingestion on the rate of folate catabolism and hepatic enzyme induction in mice, **Clin. Sci.**, 60:221-4.

Kim, C., Leo, M. A., Lowe, N. and Lieber, C. S., 1988, Effects of vitamin A and ethanol on liver plasma membrane fluidity, **Hepatology**, 8:735-41.

Klipstein, F. A., Lipton, S. D. and Schenk, E. A., 1973, Folate deficiency of the intestinal mucosa, **Am. J. Clin. Nutr.**, 26:728-37.

Leevy, C. M., 1966, Abnormalities of hepatic DNA synthesis in man, **Medicine**, 45:423-33.

Leevy, C. M., Thomson, A. and Baker, H., 1970, Vitamins and liver injury, **Am. J. Clin. Nutr.**, 23:493-8.

Leo, M. A. and Lieber, C. S., 1982, Hepatic vitamin A depletion in alcoholic liver injury, **N. Engl. J. Med.**, 307:597-601.

Leo, M. A. and Lieber, C. S., 1985, New pathway for retinol metabolism in liver microsomes, **J. Biol. Chem.**, 260:5228-31.

Lieber, C. S., 1988, Biochemical and molecular forms of alcohol-induced injury to liver and other tissue, **N. Engl. J. Med.**, 319:1639-50.

Lieber, C. S. and DeCarli, L. M., 1972, An experimental model of alcohol feeding and liver injury in the baboon, **J. Med. Primatol.**, 3:153-63.

Lelbach, W. K., 1975, Cirrhosis in the alcoholic and its relation to the volume of alcohol abuse, **Ann. N. Y. Acad. Sci.**, 252:85-105.

Lindenbaum, J. and Lieber, C. S., 1975, Effect of chronic ethanol administration on intestinal absorption in man in the absence of nutritional deficiency, **Ann. N. Y. Acad. Sci.**, 252:228-34.

Lindenbaum, J., Healton, E. B., Savage, D. G., Brust J. C. M., Garrett T. J., Podell, E. R., Marcell, P. D., Stabler, S. P. and Allen R. H., 1988, Neuropsychiatric disorders caused by cobalamin deficiency in the absence of anemia or macrocytosis, **N. Engl. J. Med.**, 318:1720-8.

Linscheer, W. G., 1970, Malabsorption in cirrhosis, **Am. J. Clin. Nutr.**, 23:488-92.

Lumeng, L., 1978, The role of acetaldehyde in mediating the deleterious effect of ethanol in pyridoxal 5'-phosphate metabolism, **J. Clin. Invest.**, 62:286-93.

Mak, K. M., Leo, M. A. and Lieber, C. S., 1984, Alcoholic liver injury in baboons: transformation of lipocytes to transitional cells, **Gastroenterology**, 87:188-200.

Marchesini G., Zoli, M., Angiolina, A., Dondi, C., Bianchi, F. B. and Pisi, E., 1981, Muscle protein breakdown in liver cirrhosis and the role of altered carbohydrate metabolism, **Hepatology**, 1:294-9.

Mason, J. B., 1990, Intestinal transport of monoglutamyl folates in mammalian systems, **in**: "Folic Acid Metabolism in Health and Disease," M. F. Picciano, E. L. R. Stokstad, and J. F. Gregory, III, eds., pp. 47-63, Wiley-Liss, New York.

McMartin, K. E. and Collins, T. D., 1983, Relationship of alcohol metabolism to folate deficiency produced by ethanol in the rat, **Pharmacol. Biochem. Behav.**, 18:257-62.

McMartin, K. E., Collins, T. D., Shiao, C. Q., Vidrine, L. and Redefski, H. M., 1986, Study of dose-dependence and urinary folate excretion produced by ethanol in humans and rats, **Alcoholism**, 10:419-24.

Mekhjian, H. S. and May, E.S., 1977, Acute and chronic effects of ethanol on fluid transport in the human small intestine, **Gastroenterology** 72:1280-6.

Mendenhall, C. L., Anderson, S., Weesner, R. E., Goldberg, S. J. and Crolic, K. A., 1984, Protein-calorie malnutrition associated with alcoholic hepatitis, **Am. J. Med.**, 76:211-22.

Mezey, E., Jow, E., Slavin, R. E. and Tobon, F., 1970, Pancreatic function and intestinal absorption in chronic alcoholism, **Gastroenterology**, 59:657-64.

Mezey, E., Kolman, C. J., Diehl, A. M., Mitchell, M. and Herlong, H. F., 1988, Alcohol and dietary intake in the development of chronic pancreatitis and liver disease in alcoholism, **Am. J. Clin. Nutr.**, 48:148-51.
Morgan, M. Y. and Levine, J. A., 1988, Alcohol and nutrition, **Proc. Nutr. Soc.**, 47:85-98.
Morgan, M. Y., Marshall, A. W., Milstrom, J. P. and Sherlock, S., 1982, Plasma amino acid patterns in liver disease, **Gut**, 23:363-70.
Naughton, C. A., Chandler, C. J., Duplantier, R. B. and Halsted, C. H., 1989, Folate absorption in alcoholic pigs: in vitro hydrolysis and transport at the intestinal brush border membrane, **Am. J. Clin. Nutr.**, 50:1436-41.
Neville, J., Eagles, J. A., Samson, G. and Olson, R. E., 1968, Nutritional status of alcoholics, **Am. J. Clin. Nutr.**, 21:1329-40.
Owen, O. E., Trapp, V. E., Reichard, G. A., Mozzoli, M. A., Moctezuma, J., Paul, P., Skutcher, C. C. and Boden, G., 1983, Nature and quality of fuels consumed in patients with alcoholic cirrhosis, **J. Clin. Invest.**, 72:1821-32.
Patek, A. J., Toth, I. G., Saunders, M. G., Castro, G. A. M. and Engel, J. J., 1975, Alcohol and dietary factors in cirrhosis, **Arch. Intern. Med.**, 135:1053-7.
Perlow, W., Baraona, E. and Lieber, C. S., 1977, Symptomatic intestinal disaccharidase deficiency in alcoholics, **Gastroenterology**, 77:680-4.
Pirola, R. C. and Lieber, C. S., 1972, The energy cost of the metabolism of drugs, including ethanol, **Pharmacology**, 7:185-96.
Pratt, R. F. and Cooper, B. A., 1971, Folates in plasma and bile of man after feeding [^{3}H] folic acid and 5-formyltetrahydrofolate (folinic acid), **J. Clin. Invest.**, 50:455-62.
Reisenauer, A. M., 1990, Affinity labelling of the folate-binding protein in pig intestine, **Biochem. J.**, 267:249-52.
Reisenauer, A. M., Chandler, C. J. and Halsted, C. H., 1986, Folate binding and hydrolysis by pig intestinal brush border membranes, **Am. J. Physiol.**, 251:G481-6.
Reisenauer, A. M., Buffington, C. A. T., Villanueva, J. A. and Halsted, C. H., 1989, Folate absorption in alcoholic pigs: in vivo intestinal perfusion studies, **Am. J. Clin. Nutr.**, 50:1429-35.
Robles, C. A., Mezey, E., Halsted, C. H. and Schuster, M. M., 1974, Effect of ethanol on motility of the small intestine, **Johns Hopkins Med. J.**, 135:17-24.
Roggin, C. M., Iber, F. L. and Linscheer, W. G., 1972, Intraluminal fat digestion in the chronic alcoholic, **Gut**, 13:107-11.
Romero, J. J., Tamura, T. and Halsted, C. H., 1981, Intestinal absorption of [^{3}H] folic acid in the chronic alcoholic monkey, **Gastroenterology**, 80:99-102.
Rothschild, M. A., Oratz, M. and Schrieber, S., 1987, Effects of ethanol on protein synthesis, **Ann. N. Y. Acad. Sci.**, 492:233-43.
Rubin, E., Ryback, B. J., Lindenbaum, J., Gerson, C. D., Walker, G. and Lieber, C. S., 1972, Ultrastructural changes in the small intestine induced by ethanol, **Gastroenterology**, 63:801-14.
Rueler, J. B., Girard, D. E. and Cooney, T. G., 1985, Wernicke's encephalopathy, **N. Engl. J. Med.**, 312:1035-9.
Russell, R. M., Rosenberg, I. M., Wilson, P. D., Iber, F. L., Oaks, E. V., Giovetti, A. C., Otradovec, C. L., Karwoski, P. A. and Press, A. W., 1983, Increased urinary excretion and prolonged turnover time of folic acid during ethanol ingestion, **Am. J. Clin. Nutr.**, 38:64-70.
Savage, D. and Lindenbaum, J., 1986, Anemia in alcoholics, **Medicine (Baltimore)**, 65:322-8.
Shaw, S., Jayatilleke, E., Herbert, V. and Colman, N., 1989, Cleavage of folates during ethanol metabolism. Role of xanthine oxidase-generated superoxide dismutase, **Biochem. J.**, 257:277-80.

Shaw, S., Jayatilleke, E., Ross, A., Gordon, E. R. and Lieber, C. S., 1981, Ethanol induced lipid peroxidation: Potentiation by long-term alcohol feeding and attenuation by methionine, **J. Lab. Clin. Med.**, 98:417-24.

Soberon, S., Pauley, M. P., Duplantier, R., Fan, A. and Halsted, C. H., 1987, Metabolic effects of enteral formula feeding in alcoholic hepatitis, **Hepatology**, 7:1204-9.

Steinberg, S. E., Campbell, C. L. and Hillman, R. S., 1979, Kinetics of the normal folate enterohepatic cycle, **J. Clin. Invest.**, 64:83-8.

Sullivan, L. W. and Herbert, V., 1964, Suppression of hematopoiesis by ethanol, **J. Clin. Invest.** 43:2048-62.

Swanson, V. L. and Thomassen, R. W., 1965, Pathology of the jejunal mucosa in tropical sprue, **Am. J. Pathol.**, 46:511-51.

Taber, M. M. and Lakshmaiah, N., 1987, Studies on hydroperoxide-dependent folic acid degradation by hemin, **Arch. Biochem. Biophys.**, 257:100-6.

Tamura, T. and Halsted, C. H., 1983, Folate turnover in chronically alcoholic monkeys, **J. Lab. Clin. Med.**, 101:623-8.

Tamura, T., Romero, J. J., Watson, J. E., Gong, E. J. and Halsted, C. H., 1981, Hepatic folate metabolism in the chronic alcoholic monkey, **J. Lab. Clin. Med.**, 97:654-61.

Thomson, A. L., Baker, H. and Leevy, C. M., 1970, Patterns of ^{35}S-thiamine hydrochloride absorption in the malnourished alcoholic patient, **J. Lab. Clin. Med.**, 76:34-45.

Wagner, C., 1982, Cellular folate binding proteins; function and significance, **Ann. Rev. Nutr.**, 2:229-48.

Wang, D. L. and Reitz, R. C., 1983, Ethanol ingestion and polyunsaturated fatty acids: Effects on the acyl-CoA desaturases, **Alcoholism**, 7:220-6.

Waring, A. J., Rottenberg, H., Ohnishi, T. and Rubin, E., 1981, Membranes and phospholipids of liver mitochondria from chronic alcoholic rats are resistant to membrane disordering by alcohol, **Proc. Natl. Acad. Sci. USA**, 78:2582-6.

Weber, F. L., Jr. and Reiser, B.J., 1982, Relationship of plasma amino acids to nitrogen balance and portal systemic encephalopathy in alcoholic liver disease, **Dig. Dis. Sci.**, 27:103-10.

Weber, F. L., Jr., Bagby, B. S., Licate, L. and Kelson, S., 1990, Effects of branched-chain amino acids on nitrogen metabolism in patients with cirrhosis, **Hepatology**, 11:942-50.

Wilkinson, J. and Shane, B., 1982, Folate metabolism in the ethanol-fed rat, **J. Nutr.**, 112:604-9.

Williams, W. M. and Hueng, K. C., 1982, Renal tubular transport of folic acid and methotrexate in the monkey. **Am. J. Physiol.**, 242:F484-90.

Zidenberg-Cherr, S., Halsted, C. H., Olin, K. L., Reisenauer, A. M. and Keen, C. L., 1990, The effect of chronic alcohol ingestion on free-radical defense in the miniature pig, **J. Nutr.**, 120:213-7.

ALCOHOL AND PROTEIN TURNOVER

Victor R. Preedy[1], Tahir Siddiq[2], Elisabeth Cook[1], Darcey Black[3], T. Norman Palmer[4] and Timothy J. Peters[1]

Departments of [1]Clinical Biochemistry and [2]Cardiology
King's College School of Medicine and Dentistry
London, UK
[3]Department of Cellular Sciences
Glaxo Group Research
Greenford, Middlesex, UK
[4]Department of Biochemistry
University of Western Australia
Perth, Australia

INTRODUCTION

Ethanol administration induces a variety of pathological responses which in many tissues are manifested as reductions in tissue protein content. These alterations must be due to changes in protein turnover. This review is, therefore, primarily concerned with investigating the mechanisms whereby tissue protein synthesis or degradation is altered as a consequence of ethanol toxicity. Attention will be focused on the various muscle types (i.e. skeletal, cardiac and smooth muscle) and bone.

In the subsequent paragraphs, the different methods for measuring protein synthesis and degradation will be reviewed. It is tempting to cover all of the ethanol-induced pathogenic defects in protein turnover, but limitations in space preclude this option. Primary concern has, therefore, focused on *in vivo* measurements in laboratory animals.

PROTEIN TURNOVER: CONCEPTS AND COMPONENTS

Protein turnover is the process where tissue proteins are unceasingly being synthesised and degraded. In the steady state, the rates of both processes are equal, thereby

† Abbreviations used in text: S_i, specific radioactivity of free amino acid in intracellular space; S_p, specific radioactivity of free amino acid in extracellular space; S_B specific radioactivity of protein-bound amino acid; k_s, fractional synthesis rate; k_d, fractional degradation rate; k_g, fractional growth rate; k_{RNA}, RNA activity, ie amount of protein synthesis per unit RNA.

maintaining the tissue protein composition (Waterlow et al., 1978). For example, in the adult laboratory animal, the rate of tissue protein synthesis will generally equal the rate of tissue protein degradation, thereby maintaining tissue protein composition. In reality this is an over-simplification as tissue protein synthesis is continually responding to stimuli, such as feeding, muscular activity and diurnal fluctuations in hormonal status or sensitivity. Even though most studies in the field of alcohol and protein turnover have concentrated solely on the measurement of protein synthesis, it is important to emphasize that physiological or biochemical perturbation that cause alterations in synthetic pathways may also cause concomitant disturbances in the rate of protein degradation. Another important point is that ethanol-induced changes in tissue protein composition will depend not only on whether the rates of protein synthesis or protein degradation are being altered, but also on the magnitude of these changes. Reductions in tissue protein composition will occur either by (a) a reduction in protein synthesis, (b) an increase in protein degradation, (c) an increase in both synthesis and degradation, with the increase in degradation being greater than the increase in synthesis or (d) a decrease in both synthesis and degradation, with the decrease in synthesis being greater than the decrease in degradation. Thus, to ascertain the mechanisms responsible for ethanol-induced changes in composition, it is essential that the rates of both processes are accurately measured.

MEASUREMENTS OF PROTEIN SYNTHESIS

In laboratory animals, rates of tissue protein synthesis are usually measured with radiolabelled amino acids. Accurate rates of protein synthesis can only be reliably determined if the specific radioactivities of both the precursor (i.e., the free amino acid at the site of protein synthesis) and the product (i.e., the protein-bound amino acid) are accurately measured throughout the labelling period (Waterlow et al., 1978).

The measurement of the specific radioactivity of the free amino acid at the site of protein synthesis, i.e., the aminoacyl-tRNA, is technically difficult. Tissue tRNA only occurs in relatively small quantities and may be compartmentalized. It is also extremely labile. Although some studies on the effects of ethanol on tissue protein synthesis (i.e., Baraona et al., 1977) have measured the specific radioactivity of the aminoacyl-tRNA, the numerous practical limitations in its extraction and isolation have shown it is not amenable for routine determination on a large number of samples. One way to resolve this is to measure the specific radioactivity of the free amino acid in either the intracellular (S_i) or extracellular (S_p) compartment on the assumption that they reflect the specific radioactivity of the aminoacyl-tRNA. S_i and S_p are represented by acid-soluble portions of tissue homogenates or plasma, respectively. However, in many situations, it is not known whether the free amino acids incorporated into the tissue protein are derived mainly from the intracellular or extracellular free amino acid pools. To circumvent this problem, calculations are based on either S_i or S_p. It follows that uncertainty arises as to the true synthesis rate of the tissue protein if there are marked differences between S_i and S_p.

There are other criteria which are also important in the measurement of protein synthesis, although they do not make the data any more accurate. These are (a) use of a minimum number of animals, (b) measurement of protein synthesis over a few minutes rather than a few hours, (c) that the technique of protein synthesis measurement itself does not influence protein synthesis, (d) that the laboratory procedures for the isolation of labelled amino acids are routine and/or can be automated, and (e) that one should be able to define precursor and product specific radioactivities relatively easily, with a small number of measurements.

The three basic methods used to measure the rate of tissue protein synthesis are (i) single pulse injection of a tracer amount of labelled amino acid, (ii) constant infusion of a tracer amount of labelled amino acid, and (iii) single injection of a large "flooding" dose of labelled amino acid. The merits of each of these methods have been extensively reviewed by Waterlow et al (1978) and Garlick (1980), but are also briefly summarised below.

The pulse injection of a trace amount of labelled amino acid causes rapid alterations in the value of S_i and S_p, between which there is much disparity. The qualitative changes in S_i and S_p are also complex and a large number of animals are required to accurately define the specific radioactivities of precursor and product amino acids. For example, in the study of Baraona et al (1977) a very large number of rats were required to measure protein synthesis in one treatment group.

The constant infusion technique is perhaps one of the better known methods for measuring protein synthesis (Garlick et al., 1973; Waterlow et al., 1978). The major advantage of the method is that rates of protein synthesis can be measured in single animals. The disadvantage of this method, however, is that animals have to be infused for at least 3-6 hours and the immobilisation may perturb synthesis (Preedy and Garlick, 1988). There may also be marked differences between S_i and S_p in some tissues (such as the intestine), which may be exacerbated by experimental conditions (such as starvation). The constant infusion technique is not amenable for investigating acute regulatory factors. Nevertheless, the data derived from this technique are generally reliable and have been used in various investigations studying the ethanol-induced regulation of protein turnover (i.e., Tiernan and Ward, 1986).

Perhaps the most accurate method for measuring protein synthesis in use today is the "flooding dose" technique. This was originally characterised and described in depth by McNurlan et al (1979), who used labelled leucine. In this technique, rats are injected with a large amount of labelled amino acid which floods all endogenous free amino acid pools. This ensures that the specific radioactivities of the free amino acids in intracellular and extracellular compartments are generally equal and reflect the specific radioactivity of the aminoacyl-tRNA (McNurlan et al., 1979). Thus, accurate rates of protein synthesis can be measured without the large errors incurred by the compartmentation of the free amino acid pools. Synthesis rates can be determined in periods as little as 10 minutes, which is amenable for determining acute metabolic effects or responses (McNurlan et al., 1979). This acute period also minimises the errors involved in radiolabelled proteins with short half lifes being degraded during the experimental period, which can underestimate the rate of protein synthesis.

Concern that leucine may be a potent regulator of tissue protein synthesis led to a modification of the "flooding" dose technique, in which leucine was substituted with phenylalanine (Garlick et al., 1980). Inherent in this was the development of a test-tube assay technique for phenylalanine specific radioactivities, which did not require the lengthy and expensive chromatographic separation procedures associated with the determination of leucine.

DETAILS OF THE PHENYLALANINE FLOODING DOSE TECHNIQUE

The measurement of protein synthesis using the phenylalanine "flooding dose" technique is as follows: the rats are injected with 1 ml of 150 mmol/l L-[4-^{3}H]- or L-[U-^{14}C] phenylalanine and are killed at either 2 or 10 minutes. These two time points are necessary to accurately define the specific radioactivity of the free phenylalanine (McNurlan et al., 1979; Garlick et al., 1980). For some tissues, particularly those with low synthetic rates such as skeletal muscle, it is only necessary to sacrifice animals at 10 minutes.

In the original method of Garlick et al., (1980) the isotope was injected via the intravenous route, i.e., lateral tail vein. Because of technical difficulties in locating and cannulating a lateral tail vein, some modifications have been devised for the administration of the isotope via the intraperitoneal route (for example, Martinez, 1987). We have also used this approach in some studies but recently we have shown that for the intestine, ethanol inhibits the transfer of [^{3}H]phenylalanine from the extracellular to the intracellular compartments, and as a consequence, phenylalanine is now administered intravenously (V.R. Preedy and T.J. Peters, unpublished observations).

After sacrifice of the rats at the specified time points, blood is collected for subsequent extraction of plasma (to obtain S_p). The dissected tissues are processed to obtain the specific radioactivities of the tissue free phenylalanine (S_i) and, after hydrolysis, protein-bound phenylalanine (S_B). Fractional rates of protein synthesis, i.e., the percentage of tissue protein renewed each day (k_S, %/day) are from the formula:

$$k_S = \frac{S_B \times 100}{(S_i \text{ or } S_p) \times t} \quad \text{\%/day}$$

where 't' is the radiolabelling period.

The basis of these measurements are discussed in greater detail by Garlick et al (1980). The phenylalanine injection technique has been used to investigate the regulation of protein synthesis in a variety of pathological states, including anaesthesia (T. Siddiq and V.R. Preedy, unpublished results), surgery (Preedy et al., 1988a), starvation (Preedy et al., 1988a) and immobilisation (Preedy and Garlick, 1988).

MEASUREMENT OF PROTEIN DEGRADATION

There are no reliable direct methods for measuring the fractional rate of mixed protein degradation i.e., the percentage of tissue protein pool degraded each day (k_d) in intact animals. Nevertheless, various methods have been developed for the assessment of tissue protein degradation (albeit with methodological reservations) in experimental studies. Isotopic techniques are based on the rate of change in the decay of protein-bound specific radioactivities. These are invariably subject to errors because of the re-incorporation (via protein synthesis) of previously protein-bound amino acids or the decay of pre-labelled proteins may not be exponential (Waterlow et al., 1978). Recycling can partially be overcome with ^{14}C-bicarbonate, and subsequent measurement of the carboxyl carbon of protein-bound glutamate and aspartate; though this method is not appropriate for acute studies (Waterlow et al., 1978). Another way is to measure the arterio-venous differences in the concentration of non-metabolised amino acids, for example, tyrosine or phenylalanine release from muscle (Barrett et al., 1987). Use of this technique has to take into account possible effects of surgical treatments to cannulate blood vessels (Preedy et al., 1988a), or the possibility that amino acids derived from skin and bone may alter the amino acid profile of the venous return (Preedy et al., 1983). The urinary excretion of 3-methylhistidine has also been used to measure the degradation of myofibrillar proteins (Tomas et al., 1984). However, 3-methylhistidine may also be derived from non-muscle tissues, such as skin and gastro-intestinal tissue (Millward et al., 1980). Similar criticisms may also be applied to the use of urinary hydroxyproline excretion to assess collagen degradation (Robins, 1982), as hydroxyproline is also derived from elastin and complement and, in addition, a relatively large proportion (75-90%) of the hydroxyproline released from the breakdown of collagen is oxidised prior to entering the urinary pool (Robins, 1982). In contrast, the pyridinium cross-link deoxy-pyridinoline (hydroxylysyl pyridinoline) is considered to be a more specific marker of bone Type I collagen (Robins, 1982; Black et al., 1988). Its urinary excretion therefore provides a suitable means for assessing protein degradation in skeletal tissue. Pyridinoline (lysyl-pyridinoline) is another urinary marker of pyridinium cross links, derived from Type II and IX collagens of cartilage and to a lesser extent, Type I collagen of bone (Robins, 1982; Black et al., 1988). Neither pyridinoline nor deoxy-pyridinoline are found in skin.

In vitro techniques for the estimation of muscle protein degradation include the release of non-degradable amino acids, such as tyrosine or phenylalanine (Baracos and Goldberg, 1986). However, tyrosine and phenylalanine (but not 3-methylhistidine) may be re-incorporated into protein though this can be prevented by suitable inhibitors of protein synthesis, such as cyclohexamide. In skeletal and cardiac muscle tissue, 3-methylhistidine is neither degraded or re-incorporated into protein and has been utilized as an index of contractile protein degradation *in vitro* (e.g, Smith and Sugden, 1986). However, the measurement of k_d by the release of amino acids is not applicable to all isolated tissues. In perfused hemicorpuses or hind-limb preparations consideration has also to be given to the contribution of amino acid derived from skin and bone, which may comprise of 30% of the preparation weight (Preedy and Garlick, 1981). Moreover, it must be remembered that the release of individual amino acids only represents the final phase in the multi-step processes involved in protein degradation. It is conceivable that removing a tissue from its normal environment *in vivo* may impose metabolic restraints at various sites in the protein degradative pathway.

The measurement of the activities of specific proteolytic enzymes (i.e., Cathepsins) may be useful in determining the potential or capacity for tissues to degrade proteins. Goldspink and Lewis (1985) showed there was a good qualitative correlation between the activities of various Cathepsins (i.e. B and D) and release of non-metabolisable amino acids (i.e., tyrosine) into the incubation medium of isolated muscles.

Perhaps one of the most widely used methods for measuring the rate of degradation *in vivo* is the indirect technique, based on the difference between the fractional rate of growth (k_g) and the fractional rate of protein synthesis (k_s) (Waterlow et al., 1978). If the rates of both these processes have been accurately determined, then it follows that the rate of protein degradation (k_d) can be calculated with a high degree of confidence, i.e., $k_d = k_s - k_g$ %/day.

EXPERIMENTAL ETHANOL TOXICITY STUDIES IN THE RAT

In our chronic ethanol toxicity studies, rats were fed a nutritionally-adequate diet containing 36% of total energy content as ethanol. Control rats were provided with a glucose-containing liquid diet that was iso-nitrogenous, iso-lipidic and iso-caloric, as described by Lieber and DeCarli (1982). Either young (0.1 kg body wt.) or old (0.3 kg body wt.) rats were fed for up to 6 weeks on this diet. In acute ethanol studies, young rats were injected intraperitoneally with a single dose of ethanol (75 mmol/kg body weight), and measurements made at the end of 2.5 hours. In general, plasma ethanol levels were maintained throughout this period (Tiernan and Ward, 1986). For these acute studies, we avoided treating controls with glucose, as this would have caused secondary reactions, such as increased plasma insulin concentrations, which is known to regulate protein turnover (Garlick et al., 1983).

EFFECTS OF ETHANOL ON THE HEART

Numerous authors have comprehensively investigated the *in vivo* effects of experimental ethanol toxicity, using *in vitro* techniques to measure rates of cardiac protein synthesis (e.g., Schreiber et al., 1974; 1982). Although *in vitro* studies provide an excellent means of dissecting out biochemical pathways, studies of this kind are based on the primary assumption that the isolated perfused heart always reflects the biochemical activity of the heart *in vivo*. These premises have been comprehensively studied by Preedy et al (1984; 1985a; 1985b) in investigating the response of the heart to metabolic disturbances, such as fasting or hypoxia. There was an excellent agreement between qualitative changes in the rates of protein synthesis *in vivo* and *in vitro* (Preedy et al., 1984; 1985a; 1985b). In contrast, some conflicting observations were obtained for the rates of protein breakdown, and this appeared to be due to whether or not there was supplementation of the perfusate with anabolic factors such as insulin (Preedy et al., 1984). Thus, there is no reliable substitution for carrying out measurements of protein turnover *in vivo*.

We have shown that in response to acute ethanol toxicity, k_s of mixed cardiac protein was reduced by approximately 20% (Preedy and Peters, 1990a). In chronic ethanol studies on young rats, the synthetic rates of mixed cardiac proteins were not significantly altered, though there was a slight reduction in protein content, suggestive of an adaptive response. Fractionation of the heart into different proteins showed that chronic ethanol feeding preferentially reduced the amount of myofibrillar (i.e., contractile fraction) proteins and that the changes in the cardiac sarcoplasmic (i.e., soluble fraction) and stromal (i.e., insoluble fraction) proteins were not significantly affected (Preedy and Peters, 1989a). Similar reductions in the amount of cardiac contractile apparatus have been noted by microscopic analysis of human cardiac muscles biopsies (Urbano-Marquez et al., 1989). In our study, the reductions in rat heart myofibrillary proteins were small (approx. 10%), but they occurred in spite of an increase (approx. 15%) in the rate of myofibrillar protein synthesis (Preedy and Peters, 1989a). Chronic ethanol feeding had no overt effect on the synthesis rates of the cardiac sarcoplasmic and stromal proteins (Preedy and Peters, 1989a). By implication, it is apparent that the degradation rate of contractile apparatus proteins increased.

The above data introduces some important facets of protein turnover. Although changes in the fractional rate of protein synthesis can appear to be relatively inconsequential, they may, nevertheless, have deleterious accumulative effects on tissue protein composition. Thus, it is interesting to speculate on whether much greater reductions in cardiac myofibrillar protein content would have occurred if the experimental ethanol feeding was extended to beyond 6 weeks. Furthermore, relatively minor changes in protein composition may have important physiological or biochemical consequences. For example, reductions in the relative amount of cardiac contractile proteins may influence ventricular function and other haemodynamic indices. Lastly, in examining whole organs, consideration must be given to the fact that perturbations in protein synthesis may selectively affect one particular protein or group of proteins. This may occur either because individual species of mRNA are being translated at different rates, i.e., translational control, or because there is a selective effect on the amount of individual mRNA pertaining to the protein or group of proteins or because there is a change in the proportion of mRNA being translated, i.e., transcriptional control (Waterlow et al., 1978).

We have recently investigated the possibility that ethanol-induced reductions in protein synthesis may occur secondary to acetaldehyde formation. Unfortunately, acetaldehyde is exceedingly toxic *in vivo* and this imposes severe experimental constraints. To test the role of acetaldehyde we, therefore, used inhibitors of alcohol dehydrogenase (4-methylpyrazole) and aldehyde dehydrogenase (cyanamide) in acute ethanol-dosage experiments (T. Siddiq and V.R. Preedy, unpublished results). The results showed that acute ethanol dosage reduced the fractional synthesis rates of myofibrillary proteins by 20%, from 28±1 %/day (saline control, all data as mean±SEM, n=7-9) to 22±1 %/day (ethanol treated, $P<0.001$). A similar inhibition of cardiac myofibrillary protein synthesis occurred when ethanol-dosed rats were pre-treated with 4-methylpyrazole (i.e., k_s was 20±2, %/day, $P<0.001$ versus saline control), implying that ethanol can suppress protein synthesis in the absence of acetaldehyde formation. Moreover, a 80% inhibition of myofibrillar protein synthesis occurred when ethanol-dosed rats were pre-treated with cyanamide; k_s was 6±1 %/day, which was markedly lower than the rate in rats dosed with ethanol alone ($P<0.001$). This is strong evidence that acetaldehyde is a potent myocardial toxin and overall the data imply that ethanol inhibition of muscle protein synthesis takes place by at least two steps (T. Siddiq and V.R. Preedy, unpublished results). At the present we do not know whether all cardiac proteins (i.e., sarcoplasmic and stromal) are similarly affected.

EFFECT OF CHRONIC ETHANOL FEEDING ON THE SMALL INTESTINE

Our studies into the relationship between ethanol toxicity and intestinal protein synthesis originated from the observation that alcohol disturbs intestinal motility (Robles et al., 1974). We postulated that ethanol abuse may cause smooth muscle contractile protein dysfunction. This was supported by our initial studies into the effects of acute ethanol dosage on intestinal protein synthesis. In the combined duodenum and proximal jejunum, we showed that there was a marked reduction in mixed intestinal protein synthetic rates of the order of 20% (Preedy et al., 1988b). This was similar in magnitude to the change in the heart described above. The proportionate change in the rate of myofibrillar protein synthesis was greater than the fall in soluble intestinal protein synthesis (Preedy et al., 1988b). This suggested that ethanol exerted a regulatory effect on protein synthesis by modulating translation and possibly transcription as well; the latter to account for the effect on contractile proteins.

In chronically-treated young rats we found that ethanol reduced the content of mixed proteins in the entire small intestine by approx 20% (Preedy and Peters, 1989b). More detailed investigations showed that there was a selective effect on the myofibrillar proteins, i.e., 32% reduction, compared to the 18% decrease in the soluble protein fraction (Preedy and Peters, 1990b). Paradoxically, we were unable to detect any significant changes in the fractional rates of either myofibrillar or soluble protein synthesis (Preedy and Peters, 1990b). This refractory response of the small intestine may possibly have reflected an adaptive responsive and/or that protein degradation had increased.

The profound reductions in the amount of intestinal smooth muscle contractile protein of ethanol-fed rats have important implications. If similar pathogenic alterations were to occur clinically, then ethanol abuse may cause a hitherto unrecognised form of visceral myopathy. Unfortunately, most studies on the small intestine of chronic ethanol abusers have been concerned with mucosal pathology. But clearly, our experimental results also implicate contractile protein abnormalities as a cause of some or many of the gastro-intestinal disturbances associated with ethanol abuse and this merits further detailed investigation.

EFFECT OF ETHANOL ON SKELETAL MUSCLE

In man, skeletal myopathy is a common feature of chronic ethanol misuse. One of the primary characteristics of this disease is a preferential reduction in the diameter of Type II (anaerobic) skeletal muscle fibres, whilst Type I (aerobic) skeletal muscle fibres are relatively unaffected (Martin et al., 1985; Urbano-Marquez et al., 1989). In order to study the pathogenesis of this disease in the laboratory rat, we selected anatomically-distinct skeletal muscles with predominant proportions of either Type II or Type I skeletal muscle fibres, namely the plantaris and soleus muscles, respectively. To represent the muscle mass as a whole we also selected the gastrocnemius muscle as it contains a mixture of both Type I and Type II skeletal muscles (Preedy et al., 1988c, 1988d, 1989a, 1989b. 1989c; Preedy and Peters, 1988a, 1988b, 1988c, 1989c). Our results showed that chronic ethanol feeding reduced the weight and protein content of the entire skeletal musculature, which was principally due to the selective susceptibility of Type II fibres: the Type I fibres were relatively unaffected. The results of these studies have recently been summarised by Preedy and Peters (1990c).

Defects in protein synthesis have been implicated in the pathogenesis of the myopathic lesion. Laboratory studies in the rat demonstrated that acute ethanol dosage profoundly reduced the fractional synthetic rate of skeletal muscle proteins (Preedy and Peters, 1988a; Preedy et al., 1988d). In the Type II fibre-preponderant plantaris k_s fell by 30%, whilst the rate of protein synthesis in the Type I fibre-preponderant soleus was reduced by 22% (Preedy and Peters 1988a). The susceptibility of the Type II muscle was therefore greater than the Type I muscle. This selective sensitivity of Type II fibres is similar to clinical observations (Martin et al., 1985), and re-affirms the suitability of the animal model.

We postulated that some of the muscle weakness observed in chronic ethanol abusers may be due to discriminating effects on the contractile apparatus proteins. This supposition was not, however, supported by investigations into the gastrocnemius muscle (Type I and II fibres), which showed that in acute ethanol-dosage studies on young rats,the synthesis rates of sarcoplasmic, myofibrillar and stromal protein fractions decreased by approximately 30%, and all fractions responded with equal sensitivity (Preedy and Peters, 1988a; Preedy et al., 1988d). Similar analyses were carried out on the chronic ethanol-feeding model, where we also compared both young and mature rats. After 6 weeks, the protein contents of soluble, myofibrillar and stromal fractions in gastrocnemius muscles were decreased. Greater changes were observed in small than in large rats. Fractional synthesis rates of soluble, myofibrillar and stromal proteins of gastrocnemius were all decreased by ethanol treatment. All fractions responded similarly, though percentage decreases in large rats were greater than in small rats (Preedy and Peters, 1988b). Thus these results tentatively affirm, that in skeletal muscle, the myofibrillary protein fraction behaves similarly to all the other proteins. This suggests that for skeletal muscle the regulation of protein synthesis is by translational control mechanisms. Nonetheless, three lines of evidence implicate more complex mechanisms. Firstly, it must be emphasised that the subcellular isolation of muscle proteins by differential solubilisation is relatively crude and individual proteins of the contractile apparatus are not identified (although this is now an active area of research within our group). Nevertheless, electrophoretic fractionation of skeletal muscle proteins from the hind-limbs of young alcohol-fed have indeed revealed striking losses of specific proteins (Macpherson et al., 1989). This implies that ethanol was either exerting differential effects on the translation of individual mRNAs for these proteins or that there was a change in the relative amounts of the mRNA species. Secondly, a transcriptional regulatory mechanism is supported by the observation that 6 weeks chronic ethanol feeding markedly reduced muscle RNA content (an indicator of the capacity for protein synthesis) by 20-40%. The greater reductions were observed in Type II muscle (Preedy and Peters, 1988b; Preedy et al., 1989b, 1989c; Marway et al., 1990). Lastly, studies with inhibitors of alcohol and aldehyde dehydrogenases have implicated both ethanol and acetaldehyde in reducing skeletal muscle protein synthesis. But as yet it is not known whether ethanol or acetaldehyde are primarily responsible for the disturbances in RNA metabolism.

We have recently made extensive investigations into the contribution of degradative pathways in the development of alcoholic skeletal myopathy. The first study we carried out was the indirect estimation of protein breakdown, as measured from the fractional rate of protein accretion and the fractional rate of skeletal muscle protein synthesis, using the formula: $k_d = k_s - k_g$ %/day. In both young and mature rats, the fractional rate of soluble, myofibrillar and stromal proteins were reduced by approx 10-20% (Preedy and Peters, 1989c). These perturbations in k_d were in direct contrast to the increased rate of skeletal muscle protein breakdown, as determined by urinary 3-methylhistidine excretion (Tiernan and Ward, 1984). The differences in responses to chronic ethanol feeding may possibly be methodological, as changes in urinary 3-methylhistidine reflect alterations in other tissues such as the gut or skin as well as skeletal muscle (see Millward et al., 1980). In man, Martin and Peters (1985) showed that urinary 3-methylhistidine was relatively unaltered in chronic alcoholics with skeletal muscle myopathy. But when non-skeletal muscle sources of 3-methylhistidine were taken into consideration as described by Afting et al (1981), 3-methylhistidine excretion was significantly reduced in alcoholics with myopathy compared to those without myopathy (Martin and Peters, 1985).

Martin and Peters (1985) also estimated the activities of neutral protease in skeletal muscle biopsies of control and myopathic alcoholics, and found that the activities were similar in both groups. It is not known whether neutral protease activity makes a relatively minor or major contribution to the overall rate of muscle protein degradation, though possibly they may be involved in initial hydrolysis steps. In contrast, the lysosomal proteases, such as the acid hydrolases Cathepsins B and D are probably important contributors to the overall rate of skeletal muscle protein degradation. For example, Millward (1980) has suggested that Cathepsins D and B are capable of degrading all of the myofibrillar proteins.

We have assayed *in vitro* gastrocnemius muscle neutral proteases and Cathepsins B and D activities of our animal model of chronic alcoholic myopathy (E. Cook, T.N. Palmer, T.J. Peters and V.R. Preedy, unpublished results). The data showed that in young rats ethanol-feeding caused a small reduction (15%) in skeletal muscle neutral protease activity, but the changes did not achieve statistical significance. Addition of ethanol to the muscle homogenates of control animals also caused a small but significant decrease in neutral protease activity (20% reductions, $P<0.01$). In contrast, the activities of Cathepsin B were increased by chronic ethanol feeding by approx 25% ($P<0.01$). Furthermore, Cathepsin B activities were increased by the addition of ethanol to homogenate of control muscle, by approx 10% ($P<0.05$). However there were no significant effects of either *in vivo* or *in vitro* ethanol treatments on skeletal muscle Cathepsin D activities (E. Cook, T.N. Palmer, T.J. Peters and V.R. Preedy, unpublished results).

To a certain extent, these results are to be somewhat expected, as consideration must be given to the fact that skeletal muscle is a composite mixture of proteins, each with its own characteristic regulatory mechanisms. The synthesis rates of sarcoplasmic proteins were higher than the myofibrillar proteins (Preedy and Peters, 1988a; 1988b) and yet after six weeks chronic ethanol feeding, their relative composition was maintained, even though the synthesis rates of the different fractions were affected equally by acute and chronic ethanol toxicity. Thus there must be a generalised differential effect of ethanol on muscle proteolysis. However, it is difficult to assign the relative contributions of each of the proteases to the "grand scheme" of protein breakdown. Protease activity only gives the potential or capacity for tissue to degrade protein and the data described above also includes proteolytic activities contained in the subcellular organelles of muscle, i.e., those contained in the sarcotubular system, as well as those in the cytoplasmic "milieu".

EFFECT ON BONE

Alcoholic bone disease, or osteopathy, is a frequent feature of chronic ethanol abuse (see Rico, 1990). To examine the pathogenesis of the increased bone fragility, we have analysed tibia from ethanol-fed rats. The results of these studies showed that chronic ethanol feeding reduced the total tibial collagen content after 4-6 weeks, by approximately 20% (Preedy et al., 1990a): this must have been due to defects in collagen synthesis and/or degradation. Unfortunately, the measurement of collagen synthesis *in vivo* is technically difficult and our attentions have been targeted towards the role of degradative mechanisms. We showed that urinary hydroxyproline was markedly increased, by approx 40-50%, in chronic ethanol-fed rats compared with glucose-fed controls (Preedy et al., 1990a). However, as mentioned earlier, it is difficult to unequivocally ascribe these changes to bone, as collagen is ubiquitous and occurs in most mammalian tissues to varying degrees. We, therefore, studied the urinary pyridinium cross-links of collagen, pyridinoline and deoxy-pyridinoline (Preedy et al., 1990b). Pyridinoline is found predominantly in the Type II and IX collagens of cartilage and in Type I collagen of bone and, to a lesser extent, in other tissues (Robins, 1982; Black et al., 1988). Deoxy-pyridinoline is found only in Type I collagen of bone and dentine and thus, a selective marker of skeletal collagen turnover. Neither pyridinoline nor deoxy-pyridinoline is found in skin (Robbins, 1972; Black et al. 1988). After six weeks of chronic ethanol feeding, the 24-hr urinary excretion of total pyridinoline was slightly reduced, by 15% (Preedy et al., 1990b). The excretion of free and conjugated form of pyridinoline were not significantly altered (Preedy et al., 1990b). In contrast, significant reductions were obtained for the 24-hr urinary excretion of total (45% reduction), free (25% reduction) and conjugated (55% reduction) forms of deoxy-pyridinoline (Preedy et al., 1990b). The mechanism for the reduced urinary deoxy-pyridinoline excretion in alcohol-fed rats is unknown and without data on carcass deoxy-pyridinoline composition, we can only conclude that the absolute rates of bone collagen were reduced; and no presumptions can be forwarded as to whether the fractional rate of collagen degradation was similarly altered.

Thus the conclusions derived from the urinary hydroxyproline and pyridinium-crosslink analyses were apparently contradictory. However, interpreting urinary hydroxyproline data is complicated by the fact that the origins of this imino acid are ill-defined (Millward et al., 1980). Changes in urinary hydroxyproline may possibly reflect dermatological and other soft tissues perturbations. For example, although we have not measured skin k_d in chronically ethanol dosed rats, we have nevertheless shown that protein synthesis in skin *in vivo* is markedly sensitive to short-term ethanol administration (Preedy et al., 1990c). In contrast, protein synthesis in the liver is relatively insensitive to the effects of acute ethanol injection (Preedy et al., 1988d). The above observations clearly demonstrate some of the pitfalls associated with urinary analysis and emphasises the need to be cautious in selecting analytes to measure degradative pathways.

THE CONTRIBUTION OF DIFFERENT TISSUES TO WHOLE BODY PROTEIN SYNTHESIS

There is considerable variability in tissue synthesis rates. The lowest rates are observed for skeletal muscle, whilst the highest rates are observed in the hepato-gastro-intestinal tract. In elucidating the significance of these protein synthetic rates, consideration must be given to the contribution of each tissue to the total-body nitrogen pool. Thus, combined bone and skin, skeletal muscle, liver and the small intestine each contribute to approximately 20% of whole-body rates of protein synthesis (McNurlan and Garlick, 1980; Preedy et al 1983; 1988a; 1990c; Preedy and Peters, 1989b). Tissues such as heart are only minor contributors to whole body nitrogen homoeostasis. We have shown that ethanol toxicity reduces protein synthesis in bone, skin, skeletal muscle and the small intestine. Therefore, one might predict that administration of ethanol would affect whole body nitrogen homoeostasis. This supposition has been confirmed by numerous studies, both in laboratory rats and man (Rodrigo et al., 1971; McDonald and Margen, 1976; Bunout et al., 1987; Reinus, et al., 1989). Thus, not only does ethanol cause tissue specific changes in protein metabolism, but also whole-body effects. Enhanced nitrogen excretion may conceivably contribute to the weight loss that occurs when ethanol is administered for relatively chronic periods. This would be in addition to the inefficient wastage of energy as a result of the activation of the MEOS (microsomal ethanol oxidizing system) system of ethanol oxidation.

CONCLUSION

We have demonstrated that acute or chronic ethanol toxicity reduces the protein composition of a variety of tissues, such as skeletal muscle, heart, bone and the gastro-intestinal tract. These perturbations may be the basis of a variety of diseases, such as chronic skeletal muscle myopathy, cardiomyopathy, intestinal motility disorders and osteoporosis. The mechanisms responsible for these changes must include defects in protein turnover and ethanol must alter either protein synthesis or protein degradation, or both. Systematic investigations in this area must, therefore, be able to reliably measure the rates of both these processes. An accurate method for directly measuring the rates of protein synthesis in the rat *in vivo* is available, namely the "flooding dose" technique with [^{3}H]phenylalanine. However, there are no direct methods for measuring protein degradation *in vivo* in either skeletal muscle, heart or intestine. Therefore, indirect methods are used to assess protein breakdown, i.e., by the differences between rates of synthesis and growth or *in vitro* protease activity. Nevertheless, *in vivo* rates of bone collagen may be directly measured by monitoring specific urinary markers of skeletal collagen (ie, deoxy-pyridinoline). Using these techniques, we have shown that protein turnover in individual tissues of the rat display markedly diverse and contrasting changes in response to acute and chronic ethanol exposure. For example, fractional rates of myofibrillar protein synthesis *in vivo* in chronic ethanol-fed rats were either reduced (skeletal muscle), increased (heart) or relatively unaffected (small intestine). Whole-body collagen degradation (indicated by urinary hydroxy-proline excretion) was shown to increase in response to ethanol feeding, while bone Type I collagen degradation was reduced (indicated by urinary deoxy-pyridinoline excretion). The information gained from these studies may be applicable to understanding how protein mass is regulated in other metabolic disorders as well as ethanol abuse.

ACKNOWLEDGMENTS

We are extremely grateful for the unfailing co-operation, secretarial skills and patience of Miss Cheryl M Riley, without whom this manuscript would not be possible. We also wish to thank Ms H Mangham and Ms D Criddle (Glaxo Laboratories, U.K.) for supplying us with computerized literature searches.

REFERENCES

Afting, E. G., Bernhardt, W., Janzen, R. W. and Rothig, H. J., 1981. Quantitative importance of non-skeletal muscle N-tau-methylhistidine and creatinine in human urine. **Biochem. J.,** 200: 449.

Baraona, E., Leo, M.A., Borowsky, S. A. and Lieber C. S., 1977. Pathogenesis of alcohol-induced accumulation of protein in the liver. **J. Clin.Invest.**, 60: 546.

Baracos, V. E., Goldberg, A. L., 1986. Maintenance of normal length improves protein balance and energy status in isolated rat skeletal muscles. **Amer. J. Physiol.**, 251:C588.

Barrett, E. J., Revkin, J. H., Young, L. H., Zaret, B. L., Jacob, R. and Gelfand, R. A., 1987. An isotopic method for measurement of muscle protein synthesis and degradation *in vivo*. **Biochem. J.**, 245: 223.

Black, D., Duncan, A. and Robins, S. P., 1988. Quantitative analysis of the pyridinium crosslinks of collagen in urine using ion-paired, reverse-phased, high-performance liquid chromatography. **Anal. Biochem.**, 169: 197.

Bunout, D., Petermann, M. Iturriaga, H., 1987. Nitrogen economy in alcoholic patients without liver disease **Metabolism Clin Exp**, 36: 651.

Garlick, P. J., 1980. Protein turnover in the whole animal and specific tissues. **Comp. Biochem.**, 19B: 77.

Garlick, P. J., Millward, D. J. and James, W. P. T., 1973. The diurnal response of muscle and liver protein synthesis *in vivo* in meal-fed rats. **Biochem. J.**, 136: 935.

Garlick, P. J., McNurlan, M. A. and Preedy, V. R., 1980. A rapid and convenient technique for measuring the rate of protein synthesis in tissues by injection of [^{3}H]phenylalanine. **Biochem. J.**, 192: 719.

Garlick, P. J., Fern, M. and Preedy, V. R., 1983. The effect of insulin infusion and food intake on muscle protein synthesis in postabsorptive rats. **Biochem. J.,** 210: 669.

Goldspink, D. F. and Lewis, S. E., 1985. Age- and activity-related changes in three proteinase enzymes of rat skeletal muscle. **Biochem. J.**, 230: 833.

Kar, N. C. and Pearson, C. M., 1980. Elevated activity of a neutral proteinase in human muscular dystrophy. **Biochem. Med.**, 24: 238.

Kelley, J., Stirewalt, W. S. and Chrin, L., 1984. Protein synthesis in rat lung. Measurements *in vivo* based on leucyl-tRNA and rapidly turning-over procollagen I. **Biochem. J.,** 222: 77.

Li, J.B., Higgins, J.E. and Jefferson, L.S. 1979. Changes in protein turnover in skeletal muscle in response to fasting. **Amer. J. Physiol.,** 236: E222.

Lieber, C.S. and DeCarli, L.M., 1982. The feeding of alcohol in liquid diets: two decades of applications and 1982 update. **Alcohol: Clin. Exp. Res.**, 6: 523.

McDonald, J. T. and Margen, S., 1976. Wine versus ethanol in human nutrition. I. Nitrogen and calorie balance. **Amer. J. Clin .Nutr.,** 29: 1093.

McNurlan, M. A. and Garlick, P. J., 1980. Contribution of rat liver and gastrointestinal tract to whole-body protein synthesis in the rat. **Biochem. J.**, 186: 381.

McNurlan, M. A., Tomkins, A. M. and Garlick, PJ., 1979. The effect of starvation on the rate of protein synthesis in rat liver and small intestine. **Biochem. J.**, 178:373.

Macpherson, A. J. S., Marway, J. S., Preedy, V. R. and Peters, 1989. Fractionation of hind limb skeletal muscle proteins by electrophoresis and differential solubilisation in the young alcohol-fed rat. **Clin. Sci.**, 76: suppl 20, 18.

Martin, F. C. and Peters, T. J., 1985. Assessment *in vitro* and *in vivo* of muscle degradation in chronic skeletal muscle myopathy of alcoholism. **Clin. Sci.**, 68: 693.
Martin, F. C., Ward, K., Slavin, G., Levi, J. and Peters, T. J., 1985. Alcoholic skeletal myopathy, a clinical and pathological study. **Q. J. Med.**, 55: 233.
Martinez, J. A., 1987. Validation of a fast, simple and reliable method to assess protein synthesis in individual tissues by intraperitoneal injection of a flooding dose of [^{3}H]phenylalanine. **J. Biochem. Biophys. Methods.**, 14: 349.
Marway, J. S., Preedy, V. R. and Peters, T. J, 1990. Experimental alcoholic skeletal muscle myopathy is characterised by a rapid and sustained decrease in muscle RNA content. **Alcohol Alcohol.**, 25: 401-406.
Millward, D. J., 1980. Protein degradation in muscle and liver. **Comprehen. Biochem.**, 19B: 153.
Millward, D. J., Bates, P. C., Grimble, G. K., Brown, J. G., Nathan, M. and Rennie, M.J., 1980. Quantitative importance of non-skeletal muscle sources of N^T-methylhistidine in urine. **Biochem. J.**, 190: 225.
Preedy, V. R. and Garlick, P. J., 1981. Rates of protein synthesis in skin and bone, and their importance in the assessment of protein degradation in the perfused rat hemicorpus. **Biochem. J.**, 194: 373.
Preedy, V. R. and Garlick, P. J., 1983. Protein synthesis in skeletal muscle of the perfused rat hemicorpus compared with rates in the intact animal. **Biochem. J.**, 214: 433.
Preedy, V. R. and Garlick, P. J., 1985. The effect of glucagon administration on protein synthesis in skeletal muscles, heart and liver *in vivo*. **Biochem.. J.**, 228: 575.
Preedy, V. R. and Garlick, P. J., 1988. The influence of restraint and infusion on rates of muscle protein synthesis in the rat. Effect of altered respiratory function. **Biochem J.**, 251: 577.
Preedy, V. R., Peters, T. J., 1988a. Acute effects of ethanol on protein synthesis in different muscles and muscle protein fractions of the rat. **Clin. Sci.** , 74: 461.
Preedy, V. R. and Peters, T. J, 1988b. The effect of chronic ethanol ingestion on protein metabolism in Type I and Type II fibre-rich skeletal muscles of the rat. **Biochem. J.**, 254: 631.
Preedy, V. R. and Peters, T. J, 1988c. The effect of chronic ethanol feeding on body and plasma composition and rates of skeletal muscle protein turnover in the rat. **Alcohol Alcohol.**, 23: 217.
Preedy, V. R. and Peters, T. J., 1989a. Synthesis of subcellular protein fractions in the rat heart in response to chronic ethanol feeding. **Cardiovas. Res.**, 23: 730.
Preedy, V. R. and Peters, T. J., 1989b. Protein metabolism in the small intestine of the ethanol-fed rat. **Cell. Biochem. Funct.**, 7: 235.
Preedy, V. R. and Peters, T. J., 1989c. The effect of chronic ethanol ingestion on synthesis and degradation of soluble, contractile and stromal protein fractions of skeletal muscles from immature and mature rats. **Biochem. J.**, 259: 261.
Preedy, V. R. and Peters, T. J., 1990a. The acute and chronic effects of ethanol on cardiac muscle protein synthesis in the rat *in vivo*. **Alcohol Alcohol.**, 7: 97.
Preedy, V. R. and Peters, T. J., 1990b. Protein synthesis of muscle fractions from the small intestine in alcohol fed rats. **Gut**, 31: 305.
Preedy, V. R. and Peters, T. J., 1990c. Alcohol and skeletal muscle disease. **Alcohol Alcohol.**, 25: 97.
Preedy, V. R., McNurlan, M. A. and Garlick, P. J., 1983. Protein synthesis in skin and bone of the young rat. **Brit. J. Nutr.**, 49: 517.
Preedy, V. R., Smith, D. M., Kearney, N. F. and Sugden, P. H., 1984. Rates of protein turnover *in vivo* and *in vitro* in ventricular muscle of hearts from fed and starved rats. **Biochem. J.**, 222: 395.
Preedy, V. R., Smith, D. M., Kearney, N. F. and Sugden, P. H., 1985a. Regional

variation and differential sensitivity of rat heart protein synthesis *in vivo* and *in vitro*. **Biochem. J.**, 225: 487.

Preedy, V. R., Smith, D. M. and Sugden, P. H., 1985b. The effects of 6 hours of hypoxia on protein synthesis in rat tissues *in vivo* and *in vitro*. **Biochem. J.**, 228: 179.

Preedy, V. R., Paska, L., Sugden, P. H., Schofield, P. S. and Sugden, M. C., 1988a. The effects of surgical stress and short-term fasting on protein synthesis in diverse tissues of the mature rat. **Biochem. J.**, 250: 179.

Preedy, V. R., Duane, P. and Peters, TJ, 1988b. Acute ethanol dosage reduces the synthesis of smooth muscle contractile proteins in the small intestine of the rat. **Gut**, 29, 1244.

Preedy, V. R., Duane, P. and Peters, T. J., 1988c. Biological effects of chronic ethanol consumption: a reappraisal of the Lieber-DeCarli liquid-diet model with reference to skeletal muscle. **Alcohol Alcohol.**, 23: 151.

Preedy, V. R., Duane, P. and Peters, T. J., 1988d. Comparison of the acute effects of ethanol on liver and skeletal muscle protein synthesis in the rat. **Alcohol Alcohol.**, 23: 155.

Preedy, V. R., Bateman, C. J., Salisbury, J.B. and Peters, T. J., 1989a. Ethanol-induced skeletal muscle myopathy: biochemical and histochemical measurements on type I and type II fibre-rich muscles in the young rat . **Alcohol Alcohol.**, 24: 533.

Preedy, V. R., Marway, J. S. and Peters, T. J., 1989b. Use of the Lieber-DeCarli liquid feeding regime with specific reference to the effects of ethanol on rat skeletal muscle RNA. **Alcohol Alcohol.**, 24: 439.

Preedy, V. R., Venkatesan, S., Peters, T. J., Nott, D. M., Yates, J. and Jenkins, S. A., 1989c. Effect of chronic ethanol ingestion on tissue RNA and blood flow in skeletal muscle with comparative reference to bone and tissues of the gastrointestinal tract of the rat. **Clin. Sci.** 76: 243.

Preedy, V. R., Moniz, C., Hammond, B., Baldwin, D. and Peters, TJ. 1990a. Changes in bone collagen content after chronic ethanol feeding. **Bone**, 11: 222.

Preedy, V. R., Sherwood, R. A., Akpoguma, C. I. O. and Black, D., 1990b The effect of experimental alcoholic bone disease in the rat on the urinary excretion of pyridinoline and deoxy-pyridinoline markers of collagen degradation. **Alcohol Alcohol.**, in press.

Preedy, V. R., Marway, J. S., Salisbury, J. R. and Peters, T.J., 1990c. Protein synthesis in bone and skin of the rat are inhibited by ethanol: implication for whole body metabolism. **Alcoholism: Clin. & Exp. Res.**, 14: 165.

Reinus, J. F., Heymsfield, S. B., Wiskind, R., Casper, K. and Galambos, J. T., 1989. Ethanol: relative fuel value and metabolic effects *in vivo*. **Metabolism.**, 38: 125.

Rico, H., 1990. Alcohol and bone disease. **Alcohol Alcohol.**, 25: 345.

Robins, S. P., 1982. Turnover and crosslinking of collagen. **In**: Collagen in Health and Disease (Eds. J. B. Weiss, and M. I. V. Jayson), Churchill Livingstone, Edinburgh: 160.

Robles, E. A., Mezey, E., Halstead, C. H. and Schuster, M. M., 1974. Effect of ethanol on motility of the small intestine. **John Hopkin's Med. J.**, 135: 17.

Rodrigo, C., Antezana, C. and Baraona, E., 1971. Fat and nitrogen balances in rats with alcohol-induced fatty liver. **J. Nutr.**, 101: 1307.

Schreiber, S. S., Evans, C. D., Reff, F., Rothschild, M. A. and Oratz, M., 1982. Prolonged feeding of ethanol to the young growing guinea pig. 1. The effect on protein synthesis in the overloaded right ventricle measured *in vitro*. **Alcoholism: Clin. & Exp. Res**, 6: 384.

Schreiber, S. S., Oratz, M., Rothschild, M. A. and Evans, C., 1974. Alcoholic cardiomyopathy II. The inhibition of cardiac microsomal protein synthesis by acetaldehyde. **J. Mol. Cell. Cardiol.**, 6: 207.

Smith, D. M. and Sugden, P. H., 1986. Contrasting responses of protein degradation to starvation and insulin as measured by release of N^{T}-methylhistidine or phenylalanine from the perfused rat heart. **Biochem. J.,** 237: 391.

Tiernan, J. M. and Ward, L. C., 1984. N(tau)-methylhistidine excretion by ethanol-fed rats. **IRCS Med. Sci.**, I2: 945.

Tiernan, J. M. and Ward, L. C., 1986. Acute effects of ethanol on protein synthesis in the rat. **Alcohol Alcohol.**, 21: 171.

Tiernan, J. M., Ward, L. C. and Cooksley, W. G., 1985. Inhibition by ethanol of cardiac protein synthesis in the rat. **Inter. J. Biochem.**, 17: 793.

Tomas, F. M., Murray, A. J. and Jones, L. M., 1984. Interactive effects of insulin and corticosterone on myofibrillar protein turnover in rats as determined by N^{T}-methylhistidine excretion. **Biochem. J.**, 220: 469.

Urbano-Marquez, A., Estruch, R., Navarro-Lopez, F., Grau, J.M., Mont, L. and Rubin, E., 1989. The effects of alcoholism on skeletal and cardiac muscle. **New Eng. J. Med.**, 320:409.

Waterlow, J. C., Garlick, P. J. and Millward, D. J., 1978. Protein turnover in mammalian tissues and in the whole body. North Holland, Amsterdam.

ROLE OF THIAMINE DEFICIENCY IN THE PATHOGENESIS OF ALCOHOLIC PERIPHERAL NEUROPATHY AND THE WERNICKE-KORSAKOFF SYNDROME: AN UPDATE

Roger F. Butterworth[1], Monique D'Amour[2], Julie Bruneau[3], Maryse Héroux[1] and Suzanne Brissette[3]

[1]Lab. of Neurochemistry
André-Viallet Clinical Research Center
[2]Neurology Service and
[3]Detoxification Unit
Hôpital St-Luc (University of Montreal)
Montreal, Quebec, Canada H2X 3J4

THIAMINE DEFICIENCY AND ALCOHOLISM

Chronic alcoholism results in thiamine deficiency partly due to inadequate dietary intake of the vitamin. However, in addition, there is evidence to suggest that chronic alcohol consumption leads to impaired absorption of thiamine from the gastrointestinal tract (Thomson et al., 1970). Central nervous system neurons require a continuous supply of thiamine for metabolic and biosynthetic purposes. Transport of thiamine across the blood-brain barrier takes place by both a saturable process and (to a lesser extent) passive diffusion. Saturable uptake occurs at a rate of 0.3 µg/h/g tissue (Greenwood et al., 1982). This rate is surprisingly similar to the rate calculated for thiamine turnover in brain (Rindi et al., 1980), suggesting that thiamine transport across the blood-brain barrier may be just sufficient to meet brain requirements and that there might exist only a limited surplus capacity of this system. Brain structures such as cerebellum and brainstem have particularly high thiamine turnover rates (of the order of 0.3-0.5 µg/h/g tissue) (Rindi et al., 1980), which might explain the selective vulnerability of these brain structures to thiamine deficiency. Furthermore, ethanol ingestion appears to have direct effects on brain thiamine metabolism.

THIAMINE DEFICIENCY AND ALCOHOLIC PERIPHERAL NEUROPATHY

Peripheral neuropathy is a frequent neurological complication of chronic alcohol abuse. In a recent study, it was demonstrated that 38% of otherwise "healthy" chronic alcohol abusers admitted to a detoxification unit had at least one abnormal peripheral nerve conduction parameter (D'Amour et al., 1990). In their study of 245 cases of patients with the Wernicke-Korsakoff Syndrome, Victor et al (1989) reported that 82% showed signs compatible with a diagnosis of peripheral neuropathy; the majority of these patients were alcoholics.

Peripheral neuropathy is classically defined as a symmetrical sensory and motor nerve conduction impairment with diminished or abolished deep tendon reflexes. It is more frequently observed in the lower limbs, with earlier involvement of the distal segments, progressing proximally. The patient's symptoms often progress very slowly and among those frequently noted are weakness of the affected limbs, paresthesias and pain. The painful sensation is usually dull but some patients complain of sharp, stabbing pains for short periods of time. EMG studies are consistent with an axonal type of neuropathy (Lefebvre et al., 1976). Recovery is a slow process and is often incomplete. Early diagnosis and prompt treatment are essential in order to avoid the possible debilitating consequences of peripheral neuropathy.

The exact cause of alcoholic peripheral neuropathy still remains unclear. In one study, 86% of alcoholic patients with peripheral neuropathy had significantly decreased blood thiamine levels (Fennelly et al., 1964). Deficiencies of other vitamins have been shown to produce peripheral neuropathy in humans, particularly riboflavin, pantothenic acid and folic acid and chronic alcohol abusers are known to be deficient in these vitamins in addition to thiamine (Butterworth, 1990). Few studies have been done in man and particularly in alcoholics regarding the etiology of peripheral neuropathy. Recently, it was reported that, in a group of alcoholic patients admitted for treatment, 35% were found to be thiamine deficient, using the erythrocyte transketolase activation assay and 38% had at least one abnormal electromyographic parameter (D'Amour et al., 1990). Unfortunately, correlations between peripheral nerve conduction parameters and thiamine status were unimpressive, with the exception of the peroneal nerve where a significant association between thiamine status and conduction velocities was found. This data would seem to suggest a contributory (but not exclusive) role of thiamine deficiency in the pathogenesis of alcoholic peripheral neuropathy.

THIAMINE DEFICIENCY AND THE WERNICKE-KORSAKOFF SYNDROME

The Wernicke-Korsakoff Syndrome is a neuropsychiatric condition characterized by ophthalmoplegia, ataxia and global confusional state. Although, in the Western hemisphere, the condition most commonly results from chronic alcoholism, it is also encountered in cases of gastrointestinal carcinoma, AIDS and other conditions associated with grossly impaired nutritional status.

Genetic Factors

Two clinical observations suggest the importance of genetic factors in the Wernicke-Korsakoff Syndrome. Firstly, it develops only in a small minority of alcoholics and other malnourished patients. Secondly, it occurs more frequently in Europeans than non-Europeans consuming similar thiamine-deficient diets (Blass and Gibson, 1977). Studies of transketolase, a thiamine-dependent enzyme, in cultured skin fibroblasts from patients with the Wernicke-Korsakoff Syndrome were found to show an increased K_m for thiamine pyrophosphate (TPP) when compared to cells from age-matched controls (Blass and Gibson, 1977). The K_m abnormality persisted through serial passages in tissue culture in cells grown in medium containing excess thiamine and no ethanol suggesting that the abnormalities were genetic rather than dietary in origin.

Thiamine-Dependent Enzymes

Three important enzyme systems, involved in brain glucose utilization, are thiamine-dependent, namely, the pyruvate dehydrogenase complex (PDHC), α-ketoglutarate dehydrogenase (αKGDH) and transketolase (TK) (Figure 1).

Over 50 years ago, Peters (1936) showed that thiamine deficiency in pigeons resulted in opisthotonus and in accumulation of pyruvic and lactic acids in brainstem of affected birds. Addition of small quantities of pure thiamine to the thiamine deficient brain tissue *in vitro* resulted in a correction of the metabolic defect leading Peters to formulate the concept of "the biochemical lesion". Since that time, several studies revealed decreased activities of PDHC in the brainstem of rats maintained on a thiamine-free diet (Butterworth et al., 1985; Gibson et al., 1984). However, these effects were noted only **after** the onset of neurological symptoms. On the other hand, activities of a second thiamine-dependent enzyme, namely α-ketoglutarate dehydrogenase (αKGDH) have consistently been found to be reduced in thiamine deficiency encephalopathy (Gibson et al., 1984; Butterworth et al., 1986). αKGDH abnormalities in thiamine deficiency are found to:

a) precede the onset of neurological symptoms,
b) be particularly severe in brainstem,
c) be reversed promptly following thiamine-reversal of neurological signs of thiamine deficiency.

In the light of these findings, it is now suggested that reversible deficits of αKGDH rather than PDHC (as had initially been postulated) constitute "the biochemical lesion" in thiamine deficiency encephalopathy.

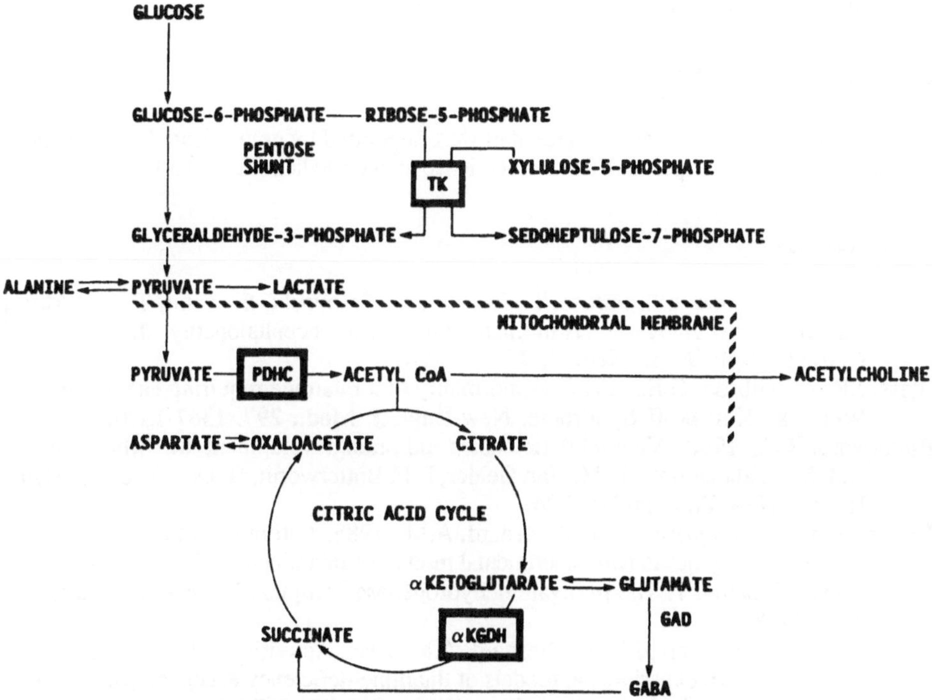

Figure 1. Thiamine-dependent enzymes (from Butterworth R.F., **Alcohol and Alcoholism**, 24: 271, 1989 with permission)

Transketolase activities in brain in thiamine deficiency are decreased early in the development of the deficiency state and are slow to reverse following thiamine supplementation (Giguère and Butterworth, 1986). However, there appears to be a substantial surplus activity of TK in brain over that required to maintain metabolic flux through the pentose shunt pathway (McCandless et al., 1976) bringing into question the possible role of decreased activities of this enzyme in the pathogenesis of thiamine-deficiency encephalopathy.

Decreased activities of αKGDH resulting from thiamine deficiency would be expected to have serious metabolic consequences for the central nervous system. Optimal activities of αKGDH measured *in vitro* are of the order of 10-15 nmol/min/mg protein (Gibson et al., 1984; Butterworth et al., 1986) Such activities are similar to the calculated flux of 3-carbon units derived from glucose *in vivo* (Gibson et al., 1984) suggesting that even moderately decreased αKGDH activities could result in decreased brain glucose oxidation. Being a tricarboxylic acid cycle enzyme, αKGDH is one of the enzymes responsible for the maintenance of mitochondrial energy metabolism. Reductions of αKGDH would be expected to result in decreased ATP synthesis. In a study of oxidative phosphorylation in brain mitochondria from thiamine-deficient rats, decreased respiration was observed when α-ketoglutarate was employed as substrate but normal respiration using succinate (Parker et al., 1984), leading to the suggestion that the permanent neurological deficits and neuropathological damage observed in thiamine deficiency were the result of decreased brain ATP synthesis. The finding of decreased respiration using α-ketoglutarate (but not succinate) is consistent with a decrease in activity of αKGDH (see Figure 1). Moreover, direct measurements of high energy phosphates in the brains of thiamine-deficient rats showed reduced brainstem ATP in symptomatic animals (Aikawa et al., 1984). It was suggested that brain energy failure results in edematous lesions observed in thiamine deficiency. Similar mechanisms could play a role in the pathophysiology of neuronal cell loss in the Wernicke-Korsakoff Syndrome in humans.

ACKNOWLEDGMENTS

Studies from the authors research unit were supported by grants from The Medical Research Council of Canada and The Alcoholic Beverage Medical Research Foundation.

REFERENCES

Aikawa, H., Watanabe, I.S., Furuse, T., Iwasaki, Y., Satoyoshi, E., Sumi, T., and Moroji, T., 1984, Low energy levels in thiamine-deficient encephalopathy. **J. Neuropathol. Exp. Neurol.** 43: 276-287.

Blass, J.P. and Gibson, G.E., 1977, Abnormality of a thiamine requiring enzyme in Wernicke-Korsakoff Syndrome. **New Engl. J. Med.,** 297, 1367-1370.

Butterworth R.F., 1990, Vitamin deficiencies and brain development, **in**: "Malnutrition and the Infant Brain". (N.M. Van Gelder, R.F. Butterworth, B. Drujan, eds). Alan R. Liss, New York, pp 207-224.

Butterworth, R.F., Giguère, J.F. and Besnard, A.M, 1985, Activities of thiamine-dependent enzymes in two experimental models of thiamine-deficiency encephalopathy. 1. The pyruvate dehydrogenase complex. **Neurochem. Res**. 10: 1417-1428.

Butterworth, R.F., Giguère, J.F. and Besnard, A.M, 1986, Activities of thiamine-dependent enzymes in two experimental models of thiamine-deficiency encephalopathy. 2. α-Ketoglutarate dehydrogenase. **Neurochem. Res.** 11: 567-577.

D'Amour M.D., Bruneau J., and Butterworth R.F, 1990, Abnormalities of peripheral nerve conduction in relation to thiamine status in alcoholic patients. **Can. J. Neurol. Sci.**, in press.

Fennelly, J., Frank O., Baker H. et al., 1964, Peripheral neuropathy of the alcoholic: I. Etiological role of aneurin and other B-complex vitamins. **Brit.Med.J.**, 2, 1290-1292.

Gibson, G.E., Ksiezak-Reding, H., Sheu, K.F.R., Mykytyn, V. and Blass, J.P., 1984, Correlation of enzymatic, metabolic and behavioral deficits in thiamine deficiency and its reversal. **Neurochem. Res.** 9: 803-814.

Giguére, J.F. and Butterworth, R.F., 1986, Activities of thiamine-dependent enzymes in two experimental models of thiamine-deficiency encephalopathy. 3. Transketolase. **Neurochem. Res.** 12, 305-310.

Greenwood, J., Love, E.R. and Platt, O.E., 1982, Kinetics of thiamine transport across the blood-brain barrier in the rat. **J. Physiol.** 327, 95-103.

Lefebvre-D'Amour M., Shahani B.T., Young, R.R. et al., 1976, Importance of studying sural conduction and late responses in the evaluation of alcoholic subjects. **Neurology**, 26, 368.

McCandless, D.W., Curley, A.D. and Cassidy, C.E., 1976, Thiamine deficiency and the pentose phosphate cycle in rats: Intracerebral mechanisms. **J. Nutr.** 106: 1144-1151.

Parker, W.D. Jnr., Haas, R., Stumpf, D.A., Parks, J., Eguren, L.A. and Jackson, C., 1984, Brain mitochondrial metabolism in experimental thiamine deficiency. **Neurology**, 34: 1477-1481.

Peters, R.A., 1936, The biochemical lesion in vitamin B_1 deficiency. Application of modern biochemical analysis in its diagnosis. **The Lancet**, i, 1161-1164.

Rindi, G., Patrini, C., Comincioli, V. and Reggiani, C., 1980, Thiamine content and turnover rates of some rat nervous regions using labelled thiamine as a tracer. **Brain Res.** 181: 369-380.

Thomson, A.D., Baker, H. and Leevy, C.M., 1970, Patterns of ^{35}S-thiamine hydrochloride absorption in the malnourished alcoholic patient. **J. Lab. Clin. Med.** 76: 34-45.

Victor M., Adams R.D., and Collins G.H., 1989, The Wernicke-Korsakoff syndrome and related neurologic disorders due to alcoholism and malnutrition, 2nd Ed. Contemporary Neurology Series, F.A. Davis, Philadelphia.

ALCOHOL AND CANCER: A CRITICAL REVIEW

Helmut K. Seitz and Ulrich A. Simanowski

Alcohol Research Laboratory
Section of Gastroenterology & Hepatology
Department of Medicine
University of Heidelberg
Heidelberg, F.R.G

INTRODUCTION

An association between alcoholism and the occurrence of certain tumors has been recognized for decades. At the beginning of this century Lamu (1910) in France showed that absinthe drinkers have an increased risk of esophageal cancer. Meanwhile a great number of epidemiologic studies has shown that there is a good correlation between chronic alcohol consumption and the occurrence of tumors in the oropharynx, in the larynx and in the esophagus (for review see Seitz and Simanowski, 1988; Garro and Lieber, 1990). In addition chronic alcohol consumption also enhances the development of primary hepatic carcinoma especially in a cirrhotic liver (for review see Seitz et al 1989). Most recently, retrospective, prospective and case-control epidemiologic studies gave evidence that there is also a correlation between alcohol intake (especially in form of beer) and rectal cancer (for review see Seitz and Simanowski, 1989). The data with respect to cancer of the breast are still controversial. The major purpose of this review is to discuss the mechanism of the ethanol-associated mechanism of tumor stimulation.

UPPER ALIMENTARY AND RESPIRATORY TRACT

Epidemiology and Animal Experiments

It has been shown that heavy drinkers of high concentrated alcoholic beverages have an approximately 10-12 fold increased risk to develop a tumor in the mouth, pharynx and larynx, while this risk was only elevated by a factor of two when wine or beer were consumed (Wynder and Bross, 1957). Furthermore heavy alcohol intake is often combined with tobacco consumption. Both factors have a synergistic effect on carcinogenesis in the upper alimentary tract. In a carefully performed French study Tuyns (1978,1979) showed that alcohol consumption of more than 80 g per day (approximately one bottle of wine) increases the risk of esophageal cancer by a factor of 18, while smoking alone of more than 20 cigarettes per day does this only by a factor of 5. Both together enhance the risk synergistically by a factor of 44! (Tuyns 1978,1979; Tuyns and Masse, 1973; Tuyns et al., 1977,1979). It has been calculated that 76% of all those cancers could be avoided by avoiding alcohol and tobacco consumption (Rothmann and Keller, 1972).

Alcoholism: A Molecular Perspective
Edited by T.N. Palmer, Plenum Press, New York, 1991

In a very recent epidemiologic study, Maier et al. (1990) found that 90% of all patients with head and neck malignancies exhibited a regular daily alcohol consumption which was almost double that in the control group. Similarly, as is the case for chronic tobacco consumption, the relative risk (RR) of developing a squamous epithelium carcinoma of the upper alimentary and respiratory tract was dose dependent. Thus, if one calculates the RR of a person with a daily alcohol consumption of 25 g as 1.0, the RR for 100 g per day is 32.4.

Animal experiments also showed a cocarcinogenic effect of chronic alcohol consumption on the upper alimentary and respiratory tract (Table 1). The results, however, depend on experimental conditions, especially on the means, duration and doses of alcohol. Factors which influence the effect of alcohol on carcinogenesis are summarized in Table 2. Alcohol per se is not a carcinogen (Ketcham et al 1963): it modulates carcinogenesis which is induced by chemical procarcinogens; it also can function as a tumor promoter (Mufti et al 1989) and/or as a cocarcinogen (McCoy et al., 1986; Castonguay et al., 1984; Gibel 1967).

Table 1. Effect of Ethanol on Upper Alimentary and Respiratory Tract Carcinogenesis in Rodents[a]

Author	Carcinogen[b]	Ethanol[c]	Organ[d]	Effect[e]
Gibel	DENA i.g.	30% i.g. with c	esophagus	+
Mufti et al	MBNA i.p.	6% .d., prior, with or after c	esophagus	-/+
Gabrial et al	MBNA i.g.	4% in d.w. continuously, zinc deficiency	esophagus	+
Newberne et al	MBNA i.g.	4% in d.w. continuously, zinc deficiency	esophagus	+
Schmähl	MPNA s.c.	25% in d.w. continuously	esophagus	0
Kohnishi et al	NNP in diet	50% i.ph. and 10% in d.w. continuously	esophagus	0
McCoy et al	NPYP/NNN i.p. with c	5% l.d. prior to and trachea	nasal cav. trachea	+/0
Griciute et al	NNN i.g.	40% i.g. with c	nasal ca.	0
Castonguay et al	NNN orally NNN s.c.	6% l.d. prior to and with c	nasal cav.	+/0
Iishi et al	MNNG orally	20% i.p. after c	stomach	+

a All studies were performed with rats except McCoy et al used hamsters

b DENA=diethylnitrosamine; MBNA=methylbenzylnitrosamine; MPNA=methylphenylnitrosamine; NNP=N-Nitrosopiperidine; NPYR=N-nitrosopyrrolidine; NNN=N-nitrosonornicotine; MNNG=N-methyl-N'-nitro-N-nitrosoguanidine i.g.=intragastrically; s.c.=subcutaneously; i.p.=intraperitoneally

c l.d.=liquid diet; d.w.=drinking water;i.ph.=intrapharyngeal; c=carcinogen

d nasal cav.=nasal cavity

e - = Inhibition; O = No effect; + = Stimulation

Table 2. Factors which Influence the Effect of Ethanol on Carcinogenesis in Animal Experiments

Type of chemical carcinogen (organ, specificity, metabolism)
Dose of carcinogen and duration of its application
Means of carcinogen application (local, oral, parenteral)
Amount, concentration, and duration of ethanol administration
Means of ethanol administration (in drinking water, as liquid diet, intragastrically, parenteral injection)
Combination of carcinogen and ethanol (ethanol application prior to, with, or after carcinogen application)
Species, strain, sex
Dietary constituents during carcinogenesis

PATHOGENETIC MECHANISMS

Ethanol can increase the susceptibility of various tissues towards chemical carcinogens by a variety of mechanisms. Among those are an enhanced activation of procarcinogens through microsomal enzyme induction, changes in the metabolism and/or distribution of carcinogens, interference with the repair-system which repairs carcinogen-induced DNA-alkylations and the immune system, stimulation of cell regeneration and alcohol mediated malnutrition (Figure 1). Depending on the target organ one or other of the mechanisms prevail.

Ethanol may facilitate the uptake of environmental carcinogens through a damaging effect on cell membranes or by changing their physicochemical properties. Furthermore, ethanol could act as a solvent leading to an enhanced penetration of carcinogenic substances into the mucosa cell. Both factors could play a role in the upper alimentary tract (Elzay, 1966, Henefer, 1966, Horie et al., 1965, Stenback, 1969), but have not been clearly investigated. Chronic alcohol consumption also leads to an atrophy and lipomatic metamorphosis of the parenchyma of the salivary glands which is associated with a striking reduction in saliva secretion (Maier et al., 1986, 1988). This decreased secretion of saliva could result in a decreased rinsing of the mucosal surface which in turn may lead to elevated concentrations of locally acting carcinogens and to a prolongation of the contact time of those substances with the mucosa. Subsequently, it has to be emphasized that various alcoholic beverages such as whiskey, vermuth, sherry, beer and wine may contain carcinogenic compounds. The significance of these compounds in alcohol associated carcinogenesis has been discussed recently in great detail (Garro and Lieber, 199O).

Animal experiments have shown that zinc deficiency which is frequent after alcohol intake (McClean and Su, 1983) may enhance chemically-induced esophageal carcinogenesis (Barch and Iannaccone, 1986; Gabrial et al., 1982). One reason for this observation is that zinc inhibits the activation of nitrosamines by microsomal cytochrome P-450 depending enzymes.

In addition to the systemic alcohol effect, there is a local damaging effect of alcohol to the mucosa by highly concentrated alcoholic beverages which is associated with a secondary compensatory hyperregeneration of the tissue. Hyperregeneration occurs in the

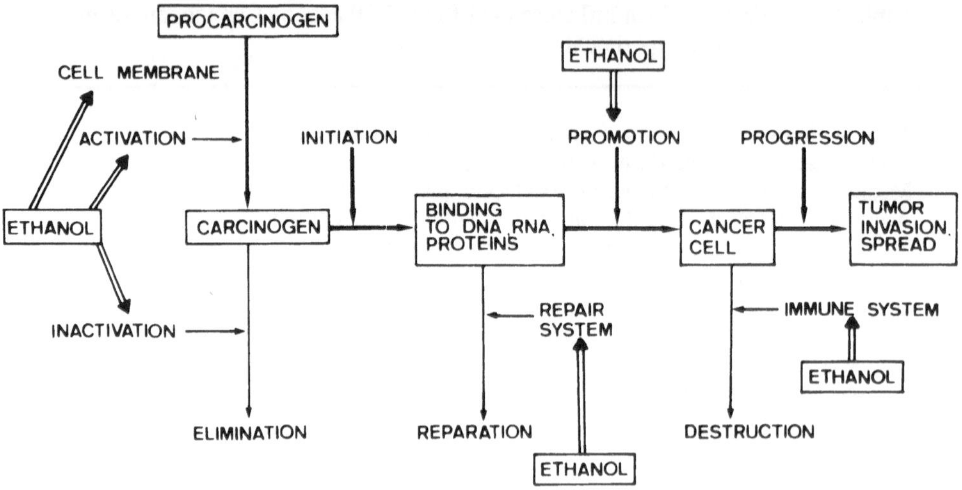

Figure 1. Simplified scheme of 2-step carcinogenesis and possible sites of action of ethanol

absence of overt mucosal injury (Mak et al., 1987). Another substance which could damage the epithelium may be acetaldehyde (AA), the first metabolite of ethanol, which is highly toxic and binds rapidly to cell macromoleules. AA can be produced from ethanol by oral bacteria (Pikkarainen et al; 1981) which may be present in excess in the alcoholic with a poor dental status. Rapidly regenerating tissues, however, are extremely susceptible to the carcinogenic effects of environmental carcinogens. The mechanisms of ethanol-associated oropharyngeal and esophageal carcinogenesis are summarized in Figure 2.

LIVER

Hepatocellular carcinoma (HCC) is frequent in the alcoholic (for review see Lieber et al., 1986; Seitz et al., 1989; Garro and Lieber, 199O). The pathogenesis of the HCC may include direct effects of alcohol during promotion and initiation, alcohol associated cirrhosis of the liver and a concomitant infection with hepatitis B virus (HBV).

The Effect of Ethanol during Initiation and Promotion of HCC

a) Animal Studies To study the effect of alcohol on the development of HCC various animals experiments have been performed (Table 3). The majority of the experiments with alcohol on hepatocarcinogenesis were performed with nitrosamines as tumor inducing agents. Only a few other procarcinogens have been investigated. Hepatocarcinogenesis with nitrosamines is increased only when alcohol is given during promotion (Takada et al., 1986, Driver and McLean, 1986) or when a concomitant methyl deficiency is present (Porta et al., 1985). When alcohol is given prior to or together with a procarcinogen the carcinogenesis is not influenced or even inhibited. Radike et al. (1977) reported a four-fold increased tumor induction by vinylchloride after chronic ethanol consumption in the rat. This was associated with histological changes and serious mitochondrial damage which where thought to be the result of a combined effect of vinylchloride and alcohol (Miller et al., 1982). These animal data are of special interest in

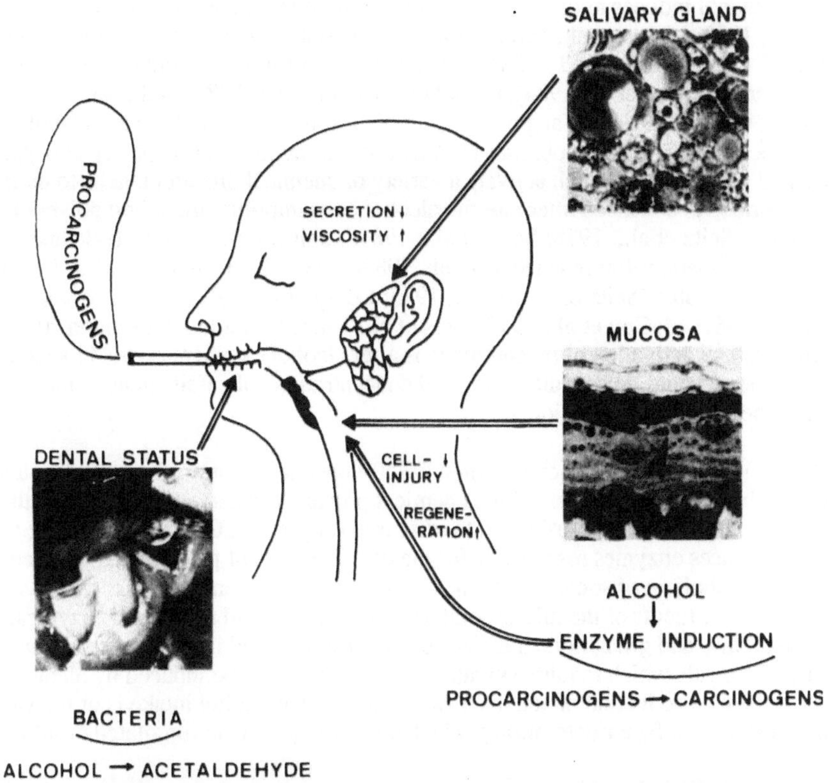

Figure 2. Possible mechanisms of ethanol-associated carcinogenesis in the oropharynx. Procarcinogens enter via smoking and may be activated in increased amounts by microsomal enzymes. In addition, ethanol itself or acetaldehyde (produced by oral bacteria) may damage the oral mucosa and may lead to morphological and functional alterations of the salivary glands.

the light of two accidents in the vinylchloride industry. One vinylchloride- exposed alcoholic developed angiosarcoma of the liver and HCC. His colleague, a nonalcoholic developed "only" angiosarcoma of the liver (Tamburro and Lee, 1981). When aflatoxin B was used as a tumor inducer, ethanol had no effect on liver carcinogenesis (Mendenhall and Chedid, 1980; Misslbeck et al., 1984), although an increased occurrence of peliosis hepatis in the rat was reported.

One epidemiologic study showed that the daily alcohol consumption of more than 24g increased the risk of HCC by the factor of 35 when 4 ug of aflatoxin Bl are ingested at the same time in the diet (Bulatao-Jayme et al., 1982).

b) Ethanol and Microsomal Metabolism of Procarcinogens Most environmental carcinogens exist in their procarcinogenic form and require metabolic activation by microsomal cytochrome P-450 dependent enzymes to become carcinogenic. The induction of microsomal enzyme activities increases the mutagenic effect of many compounds in the Ames Salmonella-Mutagenicity test (Ames et al., 1975). It is well known that ethanol induces microsomal enzymes in the liver and in other tissues (Seitz and Simanowski, 1987, Lieber et al., 1986, Lieber, 1985, Seitz, 1985). The ethanol induced cytochrome P-450 (cytochrome P-450IIEl) has preferred affinity for aniline, 7-ethoxycoumarine (Elves et al., 1984), retinol (Sato and Lieber, 1982) and also dimethylnitrosamine (DMN) (Yang et al., 1985). Thus, in the alcoholic the metabolism of a large number of drugs and xenobiotics is enhanced. Furthermore the capacity of hepatic and intestinal microsomes, which activate a variety of chemical procarcinogens to their mutagenic forms, is enhanced after chronic alcohol consumption, including polycyclic carbohydrates (Seitz et al., 1978; Seitz et al., 1981a; Seitz et al., 1983; Steele and Ionnides, 1986), 2-aminofluorene (Seitz et al., 1981b,1983; Steele and Ionnides, 1986), amino acid pyrolysates (Seitz et al., 1981b, 1983; Loury et al., 1985) and nitrosamines (Garro et al., 1981; McCoy et al., 1979; Neis et al., 1985; Smith and Guttmann, 1984).The enhanced intestinal activation of procarcinogens after alcohol can increase the bioavailability of these compounds and thus result in elevated concentrations of carcinogens in the portal vein and in the systemic circulation.

Although the microsomal cytochrome P-450-dependent biotransformation system is essential for the activation of most of the chemical procarcinogens, an induction of these enzyme system does not necessarily lead to increased cancer risk. One reason may be that alcohol also induces enzymes responsible for the detoxification of produced carcinogens. The microsomal metabolism of some compounds, such as benzopyrene, leads to various products and constituents of the microsomal enzyme system, and associated enzymes, such as epoxy hydratase and glutathione transferase, are also involved in the detoxification of many such chemicals which require activation and they also can be induced by alcohol. In this context the net production of activated carcinogens after alcohol intake is of relevance and this is one critical factor determining whether carcinogenesis is stimulated or inhibited.

In aflatoxin Bl-induced hepatocarcinogenesis, chronic alcohol consumption leads to an increased activation of the procarcinogen (Toskulkao and Glinsukon, 1986; Toskulkao et al., 1986) but does however not enhance the amount of DNA-bound of aflatoxin Bl in the liver of male F344 rats (Marinovich and Lutz, 1985). This lack of DNA-binding is in agreement with the observation that chronic alcohol consumption has no effect of aflatoxin Bl induced liver carcinogenesis.

c) Alcohol and Nitrosamine Metabolism Nitrosamines have been detected in alcoholic beverages (for review see Garro and Lieber, 1990). Since ethanol and nitrosamines are both metabolized by cytochrome P-450 dependent enzymes it is not surprising that a interaction between these compounds occurs. Ethanol and DMN are metabolized via a similar cytochrome P-450 in hepatic microsomes. Therefore alcohol is capable of inhibiting the activity of hepatic Km DMN-demethylase (Swann et al., 1984; Peng et al., 1982, Tomera et al., 1984). On the other hand chronic alcohol consumption increases microsomal DMN-demethylase activity as a result of the induction of cytochrome P-450IIEl, which has a high affinity for DMN (Garro et al., 1981). Thus, chronic alcohol ingestion increases the microsomal activation of DMN to a mutagen in the Ames Test (Garro et al., 1981). Such an increased activation was already observed with a DMN concentration of 0.3 mM or less which could be of pathophysiological importance. However no increased methylation of hepatic DNA was detected when radioactively labelled DMN was given to ethanol-fed and control rats (Kouros et al., 1983). In addition, mutagenicity in the "host mediated assay" was not increased after chronic alcohol feeding (Glatt et al., 1981).

It is noteworthy that chronic alcohol feeding does not stimulate nitrosamine- induced hepatocarcinogenesis. This lack of ethanol-associated cocarcinogenicity may possibly be

Table 3. Effect of ethanol on chemically induced hepatic carcinogens in the rat

Authors	Carcinogen[a]	Ethanol[b]	Effect[c]
Griciute et al	DMNA i.g./NNN i.g.	40% i.g. with c	-[d]
Habs & Schmäl	DMNA, orally	25% in d.w. after c	-
Teschke et al	DMNA i.p.	6% l.d. prior to c	/-[e]
Gibel	DENA i.g.	30% i.g. with c	0[f]
Porta et al	DENA i.p.	25 - 32% l.d. and methyl deficiency after c	+
Takada et al	DENA i.p., 70% hepatectomy	20% in l.d. after c	+
Driver & McLean	DENA i.p.	5% in d.w. after c	+
Mendenhall & Chedid	Aflatoxin B_1 i.g.	6% l.d. continuously	0[g]
Misslbeck et al	Aflatoxin B_1 i.g.	6% l.d. after c	0
Radike et al	Vinyl chloride, in air	5% in d.w. prior to and with c	+
Yanagi et al	3-M-4-DMAAB, orally	0-15% in d.w. prior to and with c	0
Weisburger et al	N-OH-2AAF, orally	10% in d.w. with c	0

a DMNA=dimethylnitrosamine;DENA=diethylnitrosamine;NNN=N-nitrosonornicotine; N-OH-2AAF=N-hydroxy-2-acetaminoflourene;3-M-4-DMAAB=3-methyl-4-dimethyl aminoazobenzene; i.g.=intragastrically; i.p.=intraperitoneally

b d.w.=drinking water; l.d.=liquid diet; c=carcinogen

c - = Inhibition; O = No effect; + = Stimulation

d Occurrence of olfactoric neuroepithelioma after alcohol

e Similar tumor yield, but prolonged latency period after alcohol

f Enhanced esophageal cancer after alcohol

g Stimulation of hepatic peliosis after alcohol

related to the fact that inactivating enzyme activities may also be increased after alcohol ingestion or that the presence of ethanol in the liver during procarcinogen application may inhibit hepatic activation of nitrosamines. When DMN is given orally it undergoes a first-pass metabolism in the liver up to a dose of 30 µg/kg body weight (Swann et al., 1984). At higher doses the hepatic enzymes are saturated and a methylation in other organs such as the kidney or the esophagus occurs. When ethanol is given to rats at low levels it inhibits the first pass metabolism of DMN by competing with the hepatic microsomal enzyme. As a result more nitrosamine can bypass the liver and extrahepatic organs are exposed to higher concentrations of the procarcinogen. Measurements of DMN metabolism in liver slices and esophageal epithelium suggests that the changes in alkylation of esophageal DNA can be the result of a selective inhibition of DMN metabolism in liver and kidney. These biochemical data on the interaction between ethanol and nitrosamine metabolism in the liver and extrahepatic tissues may, at least in part, explain why alcohol does not stimulate the nitrosamine-induced hepatocarcinogenesis but stimulates the development of extrahepatic tumors such as esophageal carcinoma, or tumors in the nasal cavity or in the trachea. The complex interaction between the procarcinogen activation and ethanol metabolism especially with respect to nitrosamines is demonstrated in Figure 3.

d) Alcohol, Radicals and Lipid Peroxidation Lipid peroxidation has been implicated in promoting the carcinogenic process. This suggestion is based on the antagonistic effects of dietary polysaturated fats and dietary antioxidants on carcinogensis. In general, dietary fats enhance tumorigenesis (Carroll, 1980), whereas antioxidants inhibit the process (Wattenberg, 1978). Experimentally, microsomes from ethanol-fed rats have been shown to generate reactive oxygen intermediates such as superoxides, peroxides, and hydroxyl radicals at elevated rates compared with controls (Dicker and Cederbaum, 1987,1988). This is associated with increased lipid peroxidation in animals (Videla and Valenzuela, 1985) and in man (Shaw et al., 1983).

Glutathione (GSH) plays a key role in the detoxification of electrophiles and in the reduction of lipid peroxides. Acute ethanol consumption decreases hepatic GSH levels (Fernandez and Videla, 1981). This effect of ethanol could therefore contribute both to an increase in the number of carcinogen-DNA adducts produced as a result of electrophile production from carcinogens and increased levels of lipid peroxidation.

e) Alcohol and DNA-Metabolism There are two effects of ethanol on DNA-metabolism which could be associated with cocarcinogenicity, namely the effect on DNA sister chromatid exchanges (SCE) and on the DNA repair-system. Obe and Ristow (1977) reported that AA induces SCE in cell cultures. In addition, there was an increase in chromosomal aberrations in lymphocytes of alcoholics (Obe and Ristow, 1979). The importance of this observation with respect to tumor promotion is that compounds with SCE activity can act as tumor promoters (Kinsella and Radman, 1978).

A further mechanism by which alcoholism can increase the risk of cancer is the inhibition of the cellular repair-system which repairs DNA damage. It was reported that the DMN-induced hepatic DNA-alkylations do persist for a longer time in chronically ethanol-fed rats as compared to controls (Garro et al., 1986). This effect seems to be specific for the O^6 methylguanine (MeG) repair system. The enzyme which is responsible for the O^6 MeG repair is the O^6 MeG-transferase. It was found that chronic alcohol consumption reduces the activity of this enzyme significantly (Espina et al., 1988). Since the alkylation on the O^6 position of guanine is associated with carcinogenesis and with mutation, it is obvious that reduced O^6 MeG-transferase activity in alcohol-fed animals could be an important mechanism in alcohol associated hepatocarcinogenesis. It must be emphasized, however, that in two other studies it was not possible to demonstrate an effect of alcohol on the repair of DMN-induced O^6 MeG alkylation (Schwarz et al., 1982; Belinsky et al., 1982).

f) Alcohol as a Tumor Promoter In animal experiments it was shown that alcohol can act as a tumor promoter in the liver at 8 weeks after a single injection of diethylnitrosamine (DENA) in rats which received alcohol following a 70% hepatectomy. After 32 weeks a number of visible tumor nodes in the liver and a number of gamma GT-positive areas (Takeda et al., 1986) were significantly increased after alcohol. These results underline the fact that ethanol like pentobarbital can act as a promoter in chemically-induced hepatocarcinogenesis and this may be of relevance in HBV-carriers which concomitantly drink alcohol in large amounts. In a further experiment Driver and McLean (1986) administered alcohol to rats after initiation with the hepatocarcinogen DENA. Again alcohol as a promoter stimulated carcinogenesis.

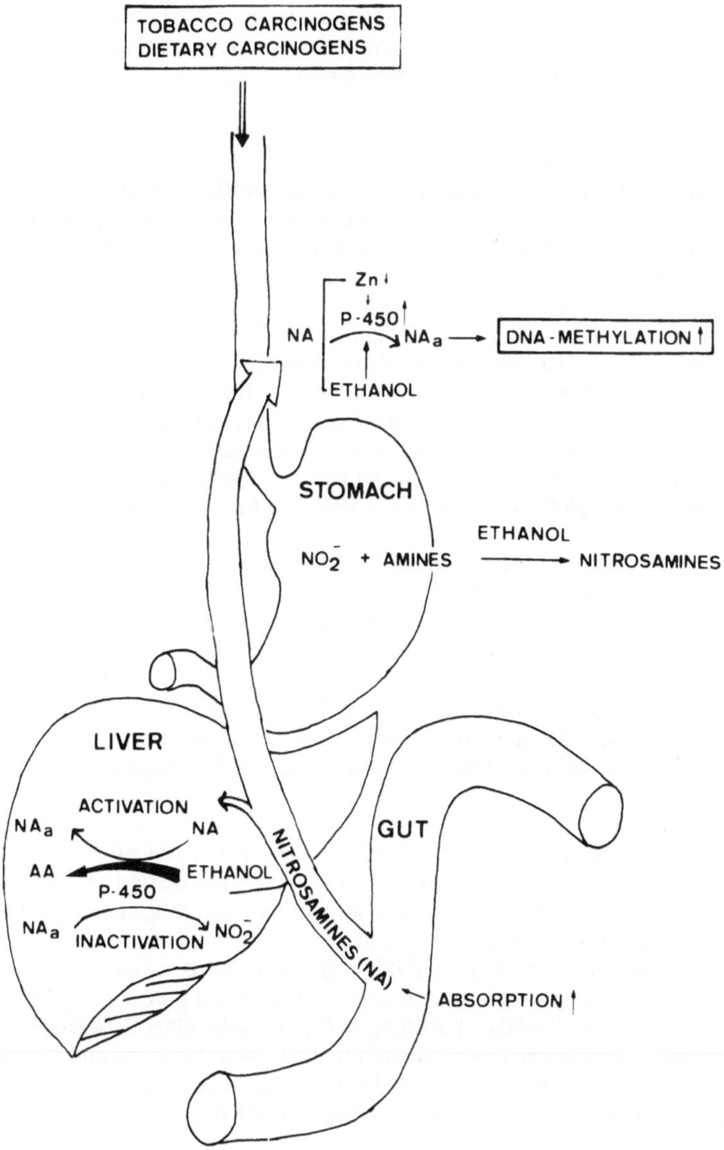

Figure 3. Effect of ethanol on nitrosamine metabolism. Tobacco or dietary nitrosamines as well as endogenously-produced (catalyzed by ethanol is the stomach) nitrosamines are absorbed and may bypass the liver since ethanol competes with the microsomal cytochrome P-450 in the liver. Extrahepatic tissues such as the esophagus are more exposed to nitrosamines which will be activated especially in zinc deficiency.

g) Alcohol and Malnutrition In the alcoholic 50% and more of the daily caloric intake consists of ethanol. Alcoholic beverages, however, lack important nutrients and therefore malnutrition is frequent (Seitz and Kommerell, 1990). With respect to hepatocarcinogenesis most recent studies in rats had shown that ethanol can exert a cocarcinogenic effect on the liver by diminishing methionine and choline. In addition, vitamin A and vitamin B6 deficiency may play a role.

It is well known that chronic alcoholism increases the need for methyl groups (Finkelstein et al., 1974; Tuma et al., 1973; Uthus et al., 1976) and dietary methyl deficiency increases the activity of some hepatocarcinogens (Rogers and Newberne, 1980; Mikol et al., 1983). It must be emphasized however that primates are less susceptible to ethanol associated lipotrope deficiency as compared to rodents (Lieber et al., 1985). Methionine which is a dietary constituent and which is also synthesized in the body by a variety of reactions is the primary precursor of S-adenosylmethionine, the primary methyl donor of the body. An interruption in methionine metabolism and in the methylation-reaction may play a role in carcinogenesis (Van der Westhuyzen, 1985). S-Adenosylmethionine is involved in the methylation of a small percentage of cytosine bases of the DNA. Most recent results underline the fact the enzymatic DNA-methylation is an important component of gene control which can suppress gene function. Some carcinogens interfere with enzymatic DNA-methylation and this may lead to an activation of oncogenes (Riggs and Jones, 1983; Hoffman, 1984; Felsenfeld and McGhee, 1982).

Vitamin A deficiency is also common in the alcoholic. Chronic alcoholism decreases vitamin A concentrations in the liver of animals (Sato and Lieber, 1981) and in man (Leo and Lieber, 1982). As a possible mechanism an increased microsomal cytochrome P-450 dependent degradation of vitamin A and an increased mobilisation of retinol as retinol esters from the liver to the circulation has to be discussed (Lieber et al., 1986). It was reported that retinol can compete with DMN for the microsomal activation in the liver. Therefore it is possible that a decrease in hepatic vitamin A may favour the activation of chemical carcinogens by a decreased inhibition. In addition, vitamin A is important for the integrity of epithelial cells.

Wynder (1976) has found that pyridoxine deficiency may be associated with an increased hepatic tumor development. Alcoholics also have pyridoxine deficiency and it is believed that AA causes an enhanced degradation of this vitamin (Lumeng and Li., 1974). In addition to its key function in hematopoesis, it was shown that vitamin B6 plays an important role in the induction of the antibody response towards various antigens (Axelrod and Trakatellis, 1964); this may influence tumor development indirectly by modifying the immune response towards the HBV infection. In the alcoholic other deficiencies of vitamins and trace elements, such as riboflavin, vitamin E and C as well as zinc and selenium, have also been reported and may also be associated with an increased extrahepatic carcinogenesis (Seitz and Simanowski, 1987, 1988; Garro and Lieber, 1990; Seitz and Kommerell, 1990).

The Importance of Alcoholic Liver Cirrhosis for HCC

In general it seems that that HCC in the alcoholic is associated with cirrhosis of the liver. Indeed the prevalence of liver cirrhosis in patients with HCC varies between 16-80% (Chan, 1975); most reports indicating approximately 55-80%. Cirrhosis of the liver by itself is a risk factor independent of alcohol for HCC. HCC in the alcoholic without cirrhosis is rather rare.

HBV Infection, Ethanol and HCC

There is some evidence that HBV infection is more frequent in the alcoholic compared to nonalcoholic, which may contribute by itself to the increased incidence of HCC. Indeed an increased prevalence of serological HBV-infection markers has been reported in alcoholics (Mills et al., 1979). These results have been confirmed in alcoholics in the same or different geographic areas (Hislop et al., 1981; Orholm et al., 1981; Gluud et al., 1982; Chevilotte et al., 1983). An increased prevalence of anti-HBV-antibodies was also found (Gluud et al., 1984). Brechot et al (1982) reported that 19 of 51 patients with various stages of of alcoholic liver disease revealed several HBV-markers in the serum; in 8 patients there was HBV DNA in the liver; in 5 patients there was DNA integrated in the genome. Integrated HBV-DNA sequences in the liver was reported for various individuals especially in chronic HBV carriers (Brechot et al., 1982; Shafritz et al., 1981). Brechot et al. (1982) studied 20 patients with alcoholic cirrhosis of the liver and HCC and all patients had HBV-DNA integrated in the genome of the neoplastic liver cells. 9 patients had serologic markers of a HBV-infection. These results are in good agreement with the data of Shafritz et al. (1981) and of Ohnishi et al. (1982) who found that hepatocarcinogenesis in HbsAg-carriers who continued to drink alcohol was increased. On the other hand autopsy studies and prospective epidemiologic studies could not found an association between alcoholism and HBV-infection (Omata et al., 1979; Goudeau et al., 1981; Yarrish et al., 1980). Whether the increased incidence of HBV-infection in the alcoholic reflects the social and economic status of the alcoholic, is a consequence of increased blood transfusions in these patients or is the result of an enhanced susceptibility towards HBV needs to be determined.

RECTUM

Epidemiology

In 1974, Breslow and Enstrom reported an association between beer consumption and the development of rectal cancer. Most of the subsequent retrospective and case-control studies on rectal cancer and alcohol have confirmed these data (for review see Seitz and Simanowski, 1989). Two case-control studies are worth reporting: In one study in Copenhagen (Jensen, 1979) there was no effect of beer on rectal cancer in brewery workers, while in Dublin a two-fold increased risk was observed after chronic beer consumption (Dean et al., 1979). A re-evaluation of the Danish study showed, however, that there was a elevated risk for a certain social group of brewery workers with alcohol consumption (Potter et al., 1982). Another cause underlying the difference in the results between the Copenhagen and the Dublin study may be that Danish beer had a significantly lower DMN content compared to Irish beer (Walker et al., 1979).

Two prospective studies were performed in the US which investigated the effect of chronic beer consumption on rectal cancer. In Hawaii 8006 persons were followed up for 14 years. In this study it was shown that there was a 3-fold increased risk of rectal cancer in the chronic beer drinkers. However, colon cancer was not affected (Pollack et al., 1984). Similar results were found in Southern California where 11888 inhabitants of a retirement community were observed for 1 - 2 years. Here the alcohol drinkers revealed a 2-fold increased risk for rectal cancer (Wu et al., 1987). Most recently it has been reported that chronic alcohol intake also increases the risk for the development of polyps in the large intestine (Kikendall et al., 1989).

In summary, all these retrospective, prospective and case-control studies show a small but significant increased risk for rectal cancer but not for colon cancer after chronic alcohol consumption. This increased risk seems mainly to be associated with beer drinking, although alcohol itself may influence rectal carcinogenesis.

Animal Experiments

To date a total of 8 studies on the effect of alcohol on colorectal cancer have been performed. Two of these studies used alcohol in the drinking water and therefore the results are questionable (Howarth and Pihl, 1985; Nelson and Samuelson, 1985). In both studies no effect of alcohol on chemical induced carcinogenesis was found. When the procarcinogens 1,2-dimethylhydrazine (DMH) and azoxymethane (AOM) were used for tumor induction controversial results appeared which depended on the experimental conditions, especially on the means and time of alcohol application (Hamilton et al., 1987a,c; Seitz et al., 1984; McGarrity et al., 1986). The conclusion of all those studies are as follows:

- Modulation of chemical induced colorectal carcinogenesis is due to alcohol and not to beer.
- Chronic alcohol consumption influences tumor development in the right and left colon differently. Increased alcohol doses (18-33% of total calories) inhibits carcinogenesis in the right colon and has no effect on tumor development in the left colon. Lower alcohol doses (9-12% of total calories) enhance carcinogenesis in the left colon without any effect on the right colon.
- Alcohol affects colorectal carcinogenesis in the preinduction and/or induction phase but is without effect during promotion.
- The interaction between alcohol and procarcinogen metabolism influences tumor incidence.

It has to be emphasized in one study that alcohol increases tumor incidence in the rectum but not in the remaining colon (Seitz et al., 1984). These results could not be confirmed in a similar study by McGarrity et al. (1986). Recent data with the primary carcinogen acetoxymethylmethylnitrosamine (AMMN), which is locally applied to the rectal mucosa, showed a significant earlier occurrence of carcinoma during alcohol application (Seitz et al., 1990; Garzon et al., 1987). Since AMMN does not need metabolic activation it is concluded from these results that alcohol enhances tumor development by local mechanisms in the rectal mucosa.

Possible Mechanisms in Carcinogenesis

a) Procarcinogen Activation As already pointed out chronic alcohol consumption results in an induction of the specific cytochrome P-450, which also leads to an activation of various procarcinogens. It has been shown that ethanol inhibits the hepatic microsomal activation of AOM (Sohn et al., 1987) and DMN (Sohn et al., 1987; Peng et al., 1982) while the activation of these two compounds is enhanced after chronic alcohol consumption when alcohol is absent. The conversion of AOM to methylazoxymethanol (MAM) is catalyzed by a microsomal cytochrome P-450 dependent N-hydroxylase in the liver (Fiala, 1977) and in the colon (Wargovich and Felkner, 1982; Glauert and Bennink, 1983). Pretreatment of animals with a microsomal enzyme inducer such as phenobarbital, chrysen (Fiala, 1977) or alcohol (Sohn et al., 1987) leads to an increased metabolism of AOM, probably by induction of the microsomal enzyme. On the other hand compounds which inhibit DMH-metabolism also inhibit DMH-induced colorectal carcinogenesis in vivo (Wattenberg, 1975). It seems therefore possible that the alcohol effect, which is observed in the animals treated with DMH and AOM, is at least in part due to alcohol-induced changes in the metabolism of procarcinogens. In the light, it is understandable why high alcohol intake, which leads to relatively high alcohol blood levels, results in an inhibition of colorectal carcinogenesis, while lower alcohol intake has no effect. In the presence of alcohol the activation of procarcinogens during tumor induction is inhibited and so is carcinogenesis (Hamilton et al., 1987b), while after alcohol withdrawal with associated enzyme induction or after low alcohol intake, which is high enough to induce the enzyme, the carcinogenic process is enhanced (Seitz et al., 1984). However, it is unclear why alcohol under certain experimental conditions stimulates tumor development in the left colon and rectum but not in the right colon (Hamilton et al., 1987a; Seitz et al.,1984).

b) Local Mechanisms The fact that chronic alcohol consumption also stimulates rectal carcinogenesis induced by the local application of the primary carcinogen AMMN raises the question whether alcohol may exert its stimulating effect by a local mechanism in the rectal mucosa and not only via an increased activation of procarcinogens (Garzon et al., 1987; Seitz et al.,1990).

An important factor in intestinal carcinogenesis is the state of regeneration of the mucosa. In animal experiments cell proliferation in the rectal mucosa is strikingly increased after chronic alcohol feeding compared to controls (Simanowski et al., 1986). A similar stimulating effect of alcohol on cell regeneration was also observed for the esophagus where alcohol as already mentioned acts as a cocarcinogen (Mak et al., 1987). A concomitant increase in the proliferative compartment size of the rectal crypt after alcohol intake has also been observed. Such a hyperproliferation and expansion of the proliferative compartment of the crypts towards the rectal lumen seems to be predictive for an increased susceptibility towards chemical carcinogens. Cell regeneration is triggered by polyamines and the synthesis of polyamines is regulated by the activity of the trigger enzyme ornithine decarboxylase (ODC). It was found that the activity of this enzymes is significantly increased after chronic alcohol consumption but rapidly returns to normal after alcohol withdrawal (Hamilton and Luk, 1987; Seitz et al., 1990).

The rectal hyperproliferation after chronic alcohol consumption may be of a secondary compensatory nature since light microscopy of the rectal mucosa in alcoholics shows superficial cell damage which is completely reversible after two weeks of alcohol abstinence (Brozinski et al., 1979). Furthermore the life time of cryptal cells in the functional compartment of the rectal crypt is decreased (Simanowski et al., 1986).

It was believed that AA may cause damage of the rectal mucosa in the alcoholic. Significantly elevated AA concentrations were found in the distal colon after alcohol application (Seitz et al., 1990). These AA-concentrations related to organ weight were significantly elevated compared with the proximal colon and liver. In addition, there was also increased alcohol dehydrogenase (ADH) activity in the mucosa of the distal compared to the proximal colon (Seitz et al., 1984), which may favour AA accumulation through ethanol oxidation. It seems, however, impossible that the rather low activities of colon ADH can be responsible for the observed accumulation of AA in the rectum. It was therefore proposed that bacterial production of AA may play a role. This theory is supported by the fact that various aldehydes including AA have been detected in vitro after incubation of feces with alcohol (Levitt et al., 19821).

Most recent data on the effect of alcohol on AMMN-induced rectal carcinogenesis support the concept that AA is involved in ethanol-associated carcinogenesis. Animals (germ-free rats) which received ethanol and cyanamide, a potent AA-dehydrogenase inhibitor, exhibited after induction by AMMN an earlier occurrence of rectal tumors compared to animals which received alcohol alone (Seitz et al., 1990). A positive correlation between the AA-concentrations in the mucosa and the number of fecal bacteria was found. Both the numbers of bacteria and AA-concentration increased from the cecum to the rectum. In summary, there is some evidence that alcohol can be metabolized by fecal bacteria, especially in the rectum where more bacteria are present compared to the cecum and that the produced AA damages the rectal mucosa. This is followed by a compensatory secondary hyperregeneration of the crypt cells which by itself favours carcinogenesis.

Most recently we observed that the increased cell regeneration after alcohol in the rectum is strikingly enhanced with age which may experimentally support the fact that rectal cancer is especially high in the elderly (Simanowski et al., 1990).

VARIOUS OTHER ORGANS

The data obtained on a possible correlation between alcohol consumption and cancer in other organs are partly controversial. This includes mainly the pancreas and the breast (Burch and Ansari, 1968; IARC, 1973; Lin and Kessler, 1981; Durbec et al., 1983; Heuch et al., 1983; Breslow and Enstrom, 1974; Rosenberg et al., 1982; Begg et al., 1983; Hiatt and Bawol, 1984; Le et al., 1984; Talamini et al., 1984; La Vecchia et al., 1985; Harvey et al., 1987; O'Connell et al., 1987; Willett et al., 1987; Hiatt et al., 1988; Rohan and McMichael, 1988; Schatzkin et al., 1987). As already mentioned with respect to the rectum there are studies which have not shown a correlation between alcohol and the appearance of mammary (Byers and Func, 1982; Paganini-Hill and Ross, 1983; Webster et al., 1983; Harris and Wynder, 1988; Schatzkin et al., 1989) and pancreatic tumors (Wynder et al., 1977; Mack et al., 1986; Norell et al., 1986).

Animal experiments on the pancreas are also very limited. Alcohol exclusively applied in the drinking water when nitroso (2-oxopyrol)amine (BOP) was used as a tumor inducer produced either an inhibition or no effect on carcinogenesis (Tweedie et al., 1981;Pour et al., 1983; Woutersen et al., 1986). When azaserin was used as a tumor-inducing compound a stimulation of carcinogenesis after alcohol intake was found in the rat (Woutersen et al., 1986).

Schrauzer et al. (1979) showed that chronic alcohol consumption shortens the latency period and increases the tumor volume of spontaneous adenocarcinoma of the breast of female C3H/St mice. Ethanol administered by gavage for 3-5 weeks prior to carcinogen administration was also reported to enhance the initiation of mammary cancers induced either with dimethylbenzanthrazene or methylnitrosourea (Grubbs et al., 1988).

The reported association between moderate alcohol intake (1-2 drinks per day) and breast cancer has been of major interest in recent years. The majority of the case-control studies and 4 of the 5 cohort studies, as well as the most recently performed meta-analyses (Longnecker et al., 1988), showed a significantly positive correlation between alcohol and breast cancer. Since a great number of women consume alcohol moderately and since breast cancer is the most frequent cancer in women in western Europe and North America, a correlation between alcohol consumption and the occurrence of this tumor, if it holds true, will be of great importance, especially in women who have an increased risk for breast cancer because of other reasons.

REFERENCES

Ames, B.N., McCann, J. and Yamasaki, E., 1975. Methods for detecting carcinogens and mutagens with the salmonella/mammalian-microsomes mutagenicity test. **Mutat Res.** 31: 374-364.

Axelrod, A.E. and Trakatellis, A.C., 1964. Relationship of pyridoxine to immunologic phenomenon. **Vitamin Horm.** 22: 591-607.

Barch, D.H. and Iannaccone, P.M., 1986. Role of zinc deficiency in carcinogenesis. In: **Essential nutrients in carcinogenesis** (Hrsg. LA Poirier, PM Mewberne, MW Pariza) Plenum Publ. Coop. 1986: 517-527.

Begg, C.B., Walker, A.M., Wessen, B. and Zelen, M., 1983. Alcohol consumption and breast cancer. **Lancet** 1: 293-294.

Belinsky, S.A., Bedell, M.A. and Swenberg, J.A., 1982. Effects of chronic ethanol diet on the replication, alkylation and repair of DNA from hepatocytes and nonparenchymal cells following dimethylnitrosamine administration. **Carcinogenesis** 3: 1293-1297.

Brechot, C., Nalpas, B., Courouce, A.M., Duhamel, G., Gallard, P., Cornot, F., Tiollais, P. and Berthelot, P., 1982. Evidence that hepatitis B virus has a role in the liver cell carcinoma in alcoholic liver disease. **N Engl J Med.** 306: 1384-1387.

Breslow, N.E. and Enstorm, J.E., 1974. Geographic correlations between mortality rates and alcohol, tobacco consumption in the United States. **JNCI.** 53: 631-639.

Brozinski, S., Fami, K., Grosberg, J.J., 1979. Alcohol ingestion-induced changes in the human rectal mucosa: light and electron microscopic studies. **Dis Colon Rectum.** 21: 329-335.

Bulatao-Jayme, J., Almero, E.M., Castro, C.A., Jardeleza, T.H.and Salamat, L.A., 1982. A case controlled dietary study of primary liver cancer risks from aflatoxin exposure. **Int J Epidemiol.** 11: 112-119.

Burch, C.E.and Ansari, A., 1968. Chronic alcoholism and carcinoma of the pancreas: a correlative hypothesis. **Arch Intern Med.** 122: 273-275.

Byers, T. and Funch, D.P., 1982. Alcohol and breast cancer **Lancet** 1: 799-800.

Capel, I.D., Turner, M., Pinock, M.H. and Williams, D.C., 1978. The effect of chronic alcohol intake upon the hepatic microsomal carcinogen activation system. **Oncology** 35: 168-170.

Carroll, K.K., 1980. Lipids and carcinogenesis **J Environ Pathol Toxicol.** 3: 253-271.

Castonguay, A., Rivenson, A., Trushin, N., Reinhardt, J. and Spathopoulos, S., 1984. Effect of chronic ethanol consumption on the metabolism and carcinogenicity of N'-nitrosonornocotine in F344 rats. **Cancer Res.** 44: 2285-2290.

Chan, C.H., 1975. Primary carcinoma of the liver. **Med Clin North Am.** 59: 989-994.

Cheviolotte, G., Durbec, J.P., Gerolami, A., Berthezene, P., Bidart, J.M. and Camatte, R., 1983. Interaction between hepatitis B virus and alcohol consumption in liver cirrhosis. **Gastroenterology** 85: 141-145.

Dean, G., MacLennan, R., McLoughlin, H. and Shelley, E., 1979. The cause of death of blue collar workers at a Dublin brewery 1954-1973. **Br J Cancer.** 40: 581-598.

Dicker, E., Cederbaum, A.I., 1987. Hydroxyl radical generation by microsomes after chronic ethanol consumption. **Alcohol: Clin Exp Res.** 11: 309-314.

Dicker, E. and Cederbaum, A.I., 1988. Increased oxygen radical-dependent inactivation of metabolic enzymes by liver microsomes after chronic ethanol consumption. **FASEB J.** 2: 2901-2906.

Driver, H.E. and McLean, A.E., 1986. Dose-response relationship for initiation of rat liver tumors by diethylnitrosamine and promotion by phenobarbitone or alcohol. **Food Chem Toxic.** 24: 241-245.

Durbec, J.P., Chevilotte, G., Bidart, J.M., Berthezene, P.and Sarles, H., 1983. Diet, alcohol, tobacco and risk of cancer of the pancreas: a case control study. **Br J Cancer.** 47: 463-470.

Elves, R.G., Ueng, T.H. and Alvares, A.P., 1984. Comparative effects of ethanol administration on hepatic monooxygenase in rats and mice. **Arch Toxicol.** 55: 258-264.

Elzay, R.P., 1966. Local effect of ethanol in combination with DMBA on hamster cheek pouch. **J Dent Res.** 45: 1788-1795.

Espina, J., Lima, V., Lieber, C.S. and Garro, A.J., 1988. *In vitro* and *in vivo* inhibitory effect of ethanol and acetaldehyde on 0^6 methylguanine transferase. **Carcinogenesis** 9: 761-766.

Felsenfeld, G. and McGhee, J., 1982. Methylation and gene control. **Nature** 296: 602-603.

Fernandez, V., Videla, L.A., 1981. Effect of acute and chronic ethanol ingestion on the content of reduced glutathione of various tissues of the rat. **Experientia** 37: 392-394.

Fiala, E.S., 1977. Investigations into the metabolism and mode of action of the colon carcinogen 1,2-dimethylhydrazine and azoxymethane. **Cancer** 40: 2436-2445.

Finkelstein, J.D., Cello, J.P. and Kyle W.E., 1974. Ethanol-induced changes in methionine metabolism in rat livers **Biochem Biophys Res Commun.** 61: 525-531.

Gabrial, G.N., Schrager, T.F. and Newberne, P.M., 1982. Zinc deficiency, alcohol, and retinoid: association with esophageal cancer in rats. **JNCI.** 68: 785-789.

Garro, A.J., Lieber, C.S., 1990. Alcohol and cancer **Ann Rev Pharmacol Toxicol.** 30: 219-249.

Garro, A.J., Seitz H.K. and Lieber, C.S., 1981. Enhancement of dimethylnitrosamine metabolism and activation to a mutagen following chronic ethanol consumption in the rat. **Cancer Res.** 41: 120-124.

Garro, A.J., Espina, N., Farinati, F., Salvagnini, M., 1986. The effect of chronic ethanol consumption on carcinogen metabolism and on 0^6-methylguanine transferase-mediated repair of alkylated DNA. **Alcohol Clin Exp Res.** 10: 73-77.

Garzon, F.T., Simanowski, U.A., Berger, M.R., Schmahl, D., Kommerell, B., Seitz, H.K., 1987. Acetoxymethyl-methylnitrosamine (AMMN)-induced colorectal carcinogenesis is stimulated by chronic alcohol consumption. **Alcohol** (Suppl) 1: 501-502.

Giebel, W., 1967. Experimentelle Untersucllungen zur Synkarzinogenese beim Ösophaguskarzinom. **Arch Geschwulstforsch.** 30: 181-189.

Glatt, H., DeBalle, L. and Oesch, F., 1981. Ethanol or acetone pretreatment of mice strongly enhanced the bacterial mutagenicity of dimethylnitrosamine in assays mediated by liver subcellular fractions, but not in host mediated assays. **Carcinogenesis** 2: 1057-1061.

Glauert, H.P., Bennik, M.R., 1983. Metabolism of 1,2-dimethylhydrazine by cultured rat colon epithelial cells **Nutr Cancer.** 5: 78-86.

Gluud, C., Gluud, B., Aldersvile, J., 1984. Prevalence of hepatic B virus infection in out-patient alcoholics. **Infection** 12: 72-74.

Gluud, C., Aldersvile, J., Henriksen, J., Kryger, P. and Mathiesen, L., 1982. Hepatitis A and B virus antibodies in alcoholic steatosis and cirrhosis. **Clin Pathol.** 35: 695-697.

Goudeau, A., Maupas, P., Dubois, F., Coursaget, P. and Bougnoux, P., 1981. Hepatitis B infection in alcoholic liver disease and primary hepatocellular carcinoma in France. **Prog Med Virol.** 27: 26-34.

Griciute, L., Castegnaro, M. and Bereziat, J.C., 1981. Influence of ethyl alcohol on carcinogenesis with N-nitrosodiethylamine. **Cancer Lett.** 13: 345-352.

Griciute, L., Castegnaro, M., Bereziat, J.C. and Cabral, J.R.P., 1986. Influence of ethyl alcohol on the carcinogenic activity of N-nitrosonornicotine. **Cancer Lett.** 31: 267-275.

Grubbs, C.J., Juliana, M.M. and Whittaker, L.M., 1988. Effect of ethanol in initiation of methylnitrosourea (MNU) and dimethylbenzanthracene (DMBA)-induced mammary cancers. **Proc Annu Meet Am Assoc Cancer Res.** 29: A590.

Habs, M. and Schmahl, D., 1981. Inhibition of the hepatocarcinogenic activity of diethylnitrosamine (DENA) by alcohol in rats. **Acta Gastroenterol.** 28: 242-244.

Hamilton, S.R., Hyland, J., McAvinchey, D., Chaudhry, Y., Hartka, L., Kim, H.T., Cichon, P., Floyd, J., Turjman, N., Kessie, G., Nair, P.P. and Dick, J., 1987a. Effects of chronic dietary beer and ethanol consumption on experimental colonic carcinogenesis by azoxymethane in rats. **Cancer Res.** 47: 1551-1559.

Hamilton, S.R. and Luk, G.D., 1987b. Induction of colonic mucosal ornithidine decarboxylase activity by chronic dietary ethanol consumption in the rat (Abstract). **Gastroenterology** 92: 1423.

Hamilton, S.R., Sohn, O.S. and Fiala, E.S., 1987c. Effects of timing and quantity of chronic dietary ethanol consumption on azoxymethane-induced colonic carcinogenesis and azoxymethane metabolism in Fischer 344 rats. **Cancer Res.** 47: 4305-4311.

Harris, R.E. and Wynder, E.L., 1988. Breast cancer and alcohol consumption: a study in weak associations. **JAMA.** 259: 2867-2871.

Harvey, E.B., Schairer, C., Brinton, L.A., Hoover, R.N. and Fraumeni, J.F., 1987. Alcohol consumption and breast cancer. **JNCI.** 78: 657-661.

Henefer, E.P., 1966. Ethanol 30% and hamster pouch carcinogenesis. **J Dent Res.** 45: 838-844.

Heuch, I., Kvale, G., Jacobsen, B.K. and Bjelke, E., 1983. Use of alcohol, tobacco and coffee and risk of pancreatic cancer. **Br J Cancer.** 48: 637-643.

Hiatt ,R.A. and Bawol, R.D., 1984. Alcoholic beverages consumption and breast cancer incidence. **Am J Epidemiol.** 120: 676-683.

Hiatt, R.A., Klatsky, A.L., Armstrong, M.A., 1988. Alcohol consumption and risk of breast cancer in a pre-paid health plan. **Cancer Res.** 48: 2284-2287.

Hislop, W.S., Follett, E.A.C., Bouchier, I.A.D. and MacSween, R.N.M., 1981. Serological markers of hepatitis B in patients with alcoholic liver disease: a multicentre survey. **J Clin Pathol.** 34: 1017-1019.

Hoffman, R.M., 1984. Altered methionine metabolism, DNA methylation and oncogene expression in carcinogenesis. A review and synthesis. **Biochem Biophys Acta.** 738: 49-87.

Horie, A., Kohchi, S. and Karatsune, M., 1965. Carcinogenesis in the esophagus II. Experimental production of esophageal cancer by administration of ethanolic solutions of carcinogens. **Gann.** 56: 429-441.

Howarth, A.E. and Phil, E., 1985. High fat diet promotes and causes distal shift of experimental rat colonic cancer - beer and alcohol do not **Nutr Cancer.** 6: 229-235.

IARC (International Agency for Research in Cancer), 1973. Alcohol and Cancer Report, Interim Report Lyon, France: **IARC.**

Ishii, H., Tatsuda ,M., Baba, M. and Taniguchi, H., 1989. Promotion by ethanol of gastric carcinogenesis by N-methyl-N-nitro-N-nitrosoguanidine in Wister rats. **Br J Cancer.** 59: 719-721.

Jensen, O.M., 1979. Cancer morbidity and cause of death among Danish brewery workers. **Int J Cancer.** 23: 454-463

Ketcham, A.S., Wexler, H. and Mantel, N., 1963. Affects of alcohol in mouse neoplasia. **Cancer Res.** 23: 667-670.

Kikendall, J.W., Bowen, P.E., Burgess, M.B,, Megnetti, C., Woodward, J. and Langenberg, P., 1989. Cigarettes and alcohol as independent risk factors for colonic adenomas. **Gastroenterology** 97: 660-664.

Kinsella, A. and Radman, M., 1978. Tumor promotor induces sister chromatid exchanges: relevance to mechanisms of carcinogenesis **Proc Natl Acad Sci USA.** 75: 6149-6153.

Konishi, N., Kitahori, Y., Shimoyama, T., Takahashi, M. and Hiasa, Y., 1986. Effects of sodium chloride and alcohol on experimental esophageal carcinogenesis induced by N-nitrosopiperidine in rats. **Gann.** 77: 446-451.

Kouros, M., Monch, W. and Reiffer, F.J., 1983. The influence of various factors on the

methylation for DNA by the esophageal carcinogen N-nitrosomethylbenzylamine. I. The importance of alcohol. **Carcinogenesis** 4: 1081-1084.
Lamu, L., 1910. Etude de statistique clinique de 131 cas de cancer de l'oesophage et du cardia. **Arch Mal Appar Dig Mal Nutr.** 4: 451.
La Vecchia, C., Decarli, A., Franceschi, S., Pampallona, S. and Tognoni, G., 1985. Alcohol consumption and the risk of breast cancer in women. **JNCI** 75: 61-65.
Le, M., Hill, C., Kramar, A., Flamant, R., 1984. Alcoholic beverage consumption and breast cancer in a french case control study. **Am J Epidemiol.** 120: 350-357.
Leo, M.A., Lieber, C.S., 1982. Vitamin A depletion in alcoholic liver injury in man. **N Engl J Med.** 307: 597-601.
Levitt, M.D., Doizaki, W. and Levine, A.S., 1982. Hypothesis: metabolic activity of the colonic bacteria influences organ injury from ethanol. **Hepatology** 2: 598-600.
Lieber, C.S., 1985. Ethanol metabolism and pathophysiology of alcoholic liver disease **In**: Seitz HK, Kommerell B (Eds) Alcohol related diseases in gastroenterology. Springer, Berlin Heidelberg New York Tokio pp 19-47.
Lieber, C.S., Garro, A.J., Leo, M.A., Mak, K.M., Worner, T.M., 1986. Alcohol and cancer. **Hepatology** 6: 1005-1019.
Lieber, C.S., Leo, M.A. and Mak, K.M., 1985. Choline fails to prevent liver fibrosis in ethanol-fed baboons but causes toxicity. **Hepatology** 5: 561-572.
Lin, R.S. and Kessler, II., 1981. Multifactorial model for pancreatic cancer in man. **JAMA** 245: 147-152.
Longnecker, M.P., Berlin, J.A., Orza, M.J. and Chalmers, T.C., 1988. A meta analysis of alcohol consumption in relation to risk of breast cancer. **JAMA** 260: 652-656.
Loury, D.J., Kado, N.Y. and Byard, J.L., 1985. Enhancement of hepatocellular genotoxicity of several mutagens from amino acid pyrolysates and broiled foods following ethanol pretreatment. **Food Chem Toxicol.** 23: 661-667.
Lumeng, L. and Li, T.K., 1974. Vitamin B6 metabolism in chronic alcohol abuse. **J Clin Invest.** 53: 693-704.
Mack, T.M., Yu, M.C., Hannisch, R. and Henderson, B.E., 1986. Pancreas cancer and smoking, beverage consumption, and past medical history. **JNCI** 76: 49-60.
Maier, H., Born, I.A., Veith, S., Adler, D., Seitz, H.K., 1986. The effect of chronic ethanol consumption on salivary gland morphology and function in the rat. **Alcoholism Clin Exp Res.** 10: 425-429.
Maier, H., Born, I.A. and Mall, G., 1988. Effect of chronic ethanol and nicotine consumption on the function and morphology of the salivary glands. **Klin Wochenschr.** 66 (suppl.XI): 140-144.
Maier, H., Dietz, A., Zielinski, D., Junemann K.H. and Heller, W.D., 1990. Risikofaktoren bei Patienten mit Plattenepithelkarzinomen der Mundhohle, des Oropharynx, des Hypopharynx und des Larynx. **DMW.** 115: 843-850.
Mak, K.M., Leo, M.A. and Lieber, C.S., 1987. Effect of ethanol and vitamin A deficiency on epithelial cell proliferation and structure in the rat esophagus. **Gastroenterology.** 93: 362-370.
Marinovich, M. and Lutz, W.K., 1985. Covalent binding of aflatoxin Bl to liver DNA in rats pretreated with ethanol. **Experientia.** 41: 1338-1340.
McCoy, G.D., Chell, C.B., Hecht, S.S., 1979. Enhanced metabolism and mutagenesis of nitrosopyrrolidine in liver fractions isolated from chronic ethanol-consuming hamsters. **Cancer Res.** 39: 793-796.
McCoy, G.D., Hecht ,S.S., Katayama, S. and Wynder, E.L., 1981. Differential effects of chronic ethanol consumption on the carcinogenicity of N-nitrosopyrrolidine and N-nitrosonornicotine in male Syrian hamsters. **Cancer Res.** 41: 2849-2854.
McClain, C.J., Su, L.C. 1983. Zinc deficiency in the alcoholic: a review. **Alcoholism Clin Exp Res.** 7: 5-10.
McGarrity, T.J., Via , E.A. and Colony, P.C., 1986. Changes in tissue sialic acid content

and staining in dimethylhydrazine (DMH)-induced colorectal cancer: effects of ethanol **Gastroenterology** 90: 1543 (Abstract).

Mendenhall, C.L. and Chedid, L.A., 1980. Peliosis hepatis: its relationship to chronic alcoholism, aflatoxin Bl and carcinogenesis in male Holtzman rats. **Dig Dis Sci.** 25: 587-592.

Mikol, Y.B., Hoover, K.L. and Creasia, D., 1983. Hepatocarcinogenesis in rats fed methyl-deficient, amino acid defined diets. **Carcinogenesis.** 4: 1619-1629.

Miller, M.L., Radike, M.J., Andringa, A. and Bingham, E., 1982. Mitochondrial changes in hepatocytes of rats chronically exposed to vinyl chloride and ethanol. **Environ Res.** 29: 272-279.

Mills, P.R., Rennington, T.H., Hay, P., MacSween, R.N.M. and Watkinson, G., 1979. Hepatitis B antibody in alcoholic cirrhosis. **J Clin Pathol.** 32: 778-782.

Misslbeck, N.G., Campbell, T.C., Roe, D.A., 1984. Effect of ethanol consumed in combination with high or low fat diets on the postinitiation phase of hepatocarcinogenesis in the rat. **J Nutr.** 114: 2311-2323.

Mufti, S.I., Becker, G. and Sipes, I.G., 1989. Effect of chronic dietary ethanol consumption on the initiation and promotion of chemically-induced esophageal carcinogenesis in experimental rats. **Carcinogenesis.** 10: 303-309.

Neis, J.M., TeBrömmelstroet , B.W.J,, VanGemert, P.J.L., Roelofs, H.M.J., Henderson, P.T., 1985. Influence of ethanol induction on the metabolic activation of genotoxic agents by isolated rat hepatocytes. **Arch Toxicol.** 57: 217-221.

Nelson, R.L., Samuelson, S.L., 1985. Neither dietary ethanol nor beer augments experimental carcinogenesis in rats. **Dis Col Rectum.** 28: 460-462.

Newberne, P.M., Charnley, G., Adams, K., Cantor, M., Roth, D. et al., 1986. Gastric and esophageal carcinogenesis: models for the identification of risk and protective factors. **Food Chem Toxicol.** 24: 1111-1119.

Norell, S.E., Ahlbom, A., Erwald, R., Jacobsen, G. and Lindberg-Navier, I., 1986. Diet and pancreatic cancer: a case control study. **Am J Epidemiol.** 124: 894-902.

Obe, G. and Ristow, H., 1977. Acetaldehyde but not alcohol induces sister chromatid exchanges in Chinese hamster cells in vitro. **Mut Res.** 56: 211-213.

Obe, G. and Ristow, H., 1979. Mutagenic, cancerogenic and teratogenic effects of alcohol. **Mut Res.** 65: 229-259.

O'Connell, D.L., Hulka, B.S., Chambless, L.E., Wilkinson, K.E. and Deubner, D.C., 1987. Cigarette smoking, alcohol consumption, and breast cancer risk. **JNCI** 78: 229-234.

Ohnishi, K., Iida, S., Iwama, S. et al., 1982. The effect of chronic habitual alcohol intake on the development of liver cirrhosis and hepatocellular carcinoma: relation to hepatitis B surface antigen carriers. **Cancer.** 49: 672-677.

Omata, M., Ashcavai, M., Liew, C. and Peters, R.L., 1979. Hepatocellular carcinoma in the USA: etiologic considerations. **Gastroenterology** 76: 280-287.

Orholm, M., Aldersvile, J., Tage-Jensen, U. Schlichting, I., Nielsen, J., Hardt, F. and Christoffersen, P., 1981. Prevalence of hepatitis B virus infection among alcoholic patients with liver disease. **J Clin Pathol.** 34: 1378-1380.

Paganini-Hill, A., Ross, R.K., 1983. Breast cancer and alcohol consumption. **Lancet** 2: 626-627.

Peng, R., Yong-Tu, Y. and Yang, C.S., 1982. The induction and competitive inhibition of a high affinity microsomal nitrosodimethylamine demethylase by ethanol. **Carcinogenesis.** 3: 1457-1461.

Pikkarainen, P.H., Baraona, E., Jauhonen, P., Seitz, H.K. and Lieber, C.S., 1979. Contribution of oropharyllx microflora and of lung microsomes to acetaldehyde in expired air after alcohol ingestion. **J Lab Clin Med.** 97: 631-636.

Pollack, E.S., Nomura, A.M.Y., Heilbrun, L.K., Stemmermann, G.N. and Green, S.B., 1984. Prospective study of alcohol consumption and cancer. **N Engl J Med.** 310: 617-621.

Porta, E.A., Markell, N. and Dorado, R.D., 1985. Chronic alcoholism enhances hepatocarcinogenicity of diethylnitrosamine in rats fed a marginally methyl-deficient diet. **Hepatology** 5: 1120-1125.

Potter, J.D., McMichael, A.J. and Hartshorne, J.M., 1982. Alcohol and beer consumption in relation to cancers of bowel and lung: an extended correlation analysis. **J Chron Dis.** 35: 833-842.

Pour, P.M., Reber, H.A. and Stepan, K., 1983. Modification of pancreatic carcinogenesis in the hamster model. XII. Dose related effect of ethanol. **JNCI** 71: 1085-1087.

Radike, M.J., Stemmer, K.L., Brown, P.B., Larson, E., and Bingham, E., 1977. Effect of ethanol and vinyl chloride on the induction of liver tumors. **Environ Health Perspect.** 21: 153-155.

Riggs, A.D. and Jones, P.A., 1983. 5-methylcytosine, gene regulation and cancer. **Adv Cancer Res.** 40: 1-30.

Rogers, A.E. and Newberne, P.M., 1980. Lipotrope deficiency in experimental carcinogenesis. **Nutr Cancer.** 2: 104-112.

Rohan, T.E. and McMichael, A.J., 1988. Alcohol consumption and risk of breast cancer. **Int J Cancer**. 41: 695-699.

Rosenberg, L., Slone, D., Shapiro, S., Kaufman, D.W., Helmrich, S.P., et al.. 1982. Breast cancer and alcoholic beverage consumption. **Lancet** 1: 267-271.

Rothmann, K.J. and Keller, A., 1972. The effect of joint exposure to alcohol and tobacco on risk of cancer of the mouth and the pharynx. **J Chron Dis.** 25: 711-716.

Sato, M. and Lieber, C.S., 1981. Hepatic vitamin A depletion after chronic ethanol consumption in baboons and rats. **J Nutr.** 111: 2015-2023.

Sato, M. and Lieber, C.S., 1982. Increased metabolism of retinoic acid after chronic ethanol consumption in rat liver microsomes. **Arch Biochem Biophys.** 213: 557-564.

Schatzkin, A., Jones, D.Y., Hoover, R.N., Taylor, P.R., Brinton, L.A., Ziegler, R.G., et al., 1987. Alcohol consumption and breast cancer in the epidemiologic follow up study of the First National Health and Nutrition Examination Survey **NEJM.** 316: 1169-1173.

Schmahl, D., 1976. Investigations of esophageal carcinogenicity by methyl-phenyl nitrosamine and ethyl-alcohol in the rat. **Cancer Lett.** 1: 215-218.

Schrauzer, G.N., McGinness, J.E., Ishmael, D. and Bell, L.J., 1979. Alcoholism and cancer: effects of long term exposure to alcohol on spontaneous mammary adenocarcinoma and prolactin levels in C3H/St mice. **J Stud Alcohol.** 40: 240-246.

Schwarz, M., Wiesbeck, G., Hummel, J. and Kunz, W., 1982. Effects of ethanol on dimethylnitrosamine activation and DNA synthesis in rat liver. **Carcinogenesis** 3: 1071-1075.

Seitz, H.K., 1985. Ethanol and carcinogenesis **In**: Seitz HK, Kommerell B (Eds) 'Alcohol related diseases in gastroenterology'. Springer, Berlin, Heidelberg, New York, Tokyo, pp 192-212.

Seitz, H.K. and Simanowski, U.A., 1987. Metabolic and nutritional effects of alcohol **In**: Hathcock JN (Ed) Nutritional toxicology, vol II. Academic, New York, pp 63-104.

Seitz, H.K. and Simanowski, U.A., 1988. Alcohol and carcinogenesis. **Ann Rev Nutr.** 8: 99-119.

Seitz, H.K., Simanowski, U.A., 1989. Ethanol and colorectal carcinogenesis **In:** Colorectal Cancel: from pathogenesis to prevention ? (Hrsg. HK Seitz, UA Simanowski, NA Wright) Springer Verlag Berlin 1989, 177-192.

Seitz, H.K., Kommerell, B., 1990. Alkoholismus als häufigste Ursache fur Mangelernahrung **Dtsch Arzteblatt** 9: 497-500 (Editorial)

Seitz, H.K., Garro, A.J., Lieber, C.S., 1978. Effect of chronic ethanol ingestion on intestinal metabolism and mutagenicity of benzo(a)pyrene. **Biochem Biophys Res Commun.** 85: 1061-1066.

Seitz, H.K., Garro, A.J., Lieber, C.S., 1981a. Sex dependent effect of chronic ethanol consumptions in rats on hepatic microsome mediated mutagenicity of Benzo(a)pyrene. **Cancer Lett.** 13: 97-102.

Seitz, H.K., Garro, A.J. and Lieber, C.S., 1981b. Enhanced pulmonary and intestinal activation of procarcinogens and mutagens after chronic ethanol consumption. **Eur J Clin Invest.** 11: 33-38.

Seitz, H.K., Garro, A.J. and Lieber, C.S., 1983. Increased activation of procarcinogens by microsomes of various tissues induced by chronic ethanol ingestion. **In**: Lieber CS (Ed) Biological Approach to Alcoholism: update 1980, pp 131-141 (Research Monograph No 11, DHHS, Publ. No. (ADM) 83-1261).

Seitz, H.K., Cygan, P., Waldherr, R., Veith, S., Raedsch, R., Kässmodel, H. and Kommerell, B., 1984. Enhancement of 1,2-dimethylhydrazine induced rectal carcinogenesis following chronic ethanol consumption in the rat. **Gastroenterology** 86: 886-891.

Seitz, H.K., Simanowski, U.A., Horner, M. and Kommerell, B., 1989. Alcohol and liver carcinoma **In**: Liver cell carcinoma (Hrsg. P. Bannasch, D. Keppler, G. Weber) Kluwer Acad Publ Dordrecht, Boston, London, 227-242.

Seitz, H.K., Simanowski, U.A., Garzon, F.Z., Rideout, J.M., Peters, T.J., Koch, A., Berger, M.R., Einecke, H. and Maiwald, M., 1990. Possible role of acetaldehyde in ethanol related rectal cocarcinogenesis in the rat. **Gastroenterology** 98: 1-8.

Shafritz, D.A., Shouval, D. and Sherman, H.I., 1981. Integration of hepatitis B virus DNA into the genome of liver cells in chronic liver disease and hepatocellular carcinoma. **N Engl J Med.** 305: 1067-1073.

Shaw. S., Rubin , K.P. and Lieber, C.S., 1983, Depressed hepatic glutathione and increased diene conjugates in alcoholic liver disease: evidence of lipid peroxidation. **Dig Dis Sci.** 28: 585-589.

Simanowski, U.A., Seitz, H.K., Baier, B., Kommerell, B., Schmidt-Gayk, H. and Wright, N.A., 1986. Chronic ethanol consumption selectively stimulates rectal cell proliferation in the rat. **Gut** 27: 278-282.

Simanowski, U.A., Suter, P., Russell, R.M., Seitz, H.K., 1990. Cell regeneration in the rat rectum after chronic ethanol ingestion is strikingly enhanced with age. **Alcoholism Clin Exp Res.** 14: 338 (abstract).

Smith, B.A. and Guttman, M.R., 1984. Differential effect of chronic consumption by the rat of microsomal oxidation of hepatocarcinogenes and their activation to mutagens. **Biochem Pharmacol.** 33: 2901-2910.

Sohn, O.S., Fiala, E.S., Puz, C., Hamilton, S.R. and Williams, G.M., 1987. Enhancement of rat liver microsomal metabolism of azoxymethane to methylazoxymethanol by chronic ethanol administration: similarity to the microsomal metabolism of N-nitrosomethylamine. **Cancer Res.** 47: 3123-3129.

Steele, C.M. and Ionnides, C., 1986. Differential effects of chronic alcohol administration to rats on the activation of aromatic amines to mutagens in the Ames test. **Carcinogenesis** 7: 825-829.

Stenback, F., 1969. The tumorgenic effect of ethanol. **Acta Pathol Microbiol Scand.** 77: 325-326.

Swann, P.F., Coe, A.M. and Mace, R., 1984. Ethanol and dimethylnitrosamine and diethylnitrosamine metabolism and disposition in the rat. **Carcinogenesis** 5: 1337-1343.

Takada, A., Neii, J., Takase, S., Matsuda, Y., 1986. Effects of ethanol on experimental hepatocarcinogenesis. **Hepatology** 6: 65-72.

Talamini, R., La Vecchia, C., DeCarli, A., Franceschi, S. and Grattoni, E., 1984. Social factors, diet and breast cancer in a northern Italian population. **Br J Cancer.** 49: 723-729.

Tamburro, C.H. and Lee, H.M., 1981. Primary hepatic cancer in alcoholics. **Clin Gastroenterol.** 10: 457-477.

Teschke, R., Minzlaff, M., Oldiges, H. and Frenzel, H., 1983. Effect of chronic alcohol consumption on tumor incidence due to dimethylnitrosamine administration. **J Cancer Res Clin Oncol.** 106: 58-64.
Tomera, J.F. Skipper, P.L., Wishnok, J.S., Tannenbaum, S.R. and Brunengraber, H., 1984. Inhibition of N-nitrosodimethylamine metabolism by ethanol and other inhibitors in the isolated perfused rat liver. **Carcinogenesis** 5: 113-116.
Toskulkao, C. and Glinsukon, T., 1986. Effect of ethanol on the in vivo covalent binding and in vitro metabolism of aflatoxin Bl in rats. **Toxicol Lett.** 30: 151-157.
Toskulkao, C., Yoshida, T., Glinsukon, T. and Kuroiwa, Y., 1986. Potentiation of aflatoxin Bl-induced hepatotoxicity in male Wistar rats with ethanol pretreatment. **J Toxicol Sci.** 11: 41-51.
Tuma, D.J., Barak, A.J., Schafer, D.F., 1973. Possible Interrelationship of ethanol metabolism and choline oxidation in the liver. **Can J Biochem.** 51: 117-120.
Tuyns, A., 1978. Alcohol and cancer. **Alcohol Health Res World.** 2: 20-31.
Tuyns, A., 1979. Epidemiology of alcohol and cancer. **Cancer Res.** 39: 2840-2843.
Tuyns, A., Masse, L.M.F., 1973. Mortality from cancer of the esophagus in Brittany. **Int J Epidemiol.** 2: 241-245.
Tuyns, A., Pequignot, G. and Jensen, O.M., 1977. Le cancer de l'esophage en Ile-et-Vilaine en fonction des niveau de consomation de alcool et de tabac. **Bull Cancer.** 64: 45-60.
Tuyns, A., Pequignot, G., Abbatucci, J.S., 1979. Esophageal cancer and alcohol consumption. Importance of type of beverage. **Int J Cancer.** 23: 443-447.
Tweedie, J.H., Reber, H., Pour, P.M. and Ponder, D.M., 1981. Protective effect of ethanol on the development of pancreatic cancer. **Surg Forum.** 32: 222-224.
Uthus, E.O., Skurdal, D.N., Cornatzer, W.E., 1976. Effect of ethanol ingestion on choline phosphotransferase and phosphatidylethanolamine methyltransferase activities in liver microsomes. **Lipids** 11: 641-644.
Van der Westhuyzen, J., 1985. Methionine metabolism and cancer. **Nutr Cancer.** 7: 179-183.
Videla, L.A. and Valenzuela, A., 1985. Alcohol ingestion, liver glutathione and lipoperoxidation; metabolic interrelations and pathological implications. **Life Sci.** 39: 3587-3591.
Walker, E.A., Castegnaro, M., Garren, L., Touissaint, G., Kowalski, B., 1979. Intake of volatile nitrosamines from consumption of alcohols. **J Natl Cancer Inst.** 63: 947-951.
Wargovich, M.J. and Felkner, I.C., 1982. Metabolic activation of DMH by colonic microsomes: a process influenced by dietary fat. **Nutr Cancer.** 4: 146-153.
Wattenberg, L.W., 1975. Inhibition of dimethylhydrazine-induced neoplasia of the large intestine by disulfiram. **JNCI** 54: 1005-1006.
Wattenberg, L.W., 1978 Inhibition of chemical carcinogenesis. **J Natl Cancer Inst.** 60: 11-18.
Webster, L.A., Layde, P.M., Wingo, P.A. and Ory, H.W., 1983. Alcohol consumption and risk of breast cancer **Lancet** 2: 724-726.
Weisburger, J.H., Yamamoto, R.S. and Pai, S.R., 1964. Ethanol and the carcinogenicity of N.hydroxy-N-2-flourenyl-acetamide in male and female rat. **Toxicol Appl Pharmacol.** 6: 363 (abstr.).
Willett, W.C., Stampfer, M.J., Colditz, G.A., Rosner, B.A., Hennekens, C.H. and Speizer, F.E., 1987. Moderate alcohol consumption and the risk of breast cancer. **NEJM** 316: 1174-1180.
Woutersen, R.A., Van Garderen-Hoetmer, A., Bax, J., Feringa A.W., Scherer, E., 1986. Modulation of putative preneoplastic foci in exocrine pancreas of rats and hamsters. I. Interactions of dietary fat and ethanol. **Carcinogenesis** 7: 1587-1593.

METABOLISM OF ETHANOL AND HIGHER ALCOHOLS PRESENT IN ALCOHOLIC DRINKS AND THEIR CORRESPONDING ALDEHYDES IN SUB-CELLULAR COMPONENTS OF RAT ESOPHAGEAL MUCOSA, AND RELEVANCE FOR ESOPHAGEAL CANCER IN MAN

V. M. Craddock

MRC Toxicology Unit
Medical Research Council Laboratories
Woodmansterne Road, Carshalton
Surrey SM5 4EF, UK

INTRODUCTION

Esophageal cancer has one of the highest world-wide frequency rates for any form of cancer, exceeding that for liver (Parkin 1988). It is also one of the most rapidly fatal cancers and one of the most unpleasant. The geographical distribution is exceptionally well demarcated, a high incidence cancer belt running from Iran to China, with other high incidence areas in South Africa, in the Transkei, in Northern France, in Switzerland and in Scotland (Day et al 1982). While the etiology must differ in different localities, in Europe and USA by far the predominant cause is consumption of alcoholic beverages. Tobacco use has a synergistic effect, but in America at least 75% esophageal cancer was estimated to depend on alcohol consumption (Rothman 1980), and in France alcohol is a much more important cause than is tobacco (Tuyns 1982).

When any chemical is incriminated as a cause of cancer, the possibility of its metabolism in the target organ becomes a leading question. In spite of the fact that consumption of alcoholic beverages has been associated with esophageal cancer for about sixty years (Stocks et al 1933), apparently no studies had been reported on the metabolism of ethanol in the esophagus.

While ethanol in any form presents a risk for esophageal cancer, many surveys have shown that, for the same amount of ethanol, the degree of risk depends on the type of beverage consumed. Spirits have been shown to present a greater risk than beer or wine in case-control studies carried out in Puerto Rico (Martinez 1969), New York (Wynder et al 1961), Washington DC (Pottern et al 1981), Los Angeles (Yu et al 1988), South Carolina (Brown et al 1988), and in France (Tuyns et al 1979).

As ethanol has not been shown to be carcinogenic in animal experiments, it has often been suggested that carcinogenicity of alcoholic beverages is due to other chemicals present in the drinks. The most hazardous drinks for esophageal cancer i.e. certain spirits, especially

Alcoholism: A Molecular Perspective
Edited by T.N. Palmer, Plenum Press, New York, 1991

apple brandy, have an outstanding characteristic, in that they contain exceptionally high levels of alcohols with molecular weights higher than that of ethanol (Postel et al 1981). These so-called 'fusel alcohols' are formed by yeast during fermentation, major components being the 2- and 3-methylbutanols and ß-phenethylalcohol. While in beer the approximate concentration of fusel alcohols is 4-5 mg/lOOml, and in wine 20-30 mg/lOOml, in Scotch it is often over 200 mg/lOOml, and in apple brandy over 300 mg/lOOml.

CYTOPLASMIC METABOLISM OF ALCOHOLS AND ALDEHYDES

The metabolism of ethanol, 3-methylbutanol, ß-phenethylalcohol, and certain other higher alcohols was studied in subcellular fractions prepared from rat esophageal mucosa and liver. The organs were homogenised in Tris buffer, pH 7.4, and subject to the usual fractional centrifugation procedures. Alcohol dehydrogenase (ADH) was determined in the cytosol fractions by measuring the substrate dependent conversion of NAD to NADH, assayed by the increase in absorption at 340 nm, using a Kontron Uvikon 860. The specific activity (SA) was calculated as the nmols NADH formed per mg protein per minute, using the amount of NADH formed during the first 30 seconds incubation. Aldehyde dehydrogenase (ALDH) was determined by measuring the substrate-dependent reduction of NAD and NADP, pyrazole being added to inhibit the reduction of the aldehydes.

Alcohol Dehydrogenases

The optimum concentration of ethanol for liver ADH was 7.5 mM, with a SA of 7.5, while for mucosa activity increased with an increase in ethanol concentration until at 2M the SA was 40 times greater than that of liver (Table 1). A similar difference between esophageal mucosa and liver was found when 3-methylbutanol and other higher alcohols were substrates. As with the liver enzyme, mucosal ADH was inhibited more effectively by 4-methylpyrazole than by pyrazole. Inhibition by 1,10-phenanthroline implied that as with other alcohol dehydrogenases the microsomal enzyme is zinc dependent.

Similar very high K_m ADH enzymes have been detected in other organs, first as a minor component in liver (Pares et al 1981), and later in the stomach of the rat (Cederbaum et al.

Table 1. ADH of liver and esophageal mucosa

	Liver		Mucosa	
Substrate	Optimum Concn. (mM)	Specific Activity[a]	Optimum Concn. (mM)	Specific Activity
Ethanol	7.5	7.5	>2000	300
3-methylbutanol	1.0	5.0	80	40
ß-phenethylalcohol	1.5	8.0	>100	40
n-butanol	2.0	8.0	200	100
12-hydroxydodecanoic acid	0.1	13.0	4.0	25

[a]SA expressed as nmol NADH formed per mg protein per minute

1975), mouse (Algar et al 1983) and baboon (Holmes et al 1986), and in the cornea (Holmes et al 1986), placenta (Pares et al 1984) and testis (Keung 1988). The role of the high K_m enzyme in organs other than those of the alimentary canal is difficult to assess, as these organs would never come into contact with sufficiently high concentrations of ethanol to be effective. The esophagus, however, is exposed to spirits with an ethanol concentration of 40%, i.e. 7M ethanol. The basal cells of the mucosa, where replication takes place, and where cancers originate, are separated by only a few cell layers from the surface of the lumen. The concentration of ethanol reaching the critical site could therefore be very high.

It has been suggested that, because solutions pass rapidly down the esophagus, direct contact with the mucosa is too transient to be important. However, our experiments on the effect of alcohols on basal cell replication have shown that a direct action has dramatic results. When ethanol containing 3-methylbutanol was given by a short intubation tube, so that it flowed through the major part of the esophagus, it triggered a large increase in basal cell replication. When given by a longer tube, so that the solution passed directly into the stomach and reached the esophagus only via the circulation, this dramatic response was not seen (Craddock and Hill, unpublished data).

Aldehyde Dehydrogenases

In contrast to the situation with alcohol metabolism, the optimum substrate concentrations for aldehyde metabolism were similar in liver and mucosa. The SA of the NAD^+ dependent mucosal acetaldehyde dehydrogenase, approx 140, was several times higher than that for liver, approx 21. The mucosa was therefore better able than liver to protect itself against this toxic metabolite by further metabolism to acetic acid and carbon dioxide. With 3-methylbutraldehyde, however, the SAs of the enzyme in two organs were similar, approx 10. Therefore although the mucosa is likely to form high levels of the 3-methylbutraldehyde from 3-methylbutanol, it is not especially able to eliminate it rapidly by further metabolism.

MICROSOMAL SYSTEMS

The MAOS (microsomal alcohol oxidising system) is of special interest in relation to esophageal cancer. This is because the only chemicals known to be potent carcinogens for the esophagus in animal experiments are the nitrosamines (Preussmann et al 1984), and human exposure to these compounds is ubiquitous. In liver, many nitrosamines are metabolised by the P450 oxidising system, ADHIIEI. As a result, alcohols act as competitive inhibitors of metabolism of these carcinogens. In the esophagus, 3-methylbutanol inhibited the metabolism of the exceptionally potent esophageal carcinogen, N-methyl-N-nitrosobenzylamine, 1000 times more effectively than did ethanol (Craddock et al 1990). The question was therefore whether the MAOS of esophageal mucosa metabolised 3-methylbutanol 1000 times more effectively than ethanol. In liver, the fusel alcohol was 100 times more effective than ethanol.

Surprisingly, it was found that the metabolism of the two alcohols showed very similar kinetics. In liver, 3-methylbutanol and ethanol had an optimum substrate concentration of around 50 mM, and a SA of 60-70. Preliminary results with the mucosal system showed that ethanol and 3-methylbutanol were oxidised at similar rates, the SA of the enzyme being lower than that for the liver MAOS. The reason why 3-methylbutanol is a much more active competitive inhibitor of nitrosamine metabolism than is ethanol remains a problem. Possibly the alcohol binds to the active site of the enzyme but is not oxidised.

HYPOTHESIS FOR ESOPHAGEAL CARCINOGENESIS

For most environmental carcinogens to be effective at the low levels of exposure, the action of a promoting factor is essential. A plausible concept is that esophageal cancer is initiated by the ubiquitous nitrosamines, exposure occurring at a time when alcohol is not being consumed, and the metabolism of the nitrosamines is therefore not inhibited . The genetically damaged initiated cells then lie quiescent, until stimulated to replicate by consumption of alcohol. Stimulation of replication is known to be a potent promoting factor in many systems (Craddock 1976). The fact that the fusel alcohols are more effective in triggering replication in the basal cells than is ethanol could explain the high risk associated with spirit consumption.

An important aspect of this work is that, if further evidence supports the idea that fusel alcohols are especially implicated in esophageal cancer, the levels present in spirits could be reduced by minor changes in fermentation procedures without decreasing the pleasurable organoleptic properties of the drinks.

ACKNOWLEDGEMENT

The author would like to acknowledge the assistance of Mr. M. Abbs throughout the course of this work, and collaboration with Professor T.J. Peters.

REFERENCES

Algar, E. M., Seeley, T. L. and Holmes, R.S., 1983, Purification and molecular properties of mouse ADH isozymes. **Eur.J.Biochem.** 137:139.

Brown, L. M., Blot, W. J., Schuman, S. H., Smith, V. M., Ershow, A.G., Marks, R. D. and Fraumeni, J. F., 1988, Environmental factors and high risk of esophageal cancer among men in coastal South Carolina. **J.Natl.Cancer Inst.**, 80:1620.

Cederbaum, A. I., Pietruszko, R., Hempel, J., Becker, F.F. and Rubin, E., 1975, Characterisation of a non-hepatic ADH from rat hepatocellular carcinoma and stomach. **Arch.Biochem.Biophys.** 171:348.

Craddock, V. M. and Henderson, A. R., 1990, Potent inhibition of esophageal metabolism of methylbenzylnitrosamine, an esophageal carcinogen, by higher alcohols present in alcoholic beverages, **in**: 'Relevance to Human cancer of N-Nitroso Compounds, Tobacco Smoke and Mycotoxins'. O'Neill, I. K., Chen, J. S. and Bartsch, H. (eds.) IARC Scientific Publication No. 105. Lyon, pp 564-567.

Craddock, V. M., 1976, Cell proliferation and experimental liver cancer, **in**: 'Liver Cell Cancer'. Cameron, H. M., Linsell, C. A. and Warwick, G.P. (eds.) Elsevier Public. 153-201.

Day, N. E., Munoz, N. and Ghadirian, P., 1982, Epidemiology of esophageal cancer: a review **in:** 'Epidemiology of Cancer of the Digestive Tract.' Correa, P. and Haenszel, W. (eds.) Martinus Nijkoff Public. 21-57.

Holmes, R. S., Courtney, Y. R. and VandeBerg, J. L., 1986, Alcohol dehydrogenase isozymes in baboons: tissue distribution, catalytic properties, and variant phenotypes in liver, kidney, stomach and testis. **Alcoholism: Clinical and Exp. Research.** 10:623.

Holmes, R. S. and VandeBerg, J. L., 1986, Ocular NAD-dependent alcohol dehydrogenase and aldehyde dehydrogenase in the baboon. **Exp.Eye Res.** 43:383.

Keung, W. M., 1988, A genuine specific alcohol dehydrogenase from hamster testis: isolation, characterisation and developmental changes. **Biochem.Biophys. Res.Com.** 156:38.

Martinez, I. 1969. Factors associated with cancer of the esophagus, mouth and pharynx in Puerto Rico. **J.Natl.Cancer Inst.**, 42:1069.

Pares, X. and Vallee, B. L., 1981, New human liver ADH forms with unique kinetic characteristics. **Biochem.Biophys.Res.Com.** 98:122.

Pares, X., Farres, J. and Vallee, B. L., 1984, Organ specific alcohol metabolism: placental χ-ADH. **Biochem.Biophys.Res.Com.** 119:1047.

Parkin, D. M., Laara, E. and Muir, C.S., 1988, Estimates of the world-wide frequency of sixteen major cancers in 1980. **Int.J.Cancer.** 41:184.

Postel, von W., Adam, L. and Jager, K. H., 1981, Herstellung und zusammensetzung von Calvados. **Die Branntweinwirtschaft,** 162-167.

Pottern, L. M., Morris, L. E., Blot, J., Ziegler, R. C. and Fraumeni, J. F., 1981, Esophageal cancer among black men in Washington DC. I. Alcohol, tobacco and other risk factors. **J.Natl. Cancer** Inst., 67:777.

Preussman, R. and Stewart, B. W. 1984. N-Nitroso Carcinogens, **in**: 'Chemical Carcinogens'. Searle, C. E. (ed.) ACS Monograph No. 182, Vol. 2. Am. Chem. Soc. Press, pp 643-828.

Rothman, K. J., 1980, The proportion of cancer attributable to alcohol consumption. **Preventive Medicine** 9:174.

Stocks, P. and Kay, M. N., 1933, A cooperative study of the habits, home life, dietary and family histories of 450 cancer patients and of an equal number of control patients. **Ann.Eugen.** 5:227.

Tuyns, A. J., 1982, Epidemiology of esophageal cancer in France, **in**: 'Cancer of the Esophagus'. Pfeiffer, C. J. (ed.) CRC Press, Florida, Vol.1, pp. 3-18.

Tuyns, A. J., Pequignot, G. and Abbatucci, J. S., 1979, Oesophageal cancer and alcohol consumption: importance of type of beverage. **Int. J. Cancer**, 23:443.

Wynder, E. L. and Bross, I. J., 1961, A study of etiological factors in cancer of the esophagus. **Cancer**, 14:389.

Yu, M. C., Garabrant, D. H., Peters, J. M. and Mack, T. M., 1988, Tobacco, alcohol, diet, occupation and carcinoma of the esophagus. **Cancer Res.**, 48:3843.

ACTIVATION BY HUMAN GASTRIC MUCOSA OF ALIMENTARY PROCARCINOGENS

R. Cardin[1], M. Zordan[1], F. Farinati[2], R.Naccarato[2] and A.G. Levis[1]

[1]Dipartimento di Biologia
Universita' degli studi di Padova via Trieste 75
35100 Padova, Italy
[2]Cattedra malattie apparato digerente
Istituto di Medicina Interna
Padova, Italy

INTRODUCTION

Alimentary Habits and Gastric Carcinogenesis

Epidemiologic evidence has been presented over the last 10 to 15 years, that the process of gastric carcinogenesis is directly correlated with alimentary habits. The first suggestion of the importance of environmental factors in the induction of gastric tumors was provided by the observation of a large variability in tumor incidence among different countries or areas that could not be explained by racial differences (American Cancer Society, 1982). Haenszel, Correa (1978), Wynder, Hiryama (1977) and McMichael (1980), reported changes in incidence for migrants from Japan via Hawaii to the mainland United States; from Eastern and Northern Europe to the United States and Canada; and from Europe generally to Australia. Migrants from any part of the world with a low tumor incidence such as Poland, South America, Japan, or southern Europe to regions with a higher risk such as the Anglo-Saxon countries, show a rapid increase in the tumor incidence.

Dietary risk factors in high-risk populations are exemplified by a high consumption of dried or salted fish, smoked fish, pickled vegetables, and a low intake of vegetables, as well as a reduced vitamin C intake (Cuello et al.,1976; Hill et al.,1983). Elevated levels of nitrates in foods and drinking water, such as occur in Colombia (Cuello et al.,1976) due to high levels in the soil and water supplies, have been correlated with the induction of gastric cancer. Another source of nitrates may arise from the use of crude salt in the preservation of certain foods such as fish (Archer,1982). Following consumption of certain foods, particularly vegetables, such as celery, spinach and beets, a dramatic increase in nitrate concentration is observed in human saliva (Spiegelhalder et al.,1976). The most relevant exposures to N-nitroso compounds (in particular N-nitrosamines) probably occurs either through the ingestion of foods or drugs contaminated with these compounds or through their in vivo intragastric synthesis. In this connection, a sequence of events has been proposed according to which alimentary amines or amides react in the stomach with nitrates whose presence is due to the reduction of nitrites in the oral cavity or in the stomach. This reduction is carried out by bacteria normally present in saliva and in the stomach.

Etiology of Gastric Cancer

Gastric cancer recognizes multiple etiopathogenetic factors, of which only a part are well identified. The possible chain of events which alter the gastric mucosa phenotypically and ultimately lead to cancer development has been accurately described. The modification stages in the so-called intestinal gastric cancer, which is the most diffuse form of gastric cancer in high-risk populations (Lauren,1965; Ming,1977), involve the shift from normal mucosa to chronic atrophic gastritis, intestinal metaplasia, epithelial dysplasia and, eventually, cancer (Correa,1988).

The occurrence of a pathological elevation in the pH of gastric juice is followed by the colonization of the stomach by bacteria that normally do not survive the acidic pH of the stomach, and this may lead to an increase in bacterial-mediated intragastric synthesis of N-nitroso compounds (Tannenbaum et al.,1977).

The quantity of N-nitroso compounds synthesized is a function not only of the amount of the amines involved, but also of nitrite concentration. For this reason chronic atrophic gastritis, through its influence on gastric pH and gastric juice nitrite concentration, may directly increase the synthesis of carcinogenic compounds.

A number of studies on gastric juice provided new evidence for the mechanisms described earlier. For example, under certain conditions, gastric juice was shown to exert a mutagenic activity, as determined by the Ames/Salmonella auxotroph reversion assay (Mirvish,1983). This was first shown in subjects living in high gastric cancer risk areas of Colombia, where the finding of mutagenic activity was correlated with the high nitrite content of gastric juice (Montes et al.,1979). Similar results were reported for subjects affected by chronic atrophic gastritis or gastric cancer (Cuello et al.,1979; Morris et al.,1984) in whom the gastric juice pH is abnormally high.

We recently confirmed these data and showed that patients with chronic atrophic gastritis and gastric cancer are also characterized by increased urinary mutagen excretion, which seems to be secondary to increased intragastric mutagen synthesis (Farinati et al.,1985).

N-NITROSO COMPOUNDS AS POSSIBLE HUMAN CARCINOGENS

N-nitroso compounds formed by nitrosation fall into two basic groups: the relatively unstable nitrosamides, such as N-methyl-N-nitro-N-nitrosoguanidine (MNNG), which exert a direct carcinogenic activity as they do not require metabolic activation in order to produce their effects, and the relatively more stable nitrosamines, such as dimethylnitrosamine (DMN), which require metabolic transformation into reactive electrophiles to exert their effects. Nitrosamines may induce a systemic carcinogenic effect, and after oral administration and activation, particularly when given in small doses over a long period of time, they induce tumors in the esophagus, the forestomach and nasal cavity, as well as in brain, lung and urinary bladder (Guttenplan.,1987). Nitrosamines, like all xenobiotics, are metabolized largely in the liver, where they undergo a first metabolic step catalyzed by one or more of the many cytochrome P-450 dependent mixed-function oxidases present in the microsomal fraction of cellular extracts. The second metabolic step may involve a number of conjugation reactions, leading to the ultimate inactivation of the compounds and facilitating their excretion.

Normally the gastric mucosa is considered to be the passive target of toxic agents. Thus the possibility that the gastric mucosa itself could play an active role in the activation of carcinogens seems particularly interesting. In this respect, detectable cytochrome P-450 activity has been reported for the rat stomach (Farinati,1985).

Our group previously demonstrated that, in rats, the mucosa of the upper digestive tract is capable of activating N-nitrosopyrrolidine (N-NPY) to a mutagenic derivative (Farinati,1985) and that rat stomach has the capacity for both enzymatic activation and detoxification (Farinati,1989). This capacity, which depends on cytochrome P-450 activity and on glutathione (GSH) and uridine-diphosphoglucuronic acid (UDPG) transferase activities, may be modified by the exposure to known modulators, such as ethanol (Farinati,1985). These findings may be of biological relevance since gastric mucosa is a site where xenobiotics can remain for varying periods of time, during which they may be at least partly absorbed and metabolized. The latter considerations are of particular relevance since, as above mentioned, under specific conditions such as chronic atrophic gastritis or gastric cancer, the stomach may contain appreciable amounts of mutagenic compounds. Thus it would seem important to evaluate the capacity of the gastric mucosa to activate N-nitrosamines to genotoxic species in order to better understand the mechanisms underlying their involvement in the process of gastric carcinogenesis.

STUDIES ON GASTRIC MUCOSA

By means of the Ames/Salmonella reversion assay (Maron et al.,1983) we tested the capacity of human gastric mucosa post-mitochondrial fractions to activate 2 carcinogenic amines: 2-aminofluorene (2- AF), an aromatic amine which is a potent mutagen/carcinogen requiring metabolic activation, and N-NPY, a cyclic nitrosamine which is a potent carcinogen occurring in food products and tobacco smoke (Hecker et al.,1979; Montesano et al.,1976).

Human gastric mucosa enzymatic fractions were found able to activate 2-AF in the Ames test on Salmonella typhimurium strain TA100 which reveals mainly base substitution mutagens, whereas negative results were obtained for N-NPY on Salmonella typhimurium strain TA98 which preferentially detects frame-shift mutagens (Zordan et al.,1988). In these experiments as a control, human liver post-mitochondrial fractions, non-induced and Aroclor induced rat liver post-mitochondrial fractions were assayed in the Ames test, in the same experimental conditions.

It is common knowledge that nitrosamine are a class of compounds that are activated in vitro with difficulty by post-mitochondrial fractions of non-induced tissues. Therefore, it should be stressed that N-NPY is a much less potent mutagen and, thus, a low capacity of activation by gastric mucosa fractions could have been missed.

The fact that gastric mucosa post-mitochondrial fractions were characterized by a lower but significant activating capacity as compared to human liver does not weaken the hypothesis of a direct involvement of the gastric mucosa in the activation of procarcinogens. Gastric mucosa is probably exposed to much higher concentrations of carcinogens than the liver and, therefore, even a much lower activation capacity may be of extreme biological relevance.

STUDIES ON HUMAN ENZYMATIC SYSTEMS

The suggestion that the gastric mucosa may be exposed to higher concentrations of xenobiotics, which might result in the induction of the enzymes involved in their metabolism, prompted us to evaluate directly in vitro the capacity of two cytochrome P-450 related enzymes, N-nitrosodimethylamine (DMN) and aminopyrine (AP) demethylase.

The determination of the enzymatic activities allows the indirect evaluation of the capacity of human tissue to activate procarcinogens to genotoxic species in vitro. Only limited quantities of tissue are required and the assays employed can measure enzymatic reaction products in the nmole range.

The activities of the two demethylases were assayed in 9000xg supernatants obtained from homogenates of 10 human gastric mucosa samples by spectrophotometric determination of formaldehyde produced in the cytochrome P-450 dependent demethylation of the two substrates (Zordan et al.,1989). DMN and AP demethylase activities were also determined in parallel in 10 human liver samples of surgical origin (as in the case of gastric specimens) and in Aroclor induced and non-induced rat liver (Zordan et al.,1988).

The mean activating capacity of the 10 gastric mucosa post-mitochondrial fractions shows that the AP demethylase and, to a lesser extent, DMN demethylase activities are similar, if not higher, in gastric mucosa as compared to human liver.

COMMENT

Although the parameters evaluated can only be considered as a partial measure of the general activating capacity toward dietary and environmental procarcinogens, these results suggest that the metabolizing capacity of human gastric mucosa is at least comparable to that of human liver, and could therefore represent a biologically relevant site of activation of ingested procarcinogens.

REFERENCES

American Cancer Society 1982: Cancer Facts and Figures. American Cancer Society, New York.

Archer, M.C., 1982, **Nutrit Toxicol**, 1:327-381.

Correa, P. and Haenszel, W., 1978, **Adv. Cancer Res.**, 26:1-141.

Correa, P., 1988, A human model of gastric carcinogenesis. **Cancer Res.** 48: 3554-3560.

Cuello, C., Correa, P., Haenszel, W., Gordillo, G., Brown C., Archer, M. and Tannenbaum, S., 1976, Studies of gastric cancer in Colombia. I. Cancer risk and suspect environmental agents. **J.Natl.Cancer. Inst**, 57: 1015-1020.

Cuello, C., Correa, P., Zorama, G., Lopez, J., Murray, J. and Gordillo, G., 1979, Histopathology of gastric dysplasias. Correlation with gastric juice chemistry. **Am.J.Surg.Pathol.**, 3: 491-500.

Farinati, F., Worner, Y.M., Lima, V., Lieber, C.S. and Garro, A.J., 1985, Mutagenic activity in gastric juice and urines of subjects with chronic atrophic gastritis, gastric cancer and alcoholism. **Gastroenterology**, 85(5): 1378.

Farinati, F., Zhou, Z.C., Bellah, J., Lieber, C.S. and Garro, A.J., 1985, Effect of chronic ethanol consumption on activation of nitrosopyrrolidine to a mutagen by rat upper alimentary tract. **Drug Metab. Dispos.**, 13: 210-214.

Farinati, F., Lieber, C.S. and Garro, A.J., 1989, Effects of chronic ethanol consumption on carcinogen activating and detoxifying systems in rat upper alimentary tract tissue. **Alcoholism Clin. Exp. Res.**, 13: 357-360.

Guttenplan, J.B., 1987, N-nitrosamines: bacterial mutagenesis and in vitro metabolism. **Mutation Res.**, 81: 134.

Hecker, L.I., Elespuru, R.K. and Farrelly, J.G., 1979, The mutagenicity of nitrosopyrrolidine is related to its metabolism. **Mutation Res.**, 62: 213-220.

Hill, M.J., 1983, **In**: Sherlock, P., Morson, B.C., Barbara, L. and Veronesi, U. (Eds.). Precancerous Lesions of the Gastrointestinal Tract. Raven Press, New York,1-22.

Lauren, P., 1965, The two histologic main types of gastric carcinoma: diffuse and so-called intestinal type carcinoma. An attempt at a histoclinical classification. **Acta Pathol. Microbiol Scand.**, 64: 31-49.

Maronm, D.and Amesm, B.N., 1983, Revised methods for the Salmonella mutagenicity test. **Mutation Res.**, 113: 173-215.

McMichael, A.J., McCall, M.G., Hartshorne, J.M. and Woodings, T.L., 1980, **Int. J. Cancer**, 25: 431-437.

Ming, S.C., 1977, Gastric carcinoma. A pathobiologic classification. **Cancer**, 39: 2475-2484.

Mirvish, S.S., 1983, The etiology of gastric cancer. Intragastric nitrosàmide formation and other theories. **J.Natl.Cancer Inst.**, 3: 631-647.

Montes, G., Cuello, C., Gordillo, G., Pelon, W., Johnson, W. and Correa, P., 1979, Mutagenic activity of gastric juice. **Cancer Lett.**, 7: 307-312.

Montesano, R. and Bartsh, H., 1976, Mutagenic and carcinogenic N- nitroso compounds: possible environmental hazards. **Mutation Res.**, 32: 179-228.

Morris, D.L., Yonngs, D., Muscroft, T.J., Cooper, J., Rojinski, C., Burdon, D.W. and Keighley, M.R.B., 1984, Mutagenicity in gastric juice. **Gut**, 25: 723-727.

Spiegelhalder, B., Eisenbrand, G. and Preussmann, R., 1976, Influence of dietary nitrate on nitrite content of human saliva: Possible relevance to in vivo formation of N-nitroso compounds. **Food Cosmet. Toxicol.**, 14: 545-548.

Tannenbaum, S.R., Archer, M., Wishnok, J.S., Correa, P., Cuello, C. and Haenszel, W., 1977, Nitrate and the etiology of gastric cancer, **In**: 'Origins of Human Cancer, Book C, Human Risk Assessment'. Hiatt H.H. Watson J.D. and Winsten J.A. (Eds.) 1609-1625.

Wynder, E.L., Hirayama, T., 1977, **Prev. Med.**, 6: 567-594.

Zordan, M., Farinati F., Nitti D., Levis, A.G. and Naccarato, R.,1988, Normal and pathologic human liver Dimethylnitrosamine (DMN) demethylase activity and capacity for 2-Aminofluorene (2-AF and Nitrosopyrrolidine) NPY activation in vitro. Joint Meeting of the British Toxicology Society and The Societa' Italiana di Tossicologia. Mechanisms of Toxicity and Repair. Venice,23-24 June 1988.

Zordan, M., Cardin, R., Farinati, F., Naccarato, R. and Levis, A.G., 1989, Dimethylnitrosamine and Aminopyrine N-demethylase activity of human gastric mucosa. Atti dell'VIII Congresso della Societa' Italiana di Tossicologia Bologna, 2-5 May 1989.

Zordan, M., Farinati, F., Cardin, R., Nitti, D., Naccarato, R. and Levis, A.G., 1988, In vitro transformation of N-nitrosodimethylamine (DMN) and N-Nitrosopyrrolidine (N-NPY to active genotoxic compounds by normal and pathologic human liver samples. **Atti Ass. Gen. It.**, 33.

EFFECT OF ETHANOL ON GASTROINTESTINAL CELL PROLIFERATION

Ulrich A. Simanowski, Burkhard Kommerell and Helmut K. Seitz

Division of Gastroenterology
Dept. of Internal Medicine, University of Heidelberg
Heidelberg, FRG

INTRODUCTION

Cell proliferation is an important measure and characteristic of tissues. Its evaluation must be quantitative as well as morpological, relating proliferative activity to histological structure. Cell proliferation is dynamic, measurement of it should include time, moreover it can even change with time. Therefore, one time point measurements are always equivocal, and even observations over time must neglect acceleration or slowing of proliferative activity, unless experiments are tremendous laborious with multiple time point observations.

In view of the before mentioned it is not surprising, that there are only sporadic and often indirect observations of cell proliferation in the G.I. tract relating to ethanol treatment. In general, there are two different entities of changes of mucosal cell proliferation. One is related to tissue damage and indicates reparative growth i.e. healing, and the other field is concerned with the acute and chronic proliferative changes during cancer development. Since ethanol itself is not a carcinogen but may rather have some tissue damaging and cocarcinogenic potential those two fields may well be connected with each other: It is evident that epithelilal hyperproliferation is often associated with enhanced tumor development. This was found to be true after ligature insertion in the rat cecum (Pozharisski 1975), after large bowel transsection and reanastomosis (Roe et al 1987), jejunoileal bypass (Bristol et al 1982), colostomy closure (Terpstra et al 1981), in *Citrobacter freundii* infected mice (Barthold and Jonas 1977), after cholic acid feeding (Cohen et al 1980), and following small bowel resection, leading to colonic hyperproliferation and hyperplasia (Williamson 1979; Williamson and Malt 1980). In humans this correlation of hyperproliferation with tumor promotion seems to be true as well: In ulcerative colitis (Bleiberg et al 1970; Collins et al 1987), coeliac disease (Wright et al 1973; Swinson et al 1983), familial colon cancer and familial polyposis (Lipkin 1981; 1984). Perhaps supporting the hypothesis that epithelial hyperproliferation leads to tumor enhancement are observations of reduced tumor incidences in defunctioned bowel segments with hypoproliferative states (Cleveland et al 1967; Wittig et al 1971; Gennaro et al 1973; Rubio 1980). However, cell proliferation is just a phenomenon of tissues, its measurement and morphological characteristics can only give indirect clues of its regulation or underlying molecular mechanisms.

Alcoholism: A Molecular Perspective
Edited by T.N. Palmer, Plenum Press, New York, 1991

Alcohol in the gastrointestinal tract can exert its action on the mucosal epithelium in two different ways. Locally from the lumen after ingestion and systemically via the blood stream after absorption. The first modus of action is only possible shortly, that is about 2 hours after oral uptake, and only from the oral cavity down to the jejunum. Thereafter, from the ileum downstream, the mucosal cells are solely exposed to ethanol via the blood stream (Halsted et al 1973).

UPPER GASTROINTESTINAL TRACT

In the oral cavity of rabbits direct mucosal damage has been reported (Mueller et al 1983), which in turn may lead to induction of reparative growth. Labelling indices were shown to be increased in the rat esophagus after chronic alcohol treatment (Mak et al 1987; Haentjens et al 1987), even without detectable mucosal damage. However birth rate measurements in the rat esophagus after vincristine was unchanged following chronic ethanol treatment (Simanowski et al 1986). This discrepancy may be due to prolongation of the S-phase, leading to increased labelling indices without effect on overall cell turnover. Approximately 12 hours after acute alcohol ingestion labelling and mitotic indices are rising above base line levels in the rat esophagus (Haentjens et al 1987). There is histological damage to the rat gastric mucosa after exposure to as little as 10% ethanol concentrations (Kvietys et al 1990). In man endoscopically diagnosed erosions and hemorrhages in the stomach and duodenum have been described after alcohol ingestion (Gottfried et al 1978), as well as histologic disturbance of duodenal structure (Lev et al 1980) and shortage of villi (Seitz et al 1985). This seems to be associated with an initial depression and subsequent (8 to 12 hours later) increase of DNA synthesis, as indicated in in vitro labelling, and incorporation studies with tritiated thymidine (Chen et al 1981; Seitz et al 1983). This sequence is thought to represent initial toxic epithelial damage followed by secondary reparative growth. In the rat intragastric ethanol administration produced hemorrhagic erosions in stomach, duodenum and proximal jejunum (Baraona et al 1974). Whereas chronic alcohol treatment induced no changes in crypt cell production rates in the rat stomach or duodenum (Simanowski et al 1986). Together with carcinogen treatment ethanol may or may not influence labelling indices in the rat stomach, depending on time of observation and part of the stomach (Iishi et al 1989).

There are numerous and sometimes conflicting reports concerning alcohol induced changes in the jejunum and ileum, covering morphology and proliferative indices like mitotic and labelling index. Most data are derived from animal experiments. After chronic ethanol feeding of rats Baraona et al (1974) demonstrated shortage of jejunal villi, jejunal and ileal crypt hyperplasia, and increased mitotic indices only in the ileum, but not in the jejunum. In that study small intestinal villus cells of chronically alcohol fed rats exhibited enzyme activities characteristic for immature cells, which may also indicate sustained mucosal injury due to ethanol. Although they did not observe small intestinal damage in light microscopy, Rubin et al (1972) demonstrated ultrastructural aberrations in electron microscopy in similarly treated animals. Others (Zucoloto and Rossi 1979) also demonstrated smaller intestinal villi after chronic ethanol feeding but smaller cryts as well and in addition decreased mitotic indices in the jejunum and ileum. Results from thymidine incorporation studies on isolated small intestinal cells would support the presence of increased epithelial proliferation after chronic alcohol feeding (Seitz et al 1982), whereas more reliable evaluation of small intestinal cell proliferation by the dynamic metaphase arrest method showed decreased crypt cell production rates following chronic ethanol ingestion in the jejunum as well as ileum (Mazzanti and Jenkins 1984). However, with a similar investigation we could not detect any change of ileal morphology and cell proliferation in chronically ethanol fed rats (Simanowski et al 1986).

Since long it was suspected that ethanol is cocarcinogenic to the rectum (for review see Seitz and Siamanowski 1989). This is strongly supported by two recent prospective epidemiologic studies (Pollack et al 1984; Wu et al 1987). Considering an adenoma-carcinoma sequence in colonic tumor development it is of special interest that there seems to be an association between alcohol (beer) consumption and the incidence of colonic adenomas (Kikendall et al 1989). In the rat alcohol model as well ethanol seems to be cocarcinogenic (Seitz et al 1984; 1990); and because tumor enhancement often is associated with epithelial hyperregeneration, we focused on the colon, investigating the influence of chronic ethanol consumption on mucosal renewal.

Labelling index and mitotic index are one time point observations and therefore state measurements. Both indices not only depend on mucosal turnover but also on the duration of

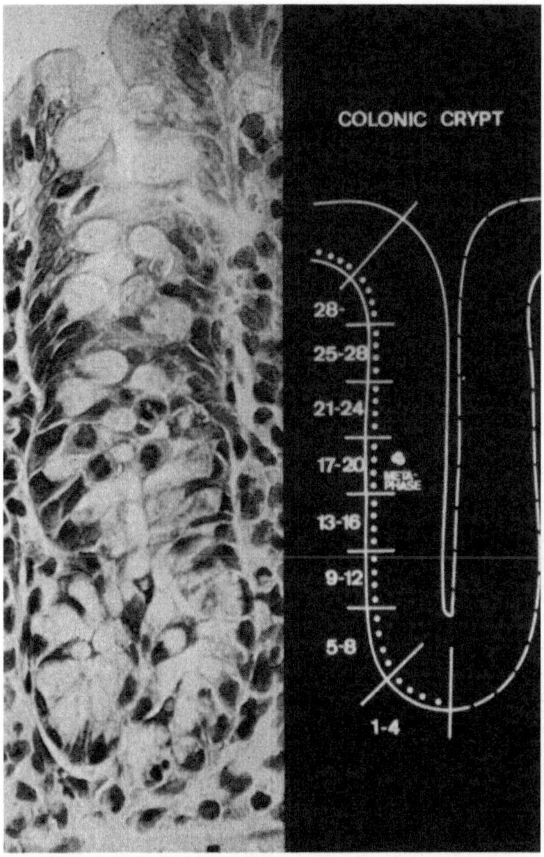

Fig.1. Histological section of colonic crypt after vincristine treatment (left hand side). There are four metaphase figures displaced towards the crypt lumen. Schematic drawing of a crypt (right hand side). Cell positions are counted from the middle of the crypt base upwards towards the intestinal lumen. Places of metaphase figures are related to positions of interphase nuclei situated close to the basement membrane.

the recorded phase. To compare groups, duration of S-phase or mitotic phase respectively have to be determined, which is difficult to achieve (Wright and Alison 1984). We therefore chose the stathmokinetic metaphase arrest technique with vincristine, which allows a dynamic measurement of epithelial renewal (Goodlad and Wright 1982). Animals are injected with 0.6 to 1.0 mg vincristine/kg body weight and seqentially sacrificed thereafter up to four hours. 8 to 16 animals per group are sufficient depending on interanimal variation. Tissues are fixed either in formalin for histological sections, or in Carnoy's fixative (6 parts ethanol, 2 parts chloroform, 1 part acetic acid) for microdissection and squash preparations. Since vincristine prohibits microtubule formation mitosis can't exceed the stage of metaphase, and metaphases accumulate with time in a linear fashion for a given cell population. From squash preparations of whole crypts the crypt cell production rate (cells/crypt/hour), and from histological sections the birth rate (fig.1) (cells/1000 cells/hour) can be calculated after linear regression analysis of accumulating metaphase numbers over time from the slope of the regression line.

In our experiments (Simanowski et al., 1986) rats were pair-fed nutritionally adequate liquid diets, thirty six percent of total calories given either as ethanol or isocaloric carbohydrates as described by Lieber and DeCarli (1970) and modified by Seitz et al (1984). Crypt cell production rates were unchanged in the proximal colon, but markedly elevated in the distal colon of the ethanol group (fig.2). This was accompanied by an extension of the proliferative compartment and reduced life span of functional epithelial cells. Because age itself increases intestinal cell proliferation in the small as well as large bowel (Holt and Yeh

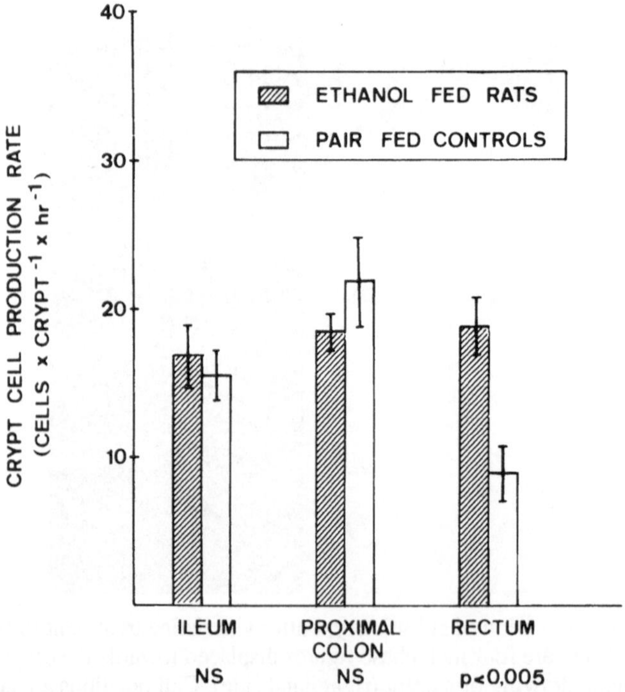

Fig.2. Crypt cell production rates in the ileum, proximal colon, and rectum of chronically ethanol fed rats and controls.

1989; Holt and Yeh 1988), and large bowel cancer is primarily a disease of advanced age, we studied large intestinal cell proliferation after chronic ethanol treatment in rats of different age (3, 14, and 23 month). Cell proliferation in the proximal colon was uneffected by age as well as ethanol. Crypt cell production rates in the rectum was increased after chronic ethanol treatment: 134% of controls in young rats, 182% in medium age, and 239% in old rats (fig.3). Therefore the ethanol related stimulation of rectal cell proliferation seems to be especially effective in old age (Simanowski et al 1990). A hint that this may be also true in humans can be derived from data showing a strong association of beer drinking and colonic adenomas in older subjects (Kikendall et al 1989).

In an attempt to elucidate the mechanism of this ethanol related rectal hyperregeneration we focused on acetaldehyde, the major ethanol metabolite. With its known toxicity (Salaspruo and Lindros 1985) acetaldehyde seems to be a good candidate for the transmission of ethanol related tissue damage. After ethanol challange we recorded surprisingly high acetaldehyde levels in the rectum, even exceeding those measured in the liver (Seitz et al 1990). From studies with germ free and conventional animals it was concluded, that some of the acetaldehyde may be produced by fecal bacteria (Seitz et al 1990).

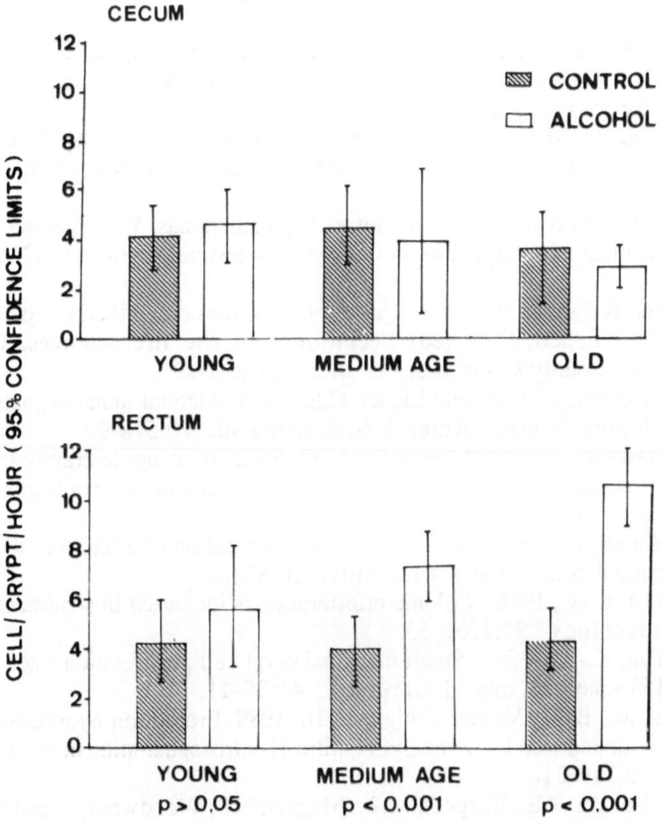

Fig.3. Crypt cell production rates in the cecum and rectum of chronically ethanol fed F344 rats and controls of different age. Young = 3 month; medium age = 14 month; old = 23 month.

Another hypothesis for alcohol/acetaldehyde related rectal cocarcinogenesis may be acetaldehyde induced local folate deficiency (Shaw et al 1989; Rosenberg 1989).

REFERENCES

Baraona, E., Pirola, R.C. and Lieber, C.S., 1974. Small intestinal damage and changes in cell population produced by ethanol ingestion in the rat. **Gastroenterology** 66:226-34.

Barthold, S.W. and Jonas, A.M., 1977. Morphogenesis of early 1,2dimethylhydrazine-induced lesions and latent period of colon carcinogenesis in mice by a variant of Citrobacter freundii. **Cancer Res.**, 37:4352-60.

Bristol, J., Wells, M. and Williamson, R.C.N., 1982. Jejunoileal bypass stimulates cell proliferation and enhances experimental carcinogenesis in the rat large bowel. (abstr.) **Proc Am Assoc Cancer Res.**, 23:233.

Bleiberg, H., Mainguet, P., Galant, P., Chretiell, J. and Dupont-Marisse, N., 1970. Cell renehal in the human rectum; in vitro autoradiographic study on active ulcerative colitis. **Gastroenterology** 58:851-5.

Chen, T., Kiernan, T. and Leevy, C.M., 1981. Ethanol and cell replication in the digestive tract. Clinics in **Gastroenterology** 10:343-54.

Cleveland, J.C., Litvak, S.F. and Cole, J.W., 1967. Identification of the route of action of the carcinogen 3,2-dimethyl-aminobiphenyl in the induction of intestinal metaplasia. **Cancer Res.**, 27:708-12.

Cohen, B.I., Raicht, R.F., Deschner, E.E., Takahashi, M., Sarwal, A.N. and Fazzini, E., 1980. Effect of cholic acid feeding on N-methyl-N-nitrosourea-induced colon tumors and cell kinetics in rats. **JNCI.**, 64:573-8.

Collins, R.H., Feldman, M. and Fordtran, J.S., 1987. Colon cancer, dysplasia, and surveillance in patients with ulcerative colitis; a critical review. **N Engl J Med.**, 316:1654-8.

Gennaro, A.R., Villaneauva, R., Sukon-Haman, Y., Vathanphas, V. and Rosemond, G.P., 1973. Chemical carcinogenesis in transposed intestinal segments. **Cancer Res.**, 33:536-41.

Goodlad, R.A. and Wright, N.A., 1982. Quantitative studies on epithelial replacement in the gut. In: Titchell, D.A. (ed) **Techniques in the life sciences. Digestive physiology**. County Clare: Elsevier Biomedical, 1-23.

Gottfried, E.B., Korsten, N.M.A. and Lieber, C.S., 1978. Alcohol induced gastric and duodenal lesions in man. **Amer J Gastroenterol.**, 70:578-92.

Haentjens, P., DeBaker, A. and Willems, G., 1987. Effect of an apple brandy from Normandy and of ethanol on epithelial cell proliferation in the esophagus of rats. **Digestion** 37:184-92.

Halsted, C.H., Robles, E.A. and Mezey, E., 1973. Distribution of ethanol in the human gastrointestinal tract. **Am J Clin Nutr.**, 26:831-4.

Holt, P.R. and Yeh, K.-y., 1988. Colonic proliferation is increased in senescent rats. **Gastroenterology** 95:1556-63.

Holt, P.R. and Yeh, K.-y., 1989. Small intestinal crypt cell proliferation rates are increased in senescent rats. **J Gerontol.**, 44:B9-11.

Iishi, H., Tatsuta, M., Baba, M. and Taniguchi, H., 1989. Promotion by ethanol of gastric carcinogenesis induced by N-methyl-N'-nitro-N-nitrosoguanidine in Wistar rats. **Br J Cancer**., 59:719-21.

Kikendall, J.W., Bowen, P.E., Burgess, M.B., Magnetti, C., Woodward, J. and Langenberg, P. 1989. Cigarettes and alcohol as independent risk factors for colonic adenomas. **Gastroenterology** 97:660-4.

Kvietys, P.R., Twohig, B., Danzell, J. and Specian, R.D.,. 1990. Ethanol-induced injury to the rat gastric mucosa. **Gastroenterology** 98:909-20.

Lev, R., Thomas, E., Parl, F.F. and Pitchumoni, C.S., 1980. Pathological and histomorphometric study of the effects of alcoholic fatty liver. **Am J Clin Nutr.**, 23:474-8.

Lieber, C.S. and DeCarli, L.M., 1970. Quantitative relationship between the amount of dietary fat and the severity of alcoholic fatty liver . **Am J Clin Nutr.**, 23 :474-8.

Lipkin, M., 1981. Early identification of population groups at high risk for gastrointestinal cancer. In: Malt, R.A., Williamson, R.C.N., (eds) **Colonic carcinogenesis.** MTP Press, Lancaster, 31-46.

Lipkin, M., 1984. Method for binary classification and risk; assessment of individuals with familial polyposis based on 3H-TdR labelling of epithelial cells in colonic crypts. **Cell Tissue Kinet.**, 17:209-22.

Mak, K.M., Leo, M.A. and Lieber, C.S. 1987. Effect of ethanol and vitamin A difficiency on epithelial cell proliferation and structure in the rat esophagus . **Gastroenterology**, 93: 362-70.

Mazzanti, R. and Jenkins,W.J., 1984. Effect of chronic ethanol ingestion on enterocyte turnover in the small intestine. Abstract. **British Society of Gastroenterology**. Autumn Meeting.

Mueller, P., Hepke, B., Meldau, U. and Raabe, G., 1983. Tissue damage in the rabbit oral mucosa by acute and chronic direct toxic action of different ethanol concentrations . **Exp Pathol.**, 24 : 171-81.

Pollack, E.S., Nomura, A.M.Y., Heilbrun, L.K., Stemmermann, G.N. and Green, S.B. 1984. Prospective study of alcohol consumption and cancer. **N Engl J Med.**, 310: 617-21.

Pozharisski, K.M., 1975. Morphology and morphogenesis of experimental epithelial tumors of the intestine. **JNCI.**, 54 :1115-35.

Roe, R., Fermor, B. and Williamson, R.C.N., 1987. Proliferative instability and experimental carcinogenesis at colonic anastomoses. **Gut** 28 : 808-15.

Rosenberg, I.H. and Mason, J.B., 1989 . Folate, dysplasia, and cancer. **Gastroenterology** 97: 502-3.

Rubin, E., Rybak, B.J., Lindenbaum, J., Gerson, C.D., Walker, G. and Lieber, C.S., 1972. Ultrastructural changes in the small intestine induced by ethanol. **Gastroenterology** 63 : 801-14.

Rubio, C.A., 1980. Experimental colon cancer in the absence of intestinal contents in Sprague Dawley rats . **JNCI.**, 64: 569-72.

Salaspuro, M. and Lindros, K., 1985. Metabolism and toxicity of acetaldehyde. In: **Alcohol related diseases in gastroenterology**. Seitz, H.K., Kommerell, B., eds., Springer Verlag, Berlin, Heidelberg.

Seitz, H.K. and Simanowski, U.A., 1989. Ethanol and colorectal carcinogenesis. In: Seitz, H.K., Simanowski, U.A., Wright, N.A. (eds) **Colorectal cancer. From pathogenesis to prevention.** Springer, Berlin, 177-89.

Seitz, H.K ., Czygan, P. and Kommerell, B., 1982 . Stimulation of thymidine incorporation in isolated rat intestinal mucosal cells by feeding an ethanol-containing liquid diet. **Digestion** 23: 65-7.

Seitz, H.K., Czygan, P., Kienapfel, H., Veith, S., Schmidt-Gayk, H. and Kommerell, B., 1983. Changes in gastrointestinal DNA synthesis produced by acute and chronic ethanol consumption in the rat: a biochemical study. **Gastroenterology**, 21: 79-84.

Seitz, H.K., Czygan, P., Waldherr, R., Veith, S., Raedsch, R., Kaessmodel, H . and Kommerell, B., 1984. Enhancement of 1, 2-dimethylhydrazine-induced rectal carcinogenesis following chronic ethanol consumption in the rat. **Gastroenterology**, 86: 886-91.

Seitz, H.K., Velasquez, D., Waldherr, R., Veith, S., Czygan, P., Weber, E., Deutsch-Diescller, O.G. and Kommerell, B., 1985. Duodenal gamma-glutamyltransferase activity in human biopsies: effect of chronic ethanol consumption and duodenal morphology. **Eur J Clin Invest.**, 15 :192-6.

Seitz, H.K., Simanowski, U.A., Garzon, F.T., Rideout, J. M., Peters, T.J., Koch, A., Berger, M.R., Einecke, H. and Maiwald, M., 1990. Possible role of acetaldehyde in ethanol-related rectal cocarcinogenesis in the rat. **Gastroenterology** 98: 406-13.

Shaw, S., Jayatilleke, E., Herbert, V. and Colmann, N., 1989. Cleavage of folates during ethanol metabolism. **Biochem J.**, 257: 277-80.

Simanowski, U.A., Seitz, H.K., Baier, B., Kommerell, B., Schmidt-Gayk, H. and Wright, N.A., 1986. Chronic ethanol consumption selectively stimulates rectal cell proliferation in the rat. **Gut** 27:278-82.

Simanowski, U.A., Suter, P., Russell, R.M., Seitz, H.K., 1990. Cell regeneration in the rat rectum after chronic ethanol ingestion is significantly enhanced with age. (abstr.) **Alcoholism Clin Exp Res.**, 14: 338.

Swinson, C.M., Salvin, G., Coles, E.C. and Booth, C.C., 1983. Celiac disease and malignancy. **Lancet** 1:111-5.

Terpstra, O.T., Peterson Dahl, E., Williamson, R.C.N., Ross, J.S. and Malt, R.A., 1981. Colostomy closure promotes cell proliferation and dimethylhydrazine-induced carcinogenesis in rat distal colon. **Gastroenterology** 81: 475-80.

Williamson, R.C.N., 1979. Hyperplasia and neoplasia of the intestinal tract. **Ann R Coll Surg Engl.**, 61:341-8.

Williamson, R.C.N. and Malt, R.A., 1980. Promotion of intestinal carcinogenesis by adaptive mucosal hyperplasia. In: Appleton, D.R., Sunter, J.P. and Watson, A.J., (eds.) **Cell proliferation in the gastrointestinal tract**. Pitman, London, 303-15.

Wittig, G., Wildner, G.P. and Zeibarth, D., 1971. Der Einfluss der Ingesta auf die Kanzerisierung des Rattendarmes durch Dimethylhydrazin. **Arch Geschwulstforsch**, 37:105-15.

Wright, N.A., Alison, M., 1984. **The biology of epithelial cell populations**. Clarendon, Oxford.

Wright, N.A., Watson, A., Morley, A., Appleton, D. and Marks, J., 1973. Cell kinetics in flat avillous mucosa of the human small intestine. **Gut** 14:701-10.

Wu, A.H., Paganini-Hill, A., Ross, R.K. and Henderson, B.E., 1987. Alcohol, physical activity and other risk factors for colorectal cancer: A prospective study. **Br J Cancer.**, 55:687-94.

Zucoloto, S. and Rossi, M.A., 1979. Effect of chronic ethanol consumption on mucosal morphology and mitotic index in the rat small intestine. **Digestion** 19:277-83.

CHRONIC ALCOHOLIC SKELETAL MYOPATHY: AN OVERVIEW

Timothy J. Peters and Victor R. Preedy

Department of Clinical Biochemistry
King's College School of Medicine and Dentistry
London UK

INTRODUCTION

It has long been recognised that chronic alcohol abusers have defects in skeletal muscle physiology and function; the most salient manifestation of which is muscle weakness. Hitherto, most of the detailed investigations into alcoholic muscle disease have concentrated on the "acute" forms of the disease process. In contrast, very little work has been carried out on the "chronic" alcoholic myopathy. Detailed studies by various research groups, including our own, have shown that the chronic form of alcoholic muscle disease occurs very much more frequently than the acute type, and may, indeed, be one of the most common metabolic myopathies in Western societies. It occurs in approximately half to two-thirds of chronic alcohol abusers (Worden, 1976; Martin et al., 1985; Urbano-Marquez et al., 1988). Thus, it is somewhat of a paradox that chronic alcoholic myopathy is one of the least researched muscle diseases, even though the causal agent is known. Furthermore, it is likely that the underlying molecular events responsible for the loss of skeletal muscle mass may be applicable to other forms of metabolic myopathy.

This review describes the clinical nature of the chronic myopathy, and the various ways in which it can be characterised, diagnosed and monitored. It will also become self-evident that there is a certain degree of uncertainty as to the molecular mechanisms responsible for the pathogenesis of alcoholic myopathy. Nevertheless, it is eventually hoped that some of the gaps in our understanding of the disease process will be filled by experiments in suitable animal models (for example, Preedy and Peters, 1988).

THE CHARACTERISTICS OF CHRONIC ALCOHOLIC MYOPATHY

A dominant feature of chronic alcoholic myopathy is a reduction in muscle mass, which occurs as a consequence of many years of alcohol misuse (Duane and Peters, 1988a). The prolonged period of time (at least 3 years) necessary for its manifestation is in part due to the fact that the rate of skeletal muscle protein synthesis in man is very low, of the order of 1%/day. In other words a 40% change in the rate of muscle protein synthesis will alter the skeletal muscle mass by 20% in 2 months, assuming that the rate of protein breakdown remains unaltered. Clinical symptoms include muscle weakness and wasting, difficulties in gait with frequent falls and muscle cramps (Martin et al., 1985). Detailed histological

Alcoholism: A Molecular Perspective
Edited by T.N. Palmer, Plenum Press, New York, 1991

analysis shows the disease is that of a 'primary myopathy', and the diameter of individual skeletal muscle fibres, particularly Type II (anaerobic, glycolytic, fast-twitch), are reduced (Hanid, et al., 1981; Martin et al., 1985; Trounce, et al., 1987). The diameter of Type I (aerobic, oxidative, slow-twitch) fibres are relatively unaffected. Plasma creatine kinase activities are not usually raised unless there is a superimposed element of acute myopathy, though they may increase in other chronic skeletal muscle myopathies (Martin et al., 1985). There may be increased lipid deposition in skeletal muscle but there is no histological evidence of myonecrosis or inflammatory responses or fibrosis (Sunnasy et al., 1983; Del Villar Negro et al., 1984; Martin et al., 1985). In contrast, acute alcoholic myopathy is characterised by rhabdomyolysis with markedly raised plasma creatine kinase activities. Histological profiles of biopsies from acute myopathic patients may reveal infiltration by inflammatory cells, with concomitant necrosis and general disruption of normal myofibre architecture. Type I fibres are predominantly affected. Symptoms of acute myopathy include muscle pain and tenderness, without evidence of muscle wasting (reviewed by Spargo, 1981; Ford et al., 1984; Preedy and Peters, 1990). Thus, patho-clinical features in the two forms of alcoholic muscle disease are strikingly different.

Analysis of the 'atrophy factor' (a weighted measure of the decrease in fibre diameter) and serum carnosinase activities have shown that chronic alcoholic skeletal myopathy is reversible (Slavin et al., 1983; Duane and Peters, 1988b). There is a considerable time period before the muscle mass returns to normal (6-12 months) and this again probably reflects the relatively slow rate of muscle protein turnover. Recent animal studies have shown that DNA accretion is reduced in ethanol-fed rats (Preedy and Peters, 1988) and it is pertinent to speculate whether chronic ethanol consumption will cause irreversible nuclear changes, particularly in young adults.

Various techniques have been used to diagnose chronic alcoholic myopathy. These include, the assessment of skeletal muscle strength (Martin et al ., 1985 ; Hickish et al ., 1989), and determination of muscle fibre diameter by morphometric analysis (Hanid, et al., 1981; Martin et al., 1985; Trounce et al ., 1987) or by measurement of serum carnosinase activities (Duane and Peters, 1988b). The evaluation of skeletal muscle strength as a diagnostic tool is prone to the criticism that it is difficult to measure and standardise and the muscle weakness may be multi-factorial in origin: Only half of the patients with histological evidence of myopathy have demonstrable muscle weakness. Determination of fibre size is more objective and determines directly if there is indeed muscle-fibre involvement; though no unequivocal information is imparted as to whether this is a direct consequence of alcohol abuse. Most of the original studies in this area utilised this method. Biopsies were fixed and stained for myosin-ATPase, to accentuate the Type II fibres and scanned using an interactive 'Magiscan' computer (Slavin et al , 1982 ; Martin et al . , 1985). An 'atrophy factor ', based on the number of individual fibres with reduced diameters was then computed. Unfortunately, analysis of muscle biopsies are time consuming and requires a minor surgical procedures, although closed needle biopsy samples are perfectly adequate for this analysis. In contrast, serum carnosinase activities require only a single blood sample and can be assayed relative rapidly. The diagnostic specificity of this estimation is not known although reduced activities have been reported in muscular dystrophies (Bando et al., 1984) and in hypothyroid myopathy, another form of Type II fibre metabolic myopathy (Bando et al., 1986). The estimation of serum creatine kinase are suitable for the appraisal of many skeletal muscle disorders as raised activities reflect skeletal muscle membrane (i.e., the sarcolemma) damage with leakage of this cytosolic enzyme. As this marker enzyme is unaltered in patients with the chronic form of alcoholic muscle disease, its use lies solely in identifying other forms of muscle disease in alcohol abusers.

MECHANISMS INVOLVED IN THE PATHOGENESIS OF THE MYOPATHY

The factors which may precipitate or cause alcoholic muscle disease may be indirect (i.e., organ dysfunction, nutrition etc.) or direct (i.e., effect on muscle carbohydrate or protein metabolism). Alcoholic myopathy may be a multi-factorial disease; for example, its development may be dependent on limitations in individual nutrient factors, but precipitated by alcohol or acetaldehyde in susceptible patients. Some of these possible pathogenic mechanisms are considered subsequently.

(i) Protein Turnover in Alcoholic Myopathy

It is important to emphasis that central to any skeletal muscle atrophy is a defect in protein turnover, regardless of whether the mechanisms are direct or indirect. This is because, in simplistic terms, the myopathic muscle fibres are smaller and, by implication, have a reduced protein content. This has recently been confirmed by direct measurement of muscle protein content in biopsies of alcoholic patients, which was highly correlated with both serum carnosinase measurements and the muscle atrophy factor (Preedy, V. R. and Peters, T.J., unpublished). The studies of Martin et al (1984) also showed that skeletal muscle biopsies from alcoholic patients with atrophy had lower myofibrillary Ca^{2+}-activated ATPase activities, compared to non-myopathic alcoholic controls, which may be suggestive of reduced contractile protein apparatus.

The concept of protein turnover implies that reductions in skeletal muscle protein content must be due to defects in protein synthesis and/or degradation. Using stable isotope methodology (^{13}C-leucine incorporation into muscle protein), Pacy et al. (1988) showed that both skeletal muscle (quadriceps) and whole-body protein synthesis rates were markedly reduced in chronic alcohol abusers, by 40% and 20%, respectively. Protein degradative pathways are clearly also important integrated components of protein turnover. Unfortunately, the data pertaining to protein degradation in chronic ethanol abusers are far from clear. This uncertainty is primarily due to the fact that there are no direct methods for reliably measuring skeletal muscle protein degradation *in vivo*. As detailed elsewhere in this volume, urinary 3-methylhistidine can be used as an indicator of myofibrillary protein degradation. This urinary analyte was used by Martin and Peters (1985) in assessing protein turnover in alcoholic patients who had consumed over 100g of ethanol/day for at least 3 years. They found no differences between control subjects and chronic alcoholics either with or without muscle wasting. This was apparently confirmed by the observation that the activities of skeletal muscle neutral proteases and lysosomal enzyme activities (Martin et al., 1984; Martin and Peters, 1985) were unaffected by chronic ethanol ingestion. These conclusions are not plausible as, if protein synthesis were reduced by 40% without alterations in protein degradation, then the entire muscle protein mass would be decreased by at least 95% in only 3 years. Calculation of myofibrillary protein breakdown rates were also made after using the correction described by Afting et al. (1981) to account for non-skeletal muscle sources of 3-methylhistidine and creatinine (Martin and Peters, 1985). These results showed that alcoholics had lower rates of muscle protein breakdown, especially those with histologically proven myopathy (Martin and Peters, 1985). Certainly, muscle protein degradative characteristics of assorted cirrhotic patients, who exhibit enhanced rates of urinary 3-methylhistidine excretion, are markedly different to those of myopathic alcoholics (Marchesini et al., 1981). This suggests that the causes of the apparent muscle wasting in severe liver disease occurs by a different mechanism. Nevertheless, it is important to further address the question of whether liver dysfunction is the principal cause of the muscle wasting seen in alcoholic subjects.

(ii) Liver Dysfunction

Skeletal muscle wasting is a common feature of chronic ethanol abusers with liver disease (Jenkins, 1984). Recent studies have indicated that patients with severe liver disease, i.e., uncompensated cirrhosis, not only had muscle wasting but also reduced rates of muscle protein synthesis (Morrison et al., 1990). This is very similar to the qualitative changes in protein turnover described above. However, in the original study of Martin et al. (1985), the prevalence of myopathy in patients with cirrhosis was not significantly different from non-cirrhotic alcoholics (although there was an increased prevalence of myopathy in subjects with severe hepatic dysfunction). Similar conclusions were obtained by Duane and Peters (1988a,b). One can tentatively infer that the patient group studied by Morrison et al. (1990) may have suffered from a muscle disorder as a consequence of ethanol toxicity per se rather than as a result of cirrhosis. This has been supported by animal studies, in which micronodular cirrhosis has been produced by carbon tetrachloride and phenobarbitone administration. After 18-20 weeks, the treatment did not cause any significant decrease in skeletal muscle protein, RNA nor DNA composition, even though there was marked reductions in plasma albumin and raised plasma enzymes activities typical of hepatic dysfunction (Preedy, V. R. and Peters, T. J, unpublished observations). In rats fed ethanol for only 6 weeks profound lesions in muscle biochemistry were observed (Preedy and Peters, 1988). One can, therefore, construe that the Type II fibre myopathy observed in chronic alcoholic abusers is non-hepatic in origin.

(iii) Nutrition

There are numerous studies showing that alcohol abusers have various nutritional deficiencies. This occurs either as a consequence of reduced intake or impairment of nutritional handling (i.e., absorption, excretion, etc.). It is also well-known that one of the most important factors that governs the rate of protein synthesis is nutritional status (for extensive reviews, see Waterlow et al., 1978; Waterlow, 1984).

The possibility that impaired nutrition was a causative factor was comprehensively explored by Duane and Peters (1988a), who showed that riboflavin, pyridoxine, thiamine, vitamin B12 and folate status in alcoholic patients with myopathy were not different to those alcoholic patients who had no skeletal muscle wasting. Similar conclusions were also obtained by Hickish et al., (1989) in examining vitamin D status, and by Urbano-Marquez et al . (1989) in examining general dietary intake. Paradoxically, Ward et al. (1988) have shown that plasma alpha-tocopherol and selenium levels in plasma of myopathic alcoholics are selectively reduced in comparison to non-myopathic alcohol abusers. Obviously, these apparently divergent set of observations require a plausible and rational explanation. At the same time, the reduction in the plasma concentrations of these micro-nutrients, together with the observation that myopathic alcoholics have reduced carnosinase activities, offer important clues as the nature and origin of alcoholic myopathy, as described below.

(iv) Free Radical Generation and Anti-Oxidant Status

Both selenium and alpha-tocopherol are important anti-oxidants, and it is considered that free radicals (oxygen reactive species) may cause deleterious changes in tissue membranes and organelles. Thus, the reduction in selenium and alpha-tocopherol could very well represent a decreased capacity in the ability of muscle to protect against oxidative and free radical damage (see also Garcia-Brunel, 1984) and there is suggestive evidence that free radical damage is implicated in the myopathic lesions. As described earlier, plasma carnosinase activities are reduced in myopathic alcoholics (Duane and Peters, 1988a). The normal biochemical and physiological functions of carnosinase (which hydrolyses the dipeptide carnosine to ß-alanine and histidine) are unknown, but it is claimed that it is an important intracellular buffering agent as well as an anti-oxidant, particularly in Type II fibres (Boldyrev, 1986; Boldyrev et al., 1988), although this view has recently been challenged (Aruoma et al., 1989). Reductions in muscle carnosine levels, as a consequence of enhanced utilization, may therefore be reflected by compensatory reductions in serum carnosinase activities. Another consideration is the fact that animal studies have shown that Type II muscles have lower concentrations of a variety of anti-oxidants (Riley et al., 1988; Ward, R. J. and Peters, T. J., unpublished) and this may explain the susceptabilty of these fibres to the damaging effects of ethanol. It is not known how oxidative damage may cause adverse changes in protein turnover. Chronic ethanol feeding reduces protein contents in skeletal muscle by deleterious effects on both translation and transcription and possibly by perturbing the rate of protein breakdown as well (Preedy and Peters, 1990). If one has to propose a unifying theory, then free radicals must effect all of these processes. A further complication is the observation that recent studies have demonstrated that both ethanol and acetaldehyde appear to be able to independently reduce protein synthesis and protein breakdown (Preedy, V. R. Cook, E. and Peters, T. J, unpublished data). One possibility is the modification of cellular proteins or protein structures such as organelles, by direct attack of radical species, may alter the fractional rates of protein turnover. This would explain why the gross architecture of the muscles of chronic alcoholics and ethanol-fed rats are relatively preserved. There is an additional line of evidence to support the theory that free radical generation may be important in the pathogenesis of chronic alcoholic myopathy. Kelly et al. (1987) and Machlin et al. (1977) have shown that selenium and vitamin E deficiency, respectively, cause skeletal muscle myopathy. The former deficiency is characterised by a selective Type II fibre atrophy. Furthermore, some muscle disorders, i.e., myotonic dystrophy, can be treated with selenium and vitamin E (Orndahl et al., 1986). Obviously, the complex relationship between protein turnover and free radicals in alcoholic muscle disease needs to be investigated in depth.

(v) Other Mechanisms

Hormonal dysfunction: Hormones are probably the most well researched factors involved in the regulation of skeletal muscle protein synthesis (Waterlow et al., 1978). It is also well-documented that alcohol ingestion may induces a variety of endocrine abnormalities, including androgens, thyroid and parathyroid hormones as well as pituitary and pancreatic hormones (see Hudgson and Hall, 1982; Van Thiel and Gavaler, 1990). In many instances, the hormonal abnormalities are also associated with Type II fibre atrophy for example in Cushing's syndrome and hypothyroidism. Excessive alcohol ingestion may also induce a pseudo-Cushing's syndrome, though published reports of this occur only very infrequently. This relationship was comprehensively studied by Duane and Peters (1987), who found that the urinary excretion and diurnal serum concentration of cortisol in alcoholic patients with myopathy were similar to those occurring in non-myopathic alcoholics. Thus, the possibility that alcoholic myopathy may be a form of pseudo-Cushing's syndrome can be confidently excluded.

Neuropathies: Peripheral neuropathies also causes skeletal muscle myopathy (Jennekens, 1982) and it is possible that this may be a contributing factor in the development of alcoholic myopathy. Mills et al. (1986) demonstrated that some alcoholics patients who had skeletal muscle myopathy, were also neuropathic. There was, however, no relationship between neuropathy and myopathy as the muscle wasting also occurred in patients without neuropathy (Mills et al., 1986). The neuro-myopathies affect distal muscle in contrast to the proximal myopathies found in chronic alcoholics. Histology and detailed neuro-physiological studies clearly indicate that alcoholic myopathy is not secondary to peripheral nerve injury.

Inactivity: Various arguments have been forwarded that the alcoholic myopathy may be a form of disuse atrophy, as a consequence of inactivity due to muscle weakness. Inactivity (i.e., leg immobilization) has been shown to reduce muscle protein synthesis (Gibson et al., 1987), similar to the qualitative changes in patients with skeletal muscle myopathy (Pacy et al ., 1988) . An argument against this is that a fundamental feature of disuse atrophy is a preferential effect on Type I skeletal muscle fibres (Gibson et al., 1987). This directly contrasts the Type II atrophy seen in chronic alcohol abusers and is strong suggestive evidence that inactivity is not the paramount cause of alcoholic myopathy. Nonetheless though, some myopathic alcoholics do have Type I fibre atrophy (albeit to an extremely limited degree), though this occurs principally in patients with very severe Type II fibre atrophy (Slavin et al ., 1983; Martin et al. , 1984). This Type I fibre atrophy may represent one end of the spectrum of muscles disorders arising as a result of oxidative damage.

CONCLUSION

Chronic skeletal muscle myopathy is a common feature of alcohol abuse and may represent one of the most common skeletal myopathies in the developed countries. Its incidence is considerably higher than that of cirrhosis. Unfortunately, the aetiological mechanisms of this disease are poorly understood, though central to the myopathy is a defect in protein turnover. In the normal individual, skeletal muscle comprises approximately 40% of body mass, and one might, therefore, expect perturbations in whole-body nitrogen metabolism to occur. This has been supported by clinical studies showing that ethanol administration markedly elevates urinary nitrogen excretion (McDonald and Margen, 1976; Reinus et al., 1989). Furthermore, it is possible that some of the biochemical mechanisms responsible for the precipitation and progression of the disease may be applicable to other disease states characterised by Type II fibre atrophy.

ACKNOWLEDGEMENT

We are extremely grateful to Miss Cheryl M. Riley for the excellent preparation of the manuscript.

REFERENCES

Afting, E.G., Bernhardt, W., Janzen, R.W. and Rothig, H.J., 1981. Quantitative importance of non-skeletal muscle N^{T}-methylhistidine and creatinine in human urine. **Biochem. J**, 200: 449.

Aruoma, O., Laughton, M.J. and Halliwell, B. 1989. Carnosine, homocarnosine and anserine: could they act as antioxidants *in vivo*? **Biochem. J.**, 264: 863.

Bando, K., Ichihara, K., Shimotsuji, T., Toyoshima, H., Koda, K., Hayashi, C. and Miya, K., 1986. Reduced serum carnosinase activity in hypothyroidism. **Ann. Clin. Biochem.**, 23: 190.

Bando, K., Shimotsuji, T., Toyoshima, H., Hayashi, C. and Miya, K., 1984. Fluorometric assay of human carnosinase activity in normal children, adults and patients with myopathy. **Ann. Clin. Biochem.**, 23: 190.

Boldyrev, A.A., 1986. Biological significance of histidine-containing dipeptides. **Biochemistry** (USSR). 51: 1651.

Boldyrev, A.A., Dupin, A.M., Pindel, E.V. and Severin, S.E., 1988. Anti-oxidant properties of histidine-containing dipeptides from skeletal muscle of vertebrates. **Comp. Biochem. Physiol.**, 89B: 245.

Del Villar Negro, A., Merino Angulo, J. and Rivera Pomar, J.M., 1984. Skeletal muscle changes in chronic alcoholic patients. A conventional, histochemical ultrastructural and morphometric study. **Acta Neurol. Scand.**, 70: 185.

Duane, P. and Peters, T.J., 1987. Glucocorticosteroid status in chronic alcoholics with and without skeletal muscle myopathy. **Clin. Sci.**, 73: 601.

Duane, P. and Peters, T.J., 1988a. Nutritional status in alcoholics with or without skeletal muscle myopathy. **Alcohol Alcohol.**, 23: 271.

Duane, P. and Peters, T.J., 1988b. Serum carnosinase activities in patients with alcoholic chronic skeletal muscle myopathy. **Clin. Sci.**, 75: 185.

Ford, C.S., Caldwell, S.H. and Kilgo, G.R., 1984. Acute alcoholic myopathy. **Am. Family Phys.**, 29: 249.

Garcia-Brunel, L., 1984. Lipid peroxidation in alcoholic myopathy and cardiomyopathy. **Medical Hypotheses** 13: 217.

Gibson, J.N.A., Halliday, D., Morrison, W.L., Stoward, P.J., Hornsby, G.A., Watt, P.W., Murdoch, G. and Rennie, M.J., 1987. Decrease in human quadriceps muscle protein turnover consequent upon leg immobilization. **Clin. Sci.**, 72: 503.

Hanid, A., Slavin, G., Mair, W., Sowter, C., Ward, P., Webb, J. and Levi, A., 1981. Fibre type changes in striated muscle of alcoholics. **J. Clin. Pathol.**, 36: 772.

Hickish, T., Colston, K.W., Bland, J.M. and Maxwell, J.D., 1989. Vitamin D deficiency and muscle strength in male alcoholics. **Clin Sci.**, 77: 171.

Hudgson, P. and Hall, R., 1982. Endocrine myopathies. **In**: "Skeletal Muscle Pathology". F.L. Mastaglia and J. Walton, Eds., pp 393-408. Churchill Livingstone, Edinburgh.

Jenkins, W., 1984. Liver disorders in alcoholism. **In**: "Clinical Biochemistry of Alcoholism", Rosalki, S.B., ed., pp 258-270, Churchill Livingstone, Edinburgh.

Jennekens, F.G.I., 1982. Neurogenic disorders of muscles. **In**: "Skeletal Muscle Pathology". F.L. Mastaglia and J. Walton, eds., pp 204-234. Churchill Livingstone, Edinburgh.

Kelly, D.A., Sherkin, A., Coe, A. and Walker-Smith, J.A., 1987. Skeletal myopathy secondary to selenium deficiency. **Gut**, 28: 75.

McDonald, J.T. and Margen, 5., 1976. Wine versus ethanol in human nutrition. I. Nitrogen and calorie balance. **Am. Clin. Nutr.**, 29: 1093.

Marchesini, G., Zoli, M., Angiolini, A., Dondi, C., Bianchi, F. B. and Pisi, E., 1981. Muscle protein breakdown in liver cirrhosis and the role of altered carbohydrate metabolism. **Hepatology**, 1: 294.

Machlin, L.J., Filipski, R., Nelson, J., Horn, L.R. and Brin, M., 1977. Effects of a prolonged vitamin E deficiency. **J. Nutr.**, 107: 1200.

Martin, F.C., Levi, A.J., Slavin, G. and Peters, T.J., 1984. Glycogen content and activities of key glycolytic enzymes in muscle biopsies from control, subjects and patients with chronic alcoholic skeletal myopathy. **Clin. Sci.**, 66: 69.

Martin, F.C. and Peters, T.J., 1985. Assessment *in vitro* and *in vivo* of muscle degradation in chronic skeletal muscle myopathy of alcoholism. **Clin. Sci.**, 68: 693.

Martin, F.C., Slavin, G., Levi, A.J. and Peters, T.J. 1984 Investigation of the organelle pathology of skeletal muscle in chronic alcoholism. **J. Clin. Pathol.**, 37: 448.

Martin, F.C., Ward, K., Slavin, G., Levi, A.J. and Peters, T.J., 1985. Alcoholic skeletal myopathy, a clinical and pathological study. **Quart. J. Med.**, 55: 233.

Mills, K.R., Ward, K., Martin, F. and Peters, T.J., 1986. Peripheral neuropathy and myopathy in chronic alcoholism. **Alcohol Alcohol.**, 21: 357.

Morrison. W. L., Bouchier, I. A. D., Gibson, J. N. A. and Rennie, M.J., 1990. Skeletal muscle and whole-body turnover in cirrhosis. **Clin. Sci.**, 78: 413.

Orndahl, G., Sellden, U., Hallin, S., Wetterqvist, H., Rindby, A. and Selin, E., 1986. Myotonic dystrophy- treated with selenium and vitamin E. **Acta Med. J. Scand.**, 219: 407.

Pacy, P.J., Read, M., Preedy, V.R., Peters, T.J. and Halliday, D., 1988. Whole body

leucine kinetics and fractional quadricep muscle synthetic rate (MPSR) in alcoholic patients. **Clin. Sci.**, supplement 19, 36p.
Preedy, V.R. and Peters, P.J., 1988. The effects of chronic ethanol ingestion on protein metabolism in Type I and II fibre-rich skeletal muscle of the rat. **Biochem. J.**, 254: 631.
Preedy, V.R. and Peters, P.J., 1990. Alcohol and skeletal muscle disease. **Alcohol Alcohol.**, 25: 177.
Reinus, J.F., Heymsfield, R.W., Casper, K. and Galambos, U.J.T., 1989. Ethanol: Relative fuel value and metabolic effects *in vivo*. **Metabolism**. 38: 125.
Riley, D.A., Ellis, S. and Bain, J.L.W., 1988. Catalase microperoxisomes in rat soleus and extensor digitorum longus muscle fibre types. **J. Histochem. Cytochem.**, 36:633.
Slavin, G., Martin, F., Ward, P., Levi, J. and Peters T. J., 1983. Chronic alcoholic excess is associated with selective but reversible injury to type 2B muscle fibres. **J. Clin. Pathol.**, 36: 772.
Slavin, G., Sowter, C., Ward, P. and Paton, K., 1982. Measurement of striated muscle fibre diameters using interactive computer-aided microscopy. **J. Clin. Pathol.,** 35: 1268.
Spargo, E. 1981. Alcohol and muscle disease. **Br. J. Alcohol Alcohol.**, 16: 124.
Sunnasy, D., Cairns, S. R., Martin, F., Slavin, G, and Peters, T. J., 1983. Chronic alcoholic skeletal myopathy: a clinical, histological and biochemical assessment of muscle lipid. **J. Clin. Pathol.**, 36: 778.
Trounce, I., Byrne, E., Dennett, X., Santamaria, J., Doery, J. and Peppard, R., 1987. Chronic alcoholic proximal wasting: physiological, morphological and biochemical studies in skeletal muscle . **Aust .N.Z. J. Med.**, 17 : 413 .
Urbano-Marquez, A. Estruch, R., Navarro-Lopez, F., Grau, J.M., Mont, L. and Rubin, E., 1989. The effect of alcoholism on skeletal and cardiac muscle. **New Engl. J. Med** 320: 409.
Van Thiel, D.H. and Gavaler, J.S., 1990. Endocrine consequences of alcohol abuse. **Alcohol Alcohol.**, 25: 341.
Ward, R.J., Jutla, J., Duane, P.D. and Peters, T.J., 1988. Reduced anti-oxidant status in patients with chronic alcoholic myopathy. **Biochem. Soc. Trans.,** 16: 581.
Waterlow, J.C., 1984. Protein turnover with special reference to man. **Quart. J. Exp. Physiol.**, 69: 409.
Waterlow, J.C., Garlick, P.J. and Millward, D.J., 1978. Protein Turnover in Mammalian Tissues and in the Whole-body. North Holland, Amsterdam.
Worden, R.E., 1976. Pattern of muscle and nerve pathology in alcoholism. **Proc. N.Y. Acad. Sci.,** 273: 351.

EFFECTS OF ETHANOL ON THE IMMUNE SYSTEM

Thomas R. Jerrells

Department of Pathology
University of Texas Medical Branch
Galveston, Texas 77550-2779, U.S.A.

INTRODUCTION

The effects of ethanol are varied and complex. A number of observations have led to the belief that ethanol abuse alters the immune system and this results in a predisposition to infectious agents and tumors. It has been shown that alcoholics have increased incidences of certain tumors and increased incidences of infections (Smith and Palmer, 1976; Adama and Jordan, 1984; Eckardt et al., 1981) that include tuberculosis and opportunistic organisms. It is clear that the effect of ethanol that produces this increased susceptibility is multifactorial but it is thought that an immunosuppression associated with alcohol abuse is important.

Of importance to many infectious diseases and malignant diseases is the observation that ethanol affects many nonspecific host defense mechanisms including natural killer cells (Meadows et al., 1989), granulocyte chemotactic responses and function measured *in vitro* (Spagnuolo and MacGregor, 1975), and clearance of particulate material by the reticuloendothelial system (Galante et al., 1982).

There are increasing numbers of observations that have shown that ethanol ingestion also affects the specific immune system. These include the observation that ethanol results in a loss of lymphoid cells from the peripheral blood, spleen and thymus (Tennenbaum et al., 1969, Jerrells et al., 1986). It has also been shown that a loss of lymphoid cell function occurs. These changes include a defect in proliferative responses of thymus-derived lymphocytes (T cells) in response to nonspecific mitogens (Jerrells et al., 1986, Roselle and Mendenhall, 1984) and to recall antigens (Jerrells et al., 1988). The immune response to antigens that require the presence of antigen-specific T cells (T dependent antigens) as measured by antibody production is also decreased in animals that have been administered ethanol (Jerrells et al., 1986; Bagasara et al., 1987). In contrast to these findings the responses to stimuli that do not require T cells for activation or antibody production (T independent stimuli) are not affected by ethanol (Jerrells et al., 1986, 1990a). The interpretation of these data has led to the suggestion that ethanol primarily effects T-cell-dependent, or cell-mediated, immune responses.

The effort in my laboratory has been focused on defining the extent and the mechanisms of the effects of ethanol on the cell-mediated immune system and to relate these changes to alterations in host defense mechanisms to infectious agents, especially agents that have well defined T cell-mediated immune responses. The present manuscript will review the study findings of my laboratory in these important areas.

EXPERIMENTAL SYSTEMS

Animals and Ethanol Feeding

In the majority of the studies that will be described the model system that was used is alcohol administration to inbred mice using a pair-feeding design (Jerrells et al., 1990a; Lieber and DeCarli, 1982). Briefly, male C57Bl/6 mice 6 to 8 weeks of age were fed either Lieber-DeCarli diet (Dyets, Bethelham, PA) containing 7% (v/v) ethanol or an amount of isocaloric diet determined by the amount of ethanol-containing diet consumed by the ethanol group. This design attempts to control for nutritional effects of the ethanol diet. Mice were maintained on these diets for varying lengths of time from 7 to 14 days and studied either with or without 1 day of withdrawal from ethanol. In the studies that have been done in my laboratory this protocol results in relatively stable blood alcohol levels, determined with the alcohol dehydrogenase technique (Sigma Chemicals, St. Louis, MO), ranging from 200 to 400 mg/dL. In the majority of studies the animals were fed the diet in the evening and blood alcohol levels are determined with sera obtained at 8:00 to 9:00 am.

Some of the data described in this review were obtained in a rat model of acute intoxication and dependency that has been described by Majchrowicz (1975). In this model male Sprague-Dawley rats were administered ethanol by gastric intubation in fractional doses to maintain intoxication. In this system each animal received from 8 to 11 g of ethanol/kg body weight per day.

To study the *in utero* effects of ethanol on the immune system, pregnant Sprague-Dawley rat dams were given either an ethanol-containing liquid diet or pair fed an isocaloric diet as described by work from Weinberg's laboratory (Gallo and Weinberg, 1986). These animals were fed the diet throughout the pregnancy and the pups were fostered on dams not exposed to ethanol. The pups from each group were raised to adults under normal conditions and their immunocompetence tested at 3 to 6 months of age.

Immunocompetence Assays

To determine the effects of ethanol on the cellular composition of the thymus and spleen, single cell suspensions were prepared from these two organs and total cell counts were performed using an electronic cell counter (Coulter Model ZM, Coulter Electronics, Hialeah, Florida). Flow cytometry was used to determine the phenotypes of cells in these organs after ethanol treatment and serial thin sections from each organ were stained with monoclonal antibodies (MABS) to cell surface markers to assess changes in the immunoarchitecture. Briefly, MABS to the pan T cell marker Thy 1.2, CD4 (L3T4), CD8 (Ly 2), or anti-IgM (all obtained from Becton-Dickinson, Mountain View ,California) were used as fluorochrome-labelled reagents for flow cytometry or in a system of detection by peroxidase-labeled second antibodies for immunohistochemistry. The details of these procedures have been published (Wirt et al., 1985; and Saad, A.J. and Jerrells, T.R., 1991, Flow cytometric and immunohistochemical evaluation of ethanol-induced changes in splenic and thymic lymphoid cell populations, Alcoholism: Clin. Exp. Med., in press).

Functional assays of T and B cell function included *in vitro* proliferation to the T cell mitogen concanavalin A (con A) or the B cell mitogen bacterial lipopolysaccharide (LPS) or the allogeneic cells in a one-way mixed lymphocyte response (MLR) using spleen cells prepared and cultured as previously described (Jerrells et al., 1986, 1990a). The levels of the T cell growth factor interleukin-2 (IL-2) in supernatants from con A-stimulated cells after 18 to 20 hours of culture were determined using CTLL cells, an IL-2-dependent cell line (Smith, 1980). The response to IL-2 was determined using blast cells isolated from con A-stimulated cell cultures after 72 to 96 hours of culture (Jerrells et al., 1990b).

In vivo antibody responses to T dependent and T independent antigens were assessed after immunization with sheep erythrocytes (SRBC) or TNP-ficoll respectively. The response to these antigens was measured by quantitating single antibody producing cells with a modified plaque assay (Cunningham and Szenberg, 1968).

RESULTS AND DISCUSSION

As described by earlier work from my laboratory (Jerrells et al., 1986;1988,1990a) and others (Tennenbaum et al., 1969) administration of ethanol to experimental animals as adults results in a loss of cell from the thymus and spleen. In our studies (Saad, A.J. and Jerrells, T.R., 1991, Flow cytometric and immunohistochemical evaluation of ethanol-induced changes in splenic and thymic lymphoid cell populations, Alcoholism: Clin. Exp. Res., in press) 7 days of ethanol feeding to mice resulted in 72% decrease of thymus cells and 50% decrease in spleen cell numbers. After 14 days of ethanol feeding the effects were even more pronounced and 94% of the thymic cells and 64% of the spleen cells were lost.

To determine if any population of cells were preferentially depleted by ethanol flow cytometric studies were performed. These studies revealed that the immature thymocytes ($CD4^+$ and $CD8^+$) were lost in greatest number from the thymus. With prolonged feeding of ethanol (14 days) mature thymocytes ($CD4^+$ and $CD8^-$ or $CD4^-$ and $CD8^+$) were also lost from the thymus and apparently the $CD8^+$ cells were lost in greater numbers. These finding were confirmed by examining sections of thymus tissue stained by immunohistochemical techniques. This technique revealed that the thymus from ethanol-treated mice lacked cortical structures and after 14 days of ethanol both the cortex and medulla were dramatically altered in morphological detail.

As described above, spleen cells were also depleted after 7 and 14 days of ethanol feeding (50% and 64% decrease respectively). Flow cytometry analyses revealed that the cells showing surface IgM staining (B cells) were lost in greatest numbers from the spleen. T cells were also lost but to a lesser extent and no preferential loss of $CD4^+$ or $CD8^+$ cells was noted in our studies. The immunoarchitecture of the spleen was not markedly changed although it was apparent that the T cell and B cell areas of the spleen were smaller in spleens from ethanol-treated mice.

It was surprising to see the somewhat preferential loss of spleen B cells when our functional data were considered. In published studies (Jerrells et al., 1986, 1990a) using both the inbred mouse model and the rat model we have shown that the T cell function is most affected by ethanol. These functional changes have included proliferation of spleen cells to the T cell mitogen concanavalin A, allogeneic cells in the MLR (presumably a measure of T cell receptor-mediated proliferation), and, recently, in response to monoclonal antibody to the CD3 complex. Stimulation of T cells through the CD3 complex is thought to approximate signal transduction via the T cell receptor. The antibody response to the T-cell-dependent antigen SRBC has also been shown to be influenced by ethanol administration or ingestion (Jerrells et al., 1990a). In contrast to these results, we and others have shown that ethanol does not markedly influence the B cell response to LPS or to the immune response to the T-cell-independent antigen TNP-ficoll (Bagasara et al., 1987; Jerrells et al., 1990a). The one exception to this were the results of a study that was done in my laboratory in which rats were administered large amounts of ethanol by gastric intubation (Jerrells et al., 1986).

It is known that ethanol administration to experimental animals results in the production of corticosteroids and withdrawal from ethanol induces large amounts of corticosteroids (Tabajoff et al., 1978; Jerrells et al., 1990c). Corticosteroids have been shown to deplete cortical thymocytes and markedly influence the immunocompetence measures that have been employed in our studies (Cupps and Fauci, 1982). To assess the role of corticosteroids in the phenomena associated with ethanol in our systems we have studied animals that were rendered incapable of producing corticosteroids by adrenalectomy (Jerrells et al., 1990c). These animals (ADX) were given ethanol and studied essentially as before. It was found that ADX only partially changed the effects of ethanol and this was most evident in the thymus. The loss of thymocytes was still evident in ADX mice and rats; however the magnitude of the cell loss was less. Importantly, the ability of both mice and rats that were ADX and given ethanol to develop antibody-producing cells to SRBC was similar regardless of whether they had adrenal glands or not. We have interpreted these data to mean that corticosteroids are only partially responsible for the changes in the immune system noted in ethanol-treated animals and that other, unknown, mechanisms are also important.

It is known that the T cell proliferative response to either antigens or mitogens is dependent on the production of the autocrine growth factor IL-2 and receptors on the cell surface for IL-2 (Smith, 1980). The interaction of IL-2 with the high-affinity IL-2 receptor results in a series of intracellular events that ultimately results in cell division. It seemed possible that the effect ethanol had on T cell proliferation might involve alterations in either IL-2 production or IL-2 receptor expression. We assayed cultures of purified T cells for levels of IL-2 following stimulation with con A using a factor dependent cell system (Jerrells et al., 1990b) and found that cells from ethanol-treated animals generally produced equivalent or sometimes higher levels of IL-2 than cells from control animals. To test if cells were capable of responding to IL-2 we isolated blast cells from the same cultures and restimulated these IL-2 receptor-bearing cells with recombinant IL-2. It was found that cells from ethanol-treated animals were repeatedly less responsive to IL-2 than cells from control animals. In subsequent studies, we have shown that the cells isolated from these cultures had equivalent levels of IL-2 receptor as measured by flow cytometry and that equivalent amounts of IL-2 were bound to cells from each treatment group.

A similar defect was noted in adult rats that were exposed to ethanol only *in utero*. However, only the male offspring of dams given ethanol during pregnancy showed a defect in proliferation to con A and the subsequent defect in response of blast cells to IL-2 (Weinberg, J. and Jerrells, T.R., 1991, Suppression of immune responsiveness: sex differences in prenatal ethanol effects, Alcoholism: Clin. Exp. Med., in press). Other aspects of immunocompetence were not affected by ethanol exposure *in utero*.

SUMMARY

The work in my laboratory has established that the T cell and T-cell-dependent immune responses are most affected by ethanol exposure either *in utero* or in adult animals. This functional change in lymphocyte action is also accompanied by a marked loss of cells from the spleen and the thymus. It is not known if the cells that replace the thymic cells are altered in function or if the T cell repertoire for antigen responsiveness is altered in these systems. The immature cells of the thymus appear to be most sensitive to the effects of ethanol and are the cells most sensitive to corticosteroids. By studying adrenalectomized animals the role of mediators or factors other than corticosteroids has been established, although it is obvious that corticosteroids are an important contributory factor in the immunosuppression associated with ethanol especially during withdrawal. Current studies in my laboratory are designed to correlate these study findings with increases in susceptibility to infectious organisms.

ACKNOWLEDGMENTS

This work was supported by a grant from The National Institute of Alcohol Abuse and Alcoholism AA 07731. The contributions by my collaborators, A. Joseph Saad, Rana Domiati-Saad, Wiley Smith, David Perrit, Joanne Weinberg, and my special friend Michael J. Eckardt are greatly appreciated and the work would not have been possible without their efforts.

REFERENCES

Adams, H.G. and Jordan, C., 1984, Infections in the alcoholic, **Med. Clin. North America**, 68:179.

Bagasara, O., Howeedy, A., Dorio, R. and Kajdacxy-Balla, A., 1987, Functional analysis of T-cell subsets in chronic experimental alcoholism, **Immunology**, 61:63.

Cunningham, A.J. and Szenberg, A., 1968, Further improvements in the plaque technique for detecting single antibody-forming cells, **Immunology**, 14:499.

Cupps, T.R. and Fauci, A.S., 1982, Corticosteroid-mediated immunoregulation in man, **Immunol. Rev.**, 65:133.

Eckardt, M.J. , Hartford, T.C., Kaelber, C.T. , Parker, E.S., Rosenthal, L.S., Ryback, R.S., Slamoiraghi, G.C. and Vanderveen, E., 1981, Health hazards associated with alcohol consumption, **JAMA**, 246:648.

Galante, D., Andreana, A., Perna, P., Utili, R. and Ruggiero, G., 1982, Decreased phagocytic and bactericidal activity of the hepatic reticuloendothelial system during chronic ethanol treatment and its restoration by levamisole, **J. Reticuloendothelial Soc.** , 32: 179.

Gallo, P.V. and Weinberg, J., 1986, Organ growth and cellular development in ethanol-exposed rats, **Alcohol**, 3:261.

Jerrells, T.R. , Marietta, C.A., Eckardt, M.J., Majchrowicz, E. and Weight, F.F., 1986, Effects of ethanol administration on parameters of immunocompetency in rats, **J. Leukocyte Biol.,** 39:499.

Jerrells, T.R. , Marietta, C.A., Bone, G., Weight, F.F. , and Eckardt, M.J., 1988, Ethanol-associated immunosuppression, **in**: 'Psychological, neuropsychiatric, and substance abuse aspects of AIDS', (ed) T.P. Bridge, Raven Press, New York, page 173.

Jerrells, T.R., Smith, W. and Eckardt, M.J., 1990a, Murine model of ethanol-induced immunosuppression, **Alcoholism: Clin. Exp. Res.**, 14:546.

Jerrells, T.R., Perritt, D., Eckardt, M.J. and Marietta, C.A., 1990b, Alterations in interleukin-2 utilization by T-cells from rats treated with an ethanol-containing diet, **Alcoholism: Clin. Exp. Res.,** 14:245.

Jerrells, T.R. , Marietta, C.A. , Weight, F.F. and Eckardt, M.J., 1990c, Effect of adrenalectomy on ethanol-associated immunosuppression, **Int. J. Immunopharmac.,** 12:435.

Lieber, C.S. and DeCarli, L.M., 1982, The feeding of alcohol in liquid diets: two decades of applications and 1982 update, **Alcoholism: Clin. Exp. Res.** , 6:523.

Majchrowicz, E., 1975, Induction of physical dependence upon ethanol and the associated behavioral changes in rats, **Psychopharmacologia**, 43:245.

Meadows, G.G., Blank, S.E. and Duncan, D.D., 1989, Influence of ethanol consumption on natural killer cell activity in mice, **Alcoholism: Clin. Exp. Res.,** 13: 476.

Roselle, G.A. and Mendenhall, C.L., 1984, Ethanol-induced alterations in lymphocyte function in the guinea pig, **Alcoholism: Clin. Exp. Res.** 8:62.

Smith, K.A., 1980, T-cell growth factor, **Immunol. Rev.,** 51:337.

Smith, F.E. and Palmer, D.L., 1976, Alcoholism, infection and altered host defenses: a review of clinical and experimental observations, **J. Chron. Dis.** 29: 35.

Spagnuolo, R.J. and MacGregor, R.R., 1975, Acute ethanol effect on chemotaxis and other components of host defense, **J. Lab. Clin. Med.,** 86:24.

Tabakoff, B., Jaffe, R.C. and Ritzmann, R.F., 1978, Corticosterone concentrations in mice during ethanol drinking and withdrawal, **J. Pharm. Pharmac.,** 30:371.

Tennenbaum, J.I., Ruppert, R.D., St. Pierre, R. and Greenberger, N.J., 1969, The effect of chronic alcohol consumption on the immune responsiveness of rats, **J. Allergy,** 44:272.

Wirt, D.P., Grogan, T.M., Jolley, C.S., Rangel, C.S., Payne, C.M. , Hansen, R.C., Lynch, P.J. and Schuchardt, M., 1985, The immunoarchitecture of cutaneous pseudolymphoma, **Human Pathology,** 16:492.

FUTURE DIRECTIONS IN BIOMEDICAL RESEARCH ON ALCOHOLISM

A Workshop organised by Jean-Pierre von Wartburg

This book is based on the proceedings of the NATO Advanced Study Institute (ASI) on 'The Molecular Pathology of Alcoholism' held at Il Ciocco in Italy on August 26th - September 6th 1990. An important part of this ASI was a one-day Workshop organised by Jean-Pierre von Wartburg on the theme 'Future Directions in Biomedical Research in Alcoholism'. The purpose of this Workshop was to allow a few distinguished researchers in the field of alcoholism to informally muse on the directions that future research in this important area may take. Selected contributions to this Workshop are included in this volume.

INTRODUCTORY REMARKS

Jean-Pierre von Wartburg

Institut für Biochemie und Molekularbiologie
der Universität Bern
CH-3012 Bern, Switzerland

These short introductory remarks to this Workshop on future trends in alcohol research try to see some of the current difficulties characteristic to this field of research in a historical perspective, in order to stimulate discussion on possibilities to overcome some of them.

REPUTATION OF ALCOHOL RESEARCH

It was in the year 1956, when Hugo Aebi, late Professor of Biochemistry at the University of Berne, suggested to me, to use the shortly available ^{14}C-labelled ethanol to investigate the effects of the newly discovered compound tetraethylthiuram disulfide (antabuse, disulfiram) on alcohol metabolism in rats. After completion of my thesis I decided to continue this line of research and to use tissue slices and the manometric technique of Warburg in order to find out more about the respective roles of alcohol dehydrogenase and catalase in alcohol metabolism. At that time a famous German professor warned me that alcohol research would be considered as second rate and not good enough for an academic career in Biochemistry, and he advised me to change the topic of my research. However, my belief, that basic biomedical research on alcoholism would ultimately lead to a better understanding and cure of that disease, made me continue until the present day.

SUPPORT FOR ALCOHOL RESEARCH

It is quite obvious that the last thirty years have brought many changes in attitudes toward alcohol research. The field has improved its quality and earned more recognition in the scientific community. National Institutes with research activities, such as NIAAA have been established, and international organizations like WHO have developed more interest in the field. Nevertheless, we have to recognize that alcohol research gets no or only little support in many countries, especially in Europe. It is still much more difficult to obtain governmental and non-governmental financial support for alcohol research than for basic research for instance on cancer or recently on AIDS. Alcohol researchers have developed their own means of communication, such as the congresses sponsored by ISBRA, RSA or ESBRA or the specialized journals in the field. However, some of the best contributions are still published in the most prestigeous journal of the corresponding disciplines, i.e. in biochemical, pharmacological or neurochemical journals. Finally, an astonishing discrepancy exists between the importance of alcoholism and the lack of interest in this disease on the part of the pharmaceutical industry. As recently pointed out by Enoch Gordis at the fifth ISBRA Congress in Toronto, the choice of research topics depends on moral and ethical values and is subject to social and political pressures, and this is especially the case for alcohol research. An analysis of the reasons for this phenomenon is of primordial importance in order to allow a definitive emancipation of the research on alcoholism.

PROGRESS AND STAGNATION

Many improvements can be noted concerning the diagnosis and treatment of alcoholism. Most of the progress concerns alcohol-induced damage to organs. Much knowledge has also been gained on the molecular mechanisms, which underlie this organ pathology. As pointed out by Charles Lieber, the recognition of the direct toxic effects of alcohol, although probably in interaction with some genetically-predisposing or environmental factors such as nutrition, has also helped to promote the disease concept of alcoholism. On the other hand, a certain stagnation in the development of new pharmacotherapeutic approaches to combat alcoholism must be noted. The pharmacological arsenal is still mainly restricted to compounds inhibiting aldehyde dehydrogenase, such as antabuse, which lead to a hypersensitivity to alcohol due to the acute toxicity of very high acetaldehyde levels, which consequently have an aversive effect. These drugs, however, do not lead to a true change in the positively reinforcing psychotropic action of alcohol. On the contrary, it has been shown that only slight elevations in acetaldehyde by a small inhibition of aldehyde dehydrogenase may enhance the reinforcing effects of alcohol, and that some alcoholics under antabuse treatment have learned to enjoy this effect by consuming relatively small amounts of alcohol.

RESEARCH ON THE PSYCHOTROPIC EFFECTS OF ALCOHOL

The rapid progress in neurobiological research on the actions of alcohol on the brain is most impressive. However, it is still difficult to correlate specific ethanol-induced effects on specific neuronal mechanisms in specific brain regions with a corresponding psychopharmacological action of alcohol, such as the anxiolytic, the hypnotic/sedative, or the stimulating/euphoric effects. Furthermore, only little is known on the individual and racial differences in the pattern of psychotropic effects, or the prevalence of certain patterns in subtypes of alcoholics. The most recent finding by T.-K. Li and his group, that alcohol-preferring rats have a higher density of the GABA-benzodiazepine-chloride-ionophore complex in the nucleus accumbens than the alcohol-nonpreferring line, represents a true breakthrough in this context, especially because this brain region has been implicated as an important component of the reward system. As shown by Boris Tabakoff and his collaborators, certain receptor systems (5HT3, GABA, NMDA, VSCC, vasopressin) are involved in the mechanisms underlying reinforcement, motoric impairment, tolerance, or physical dependence. These findings provide a better understanding of the role of the psychotropic actions of alcohol in the development of alcohol abuse and alcoholism. A totally new concept in the pharmacotherapy of alcoholism could consist of trials to modify the psychotropic effects, e.g. the euphorising or any other reinforcing component, with antagonists or partial inverse antagonists, in order to make the drug ethanol an uninteresting compound in the addictive process. The possibility to develop new drugs, which will eventually influence the craving and the alcohol- seeking behaviour of alcoholics, seems to be within reach. High priority should be given to these lines of research, because it seems essential to supplement or replace pharmacological palliative therapies by more causal therapies.

MULTIDISCIPLINARY AND INTERDISCIPLINARY RESEARCH

In many cases advances in biomedical research on alcohol abuse and alcoholism have depended on general progress and development of new technologies and methodologies akin to single disciplines, such as biochemistry, physiology, pharmacology, molecular biology, genetics or neurobiology, which subsequently could be applied to alcohol-related problems. Similar to the research on other complex diseases these research efforts have become increasingly multidisciplinary. However, the multifactorial nature of alcoholism and its subtypes, as well as the interplay between genetic and environmental factors, necessitates a truly interdisciplinary approach with wholistic models. It will be a fascinating challenge for alcohol researchers to overcome language barriers and intensify the dialogue between researchers from biomedical and psychosocial disciplines. There are still many questions, which can only be addressed in such a interdisciplinary manner. It is for instance only poorly understood how biological, psychological and social factors interact to bring about a Ledermann distribution of the consumption of alcoholic beverages in a population. Another topic concerns the beneficial effects of alcohol, which unfortunately has not obtained the necessary attention by alcohol researchers. What are the possible contributions of alcohol to the individual and collective quality of life? How much alcohol is beneficial and how much alcohol is too much?

SOCIAL IMPACT OF ALCOHOL RESEARCH

In any case, such issues imply another obligation on the community of alcohol researchers in all disciplines, that is to promote the transfer of the overwhelming knowledge already gained by basic research in the field of alcohology and to strengthen the impact of alcohol research on the attitudes of the general public toward alcohol-related problems and governmental alcohol policy.

BIOLOGIC DISORDERS IN ALCOHOLISM AS A BASIS FOR A PUBLIC HEALTH APPROACH: POLICY IMPLICATIONS

Charles S. Lieber

Bronx VA Medical Center and Mount Sinai School of Medicine
130 West Kingsbridge Road
Bronx, New York 10468, USA

There is hardly any tissue in the body that is spared by alcohol. Knowing the pathophysiology of alcohol-induced lesions is understanding the pathologic responses of most organs. This view is not meant to downgrade the social and behavioural aspects of alcoholism, but there is a need to emphasize that in terms of human suffering and economic costs, the disorders of internal medicine associated with alcoholism represent a major part of the problem. In surveys of subjects hospitalized in medical services or treated in medicine or family practice clinics, it is common to find that 30-40% of all individuals are afflicted by some alcohol-related disorders, and these are not trivial diseases. For instance, one of these disorders, namely cirrhosis of the liver (usually as a complication of alcoholism), is the fourth most frequent cause of death among those 25 to 64 years of age in urban areas (Summary of Vital Statistics, 1986). In spite of the enormity of the medical complications of alcoholism, resources to control them are usually meager and not at all commensurate with the magnitude of the problem. Indeed, it is felt in many quarters that research and treatment efforts should be focused exclusively on the determinants of excessive drinking because it is believed that if consumption is brought under control, associated complications could also be alleviated. Although superficially such an approach makes common sense, it ignores the multitude of individuals whose medical disorders due to alcoholism have reached an irreversible stage and require treatment, irrespective of the drinking stage. Furthermore, a significant segment of the alcoholic population escapes efforts to control drinking but yet might be successfully helped in terms of control or prevention of medical disorders. And finally, recent knowledge acquired on the pathogenesis of the medical disorders of alcoholism have opened up the possibility of the detection of the precursor stages of some of the major complications that in turn might enable early and more successful intervention. There is also hope that application of techniques of molecular biology to the field may eventually yield sufficient information for the identification of those individuals susceptible to develop the more severe complications of alcohol abuse even prior to the onset of alcoholism, thereby providing a handle for focusing therapeutic efforts on that subset of the population that is at the highest risk for the development of severe medical disorders.

The biological aspects of alcoholism extend to different areas of internal medicine: major and minor gastrointestinal disorders, cardiovascular and neurological abnormalities, endocrine and hematologic problems, etc. For the sake of brevity, I shall focus mainly on the liver, to exemplify the benefits that can be derived from the application of newly acquired knowledge in biochemistry and molecular biology (Lieber, 1988a). Liver disease, one of the most devastating complications of alcoholism, was formerly attributed exclusively to associated malnutrition. This view was most strongly expressed by Best, the co-discoverer of insulin, who stated that "there is no more evidence of a specific toxic effect of pure ethyl alcohol upon liver cells than there is for one due to sugar" (Best et al., 1949). Indeed, nutritional deficiencies are not uncommon in the alcoholic (Lieber, 1988b) and, when present, should be corrected, but such efforts were found to be ineffective in fully preventing liver disease. Indeed, studies conducted in the past two decades on human volunteers and subhuman primates have shown that either the initial liver lesion--the fatty liver (Lieber et al., 1965)--or the ultimate stage of cirrhosis (Lieber and DeCarli, 1974) can be produced by alcohol in the absence of dietary deficiencies. Striking changes were generated even in the presence of protein-, vitamin- and mineral-enriched diets (Lieber and Rubin, 1968; Lieber and DeCarli, 1989). The realization, by the primary care physician and the internist, of the toxicity of ethanol shifted the emphasis of treatment from correction of nutritional deficiencies to control of alcohol consumption. This recognition of the direct toxic action of alcohol and the subsequently acquired insight into the biochemistry of alcohol's effect have also provided new approaches to both pathogenesis and prevention. Elucidation of the redox changes associated with ethanol oxidation have furthered our understanding of associated disorders in carbohydrate, lipid and protein metabolism (Lieber, 1982). More recently, emphasis has also been placed on the role of the cytochrome P-450-dependent microsomal ethanol-oxidizing system (MEOS). The discovery, 20 years ago, of this accessory but adaptive pathway of ethanol metabolism (Lieber and DeCarli, 1968) was met with more than a decade of scepticism, but it now has gained general acceptance. The associated microsomal induction provides an explanation for the vulnerability of heavy drinkers to the hepatotoxicity of industrial solvents, anaesthetics, chemical carcinogens, commonly used drugs and even over-the-counter analgesics and vitamins (Lieber, 1988a). The induction of this pathway also leads to increased production of acetaldehyde, a very reactive compound. Acetaldehyde, in turn, causes injury through the formation of adducts with proteins, resulting in antibody formation, enzyme inactivation, decreased DNA repair and alterations in microtubules, mitochondria and plasma membranes. Acetaldehyde also promotes glutathione depletion, toxicity mediated by free radicals, lipid peroxidation and hepatic collagen synthesis (Lieber, 1988c). Indeed, acetaldehyde was discovered to play a key role in the production of scar tissue, one of the basic pathologic features of alcoholic cirrhosis (Savolainen et al., 1984; Moshage et al., 1990). Furthermore, acetaldehyde was incriminated in processes leading to, or favouring, alcohol dependence and, possibly, some of the manifestations of the fetal alcohol syndrome (Karl et al., 1988).

Since it is now clear that many pathologic complications of alcoholism, and perhaps alcoholism itself, are linked to the metabolism of alcohol, elucidation of the metabolic pathways and their genetic determinants has become crucial to our better understanding and management of these disorders. For other drugs, genetic abnormalities in their metabolism have been invoked to explain some of the variability in the vulnerability to toxicity. For instance, polymorphism has been demonstrated for other cytochrome P-450-dependent substrates, such as the antihypertensive drug debrisoquine sulfate: this and a score of related commonly prescribed compounds have toxic effects in certain persons who harbor a mutant gene that results in a deficiency of the corresponding active cytochrome P-450 (Gonzalez et al., 1988). It has not been determined as yet whether polymorphism for P-450IIE1 (the cytochrome P-450 involved in microsomal ethanol metabolism) exists. If it does, it might provide a mechanism for the unusual susceptibility to ethanol (and related xenobiotics) of a subset of the population. Clear cut genetic differences have already been elucidated for the alcohol dehydrogenase (ADH) pathway and for the disposition of acetaldehyde. A corresponding genetically determined enzyme deficiency results in intolerance to alcohol (Goedde et al., 1985) and may thereby "protect" against the development of cirrhosis. Elucidation of how different individuals may vary biologically will enhance our understanding of the medical complications of alcoholism and perhaps of the propensity to develop alcoholism itself. Since alcoholism and its complications do not develop in all heavy drinkers, it has been obvious that some factors of heredity or individual susceptibility might play an important role. Studies carried out over the last decade have shown significant factors of heredity, including biochemical differences between alcoholics and nonalcoholics. We may now be able to identify genes underlying alcohol abuse, as suggested by the association of the dopamine D2 gene and alcoholism (Blum et al., 1990). Genetic differences may also serve as trait or risk markers, which in turn provide new tools for prevention (vide infra).

The biochemical approach is also beginning to have an impact on the treatment of alcohol related disorders. For instance, key cellular lesions resulting from alcohol consist of striking membrane alterations, including those of the plasma membranes of the hepatocytes (Yamada et al., 1985). Phospholipids are the major constituents of these membranes. Supplementation with a specific type of phospholipid (polyunsaturated lecithin) was recently found to effectively attenuate the capacity of alcohol to produce experimental scarring or cirrhosis of the liver (Lieber et al., 1990). This protection against a major complication of alcoholism, if verified in man, may finally provide us with an effective tool to attenuate one of the most important causes of mortality in the alcoholic.

Biochemical tools are not only crucial for the elucidation of the pathophysiology and the elaboration of a rational therapy of alcoholism and its complications, but they have also begun to serve as instruments for prevention through a public health approach (Lieber, 1982b). Indeed, there are many similarities between alcoholism and some other major public health problems which have confronted our society over the last century, such as for instance tuberculosis. Of course, there are many dissimilarities between tuberculosis and alcoholism which will not be discussed here. The focus will be on some of the common aspects which might suggest possible avenues for successful preventative intervention in the field of alcoholism. For both tuberculosis and alcoholism the etiological agents are known: namely the Koch bacillus and ethanol respectively. Both agents can affect most tissues of the body, but the most severe adverse effects usually occur in one organ, namely the lungs for tuberculosis and the liver for alcohol. In both instances, exposure to the agent was or is widespread: a majority of the population had a positive reaction to tuberculin indicating contact with the Koch bacillus; similarly, the majority of the population consumes alcohol. However, only a minority of those exposed have complications. One of the successful approaches to the problem of tuberculosis has been the early detection of the affected subjects through appropriate laboratory tests. Alcoholism could lend itself to a similar approach: it is obvious that among the alcohol users there is a subpopulation of very heavy drinkers who are particularly at risk to develop alcoholism and its complications. One goal should be the early detection of those people prior to their social or medical disintegration. This might be achieved through some state marker of heavy drinking, such as blood tests that could be performed by automated procedures on a large scale. There is reasonable hope that ongoing studies might be successful in the near future and might provide us with sensitive and specific tools for the early detection of heavy drinking and the screening of populations at risk. In addition, the availability of such state markers for drinking can be expected to be useful to detect relapses in patients who have been rehabilitated, thereby facilitating their early treatment. Such state markers might also provide an objective means of assessing and comparing various treatment modalities.

Another cornerstone of the public health approach to tuberculosis has been the early detection of its major complication through mass screening for pulmonary lesions. A similar approach might also be feasible in alcoholism through improved means to detect somatic changes at an early stage (i.e., liver lesions, hematological changes, etc.). Unfortunately, commonly available tests are not very satisfactory. For instance, the severity of alcoholic liver injury is not reflected in standard liver-function tests; serum aspartate transaminase (AST) and serum alanine transaminase (ALT) levels are not always strikingly elevated, even in subjects with severe liver injury such as alcoholic hepatitis; in fact the clinical and laboratory picture may occasionally mimic that of obstructive jaundice. However, better blood tests for the detection of alcoholic liver disease are being developed (Lieber, 1985) and further improvements can be expected.

Because major complications (such as cirrhosis) do not develop in all heavy drinkers, one major task in an effort at prevention is to better define the susceptible population. We have now learned to recognize lesions in the liver which, already at a very early precirrhotic stage, enable us to predict the rapid progression to the cirrhotic stage upon continuation of drinking (Worner and Lieber, 1985). At present, liver biopsies are required to detect these precirrhotic lesions. Of course, this does not represent a practical tool for disease control on a large scale. It is not unreasonable to anticipate, however, that eventually blood tests might be developed that serve the same purpose of screening individuals who, for some reason, show a greater propensity to rapidly develop cirrhosis when exposed to the offending agent alcohol. Of interest in that respect is the recent observation of polymorphism in the collagen gene reported to be associated with alcoholic cirrhosis (Weiner et al., 1988). Similarly, some of the central nervous system complications of alcoholism may occur preferentially in predisposed individuals, as illustrated by the preponderance of the Wernicke-Korsakoff syndrome in individuals with pre-existing enzyme abnormalities (Blass and Gibson, 1977). Detection of such or other pre-existing abnormalities in high-risk groups, e.g., heavy drinkers, may allow for selective intensive treatment efforts in such subpopulations, thereby enhancing the cost effectiveness of our therapeutic interventions. Thus, further refinement of what makes an alcoholic different from a nonalcoholic not only in psychological but also in biochemical terms may be extremely useful as a basis for a biological approach to alcoholism. Eventually, application of these ongoing studies on trait markers, may provide us with some tools to detect those individuals prone to develop alcoholism prior to their exposure to alcohol.

In summary, it is being advocated that a disease control strategy toward alcoholism and its medical complications be added to our main approach based on treatment of underlying psychological and behavioural problems. This public health approach would be based on early detection of alcoholism (utilizing, in part, biochemical markers of heavy drinking), screening, among heavy consumers, for signs of medical complications (through the use of improved blood tests), and reducing the task of treatment to a manageable size by focusing major therapeutic efforts on susceptible subgroups for instance, those recognized to be prone to the development of major sequelae (i.e., cirrhosis of the liver or brain damage) upon continued heavy drinking.

The benefits of this `disease control' approach are multiple. It offers research goals which are realistic and achievable in the foreseeable future and which, though they may not solve the fundamental problem of alcoholism, would nevertheless clearly alleviate the suffering of the alcoholic and the public health impact of alcoholism on our society. This approach would also legitimize much of the research which is currently carried out in this field and is supported by public agencies. A major advantage of the disease control concept is a potential for reconciliation between classic advocates of public health approaches who wish to focus on the agent, namely alcohol and those who feel strongly that all efforts should be directed towards the susceptible host, namely the alcoholic. The antagonism between the two groups has been exacerbated, in part, by efforts on behalf of public health advocates to control alcohol consumption through measures that involve the entire population. The approach proposed here, namely the application of public health measures primarily to the alcoholic population itself may help reintegrate public health advocates into the alcoholism movement without evolving fears of 'neoprohibitionism'. An 'ecumenical' approach may hopefully provide alcoholism research with the backing needed to achieve status, stature, and support equal to that enjoyed by other public health problems.

ACKNOWLEDGEMENTS

Original studies reviewed were supported by the Department of Veterans Affairs and

DHHS grants AA03508, AA05934, AA07802 and DK 32810, and the Alcoholic Beverage Medical and Kingsbridge Research Foundations.

REFERENCES

Best, C. H., Hartroft, W. S., Lucas, C.C. and Ridout, J. H., 1949, Liver damage produced by feeding alcohol or sugar and its prevention by choline, **Brit. Med. J.**, 2:1001.

Blass, J. P., and Gibson, G.E., 1977, Abnormality of a thiamine-requiring enzyme in patients with Wernicke-Korsakoff syndrome, **New Engl. J. Med.**, 297:1367.

Blum, K., Noble, E. P., Sheridan, P. J., Montgomery, A., Ritchie, T., Jagadeeswaram, P.,Nogami, H., Briggs, A. H., Cohn, A. H., and Cohn, J.B., 1990, Allelic association of human dopamine D2 receptor gene in alcoholism, **JAMA**., 263:2055.

Goedde, H. W., Agarwal, D. P., Eckey, R., and Harada, S., 1985, Population genetic and family studies on aldehyde dehydrogenase deficiency and alcohol sensitivity, **Alcohol**, 2:283.

Gonzalez, F. J., Skoda, R. C., Kimura, S., Umeno, M., Zanger, U. M., Nebert, D. W., Gelboin,H. V., Hardwick, J. P., and Meyer, U. A. , 1988, Characterization of the common genetic defect in humans deficient in debrisoquine metabolism, **Nature**, 331:442.

Karl, P. I., Gordon, B. H. J., Lieber, C. S., and Fisher, S. E., 1988, Acetaldehyde production and transfer by the perfused human placental cotyledon, **Science**, 242:273.

Lieber, C. S., 1982a, **in:** Medical Disorders of Alcoholism: Pathogenesis and Treatment, C.S.Lieber, ed., WB Saunders, Philadelphia, PA.

Lieber, C. S., 1982b, A public health approach for the control of the disease of alcoholism, **Alcoholism: Clin. Exp. Res.**, 6:171.

Lieber, C. S., 1985, Early detection of precursor lesions of liver cirrhosis, **in:** Early Identification of Alcohol Abuse, N. Chang, and H. M. Chao, eds., Research Monograph-17, DHHS Publication No. (ADM) 85-1285, Superintendent of Documents, US Government Printing Office, Washington DC, pp. 168.

Lieber, C. S., 1988a, Biochemical and molecular basis for alcohol-induced injury to liver and other tissues. **New Engl. J. Med.**, 319:16 39.

Lieber, C. S., 1988b, The influence of alcohol on nutritional status, **Nutr. Rev.**, 46:241.

Lieber, C. S., 1988c, Metabolic effects of acetaldehyde, **Biochem. Soc. Trans.**, 16:241.

Lieber, C. S. and DeCarli, L. M., 1968, Ethanol oxidation by hepatic microsomes: adaptive increase after ethanol feeding, **Science**, 162:917.

Lieber, C. S. and DeCarli, L. M., 1974, An experimental model of alcohol feeding and liver injury in the baboon, **J. Med. Primatol.**, 3 :153.

Lieber, C. S. and DeCarli, L. M., 1989, Effects of mineral and vitamin supplementation on the alcohol-induced fatty liver and microsomal induction, **Alcoholism: Clin. Exp. Res.**, 13:142.

Lieber, C. S. and Rubin, E., 1968, Alcoholic fatty liver in man on a high protein and low fat diet, **Am. J. Med.**, 44:200.

Lieber, C. S., Jones, D. P., and DeCarli, L. M., 1965, Effects of prolonged ethanol intake:production of fatty liver despite adequate diets, **J. Clin. Invest.**, 44:1009.

Lieber, C. S., DeCarli, L. M., Mak, K. M., Kim, C-I., and Leo, M. A., 1990, Attenuation of alcohol-induced hepatic fibrosis by polyunsaturated lecithin, **Hepatology**, 12, 1390.

Moshage, H., Casini, A., and Lieber, C.S., 1990, Acetaldehyde selectively stimulates collagen production in cultured rat liver fat-storing cells but not in hepatocytes, **Hepatology**, 12, 511.

Savolainen, E-R., Leo, M. A., Timpl, R., and Lieber, C. S., 1984, Acetaldehyde and lactate stimulate collagen synthesis of cultured baboon liver myofibroblasts, **Gastroenterology** 87:777.
Summary of Vital Statistics, 1986, New York: Department of Health, City of New York, Bureau of Health Statistics and Analysis.
Weiner, F. R., Eskreis, D. S., Compton, K. V., Orrego, S., and Zern, M. A., 1988, Haplotype analysis of type I collagen gene and its association with alcoholic cirrhosis in man, **Mol. Asp. Med.**, 10:159.
Worner, T. M., and Lieber, C. S., 1985, Perivenular fibrosis as precursor lesion of cirrhosis, **JAMA.**, 254:627.
Yamada, S., Mak, K. M., and Lieber, C. S., 1985, Chronic ethanol consumption alters rat liver plasma membranes and potentiates release of alkaline phosphatase, **Gastroenterology**, 88:1799.

WHAT CAN WE EXPECT FROM ANIMAL MODELS IN ALCOHOLISM RESEARCH? LIMITATIONS AND RELEVANCE TO THE HUMAN CONDITION

Ting-Kai Li

Indiana University School of Medicine and the VA Medical Center
Indianapolis, IN 46202

Animal models are an invaluable tool in elucidating the normal and abnormal functions of the human body and behavior. The perfect model is isomorphic to the human condition, but this is rarely accomplished except with disorders arising from single gene mutations. Models become useful, however, when it can be shown that they have relevant analogy, i.e., when they reveal some aspects of a complex process to yield understanding about the human condition through hypothesis-testing in humans. In this context, the choice of the appropriate animal species (practical issues such as cost also pertain), its genetic makeup, and the experimental (environmental) variables to which it is exposed to simulate the human condition are all important considerations.

There now exists a substantial number of very useful to potentially useful models for elucidating the acute and chronic effects of ethanol on organ structure and function, and for the study of ethanol consummatory behavior. Because most common laboratory animals do not voluntarily drink substantial amounts of ethanol, methods to forcibly administer pharmacologically meaningful amounts of ethanol have been developed. However, ethanol consummatory behavior is, in part, genetically controlled, and animal lines with high and low alcohol-drinking preference have been established through selective breeding. In like manner, animal lines that differ genetically in sensitivity or susceptibility to the effects of ethanol have been raised. Such animals are useful not only for elucidating the genetic nature of individual differences (e.g., number of significant genes influencing a trait), but also as tools for studying underlying biological mechanisms.

ANIMAL MODELS FOR STUDYING EFFECTS OF ALCOHOL

Although non-human primates, miniature swine and guinea pigs have been used, the most commonly employed animals are mice and rats. In most chronic studies, alcohol is administered forcibly as part of a liquid diet that serves as the sole source of essential nutrients, calories and the fluid requirements of the animals. This ethanol-containing liquid diet can be given by intragastric intubation/infusion or be consumed orally (Lieber and DeCarli, 1982; Miller et al., 1980). The advantage with this kind of regimen is nutritional adequacy of the experimental animals as well as having nutritionally similar control animals for comparison. In some chronic studies, vaporized ethanol has been administered by inhalation (Goldstein and Pal, 1971). The advantage of this approach is that large and relatively constant doses can be administered over a relatively short period of time. Disadvantages are the lack of nutritional control and the need, in some instances, to use concurrently an inhibitor of alcohol metabolism in order to elevate blood alcohol to desired levels.

In most acute studies, ethanol is given to rodents by intraperitoneal injection. The advantage is simplicity of the technique; the disadvantage is that the route is unphysiological and blood alcohol concentration rises unrealistically fast. To study the acute and chronic effects of ethanol, the forcible administration of ethanol is acceptable experimental design; oral or intragastric administration would appear to be more desirable than other routes of administration for the reasons stated above.

The liquid-diet technique has proven over the years to be the most practicable, and has been successfully employed in the study of alcoholic liver disease in baboons and rats (Lieber, 1988; Tsukamoto et al., 1985), and in the study of fetal alcohol syndrome/effects (Randall, 1987) and chronic brain damage (Walker et al., 1981). Ethanol vapor inhalation has been used to produce physical dependence and withdrawal signs in rats and mice. It has been used successfully in the selective breeding of mouse lines that have high and low measures of withdrawal severity (vide infra).

GENETICALLY DEVELOPED EXPERIMENTAL ANIMALS

Rats and mice are the most commonly used laboratory animals, and lines and strains with different characteristics have been developed and maintained by commercial suppliers and by research institutions. The animals can be genetically homogeneous (inbred) or heterogeneous (outbred). Inbred strains are developed by successive generations of sibmating. The advantage of the inbred strains is their group stability and uniformity over time, and within-strain variations in a trait would be largely environmental in origin. Differences among inbred strains in a trait is *prima facie* evidence of a genetic influence on that trait. Outbred stocks of animals, on the other hand, offer populations of choice for normative studies and studies of correlations between and among phenotypes, since individual animals within the population are genetically different to different degrees. Unfortunately, most commercially available and commonly employed outbred animals have not been systematically maintained with respect to genetic heterogeneity. Heterogeneous stock animals of defined genetic background, however, can be developed by the systematic crossing of multiple inbred strains (Hansen and Spuhler, 1984).

Animal lines with specific phenotypes of interest can also be developed by selective breeding which, in contrast to inbreeding, is directional. This process systematically mates individuals in a heterogeneous population that exhibit the most extreme levels of the chosen phenotype (i.e., high or low measures of a trait). The end result is that the selected lines would have a high or low frequency of the genes that impact on that trait, while the frequencies for genes not affecting that trait should be unchanged. These pharmacogenetically different animals provide useful tools for investigating mechanisms, since associated traits are likely to share common mechanisms through common gene action (pleiotropy). On the other hand, phenotypic associations in inbred strains may be entirely fortuitous because the fixation of genes is entirely random.

Differences among inbred strains, particularly inbred mouse strains, have been discovered for a large variety of alcohol-related phenotypes. These include: alcohol preference, the sensitivity of the brain to an acute dose of ethanol, acute (within session) tolerance development, the development of chronic tolerance, severity of physical dependence of withdrawal signs, and alcohol elimination rate (Li, 1985; McClearn and Erwin, 1982).

There now exist several sets of mice and rats that have been selectively bred for high and low measures of ethanol sensitivity and withdrawal severity (Table 1). Two of these, the LS/SS mice and the WSP/WSR mice have been studied quite extensively, and have shown, without any doubt, that animal pharmacogenetics is a powerful approach to the study of the biological substrates of alcohol intoxication and dependence. The best example is the demonstration in the LS/SS mice that ethanol "sleep time" is intimately related to the sensitivity of cerebellar Purkinje cells to ethanol at a physiological level (Sorensen et al., 1980) and to the sensitivity of the $GABA_A$ receptor (facilitation in LS and antagonism in SS) function to ethanol at the molecular level (Wafford et al., 1990). The value of the LS and SS lines has recently been enhanced by the successful development of recombinant inbred strains (vide infra).

ANIMAL MODELS TO STUDY ALCOHOL-DRINKING BEHAVIOR

One of the most fundamental questions in alcohol-related research is "why some people drink too much in spite of negative consequences?" There is now convincing evidence of genetic risk affecting a significant segment of the alcoholic population (Cloninger, 1986). Thus there is biological predisposition for "drinking too much" in some individuals. What this inherited propensity might be in physiological and biochemical terms is a subject of intense interest, now being pursued in high risk subjects (children of alcoholics) and families with multigenerational alcoholism. However, there are practical and ethical reasons that preclude certain kinds of experimentation in human subjects. Accordingly, studies of the biological mechanisms underlying normal and abnormal (extremes of a continuum) alcohol drinking behavior in subhuman primates and lower species has become more compelling than ever before.

Voluntary alcohol consumption (self-administration) has been studied in a variety of experimental animals, including subhuman primates, miniature swine, rats, mice, and guinea pigs (Deitrich and Melchior, 1985). The rat has been studied the most extensively and a variety of environmental and biological manipulations have been shown to increase drinking. These include scheduled availability, schedule-induced polydipsia in weight-reduced animals, secondary conditioning using operant methodology, brain electrical stimulation and the forcible prior induction of physical dependence and tolerance. However, the most successful way has been the use of selective breeding, since there exists a considerable degree of variation in ethanol preference in rats and mice, much of which is genetic in origin. At this point in time, there are several sets of alcohol-preferring and nonpreferring lines developed through selective breeding for high and low voluntary ethanol intake, when free-fed animals are given free access to water inbred strains of mice (C57BL and DBA) and rats (UChB and UChA) are also available (Table 2).

Table 1

GENETIC ANIMAL MODELS SELECTED FOR DIFFERENTIAL ETHANOL EFFECTS

Line/Species	Phenotype
Long/Short Sleep (LS/SS) Mice[a]	Duration of loss of righting reflex
High/Low Alcohol Sensitive (HAS/LAS) Rats[a]	Duration of loss of righting reflex
ALKO Tolerant/Nontolerant (AT/ANT) Rats[b]	Impairment of tilting-plane performance
COLD/HOT Mice[c]	Hypothermic response
FAST/SLOW Mice[c]	Stimulation of spontaneous motor activity
Withdrawal Seizure Prone/Resistant (WSP/WSR) Mice[c]	Handling-induced convulsions
Severe/Mild Ethanol Withdrawal (SEW/MEW) Mice[a]	Withdrawal severity on multivariate index

[a] Alcohol Research Center, University of Colorado, USA
[b] ALKO, Helsinki, Finland
[c] Portland VA Medical Center, Oregon, USA

Among the selectively-bred alcohol preferring/nonpreferring lines, the P and its counterpart NP line have been studied most extensively with regard to alcohol-related behaviors and their neurobiological associations. The P line satisfactorily meets the major suggested criteria for an animal model of alcoholism (Cicero, 1979). Obviously, psychosocial variables unique to the human cannot be met in an animal model.

It has been shown that the P rats will:

1. Voluntarily drink 10-30% ethanol solutions in quantities that elevate blood alcohol concentrations (BACs) into pharmacologically meaningful ranges (Li et al., 1979, Lumeng and Li, 1986). Blood ethanol concentrations as high as 200 mg% have been observed.
2. Develop, with chronic drinking, metabolic and neuronal tolerance (Lumeng and Li, 1986; Gatto et al., 1987a), and physical dependence as evidenced by withdrawal signs (Waller et al., 1982).
3. Work by operant responding (bar-pressing) to obtain ethanol in concentrations as high as 30%, but not because of the caloric value, taste or smell of the ethanol

Table 2

GENETIC ANIMAL MODELS SELECTED FOR VOLUNTARY ETHANOL CONSUMPTION (SELF-ADMINISTRATION)

Line/Species	Phenotype
ALKO Alcohol/Nonalcohol (AA/ANA) Rats[a]	Intake of 10% (v/v) ethanol vs. water
Preferring/Nonpreferring (P/NP) Rats[b]	Intake of 10% (v/v) ethanol vs. water
High/Low Alcohol Drinking (HAD/LAD) Rats[b]	Intake of 10% (v/v) ethanol vs. water
Sardinian Preferring/Nonpreferring (sP/sNP) Rats[c]	Intake of 10% (v/v) ethanol vs. water

[a] ALKO, Helsinki, Finland
[b] Alcohol Research Center, Indiana University, USA
[c] Department of Neuroscience, University of Cagliari, Sardinia

solutions (Penn et al., 1978; Murphy et al., 1989). The NP rats will not bar-press for ethanol in concentrations higher than 5%. Importantly, the P but not the NP rats would self-administer ethanol even by direct intragastric infusion (Waller et al., 1984). Blood alcohol concentrations as high as 400 mg% were observed.

Comparison of the P rats with the alcohol-nonpreferring NP rats has revealed differences that suggest new avenues for exploration of the neurobiological basis of alcohol-seeking behavior. Compared with NP rats, the P rats:

1. Are behaviorally stimulated by low doses of ethanol and are less affected by sedative-hypnotic doses of ethanol (Waller et al., 1986).
2. Are less sensitive to the aversive properties of ethanol (Froehlich et al., 1988).
3. Develop tolerance more quickly within a single session of exposure to a sedative-hynotic dose of ethanol (acute tolerance), and this tolerance persists (as tested by a second dose of ethanol) for a much longer period of time (Waller et al., 1983; Gatto et al., 1987b). With chronic free-choice drinking, P rats develop tolerance to the aversive properties of ethanol. Concomitantly, the free-choice drinking of ethanol increases.
4. Exhibit lower levels of serotonin (5-HT) and dopamine (DA) in certain brain regions (Murphy et al., 1982), most notably the nucleus accumbens (Murphy et al., 1987). Higher densities of 5-HT receptors in some of these regions (cerebral cortex and hippocampus) have been found in the P rats as compared with NP rats (Wong et al., 1988). Preliminary immunocytochemical studies indicate that the P rats have fewer 5-HT fibers in the affected regions as compared with NP rats. The administration of

5-HT uptake inhibitors attenuates the alcohol-seeking behavior of the P rats (McBride et al., 1988). There are also higher densities of GABAergic fiber terminals in the nucleus accumbens of the P rats as compared with NP rats (Hwang et al., 1990). Other neurotransmitter neuromodulator systems of interest are the endogenous opioid peptides, corticotrophin releasing hormone, and vasopressin.

Many of the differences in the behavioral effects of ethanol and in neurochemistry between the P and NP lines have been replicated in the HAD and LAD lines. Thus these associations of innate alcohol preference appear generalizable to some degree. It is notable that a lower level of brain serotonin has not been found in the AA rats as compared with ANA rats (Korpi et al., 1988). Although the alcohol-preferring C57BL mice have lowered brain serotonin, the mechanism appears to be mediated by a low supply of the serotonin precursor tryptophan (Badaway et al., 1989). Thus there may be many different mechanisms that can underlie ethanol preference. Furthermore ethanol drinking preference and high volumes of voluntary ethanol intake cannot, *a priori*, be equated with high alcohol-seeking behavior. This relationship needs to be tested specifically by operant methodology.

RECOMBINANT INBRED (RI) STRAINS

These are developed by crossing two inbred strains of interest and initiating sibmating from the F2 offspring. Inbreeding is continued for more than 20 generations to establish the RI strains. RI strains can also be developed from selected lines, as was recently completed from LS X SS crosses (DeFries et al., 1989). Because the F2 generation is a segregating and genetically heterogeneous generation, each RI strain is a random sample of the genetic variability present in the parental strains. RI strains are useful for obtaining evidence of the existence of a single or a major gene controlling a trait (bimodal distribution of the trait in the RI panel), and for discerning patterns of genetic correlation of traits and markers of interest. Recombinant inbred strains developed from selected lines offer an advantage over the parental lines in that chance associations which may have developed during selective breeding should be reduced or eliminated.

There are several existing panels of RI strains of mice. Some have had extensive characterization in terms of genetic markers (Lyon and Searle, 1989). A limited amount of alcohol-related work has been performed in these existing lines, looking for single-gene effects on ethanol withdrawal severity and other traits. An association of the polymorphic brain protein LTW-4, mapped to chromosome one, was found with ethanol acceptance in the B X D RI panel (Goldman et al., 1987). Interesting findings from the LS X SS RI strains should be forthcoming from ongoing studies from the Colorado Alcohol Research Center.

SUGGESTIONS FOR FUTURE RESEARCH DIRECTIONS

As much as possible, new research findings in animal models should be hypothesis-tested in humans to examine whether the model has relevant analogy. For example, if a certain neurotransmitter system or behavioral response is implicated in certain actions of ethanol, can analogous studies be performed in humans to test the hypothesis? The demonstration that the GABA-benzodiazepine-Cl- channel complex is particularly sensitive to ethanol's action is a case in point. Another potentially relevant finding is that sons of alcoholics, a high risk group, appear less sensitive to the behaviorally intoxicating effects of ethanol (compare response of P rats to ethanol).

The following is a suggested list of future directions of research using animal models.

Both methodological approaches and research questions are included. The suggestions are not comprehensive, but represent focal points for discussion and reflection.

1. Use inbred animal strains to minimize within-sample variation, and use outbred stock of known composition, if a wide range of biological variation in response is desired. In this regard, the heterogeneous stock HS mice and the N/Nih rat are recommended.
2. Selective breeding for high and low measures of certain responses to ethanol not yet in progress:
   a. acute tolerance
   b. operant responding for ethanol (psychological dependence)
   c. metabolism of ethanol and acetaldehyde

   Selective breeding programs are expensive and labor intensive. Investigators should not initiate such a program without realizing that a "lifetime" commitment is required.
3. Select for more than one response to alcohol (e.g. high preference and low tolerance; low preference and high tolerance; low preference and low tolerance) by means of tandem selection or index selection.
4. Use existing RI panels to examine strain distribution patterns, correlation of patterns of distribution of alcohol-related traits, and correlation with known genetic marker distribution patterns. The purpose would be to elucidate the quantitative nature of the inheritance of a trait, the potential genetic relationship between traits of interest, and the quantitative trait loci for the phenotypic response to alcohol of interest. The advantage of using existing RI panels is the large accumulation of genetic data already collected in these RI panels.
5. Develop new recombinant inbred strains from existing selected lines, e.g., LS X SS RI strains and others. The advantage is that the parental lines/strains are widely disparate in the selected phenotype. The disadvantage is the paucity of existing information on genetic markers, especially in the rat.
6. Study the neurobiology of ethanol reinforcement and compare with other drugs of abuse that have specific receptors. In this regard, the study of cross-tolerance and other interactions at the molecular level are important.
7. Study how mechanisms of hunger and satiety play a role in alcohol abuse. In contrast with other drugs of abuse, ethanol's metabolism qualifies it as a food substance. There are relationships to food substances, such as sugars that impinge on ethanol-drinking preference, that may have relevance to human ethanol-drinking behavior.
8. Decipher how ethanol produces reinforcement, intoxication and dependence in animal models and test relevance in humans. Neurotransmitter and neuromodulator systems of interest include: serotonin, dopamine, GABA, glutamate, opioid peptides, vasopressin, angiotensin, adenosine, among others.
9. Look for candidate genes that influence high/low ethanol preference, high/low sensitivity, high/low withdrawal severity, etc.
10. Study whether the phenomenon of craving has an analog in animal models by operant methodology. If so, the neurobiological basis of this very central element of alcohol dependence (alcoholism) can be experimentally approached at a neurobiological level.
11. Study environmental manipulations that can modify genetically different responses to ethanol. Such studies are pivotal in the development of new treatment and prevention strategies, which are conspicuously lacking.
12. Use the knowledge from the neurobiological studies above to develop new pharmacological agents for treatment of alcoholism.

REFERENCES

Badawy, A.A., Morgan, C.J., Lane, J., Dhaliwal, K., and Bradley, D.M., 1989, Liver tryptophan pyrrolase: A major determinant of the lower brain 5-hydroxytryptamine concentration in alcohol-preferring C57BL mice, **Biochem. J.** 264:597-599.

Cicero, T.J., 1979, A critique of animal analogues of alcoholism, **in**: "Biochemistry and Pharmacology of Ethanol, Vol. 2", E. Majchrowicz and E.P. Noble, eds., Plenum Press, New York, pp. 533-560.

Cloninger, C.R., 1986, Genetics of alcoholism, **Alcoholism: Clin. Exp. Res.** 9:479-482.

DeFries, J.C., Wilson, J.R., Erwin, V.G., and Petersen, D.R., 1989, LS X SS recombinant inbred strains of mice: Initial characterization, **Alcoholism: Clin. Exp. Res.** 13:196-200.

Deitrich, R.A., and Melchior, C.L., 1985, A critical assessment of animal models for testing new drugs for altering ethanol intake, **in**: "Research Advances in New Psychopharmacological Treatments for Alcoholism", C.A. Naranjo and E.M. Sellers, eds., Excerpta Medica, Amsterdam, pp. 23-43.

Froehlich, J.C., Harts, J., Lumeng, L, and Li, T.-K., 1988, Differences in response to the aversive properties of ethanol in rats selectively bred for oral ethanol preference, **Pharmacol. Biochem. Behav.** 31:215-222.

Gatto, G.J., Murphy, J.M., Waller, M.B., McBride, W.J., Lumeng, L., and Li, T.-K., 1987a, Chronic ethanol tolerance through free-choice drinking in the P line of alcohol-preferring rats, **Pharmacol. Biochem. Behav.** 28:111-115.

Gatto, G.J., Murphy, J.M., Waller, M.B., McBride, W.J., Lumeng, L., and Li, T.-K., 1987b, Persistence of tolerance to a single dose of ethanol in the selectively bred alcohol-preferring P rats, **Pharmacol. Biochem. Behav.** 28:105-110.

Goldman, D., Lister, R.G., and Crabbe, J.C., 1987, Mapping of a putative genetic locus determining ethanol intake in the mouse, **Brain Res.** 420:220-226.

Goldstein, D.B., and Pal, N., 1971, Alcohol dependence produced in mice by inhalation of ethanol; grading the withdrawal reaction, **Science** 172:288-290.

Hansen, C., and Spuhler, K, 1984, Development of the National Institutes of Health genetically heterogeneous rat stock, **Alcoholism: Clin. Exp. Res.** 8:477-479.

Hwang, B.H., Lumeng, L., Wu, J.-Y., and Li, T.-K., 1990, Increased number of GABAergic terminals in the nucleus accumbens is associated with alcohol preference in rats, **Alcoholism: Clin. Exp. Res.** 14:503-507.

Korpi, E.R., Sinclair, J.D., Kaheinen, P., Viitamaa, T., Hellevuo, K., and Kiianmaa, K., 1988, Brain regional and adrenal monoamine concentrations and behavioral responses to stress in alcohol-preferring AA and alcohol-avoiding ANA rats, **Alcohol** 5:417-425.

Li, T.-K., 1985, Genetic variability in response to ethanol in humans and experimental animals, **in**: "Proceedings: NIAAA-WHO Collaborating Center Designation Meeting and Alcohol Research Seminar", L.H. Towle, ed., U.S. Government Printing Office, Washington, D.C., pp. 50-62.

Li, T.-K., Lumeng, L., McBride, W.J., Waller, M.B., and Hawkins, D.T., 1979, Progress toward a voluntary oral-consumption model of alcoholism, **Drug and Alcohol Dependence** 4:45-60.

Lieber, C.S., 1988, Biochemical and molecular basis of alcohol-induced injury to liver and other tissues, **N. Engl. J. Med.** 319:1639-1650.

Lieber, C.S., and DeCarli, L.M., 1982, The feeding of alcohol in liquid diets: Two decades of applications and 1982 update, **Alcoholism: Clin. Exp. Res.** 6:523-531.

Lumeng, L., and Li, T.-K., 1986, The development of metabolic tolerance in the alcohol-preferring P rats: Comparison of forced and free-choice drinking of ethanol, **Pharmacol. Biochem. Behav.** 25:1013-1020.

Lyon, M.F., and Searle, A.G., eds., 1989, **in:** "Genetic variants and strains of the laboratory mouse, 2nd Edition", Oxford, England, Oxford University Press, New York.

McBride, W.J., Murphy, J.M., Lumeng, L., and Li, T.-K., 1988, Effects of Ro 15-4513, fluoxetine and desipramine on the intake of ethanol, water and food by the alcohol-preferring (P) and -nonpreferring (NP) lines of rats, **Pharmacol. Biochem. Behav.** 30:1045-1050.

McClearn, G.E., and Erwin, V.G., 1982, Mechanisms of genetic influence on alcohol-related behaviors, **in:** "Alcohol and Health Monograph I. Alcohol Consumption and Related Problems", U.S. Government Printing Office, Washington, D.C., pp. 263-289.

Miller, S.S., Goldman, M.E., Erickson, C.K., and Shorey, R.L., 1980, Induction of physical dependence on and tolerance to ethanol in rats fed a new nutritionally complete and balanced liquid diet, **Psychopharmacology** 68:55-59.

Murphy, J.M., Gatto, G.J., McBride, W.J., Lumeng, L., and Li, T.-K., 1989, Operant responding for oral ethanol in the alcohol-preferring P and alcohol-nonpreferring (NP) lines of rats, **Alcohol** 6:127-131.

Murphy, J.M., McBride, W.J., Lumeng, L., and Li, T.-K., 1982, Regional brain levels of monoamines in alcohol-preferring and -nonpreferring lines of rats, **Pharmacol. Biochem. Behav.** 16:145-149.

Murphy, J.M., McBride, W.J., Lumeng, L., and Li, T.-K., 1987, Contents of monoamines in forebrain regions of alcohol-preferring (P) and -nonpreferring (NP) lines of rats, **Pharmacol. Biochem. Behav.** 26:389-392.

Penn, P.E., McBride, W.J., Lumeng, L., Gaff, T.M., and Li, T.-K., 1978, Neurochemical and operant behavioral studies of a strain of alcohol-preferring rats, **Pharmacol. Biochem. Behav.** 8:475-481.

Randall, C.L., 1987, Alcohol as a teratogen, **Alcohol and Alcoholism**, Supplement 1:125-132.

Sorensen, S., Palmer, M., Dunwiddie, T.V., and Hoffer, B.M., 1980, Electrophysiological correlates of ethanol-induced sedation in differentially sensitive lines of mice, **Science** 210:1143-1145.

Tsukamoto, H., French, S.W., Benson, N., Delgado, G., Rao, G.A., Larkin, E.C., and Largman, C., 1985, Severe and progressive steatosis and focal necrosis in rat liver induced by continuous intragastric infusion of ethanol and low fat diet, **Hepatology** 5:224-232.

Wafford, K.A., Burnett, D.M., Dunwiddie, T.V., and Harris, R.A., 1990, Genetic differences in the ethanol sensitivity of $GABA_A$ receptors expressed in Xenopus oocytes, **Science** 249:291-293.

Walker, D.W., Hunter, B.E., and Abraham, W.C., 1981, Neuroanatomical and functional deficits subsequent to chronic ethanol administration in animals, **Alcoholism: Clin. Exp. Res.** 5:267-282.

Waller, M.B., McBride, W.J., Gatto, G.J., Lumeng, L., and Li, T.-K., 1984, Intragastric self-infusion of ethanol by the P and the NP (alcohol-preferring and -nonpreferring) lines of rats, **Science** 225:78-80.

Waller, M.B., McBride, W.J., Lumeng, L., and Li, T.-K., 1982, Induction of dependence on ethanol by free-choice drinking in alcohol-preferring rats, **Pharmacol. Biochem. Behav.** 16:501-507.

Waller, M.B., McBride, W.J., Lumeng, L., and Li, T.-K., 1983, Initial sensitivity and acute tolerance to ethanol in the P and NP lines of rats, **Pharmacol. Biochem. Behav.** 19:683-686.

Waller, M.B., Murphy, J.M., McBride, W.J., Lumeng, L., and Li, T.-K., 1986, Effect of low dose ethanol on spontaneous motor activity in the alcohol-preferring (P) and nonpreferring (NP) lines of rats, **Pharmacol. Biochem. Behav.** 24:617-625.

Wong, D.T., Lumeng, L., Threlkeld, P.G., Reid, L.R., and Li, T.-K., 1988, Serotonergic and adrenergic receptors of alcohol preferring and non-preferring rats, **J. of Neural Transm.** 71:207-218.

ALCOHOLISM RESEARCH: CRYSTAL BALL OR WISHFUL THINKING

Yedy Israel

Department of Pharmacology
University of Toronto and Primary Mechanisms Department
Addiction Research Foundation
Toronto, Ontario, Canada

INTRODUCTION

The future of research in alcoholism, at least in the short term, is in great part what we make it. Thus, when talking about the future of research we in some measure mean a wish list; one that is based on the realities of the past blended with one's bold - at times outrageous - extrapolation. There have been many outrageous proposals in our field; for example, that hyperexcitibility and delirium tremens were alcohol-withdrawal phenomena. How could anybody believe that removing the injurious agent would be actually harmful. How could anybody believe that an entity in the endoplasmic reticulum different from alcohol dehydrogenase would metabolize alcohol; that animals strains could be developed to mimick many aspects of alcoholism, that opioid peptides could mediate some alcohol actions or that a small molecule such as acetaldehyde attached to proteins could elicit an immune response.

I hope the reader disagrees with what follows; if not, I am actually talking about the past or the present, but not the future.

MARKERS OF CHRONIC ALCOHOL ABUSE

I believe the field will move quickly in this direction. In 10 years there will be 2 - 3 biochemical indicators of chronic alcohol abuse which will measure alcohol consumption in (a) the last few days, (b) the last few weeks and (c) in the last one or two months. There will be efficient batteries to diagnose alcohol abuse which will be used by primary care physicians much as blood glucose and cholesterol levels are routinely tested today. We will have, in 15 years, a generalized understanding, derived from present studies in Sweden, the U K and New Zealand, and from many other studies to come, that early identification coupled with a minimal intervention-advice can cut the toll of alcohol-related morbidities by one half.

We will also have markers to better measure the consequences of alcohol abuse in populations. New markers of acute alcohol intoxication in populations will be added to markers which indicate the total volume consumed, to assess the dual pattern with which alcohol is consumed. Chronic alcohol abuse markers will be used in medical check-ups at the workplace and by life insurance companies, much as today we measure blood pressure. As members of society we should keep a vigilant eye open to ensure that these tests are not abused. We will also be able to predict the risk of alcoholic cirrhosis in an individual. Cirrhosis prevalence and mortality will go markedly down, due to prevention and improved treatments.

TRAIT MARKERS OF ALCOHOLISM

We will start abandoning the term alcoholism in favor of two terms; alcohol abuse without - and alcohol abuse with dependence (the latter what we now understand as alcoholism). In 10 - 15 years we will have about 3 genetic markers that confer a greater susceptibility to the development of alcoholism. We will, however, not use these markers in generalized clinical settings until we can show that we can negate or greatly reduce the effects of these genes by other interventions. Some genetic combinations will be much more predictive of an alcoholism risk than others, but no combination will be fully predictive. We will discover genes that protect from alcoholism in caucasian populations.

A PERMANENT OR LONG LIVED AVERSION TO ALCOHOL

One of the important lessons of the past decade is that in Oriental populations, individuals who have a genetic deficiency in one of aldehyde dehydrogenase isozymes (ALDH negative) do not develop alcoholism, or at least are 90 - 95% protected. Thus, a single mutation in ALDH - which results in aversion to alcohol - negates most or all other genetic and environmentally predisposing influences. It is important to also note that ALDH negative individuals do not show a greater prevalence of other psychiatric problems. The rest of the Oriental population without this mutation is certainly not immune to alcoholism.

I believe that in time we will be able to induce long-lived, **reversible** aversion reactions against alcohol in individuals who do not have this characteristic at birth. Work in animals will proceed towards manners of reducing the production of aldehyde dehydrogenase, for example by anti-sense gene therapeutics. A general field will develop to study aversive mechanisms per se to alcohol and alcohol products. A small number of individuals show severe hypersensitivity reactions against alcohol. These individuals do not consume alcohol. It will be possible to produce long-lived, reversible reactions which deter animals, and perhaps eventually humans, from consuming alcohol.

ANIMAL MODELS OF ALCOHOL ABUSE

The field will proceed towards the development of two types of animal models, one model in which animals given access to alcohol for very short periods, on a daily basis consistently get intoxicated and "pay a price" for the intoxication. Another model is one in which animals consume alcohol leading to alcoholic liver disease. There are partial models of both at this time but there is also the need to expand these models. In 5 - 10 years these animal strains will become commercially available, with the help of granting agencies.

COMPLIANCE OF ANTI-ALCOHOL MEDICATION

We will finally become aware that drugs that reduce the effects of alcohol are unlikely to be complied with by the alcoholic. Also, that compliance cannot be expected for drugs that produce aversion to alcohol. We will start looking for long term interventions, where daily compliance is not an issue, and for drugs that are chosen voluntarily by animals. The major conceptual problem will be to avoid drugs that are per se addictive. Obtaining specificity for alcohol consumption will constitute the main challenge.

THE BIOLOGY OF REWARD AND REINFORCEMENT

We will conduct research to understand the nature of reward and reinforcement to alcohol and drugs in general. It will be a long uphill road which will produce rewards mostly in relation to other drugs. Alcoholism will slowly fade away due to both advances in biology and due to rejection of harmful alcoholic intoxication by society. The latter, much like the "second hand" smoke rejection syndrome, is having an important impact in cigarette consumption in North America. In thirty years alcoholism, like age-old tuberculosis or syphilis will be greatly reduced. Our main problems will remain the recreational abuse of alcohol and the "nutritional" use of alcohol in some countries. On the other hand drug abuse and drug dependence of other types will increase with the advent of new designer drugs which will be able to activate and to mimick the neurobiological mechanisms of reward.

ALCOHOL AND CANCER: WHERE DO WE STAND AND WHERE SHOULD WE GO?

Helmut K. Seitz

Alcohol Research Laboratory
Section of Gastroenterology and Hepatology
Department of Medicine
University of Heidelberg
Germany

Before discussing possible future research directions with respect to alcohol and cancer one must raise the question whether or not there is a need for such kinds of research. It has been estimated that approximately 3% of all tumors in the U.S. and in Western Europe are due to chronic alcohol consumption. This is especially true for carcinoma of the oropharynx, the larynx, the esophagus, and the liver. In addition, the prevalence of these tumors increases. Thus, in the Federal Republic of Germany, the overall death rate from cancer of the mouth, pharynx, and larynx increased from 1970 to 1987 by 129% (1688 vs. 3846 cases/year) (Maier et al., 1990). It has to be pointed out that 76% of these tumors could be avoided by avoiding alcohol consumption and smoking (Rothman and Keller, 1972). Furthermore, some evidence exists that cancer of the rectum and the breast, two organs with the highest cancer mortality rates in the Western World, seem to be associated with chronic alcohol ingestion. In this context it has to be pointed out that the response rate of those tumors to chemotherapy or radiation is rather poor and that they are diagnosed at a rather late state of the disease. Thus, it seems necessary to gather more detailed information on the pathogenesis of alcohol-associated tumor development, eventually finding possible ways to prevent cancer or to interfere with the carcinogenic process at an early stage. Thus, the answer to the introductory question is yes.

The first symposium on alcohol and cancer was held at the National Institute of Health in October 1978 (Lieber et al., 1979). At that time some epidemiologic studies existed which showed a positive association between alcohol consumption and cancer of the upper alimentary and respiratory tracts. Studies on pathogenetic mechanisms were rare. Within the last 12 years 4 main types of studies have been performed:

1. Epidemiologic studies.
2. Animal studies with alcohol and carcinogens measuring tumor yield, occurence of preneoplastic lesions or DNA-alkylation.
3. In vitro studies using biological material from chronically ethanol-fed animals to study procarcinogen metabolism.
4. In vivo and in vitro studies to investigate cell regeneration, an important factor in carcinogenesis.

What have we achieved and where do we stand?
In general several important observations can been made (for more detailed review see Seitz and Simanowski, 1988):

1. From epidemiologic studies it is clear that chronic alcohol ingestion increases the risk for upper alimentary and respiratory tract cancers and for cancer of the liver. The data for other sites are still unclear.
2. From animal experiments it is obvious that ethanol per se is not a carcinogen. However, it can act as a cocarcinogen and/or as a tumor promoter.
3. Acetaldehyde can cause chromosomal aberrations in humans.
4. Ethanol and/or acetaldehyde can cause toxic cell injury due to various mechanisms leading to hyperregeneration, a state of increased susceptibility towards chemical carcinogens.
5. In vivo and in vitro studies show that chronic ethanol ingestion leads to a change in the metabolism of various pro-carcinogens including activation, inactivation and DNA-repair.
6. Chronic ethanol consumption may create situations which are known to be associated with an increased tumor risk such as certain dietary deficiencies.

Taking all of these data together we know today that alcohol stimulates the development of certain cancers. However, we still speculate on various possible ways by which ethanol affects carcinogenesis without knowing their exact mechanisms and their overall contribution to the carcinogenic process. In addition, studies which investigated similar pathogenetic mechanisms are limited, incomplete and often contradictory in certain aspects. One major reason for the controversial results obtained in experimental studies is the fact that animal experiments are often not standardized. However, many factors modulate carcinogenesis (Seitz and Simanowski, 1986):

1. Type of carcinogen used (organ specificity, metabolism).
2. Dose of carcinogen and duration of its application.
3. Means of carcinogen application (oral, local, parenteral).
4. Amount, concentration, and duration of ethanol administration.
5. Means of ethanol administration (in drinking water, as liquid diet, intragastrically, parenterally)
6. Combination of carcinogen and ethanol (preinduction, induction, postinduction).
7. Species, strain, sex of the animals.
8. Consumption of other dietary constituents during carcinogenesis.

If the experiments are not controlled for these variables it is impossible to come to meaningful conclusions. Another factor which is important and which has not sufficiently been evaluated is the dynamics of carcinogenesis. Most of the studies report data from a single time point during carcinogenesis. However, if this time point were to be changed the results can be different.

Many studies have investigated the activation of procarcinogens by microsomes from various tissues after chronic ethanol consumption. It has been shown that increased activation does not necessarily lead to an enhanced DNA alkylation. Again, the dynamics of the process have to be taken into consideration (inactivation, DNA-repair) and sequential time points during carcinogenesis should be examined. In addition, it is very difficult, if not impossible, to extrapolate the data obtained in these animal experiments to the human situation. The data from humans are still very limited.

What do we need and where should we go? In general we should collect more data from humans, should perform adequate animal experiments by controlling for the factors mentioned and by using experimental models where the time factor is integrated. In addition, it is important to know more about the effect of ethanol on factors which are involved in human carcinogenesis. This is especially valid for upper alimentary tract carcinogenesis and may include salivary gland function, epidermal growth factor output and receptor status of the mucosa, uptake of tobacco carcinogens by mucosal cells, the role of oral bacteria in ethanol metabolism and in the generation of acetaldehyde, the role of acetaldehyde in oral carcinogenesis as well as local concentrations of vitamins and trace elements. The studies should focus more on the effect of ethanol on intracellular mechanisms associated with carcinogenesis: e.g. radical formation and detoxification, acetaldehyde adduct formation with DNA, and protooncogene activation to mention only a few.

Finally, cell regeneration studies in humans are urgently needed. If regeneration studies are performed in animals dynamic methods such as the metaphase arrest technique are preferable (Seitz and Simanowski, 1986). Computer simulation offers a new tool to enhance information gathered by experiments on cell proliferation, since it allows multiple testing of hypothesis concerning organisation and regulation of tissues either in steady state, after perturbation or growing (Sandblad and Meinzer, in press).

REFERENCES

Lieber, C.S., Seitz, H.K., Garro, A.J. and Worner, T.M. (1979) Alcohol and cancer. **Cancer Res** 39: 2863-2886.

Maier, H., Dietz, A., Zielinski, D., Junemann, K.H. and Heller, W.D., 1990, Risikofaktoren bei Plattenepithelkarzinomen der Mundhohle, des Oropharynx, des Hypopharynx und des Larynx. Deutsch Med Wschr 115: 843-850.

Rothman, E. and Keller, A.Z., 1972, The effect of joint exposure to alcohol and tobacco on risk of cancer of the mouth and pharynx **J Chron Dis** 25: 711-716.

Sandblad, B. and Meinzer, H.P., 1991, Modelling of complex control structures in biology and biomedicine. **Meth Inf Med.**, in press.

Seitz, H.K.and Simanowski, U.A., 1986, Ethanol and carcinogenesis of the alimentary tract. **Alcoholism Clin Exp Res** 10: 33S-40S.

Seitz, H.K.and Simanowski, U.A., 1988, Alcohol and carcinogenesis. **Ann Rev Nutr** 8: 99-119.

Wright, N.A.and Alison, M., 1984, The biology of epithelial cell populations. Clarendon,Oxford.

CHRONIC ALCOHOLISM:
SUBTYPES USEFUL FOR THERAPY AND RESEARCH

Otto Michel Lesch

Psychiatric Univ. Clinic Vienna
Anton Proksch Institute Kalksburg Vienna

One of the most important medico-social and the most important psycho-social problem of all industrialised nations, especially of those producing alcoholic beverages (beer, wine, strong liquors), are the costs caused by patients with alcohol-related disabilities. These costs arise not only because of sickness and family problems but also because of somatic damage. These costs are far in excess of the financial support committed to research and therapy. The last hundred years has resulted in many countries (like esp. USA) in the establishment of self-help-groups. They still dominate therapy from a moralizing point of view and thus have hampered not only qualified medical therapy but also scientific research. There are therefore two important factors, namely the strict ideology associated with self-help-groups and the lack of financial resources, that have seriously interfered with research.

Another problem is that several professions are involved in alcoholism research. They all approach the problem with preconceived opinions that dominate their research concepts in etiology, diagnosis and therapy, as well as in the interpretation of their results. This kind of research reflects the diverse concepts and aims of these professions rather than being in the interests of fundamental research as the basis for further studies.

Let me - after this rather dishcartening account of the present state of research - discuss some of the facts that our research group considers as important for alcoholism research.

In relation to the etiology of chronic alcoholism, the toxic effects of ethanol and congener alcohols, together with their metabolites, represent the common factor underlying chronic alcoholism. These toxic effects affect different people in different ways, some being genetically conditioned or having different metabolizing processes. Research on tolerance development as well as that dealing with the causes for the withdrawal syndromes has to be considered in this context.

Differences in cultural background lead to different views on alcohol consumption. Some countries have a highly permissive attitude towards alcohol, supported moreover by individual, familial as well as professional attitudes, which influence not only personal characteristics but also drinking behaviour. The description of a single causal group of factors responsible for chronic alcoholism, paying regard only to factors related to personality, surroundings and drinking behaviour, has to be viewed as too unprecise. This is because certain factors have to be taken into account in the development of the psychic syndrome. Only if all the factors described in Table 1 are considered can the etiology of this syndrome can be adequately described.

This interplay of influences leads, according to differences in social surroundings and personal characteristics, to patients becoming conspicuous at varying stages of the illness (early or later) and thereby seeking therapy and entering research programs at different stages of illness development. After delineating the factors responsible for the etiology of chronic alcoholism it becomes evident that:

Table I. Factors influencing the psychopathological syndrome of chronic alcoholic patients.

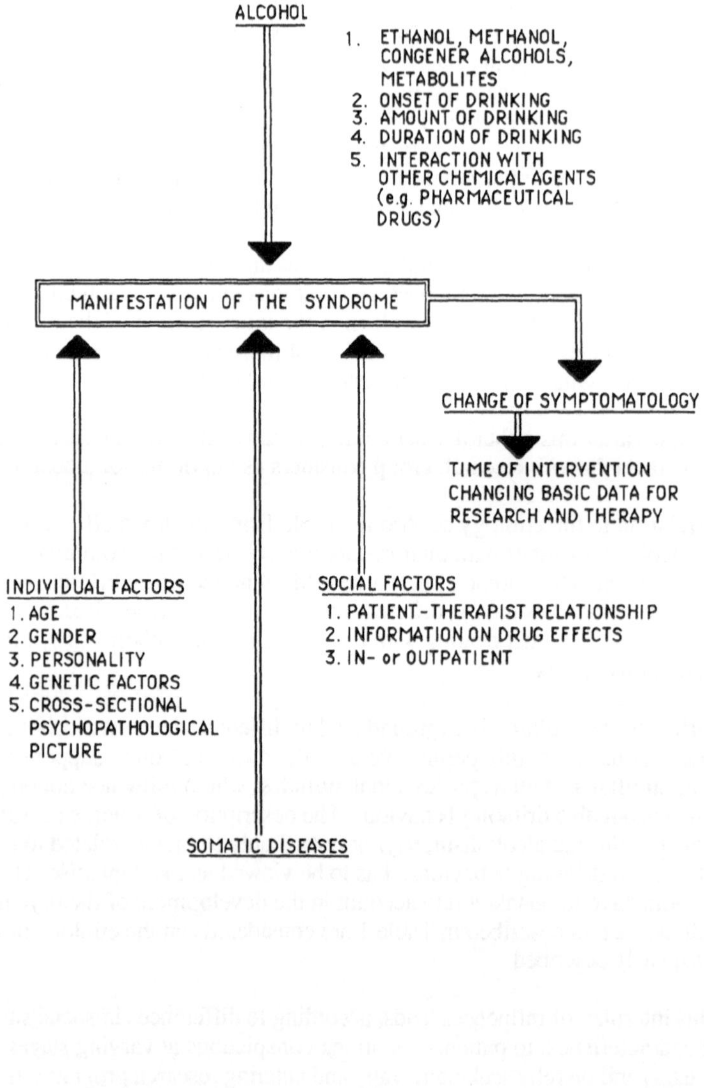

1) the border between alcohol-use, -misuse and -addiction is an artificial one and moreover is defined socially, and

2) the clinical picture of chronic alcoholism has to be a very heterogenous one.

These two points are consistent with the fact that we have extremely differing results concerning predictors (eg. gender, age) (Conley and Prioleau, 1987: Hoff, 1956: Dahlgren, 1975; Gomberg, 1974;,Guilespie, 1967; Harford and Mills, 1978) as well as attitudes towards therapy of chronic alcoholism.

This situation has led, since the introduction of the diagnosis "chronic alcoholism" into medical literature more than hundred years ago, not only to the heterogenity always being emphasised in cross-sectional (Bleuler, 1981; Raunsville et al., 1987 and others) and longitudinal investigations (Lesch 1985), but also to the definition of different subgroups according to the hypothesis of the individual investigator. A typical example of this situation is the current genetic research, in which the subgrouping method is applied, mainly using Cloninger's typology (type 1 and type 2: see Cloninger et al., 1981).

CONCLUSIONS FOR FUTURE RESEARCH

The diagnosis "chronic alcoholism", affected according to DSM-III, DSM-III-R, ICD-10 or according to other diagnostic instruments, is unsuitable not only for therapy but also for scientific research. According to the problem under investigation (in genetic, biochemical, or pharmacological research as well as psychological research), each research program should in future take into consideration the different factors of personal characteristics, personality, psychic development, social surroundings, somatic predamages and alcohol-related disabilities involved. Drinking behaviour is the only thing all alcoholic patients have in common and can therefore not be a parameter for the constitution of sub-groups.

Differences in alcoholic patients can be found - as is shown in our own typology (Lesch 1985) - in psychic development, somatic factors and social constellations. The 4 sub-groups developed by our research group can be found as pure types only in 50% of all cases. In scientific research, evaluation methods should be developed which define not only different types of chronic alcoholic patients (Lesch et al., 1990) but also evaluate the severity as well as a patient's stage in the development of chronic alcoholism.

Only by co-operation between all professions in alcoholism research will a clearer picture of this illness be developed, biological, sociological and psychological research providing the basic data for the etiology of groups of chronic alcoholic patients. Only on such an objective basis can the data provide the means for a more precise diagnosis as well as the opportunity for more successful therapy. The moralising point of view, embodying the idea that a patient who maintains absolute abstinence is a "good" patient while a patient who drinks is a "bad" patient, has to be rejected in order not to hamper research as well as progress in establishing better therapies for chronic alcoholic patients.

REFERENCES

Bleuler, E., 1981 Lehrbuch der Psychiatrie, Berlin.

Cloninger, C.R., Bohman, M. and Sigvardson, S., 1981, Inheritance of alcohol abuse - cross-fostering analysis of adopted men. **Arch. Psychiat.** 42: 1043.

Conley, J.J. and Prioleau, L.A., 1983, Personality typology of men and women alcoholics in relation to etiology and prognosis. **J. Stud. Alc.**, 44: 996.
Dahlgren, K.G., 1975, Special problems in female alcoholism. **Brit.J.Add.** 70: 18.
Gomberg, E.S., 1974, Women and alcoholism. **In**: V. Franks, V. Burtle, eds., "Women in therapy". Brunner/Mazel New York; 169.
Guillespie, D.G., 1967, The fate of alcoholics: An evaluation of alcoholism follow-up studies. **In**: D.J. Pittman (Ed.) alcoholism. Harper & Row, New York.
Harford, T.C. and Mills, G.S., 1978, Age-related trends in alcohol consumption. **J.Stud.Alc**. 39: 207.
Hoff, H., 1956, Lehrbuch der Psychiatrie, Verhütung Prognostik und Behandlung der geistigen und seelischen Erkrankungen, Basel, Stuttgart.
Koehler, K. and Sass, H., 1984, Diagnostisches und Statistisches Manual Psychischer Störungen, DMS-III. Beltz, Weinheim.
Lesch, O.M., 1985, Chronischer Alkoholismus. Typen und ihr Verlauf. Eine Langzeitstudie. Thieme Verlag.
Lesch, O.M., Kefe, J., Lentner, S., Mader, R., Marx, B., Musalek, M., Nimmerrichter, A., Presinsberger, H., Puchinger, H., Rustembegovic, A., Walter, H., 1990, Diagnosis of Chronic Alcoholism - Classificatory Problems. **Psychopathology** 23: 88.
Raunsville, B.J., Dolinsky, Z.S., Babor, T.F. and Meyer, R.E., 1987, Psychopathology as a Predictor of Treatment Outcome in Alcoholics. **Archs Psychiat**. 44: 505.

INDEX